AF833702

Physiology in Health and Disease

Published on behalf of the American Physiological Society
by Springer

This book series is published on behalf of the American Physiological Society (APS) by Springer. Access to APS books published with Springer is free to APS members.

APS publishes three book series in partnership with Springer: *Physiology in Health and Disease* (formerly *Clinical Physiology*), *Methods in Physiology*, and *Perspectives in Physiology* (formerly *People and Ideas*), as well as general titles.

Glenn McConell
Editor

Exercise Metabolism

Springer

american
physiological
society

Editor
Glenn McConell
Victoria University
Melbourne, VIC, Australia

ISSN 2625-252X ISSN 2625-2538 (electronic)
Physiology in Health and Disease
ISBN 978-3-030-94304-2 ISBN 978-3-030-94305-9 (eBook)
https://doi.org/10.1007/978-3-030-94305-9

Published on behalf of the American Physiological Society.

© The American Physiological Society 2022
This work is subject to copyright. All rights are solely and exclusively licensed by the Publisher, whether the whole or part of the material is concerned, specifically the rights of translation, reprinting, reuse of illustrations, recitation, broadcasting, reproduction on microfilms or in any other physical way, and transmission or information storage and retrieval, electronic adaptation, computer software, or by similar or dissimilar methodology now known or hereafter developed.
The use of general descriptive names, registered names, trademarks, service marks, etc. in this publication does not imply, even in the absence of a specific statement, that such names are exempt from the relevant protective laws and regulations and therefore free for general use.
The publishers, the authors, and the editors are safe to assume that the advice and information in this book are believed to be true and accurate at the date of publication. Neither the publishers nor the authors or the editors give a warranty, express or implied, with respect to the material contained herein or for any errors or omissions that may have been made. The publishers remain neutral with regard to jurisdictional claims in published maps and institutional affiliations.

This Springer imprint is published by the registered company Springer Nature Switzerland AG
The registered company address is: Gewerbestrasse 11, 6330 Cham, Switzerland

Glenn McConell
Editor

Exercise Metabolism

Springer

american
physiological
society

Editor
Glenn McConell
Victoria University
Melbourne, VIC, Australia

ISSN 2625-252X ISSN 2625-2538 (electronic)
Physiology in Health and Disease
ISBN 978-3-030-94304-2 ISBN 978-3-030-94305-9 (eBook)
https://doi.org/10.1007/978-3-030-94305-9

Published on behalf of the American Physiological Society.

© The American Physiological Society 2022
This work is subject to copyright. All rights are solely and exclusively licensed by the Publisher, whether the whole or part of the material is concerned, specifically the rights of translation, reprinting, reuse of illustrations, recitation, broadcasting, reproduction on microfilms or in any other physical way, and transmission or information storage and retrieval, electronic adaptation, computer software, or by similar or dissimilar methodology now known or hereafter developed.
The use of general descriptive names, registered names, trademarks, service marks, etc. in this publication does not imply, even in the absence of a specific statement, that such names are exempt from the relevant protective laws and regulations and therefore free for general use.
The publishers, the authors, and the editors are safe to assume that the advice and information in this book are believed to be true and accurate at the date of publication. Neither the publishers nor the authors or the editors give a warranty, express or implied, with respect to the material contained herein or for any errors or omissions that may have been made. The publishers remain neutral with regard to jurisdictional claims in published maps and institutional affiliations.

This Springer imprint is published by the registered company Springer Nature Switzerland AG
The registered company address is: Gewerbestrasse 11, 6330 Cham, Switzerland

Preface

This exercise metabolism collection provides a very comprehensive examination of the metabolic processes taking place during exercise.

It starts with a historical perspective and overview of exercise metabolism and then considers the underlying thermodynamic and bioenergetic principles underpinning metabolism during exercise. The chapters then outline anaerobic metabolism and separate chapters on muscle glycogen and blood glucose metabolism, muscle and adipose tissue fat metabolism and protein metabolism. There is also discussion on various tissues in addition to skeletal muscle, such as liver, heart and brain metabolism during exercise. In addition, the book includes chapters on other perspectives such as the effects of exercise training, age, sex, fatigue and the circadian rhythm on metabolism during exercise.

The focus was very much on metabolism during exercise rather than the effects of exercise on metabolism. Many of the chapters include more than one lead investigator to facilitate a balanced appraisal of the topic. The authors focussed on humans but included animal data to inform the human data. Although an attempt was made to avoid overlap, some topics may have been covered to some extent in more than one chapter.

Melbourne, VIC, Australia Glenn McConell

Acknowledgements

I would like to acknowledge the efforts and patience of everyone involved in this undertaking during a very disjointed and troubling time over 2020 and 2021. I would like to thank Dee Silverthorn from APS for approaching me to do this and for being happy for it to be expanded well beyond the original brief. From Springer Nature, a big thank you to Markus Spaeth for his enthusiasm and professionalism and to Bibhuti Sharma for helping me out with logistics, etc. A huge thank you to the authors who came through despite your very busy schedules and the professional and personal disruption of COVID-19. I was thrilled to have such a who's who of exercise metabolism involved, and it was a delight to interact with you all. I would also like to very much thank these exercise metabolism experts who assisted with the review of many of the chapters: David Allen, Raul Bescos, George Brooks, Andreas Fritzen, Paul Greenhaff, Mark Hargreaves, Graham Holloway, Jeffrey Horowitz, Joseph Houmard, Bente Kiens, Peter Lemon, Adriano Eduardo Lima-Silva, Troy Merry, Scott Rector, Kurt Smith, Nicole Stupka, Mark Tarnopolsky and Scott Trappe.

Contents

Chapter 1
A Brief History of Exercise Metabolism

Andrew R. Coggan and David L. Costill

Abstract The source of energy utilized during physical activity has been of intense scientific interest for at least two centuries. This chapter briefly describes how (and why) each of the three major macronutrients—i.e., protein, carbohydrate, and fat—has alternately had their moments in the sun. Specifically, although until the 1860s protein was considered to be the only fuel used during exercise, first carbohydrate, then fat, and then again carbohydrate held sway from the 1860s until World War II, from World War II until the late 1960s, and from the late 1960s to ca. 1990, respectively. It is now widely recognized, however, that contracting muscle relies upon a mixture of carbohydrate, fat, and even a small amount of protein to provide its energy needs, with the relative importance of each varying with the exercise intensity and duration, the characteristics (e.g., nutritional state, physical fitness) of the individual, etc. Thus, although substrate metabolism during exercise is now understood in greater detail than ever before, the overall picture has come full circle to that described by Zuntz at the start of the twentieth century.

Keywords Scientific history · Physical activity · Substrate oxidation · Muscle energetics · Macronutrient metabolism

1.1 Introduction

Precisely how skeletal muscle obtains the energy needed to support contractile activity, i.e., physical exercise, has been of keen interest for at least 150 years. This chapter briefly traces the historical development of our modern understanding of the answer to this question, focusing primarily upon macronutrient metabolism,

A. R. Coggan (✉)
Department of Kinesiology, School of Health and Human Sciences, Indiana University Purdue University Indianapolis, Indianapolis, IN, USA
e-mail: acoggan@iupui.edu

D. L. Costill
Human Performance Laboratory, Ball State University, Muncie, IN, USA

© The Author(s), under exclusive license to Springer Nature Switzerland AG 2022
G. McConell (ed.), *Exercise Metabolism*, Physiology in Health and Disease,
https://doi.org/10.1007/978-3-030-94305-9_1

i.e., the utilization of protein, carbohydrate, and fat (lipids) during exercise in humans. For a more detailed discussion of this and related issues, readers are referred to a previous review (Coggan 2014).

1.2 Exercise Metabolism, Late Eighteenth Century to the 1860s: Protein Reigns Supreme

Early perspectives on substrate metabolism during exercise arose in parallel with broader developments in science and especially biochemistry. In particular, the recognition of proteins as a distinct class of chemicals by de Fourcroy (1789), their careful study in the early 1800s by luminaries such as Mulder (1838), and their ultimate naming by Berzelius in 1838 (Harold 1951) initially led to the hypothesis that muscle obtained its energy primarily from consumption of its own substance. This was an extension of the common belief among chemists and physiologists of the time, such as Voit or von Haller, that "flesh becomes flesh" (Bischoff and Voit 1860; Holmes 1963). Leibig in particular asserted that protein was the only true nutrient and that carbohydrates and fats were combusted only to prevent excess O_2 from entering the body during ventilation, one major purpose of which was simply to cool the "metabolic furnace" (von Liebig 1842; Carpenter 2003). In 1862, however, Smith (1862) demonstrated that laboring for 8 h/d on a treadwheel did not increase the 24 h urea excretion of four British prisoners, thus seemingly ruling out protein as a major source of energy for muscular work. These findings were confirmed shortly thereafter by Fick and Wislicenus (1866), who measured their own urea excretion during and after a day-long mountain hike during which they consumed only protein-free snacks and beverages (including some beer and wine). They ascended nearly 2000 m during the climb, but the amount of protein that was oxidized could have provided only about half of the energy required by the vertical work that they performed, even if their muscles were 100% efficient. This led Fick and Wislicenus (1866) to conclude that "the burning of protein cannot be the only source of muscular power." Similarly, upon reviewing these and other results (e.g., von Pettenkofer and Voit 1866), Fick's brother-in-law, Edward Frankland, concluded that:

> Like every other part of the body the muscles are constantly being renewed; but this renewal is not perceptibly more rapid during great muscular activity than during comparative quiescence. After the supply of sufficient albuminized matter [protein] in the food to provide for the necessary renewal of the tissues, the best materials for the production, both of internal and external work, are non-nitrogenous material. . . . (Frankland 1866)

Thus, almost as quickly as it began, the reign of protein (whose name is derived from the Greek word πρώτειος (proteios), meaning "primary" or "of the first rank") as a major, much less the only, source of energy during exercise was over.

1.3 Exercise Metabolism, 1860s to World War II: Carbohydrate Is King

Even as protein was enjoying its brief time in power, other events were transpiring that would eventually see carbohydrate anointed—mistakenly, as it turns out—as its successor as the singular substrate fueling exercise. Major findings leading to this conclusion included the following observations:

1. Lactate levels were higher in the non-paralyzed vs. the paralyzed muscles of hunted stags (Berzelius and Berzelius 1806–1808).
2. Contractile activity resulted in an increase in water-soluble and a decrease in alcohol-soluble substances in the muscle, apparently as a result of lactate formation via degradation of glycogen (von Helmoltz 1845).
3. Muscle contractions were accompanied by tissue acidification, again presumably as a result of glycogenolysis to form lactate plus protons (du Bois-Reymond 1859; Heidenhain 1864).
4. Direct demonstration that electrically stimulated contractions resulted in a decrease in muscle glycogen content (Weiss 1871).

As discussed by Zuntz (1911), these observations led to the erroneous belief that the energy required to support muscle contractions was derived entirely from carbohydrate. Major proponents of this theory, which persisted in various forms until almost the middle of the twentieth century, included Chauveau himself and, subsequently, Archibald Vivian (A.V.) Hill.

Chauveau based his belief in part on measurement of the respiratory exchange ratio (RER; i.e., whole-body CO_2 production divided by whole-body O_2 uptake) during strenuous exercise lasting about an hour, which revealed a value close to unity, commensurate with oxidation of strictly carbohydrates (Chauveau 1896). He therefore proposed that fat could only contribute to the energy needs of contracting muscle if it were first converted to carbohydrate in the liver, a process that he estimated would increase the overall energy cost of exercise by approximately 30%.

In contrast to Chauveau, Hill's perspective was primarily based on biochemical and biophysical studies of isolated amphibian muscle, including his mentor Fletchers' definitive demonstration of lactate production during electrical stimulation (Fletcher and Hopkins 1907) and his own measurements of the effects of O_2 availability on the heat released during and after such evoked contractions (Hill 1910, 1913, 1914). These studies, for which Hill eventually shared the Nobel Prize in Physiology or Medicine with Meyerhof in 1922, demonstrated that although a fixed amount of heat was always liberated during the contractile activity, the presence of O_2 led to additional heat being released afterward. Hill interpreted these results to mean that lactate formation was directly responsible for force production by the muscle, with oxidative resynthesis of glycogen occurring during recovery.

Regardless of the somewhat different basis for their reasoning, the adamancy of Chauveau and Hill that carbohydrate was the only fuel used by muscle during exercise held considerable sway for decades. This was so even though other data available at the time, including Zuntz's own measurements of RER in exercising horses (Zuntz 1898) and humans (Zuntz and Schumberg 1901), indicated that fat could also be oxidized. As discussed previously (Coggan 2014), the refusal of Chauveau and Hill to accept this conclusion may have stemmed in part from the inability to consistently demonstrate utilization of lipids by the muscle. For example, Leathes (1906) did not find any changes in the total fat content of amphibian muscle following electrical stimulation. In contrast, Lafon (1913) found that electrical stimulation to fatigue decreased the total muscle fat content in two rabbits. Lafon also found a net uptake of fat by muscle at rest and especially during exercise in one horse and two donkeys, but not in three dogs. As emphasized by Zierler (1976), these variable results were probably due to (1) the relatively crude biochemical methods available at the time and (2) the simultaneous uptake of fatty acids by muscle and release by adipose tissue (as recognized by Lafon: "...variations could be due to the possibility that blood, at the same time in which it loses fat to muscle, replenishes [it] from reserves"). Indeed, uncertainty about the precise source and nature of the lipids oxidized by exercising muscle persisted until almost the end of the twentieth century, as will be discussed.

As a result of Chauveau's and Hill's stature as scientists, studies of exercise metabolism in the early 1900s were largely devoted to testing their hypotheses, especially Chauveau's. Often, this entailed manipulating an individual's diet in an attempt to alter bodily carbohydrate stores and then determining the effect of such an intervention on RER and efficiency during exercise (e.g., Heinemann 1901; Frentzel and Reach 1901; Benedict and Cathcart 1913; Krogh and Lindhard 1920; Marsh and Murlin 1928). Benedict and Cathcart (1913), for example, used this approach to test Chauveau's ideas, relying primarily on a professional cyclist as their subject. However, despite inducing a marked shift in substrate utilization, as evidenced by a decrease in RER of 0.10–0.15 units, Benedict and Cathcart were unable to demonstrate any significant decrease in thermodynamic efficiency. Krogh and Lindhard (1920), though, were mistrustful of these data, because Benedict and Cathcart used a mouthpiece and nose clip to collect expired air, which can lead to errors due to hypo- or hyperventilation. They therefore built and carefully validated a respiration chamber large enough to enclose a cycle ergometer, and essentially repeated Benedict and Cathcart's experiments. Unlike this prior study, Krogh and Lindhard found that subjects were 10–11% less efficient during exercise when oxidizing purely fat versus purely carbohydrate. Similarly, using a young boxing instructor as their subject, Marsh and Murlin (1928) found a 11–12% difference in efficiency after at least 3 days of a high-fat diet.

The results of Krogh and Lindhard, Marsh and Murlin, and previously also Frentzel and Reach (working in Zuntz's lab), were therefore all remarkably consistent in demonstrating a roughly 10% difference in efficiency depending upon the substrate being oxidized, which was clearly less than the 30% difference predicted by Chauveau or even the slightly lower figure calculated by Zuntz. However, Krogh and Lindhard did not consider such data to be definitive, stating that they were "...not convinced of the validity of any of these summary methods of calculating the waste of energy incidental to the conversion of fat into sugar. . . ." Marsh and Murlin, on the other hand, were somewhat more confident, concluding that "we cannot account for the lower efficiency [based] on Chauveau's theory." In contrast, Hill interpreted these results to ". . .suggest[s] strongly that the primary breakdown is of carbohydrate, and that fat is used only in a secondary manner, e.g., to restore the carbohydrate which has disappeared. . ." (Hill 1924). Thus, the question of whether muscle could use fat directly during exercise remained unsettled for some time.

Carbohydrate's grip on the battlefield as the sole substrate utilized by the contracting muscle only really began to loosen after further biochemical advances in the late 1920s and early 1930s led to what Hill himself described as a "revolution in muscle physiology" (Hill 1932). These breakthroughs, which have been reviewed in greater detail by Maruyama (1991), included:

1. The contemporaneous discovery of "phosphagen" and its identification as phosphocreatine (PCr) by Eggleton and Eggleton (1927) and Fiske and SubbaRow (1927), respectively.
2. Establishment of PCr's high heat of hydrolysis by Meyerhof and Suranyi (1927).
3. Subsequent near-simultaneous and independent discovery of "pyrophosphate" by Lohman (1929) and "adenosine triphosphonic ester" by Fiske and SubbaRow (1929), which were almost immediately recognized to be the same compound. Originally called "adenylpyrophosphate" in Meyerhof's lab where Lohman worked, this was changed to adenosine triphosphate (ATP) by Barrenscheen and Filz (1932).
4. The recognition by Lohman (1934) that ATP served as the immediate source of energy during muscle contractions, with the ATP utilized being rapidly resynthesized via the hydrolysis of PCr.

Combined with the demonstration by Lundsgaard (1930a, b, 1931, 1932) that blocking glycolysis and hence lactate production using iodoacetic acid did not prevent frog muscle from contracting, these findings finally put to rest the Hill and Meyerhof "lactic acid" theory of muscle contraction.[1]

---

[1]Even so, Hill seemed to stubbornly cling to his and Meyerhof's original beliefs, first proposing in 1933 that the "lactacid" portion of post-exercise O_2 consumption was due to the resynthesis of glycogen from lactate Margaria et al. (1933) and then in 1950 emphasizing that direct proof of ATP hydrolysis during contractions was still lacking:

> In the lactic acid era the evidence that the formation of lactic acid was the cause and provided the energy for contraction seemed pretty good. In the phosphagen era a similar attribution to phosphagen appeared even better justified. Now, in the adenosinetriphosphate era lactic acid and phosphagen have been relegated to recovery and ATP takes their place. Those of us who have lived through two revolutions are wondering whether and when the third is coming. (Hill 1950)

Hill's famous "challenge to biochemists" was only finally met in 1962, when Cain and Davies (1962) were able to demonstrate small but significant and reciprocal changes in ATP and adenosine diphosphate (ADP) in contracting frog muscle by inhibiting creatine kinase using 1-fluoro-2,4-dinitrobenzene.

Thus, in the years leading up to World War II, it gradually became accepted that along with carbohydrates, fats could also be used for fuel during exercise. However, it was still thought that this could only occur via some indirect pathway (Lundsgaard 1938; Steinhaus 1941; Gemmill 1942). For example, Lundsgaard wrote that "it is probable that the high-molecular fatty acids are not attacked, or not readily attacked oxidatively in the muscles." Instead, Lundsgaard hypothesized that fatty acids were converted into ketone bodies by the liver, which were then utilized by the muscle. In support of this hypothesis, Blixenkrone-Møller observed avid uptake of ß-hydroxybutyrate by resting and especially contracting perfused cat hindquarters (Blixenkrone-Møller 1938). Drury and coworkers also found significant ketone body extraction by muscle in various species, including humans (Barnes and Drury 1937), and shortly thereafter reported that exercise had a temporary ketone-lowering effect in both rats and humans (Drury et al. 1941). Similar results were obtained in guinea pigs and humans by Neufeld and Ross (1943). Ketone bodies therefore shared the arena with carbohydrates for approximately 15–20 years, until they were displaced by non-esterified fatty acids as described below.

1.4 Exercise Metabolism, World War II to the Late 1960s: Lipids Have Their Heyday

As described previously (Coggan 2014), research into exercise metabolism slowed during and immediately after the Second World War, due to (1) the negative impact of the conflict on the lives of important scientists and (2) a shift in emphasis in exercise physiology research toward more pragmatic studies of heat and altitude acclimation, fitness testing, ergogenic aids, etc. However, metabolic research then began to accelerate again in the 1950s, on the heels of major advances in biochemistry, driven in part by the availability of ^{14}C-labeled tracers in the new nuclear age (Krebs 1964). Such advancements soon led to the realization that long-chain fatty acids, not ketone bodies, were the plasma lipid substrate normally utilized by resting and contracting skeletal muscle. Specifically, in 1958, Fritz and colleagues used ^{14}C-labeled palmitate to demonstrate that electrical stimulation resulted in a doubling in the rate of fatty acid oxidation by isolated rat skeletal muscle (Fritz et al. 1958). Shortly thereafter, Friedberg, Estes, and coworkers reported that exercise increased the rate of clearance of a bolus of ^{14}C-labeled palmitate from plasma in humans (Friedberg et al. 1960, 1963). Friedberg and Estes also used ^{14}C-labeled palmitate to quantify the rate of $^{14}CO_2$ production across the human forearm and found that it increased during contractions (Friedberg and Estes 1961). This study therefore provided the first direct evidence that exercise increases the rate of fatty acid utilization by the human skeletal muscle. Subsequent experiments using continuous infusion of various ^{14}C-labeled fatty acids essentially confirmed this conclusion and firmly established the importance of plasma-borne fatty acids as an energy substrate during exercise (Carlson and Pernow 1961; Havel et al. 1963, 1964, 1967; Issekutz Jr et al. 1965).

This muscling aside of ketone bodies by fatty acids finally broke carbohydrate's iron-like grip as "the" source of energy during exercise. In fact, the see-saw nature of the battle between carbohydrate and fat actually seemed to swing too far the other way, with at least some scientists of the era apparently believing that the latter was equally the dominant metabolic fuel, writing that "free fatty acids are the major circulating metabolites burned by working muscle in the postabsorptive state" (Havel et al. 1963), "fatty acid oxidation is considered the primary if not the sole energy source in exercising men" (Rowell et al. 1965), and "muscular work, performed aerobically in the post-absorptive state, depends mainly on utilization of fat" (Jones and Havel 1967).[2] Glycogen was relegated to a role as an emergency substrate used only during high-intensity "anaerobic" exercise, whereas the contribution from plasma glucose as an energy source during any form of exercise was considered to be relatively unimportant (e.g., "glucose uptake from the blood [is] negligible" (Bergström and Hultman 1966b), "the rates of turnover and oxidation of plasma glucose play only a minor role in exercise metabolism" (Paul and Issekutz Jr 1967), "the amount of glucose extracted [...] does not amount to more than 5-6% of all energy production of the skeletal muscles at submaximal or maximal work levels" (Keul et al. 1967)).

1.5 Exercise Metabolism, Late 1960s to ca. 1990: Carbohydrates Mount a Comeback

Just when it seemed that fatty acids had prevailed in the struggle for supremacy as the most important fuel during exercise, carbohydrates launched a counterattack. This was largely the result of the resurrection of Duchenne's percutaneous needle muscle biopsy technique (Charrière and Duchenne 1865)[3] by Bergström and Hultman (1966a). Application of this "new" method quickly revealed that the glycogen content of the *v. lateralis* declined significantly even during low intensity exercise (i.e., 30 min of supine cycling at 50 W) (Bergström and Hultman 1966b). Even greater utilization of glycogen was observed during more prolonged or intense cycling (Ahlborg et al. 1967; Bergström et al. 1971), Nordic skiing (Bergström et al.

---

[2] Although studies using ^{14}C-labeled fatty acids highlighted their own importance as an energy source during exercise, they also resurrected the long-standing question of the role played by tissue (muscle) lipid stores. Specifically, oxidation of plasma fatty acids was generally found to account for only about half of the total amount of fat oxidized during exercise, as determined via indirect calorimetry (e.g., Havel et al. 1967). However, similar to earlier studies (Leathes 1904, Lafon 1913), attempts in the 1950s and 1960s to directly demonstrate utilization of muscle lipids during contractions were met with mixed success (Volk et al. 1952; George and Naik 1958; Neptune et al. 1960; Masoro et al. 1966; Carlson 1967).

[3] For a full history of the method, see Waclawik and Lanska (2019).

1973) or distance running (Costill et al. 1971, 1973). Perhaps most importantly, it was shown that an individual's time to fatigue during moderate-intensity exercise was highly correlated with their initial muscle glycogen level, as manipulated using variations in activity and diet (Bergström et al. 1967). These and other studies therefore cemented the importance of muscle glycogen as an energy source during exercise, especially at higher intensities typical of many athletic competitions.

During this same period, a number of studies also showed that contrary to previous suggestions, plasma glucose could also be an important source of energy during exercise. Using the arteriovenous balance approach, for example, Ahlborg et al. (1974) found that during prolonged, low-intensity cycling glucose uptake by the legs could account for 30–40% of total energy expenditure. Even though muscle glycogen utilization increases at higher exercise intensities, plasma glucose was found to account for a similar fraction of overall energy production during exercise at 60–70% of VO_2max (Wahren et al. 1971; Martin et al. 1978). Comparable results were subsequently obtained using other methods, i.e., isotopic tracer infusion or the glucose clamp technique (see Coggan 1991 for review). It was also demonstrated that fatigue during prolonged, moderate-intensity exercise was often the result of both muscle glycogen depletion and hypoglycemia, and not just glycogen depletion alone (Coyle et al. 1986; Coggan and Coyle 1987), thus aiding carbohydrate's revival as a key substrate.

This reemergence of carbohydrates forced lipids to surrender their position as the chief, if not only, metabolic fuel during exercise. Nonetheless, they retained a position of importance, especially during lower-intensity, more prolonged exercise. Indeed, the period from the 1960s to approximately 1990 was marked by an increasingly sophisticated understanding of the overall pattern of substrate utilization during exercise. For example, it was during this period that it was finally accepted that along with plasma-borne fatty acids, intramuscular triglycerides could also be an important source of energy during exercise (Watt et al. 2002). Notably, this was not the result of any single study, but rather was simply due to the accumulation of evidence over the previous 100 years. This deeper appreciation of the nuanced nature of substrate choice by contracting muscle was aided along the way by the development and application of newer, less invasive methods than muscle biopsy or arteriovenous balance sampling, such as stable isotopic tracer techniques (see Coggan 1999a, b for review) and ^{31}P, ^{1}H, and ^{13}C magnetic resonance spectroscopy, used to assess high-energy phosphate, muscle triglyceride, and muscle and liver glycogen metabolism, respectively (see Kemp and Radda 1994 for review).

The period from the 1960s to ca. 1990 was also marked by an explosion of research into the biochemical responses and adaptations to exercise at the cellular level. Spearheaded by work from the laboratory of John Holloszy (Hagberg et al. 2019), such studies demonstrated how muscle mitochondrial respiratory capacity plays a key role in determining the rates of muscle and liver glycogen utilization during exercise (Fitts et al. 1975). Research using rats by Holloszy and coworkers (Holloszy 1967; Holloszy et al. 1970; Molé and Holloszy 1971; Baldwin et al. 1972) and others such as Pette (reviewed by Pette and Vrbová 2017) and using humans by

Varnauskas et al. (1970), Morgan et al. (1971), Gollnick et al. (1972, 1973), etc. also revealed the remarkable adaptability of skeletal muscle in response to changes in demand imposed by voluntary or involuntary (i.e., electrically stimulated) contractile activity. These findings paved the way for subsequent more in-depth studies of the molecular underpinnings of the metabolic responses to acute and chronic exercise in the 1990s and beyond, as discussed previously (Coffey and Hawley 2007; Röckl et al. 2008; Hawley et al. 2015).

1.6 Exercise Metabolism, ca. 1990 to Present: Détente Prevails

Approaching the end of the twentieth century, the overall pattern of substrate utilization during exercise, at least in young, healthy male subjects, had been well described. Thus, after this point, research into exercise metabolism became increasingly focused on additional factors that might modify this pattern, as previously noted (Coggan 2014). Although space precludes a detailed discussion of the impact of such "special circumstances" on the utilization of various fuels, a brief discussion of some of them is provided below. For additional details, readers are again referred to previous reviews (e.g., Holloszy and Coyle 1984, Coggan and Williams 1995, Coggan 1996, 1999a, b, Tarnopolsky and Ruby 2001, Mittendorfer and Klein 2001, Devries 2016, etc.).

1.6.1 Endurance Training

It was widely recognized in the early years of the twentieth century that trained athletes are less dependent than untrained individuals on carbohydrate metabolism during exercise, as evidenced by their lower RER and blood lactate levels. However, it was generally assumed that this was due to an athlete's muscles being less "anaerobic," or more vaguely ascribed to their greater "skill," with the only formal study of this adaptation being a longitudinal training of three men conducted by McNelly (1936). The previously described boom in exercise/metabolic research in the 1960s and thereafter, however, brought renewed attention to this question. Key *human* experiments during this period included the first cross-sectional (Hermansen et al. 1967; Evans et al. 1979) and longitudinal (Saltin et al. 1976; Karlsson et al. 1974) studies demonstrating that training reduces the rate of muscle glycogen utilization during exercise. Subsequently, it was also shown that training reduces utilization of plasma glucose as well (Coggan et al. 1990; Mendenhall et al. 1994), with this lesser demand associated with slower rates of hepatic glycogenolysis and gluconeogenesis (Coggan et al. 1995a). Conversely, the rate of intramuscular triglyceride utilization was shown to be increased (Hurley et al. 1986). All of these adaptations were found to be evident not only at the same absolute exercise intensity but even at the same relative intensity, i.e., the same percentage of VO_2max (Coggan et al. 1995b, 2000). On the other hand, utilization of plasma free fatty acids, which was initially thought to not only compensate for the slower rate of carbohydrate utilization but also to at least partially cause it via the glucose-fatty acid cycle (Holloszy 1973), was found to actually be lower during exercise at the same absolute intensity after training (Martin 3rd et al. 1993). This is due to a slower rate of sympathetically mediated adipose tissue lipolysis in the trained state, as training increases not only the maximal capacity of muscle to oxidize fatty acids but also the maximal capacity for inward transport as well (Talanian et al. 2010). Consequently, during exercise at the same relative intensity, when activation of the sympathetic nervous system is comparable, utilization of plasma free fatty acids is higher in the trained state (Coggan et al. 2000).

1.6.2 Aging

Measurement of RER and blood lactate levels in older vs. younger subjects during incremental exercise provided the first evidence that aging results in an increase in carbohydrate utilization and a decrease in fat utilization during exercise (Robinson 1938; Durnin and Mikulicic 1956; Åstrand 1958). It was only many years later, however, that this issue was studied in any great detail. Specifically, using stable isotopic tracers, Sial et al. (1996) determined the rates of whole-body glucose and free fatty acid turnover while also measuring the overall rates of carbohydrate and fat oxidation via indirect calorimetry in older and younger men and women exercising at both the same absolute and the same relative intensity. At the same absolute intensity, carbohydrate oxidation was higher in older subjects, with this difference being apparently due to a higher rate of muscle glycogen utilization (as glucose kinetics were similar). Conversely, the overall rate of fat oxidation was lower, despite a higher turnover and especially a higher concentration of free fatty acids in the older subjects. Differences in relative rates of substrate oxidation were minimized during exercise at the same relative intensity, but free fatty acid concentrations were still higher in older subjects, due to their much lower rate of free fatty acid clearance. These age-related changes in substrate metabolism are undoubtedly the result of a complex interaction of a host of factors, including decreases in VO_2max with aging, which confounds the basis on which subjects of varying fitness have conventionally been compared. Another important factor, however, is age-related changes in the skeletal muscle itself, especially a decline in mitochondrial content and hence in respiratory capacity (Coggan et al. 1992), which results in a greater disturbance in muscle energetics during contractile activity, even when accounting for the age-related decrease in muscle mass (Coggan et al. 1993).

1.6.3 Sex

Due to societal norms that discouraged mass participation of women in physical activity and sports, possible sex-related differences in substrate metabolism during exercise were the last battlefront to be confronted. In 1979, however, Costill and colleagues studied male and female distance runners matched for both training volume and VO_2max and found no difference in RER during submaximal exercise (Costill et al. 1979). Similar results were obtained by Powers et al. (1980) and Helgerud et al. (1990), whereas Froberg and Pedersen (1984) and Blatchford et al. (1985) found RER to be lower in women. The first in-depth study, however, was performed by Tarnopolsky et al. (1990), who used indirect calorimetry and the muscle biopsy technique to compare substrate metabolism in men and women matched for training and volume and VO_2max while also controlling for menstrual status and diet. Supporting a sex-related difference in substrate preference, they found both RER and muscle glycogen utilization to be lower in the women. Although follow-up studies from the same group as well as others have yielded somewhat mixed results (Phillips et al. 1993; Tarnopolsky et al. 1995, 1997), the bulk of the evidence indicates that women do rely somewhat more on fat as fuel during exercise. As with the effects of aging, multiple mechanisms almost certainly contribute to this subtle difference, including sex-related differences in gonadal and catecholamine hormone levels, type II muscle fiber volume, muscle glycolytic/ glycogenolytic enzyme activities, etc. (Coggan 1999a, b).

1.6.4 Obesity and Type 2 Diabetes

Also reflecting societal trends, a number of relatively recent studies have determined the effects of obesity or diabetes on substrate metabolism, primarily at rest but also in response to exercise. As reviewed by Houmard (2008), these studies have demonstrated that although the overall rate of fatty acid uptake by muscle is increased, the capacity to oxidize such fatty acids is actually reduced. This leads to intramuscular accumulation of lipid metabolites (triglycerides, ceramides, diacylglycerol, etc.) that are thought to play a leading role in the muscle insulin resistance in such disease states. Interestingly, this reduced capacity to oxidize fatty acids is not reversed by weight loss (Thyfault et al. 2004), suggesting that genetics may play a significant role in its etiology.

1.7 Summary

During the last 150 years, tremendous advances have been made in understanding the pattern of substrate metabolism during exercise. The picture that has emerged is that of muscle as a metabolic omnivore, capable of utilizing carbohydrates, fats, and even to some extent protein to fulfill its energy needs, with the precise mixture of substrates being oxidized depending upon the exercise intensity and duration and the individual's diet, fitness, age, sex, etc. Future studies will undoubtedly provide even greater detail and lead to an even deeper mechanistic basis for these findings.

References

Ahlborg B, Bergström J, Ekelund L-G, Hultman E (1967) Muscle glycogen and muscle electrolytes during prolonged physical exercise. Acta Physiol Scand 70:129–140

Ahlborg G, Felig P, Hagenfeldt L, Hendler R, Wahren J (1974) Substrate turnover during prolonged exercise in man. Splanchnic and leg metabolism of glucose, free fatty acids, and amino acids. J Clin Invest 53:1080–1090

Åstrand I (1958) The physical work capacity of workers 50-64 years old. Acta Physiol Scand 42: 73–86

Baldwin KM, Klinkerfuss GH, Terjung RL, Molé PA, Holloszy JO (1972) Respiratory capacity of white, red, and intermediate muscle: adaptive response to exercise. Am J Phys 222:363–378

Barnes RH, Drury DR (1937) Utilization of ketone bodies by the tissues in ketosis. Proc Soc Exp Biol Med 36:350–352

Barrenscheen HK, Filz W (1932) Untersuchungen zur Frage der Co-Fermentwirkung. II. Mitteilung: Zur Chemie der Adenosintriphosphorsäuren. Biochem Z 250:281–304

Benedict FG, Cathcart EP (1913) Muscular work. A metabolic study with special references to the efficiency of the human body as a machine. Carnegie Institution, Washington, DC

Bergström J, Guarnieri G, Hultman E (1971) Carbohydrate metabolism and electrolyte changes in human muscle tissue during heavy work. J Appl Physiol 30:122–125

Bergström J, Hermansen L, Hultman E, Saltin B (1967) Diet, muscle glycogen, and physical performance. Acta Physiol Scand 71:140–150

Bergström J, Hultman E (1966a) Muscle glycogen synthesis after exercise: an enhancing factor localized to the muscle cells in man. Nature 210:309–310

Bergström J, Hultman E (1966b) The effect of exercise on muscle glycogen and electrolytes in normals. Scand J Clin Lab Invest 18:16–20

Bergström J, Hultman E, Saltin B (1973) Muscle glycogen consumption during cross-country skiing (the vasa ski race). Int Z Angew Physiol 31:71–75

Berzelius M, Berzelius JJ (1806–1808) Föreläsningar i djurkemien. C Delén, Stockholm

Bischoff TLW, Voit C (1860) Die Gesetze der Ernährung des Fleischfressers durch neue Untersuchungen festgestellt. CF Winter'sche Verlagahadlung, Leipzig and Heidelberg

Blatchford FJ, Knowlton RG, Da S (1985) Plasma FFA responses to prolonged walking in untrained men and women. Eur J Appl Physiol 53:343–347

Blixenkrone-Møller N (1938) Über den Abbau von Ketonkörpern. Z Physiol Chem 253:261–275

Cain DF, Davies RE (1962) Breakdown of adenosine triphosphate during a single contraction of working muscle. Biochem Biophys Res Commun 8:361–366

Carlson LA (1967) Lipid metabolism and muscular work. Fed Proc 26:1755–1759

Carlson LA, Pernow B (1961) Studies on blood lipids during exercise. II. The arterial plasma-free fatty acid concentration during and after exercise and its regulation. J Lab Clin Invest 58:673–681

Carpenter KJ (2003) A short history of nutritional science: part 1 (1785–1885). J Nutr 133:638–645

Charrière M, Duchenne GB (1865) Emporte pièce histologique. Bull Acad Natl Med 30:1050–1051

Chauveau A (1896) Source et nature du potential directment utilisé dans la travail musculairs d'asprès les exchanges respiratories, chez l'hormme en état d'abstinence. CR Acad Sci Paris 122:1163–1221

Coffey VG, Hawley JA (2007) The molecular bases of training adaptation. Sports Med 37:737–763

Coggan AR (1991) Plasma glucose metabolism during exercise in humans. Sports Med 11:102–124

Coggan AR (1996) Effect of endurance training on glucose metabolism during exercise: stable isotope studies. In: Maughan RJ, Shirreffs SM (eds) Biochemistry of exercise IX. Human Kinetics, Champaign IL, pp 26–36

Coggan AR (1999a) Use of stable isotopes to study carbohydrate and fat metabolism at the whole-body level. Proc Nutr Soc 58:953–961

Coggan AR (1999b) Effects of gender and aging on substrate metabolism during exercise. In: Lamb DR, Murray R (eds) Perspectives in exercise science and sports medicine, The metabolic basis of performance in exercise and sport, vol 12. Cooper, Carmel, IN, pp 355–387

Coggan AR (2014) Metabolic systems: substrate utilization (1910-2010). In: Tipton CM (ed) History of exercise physiology. Human Kinetics, Champaign IL, pp 423–446

Coggan AR, Abduljalil AM, Swanson SC, Earle MS, Farris JW, Mendenhall LA, Robitaille P-M (1993) Muscle metabolism during exercise in young and older untrained and endurance-trained men. J Appl Physiol 75:2125–2133

Coggan AR, Coyle EF (1987) Reversal of fatigue during prolonged exercise by carbohydrate infusion or ingestion. J Appl Physiol 63:2388–2395

Coggan AR, Kohrt WM, Spina RJ, Bier DM, Holloszy JO (1990) Endurance training decreases plasma glucose turnover and oxidation during moderate intensity exercise in men. J Appl Physiol 68:990–996

Coggan AR, Raguso CA, Gastaldelli A, Sidossis LS, Yeckel CW (2000) Fat metabolism during high intensity exercise in endurance-trained and untrained men. Metabolism 49:122–128

Coggan AR, Raguso CA, Williams BD, Sidossis LS, Gastaldelli A (1995a) Glucose kinetics during high-intensity exercise in endurance-trained and untrained humans. J Appl Physiol 78:1203–1207

Coggan AR, Spina RJ, Rogers MA, King DS, Brown M, Nemeth PM, Holloszy JO (1992) Histochemical and enzymatic comparison of the gastrocnemius muscle of young and elderly men and women. J Geront 47:B71–B76

Coggan AR, Swanson SC, Mendenhall LA, Habash DL, Kien CL (1995b) Effect of endurance training on hepatic glycogenolysis and gluconeogenesis during prolonged exercise in men. Am J Phys 268:E375–E383

Coggan AR, Williams BD (1995) Metabolic adaptations to endurance training: substrate metabolism during exercise. In: Hargreaves M (ed) Exercise Metabolism. Human Kinetics, Champaign IL, pp 177–210

Costill DL, Bowers R, Branam G, Sparks K (1971) Muscle glycogen utilization during prolonged exercise on successive days. J Appl Physiol 31:834–838

Costill DL, Fink WJ, Getchell LH, Ivy JL, Witzmann F (1979) Lipid metabolism in skeletal muscle of endurance-trained males and females. J Appl Physiol 47:781–791

Costill DL, Gollnick PD, Jansson ED, Saltin B, Stein EM (1973) Glycogen depletion pattern in human muscle fibres during distance running. Acta Physiol Scand 89:374–383

Coyle EF, Coggan AR, Hemmert MK, Ivy JL (1986) Muscle glycogen utilization during prolonged strenuous exercise when fed carbohydrate. J Appl Physiol 61:165–172

de Fourcroy A (1789) Extrait dun mémoire la ayant pour titre recherches pour servir à l'historie du gaz azote ou de la mofette, comme principe des matières animals. Ann Chim 1:40–46

Devries MC (2016) Sex-based differences in endurance exercise muscle metabolism: impact on exercise and nutritional strategies to optimize health and performance in women. Exp Physiol 101:243–249
Drury DR, Wick AN, MacKay EM (1941) The action of exercise on ketosis. Am J Phys 134:761–768
du Bois-Reymond EH (1859) On the supposed acid reaction of muscular fibre. Philos Mag Series 4(18):544–546
Durnin JVGA, Mikulicic V (1956) The influence of graded exercise on oxygen consumption, pulmonary ventilation and heart rate of young and elderly men. Q J Exper Physiol 41:422–452
Eggleton P, Eggleton GP (1927) XXV. The inorganic phosphate and a labile form of organic phosphate in the gastrocnemius of the frog. Biochem J 21:190–195
Evans WJ, Bennett AS, Costill DL, Fink WJ (1979) Leg muscle metabolism in trained and untrained men. Res Q 50:350–350
Fick A, Wislicenus J (1866) LXX. On the origin of muscle power. Philos Mag Series 4 31:485–503
Fiske CH, SubbaRow Y (1927) The nature of the "inorganic phosphate" in voluntary muscle. Science 65:401–403
Fiske CH, SubbaRow Y (1929) Phosphorous compounds of muscle and liver. Science 70:381–382
Fitts RH, Booth FW, Winder WW, Holloszy JO (1975) Skeletal muscle respiratory capacity, endurance, and glycogen utilization. Am J Phys 228:1029–1033
Fletcher WM, Hopkins FG (1907) Lactic acid in amphibian muscle. J Physiol Lond 35:247–309
Frankland E (1866) On the source of muscular power. Proc Roy Inst Great Brit 4:661–685
Frentzel J, Reach F (1901) Untersuchungen zur Frage nach der Quelle der Muskelkraft. Pflüger's Arch 83:447
Friedberg SJ, Estes EH (1961) Direct evidence for the oxidation of free fatty acids by peripheral tissues. J Clin Invest 41:677–681
Friedberg SJ, Harlan WR Jr, Trout DL, Estes EH (1960) The effect of exercise on the concentration and turnover of plasma nonesterified fatty acids. J Clin Invest 39:215–220
Friedberg SJ, Sjer PB, Bogdonoff MD, Estes EH (1963) The dynamics of plasma free fatty acid metabolism during exercise. J Lipid Res 4:34–38
Fritz IB, Davis DG, Holtrop RH, Dundee H (1958) Fatty acid oxidation by skeletal muscle during rest and activity. Am J Phys 194:379–386
Froberg K, Pedersen PK (1984) Sex differences in endurance capacity and metabolic responses to prolonged heavy exercise. Eur J Appl Physiol 52:446–450
Gemmill CL (1942) The fuel for muscular exercise. Physiol Rev 22:32–53
George J, Naik RM (1958) Relative distribution and chemical nature of the fuel store of the two types of fibres in the pectoralis major muscle of the pigeon. Nature 181:709
Gollnick PD, Armstrong RB, Saltin B, Saubert CW 4th, Sembrowich CL, Shepherd RE (1973) Effect of training on enzyme activity and fiber composition of human skeletal muscle. J Appl Physiol 34:107–111
Gollnick PD, Armstrong RB, Saubert CW IV, Piehl K, Saltin B (1972) Enzyme activity and fiber composition in skeletal muscle of untrained and trained men. J Appl Physiol 33:312–319
Hagberg JM, Coyle EF, Baldwin KM, Cartee GD, Fontana L, Joyner MJ, Kirwan JP, Seals DR, Weiss EP (2019) The historical context and scientific legacy of John O. Holloszy J Appl Physiol 127:277–305
Harold H (1951) Origin of the word "protein". Nature 168:244
Havel RJ, Carlson LA, Ekelund L-G, Holmgren A (1964) Turnover rate and oxidation of different free fatty acids in man during exercise. J Appl Physiol 19:613–618
Havel RJ, Naimark A, Borchgrevink CF (1963) Turnover rate and oxidation of free fatty acids of blood plasma in man during exercise: studies during continuous infusion of palmitate-1-C14. J Clin Invest 42:1054–1063
Havel RJ, Pernow B, Jones NL (1967) Uptake and release of free fatty acids and other metabolites in the legs of exercising men. J Appl Physiol 23:90–99

Hawley JA, Maughan RJ, Hargreaves M (2015) Exercise metabolism: historical perspective. Cell Metab 22:12–17
Heidenhain R (1864) Mechanische Leistung. Stoffumsatz bei Muskeltätigkeit–Buch neu kaufen, Wärmeentwicklung und
Heinemann HN (1901) Experimentelle Untersuchung am Menschen über den Einfluss der Muskelarbeit auf den Stoffverbrauch und die Bedeutung der einzelnen Nährstoffe als Quelle der Muskelkraft. Pflüger's Arch 83:441
Helgerud J, Ingjer F, Strömme SB (1990) Sex differences in performance-matched marathon runners. Eur J Appl Physiol 61:433–439
Hermansen LE, Hultman E, Saltin B (1967) Muscle glycogen during prolonged severe exercise. Acta Physiol Scand 71:129–139
Hill AV (1910) The heat produced in contracture and muscle tone. J Physiol Lond 40:389–403
Hill AV (1913) The energy degraded in recovery processes of stimulated muscles. J Physiol Lond 46:28–80
Hill AV (1914) The oxidative removal of lactic acid. J Physiol Lond 48:x–xi
Hill AV (1932) The revolution in muscle physiology. Physiol Rev 12:56–67
Hill AV (1950) A challenge to biochemists. Biochim Biophys Acta 4:4–11
Holloszy JO (1967) Biochemical adaptations in muscle. Effects of exercise on mitochondrial oxygen uptake and respiratory enzyme activity in skeletal muscle. J Biol Chem 242:2278–2282
Holloszy JO (1973) Biochemical adaptations to exercise: aerobic metabolism. In: Wilmore JH (ed) Exercise and sport sciences reviews, vol 1. Academic Press, New York, pp 45–71
Holloszy JO, Coyle EF (1984) Adaptations of skeletal muscle to endurance exercise and their metabolic consequences. J Appl Physiol 56:831–838
Holloszy JO, Oscai LB, Don IJ, Molé PA (1970) Mitochondrial citric acid cycle and related enzymes: adaptive response to exercise. Biochem Biophys Res Commun 40:1368–1373
Holmes FL (1963) Elementary analysis and the origins of physiological chemistry. Isis 54:50–18
Houmard JA (2008) Intramuscular lipid oxidation and obesity. Am J Phys 294:R1111–RR116
Hurley BF, Nemeth PM, Martin WH 3rd, Hagberg JM, Dalsky GP, Holloszy JO (1986) Muscle triglyceride utilization during exercise: effect of training. J Appl Physiol 60:562–567
Issekutz B Jr, Miller HI, Paul P, Rodahl K (1965) Aerobic work capacity and plasma FFA turnover. J Appl Physiol 20:293–296
Jones NL, Havel RJ (1967) Metabolism of free fatty acids and chylomicron triglycerides during exercise in rats. Am J Phys 213:824–828
Karlsson J, Nordesjo L-O, Saltin B (1974) Muscle glycogen utilization during exercise after physical training. Acta Physiol Scand 90:210–217
Kemp GJ, Radda GK (1994) Quantitative interpretation of bioenergetic data from ^{31}P and ^{1}H magnetic resonance spectroscopic studies of skeletal muscle: an analytical review. Magn Reson Q 10:43–63
Keul J, Doll E, Keppler D (1967) The substrate supply of the human skeletal muscle at rest, during and after work. Experientia 23:974–979
Krebs HA (1964) The citric acid cycle. Nobel lectures, physiology or medicine 1942–1962. Elsevier, Amsterdam, p 399
Krogh A, Lindhard J (1920) The relative value of fat and carbohydrate as sources of muscular energy. Biochem J 14:290–363
Lafon G (1913) Sur la consommalion des graisses dans l'organisme animal. CR Acad Sci Paris 156: 1248–1250
Leathes JB (1906) Problems in animal metabolism. A course of lectures given in the physiological Laboratory of the London University at South Kensington in the summer term, 1904. Blakiston's Son & Co, Philadelphia, p 99
Lohman K (1929) Über die Pyrophosphatfraktion im Muskel. Naturwissenschaften 17:624–625
Lohman K (1934) Über die enzymatische Aufspaltung der Kreatinphosphorsäure zugleich ein Beitrag zum Chemismis der Muskelkontraktion. Biochem Z 271:264–277

Lundsgaard E (1930a) Untersuchungen über Muskel-Kontraktionen ohne Milchsäurebildung. Biochem Z 217:162–177
Lundsgaard E (1930b) Weitere Untersuchungen über Muskelkontraktionen ohne Milchsäurebildung. Biochem Z 227:51–83
Lundsgaard E (1931) Über die Energetik der anaeroben Muskelkontraktion. Biochem Z 233:322–343
Lundsgaard E (1932) Betydningen af fænomenet "mælkesyrefri muskelkontraktioner" for opfattelsen af muskelkontraktionens kemi. Hospitalstidende 75:84–95
Lundsgaard E (1938) The Pasteur-Meyerhof reaction in muscle metabolism. Bull NY Acad Sci 14: 163–182
Margaria R, Edwards HT, Dill DB (1933) The possible mechanisms of contracting and paying the oxygen debt and the role of lactic acid in muscular contraction. Am J Phys 106:689–615
Marsh ME, Murlin JR (1928) Muscular efficiency on high carbohydrate and high fat diets. J Nutr 1: 105–137
Martin WH 3rd, Dalsky GP, Hurley BF, Matthews DE, Bier DM, Hagberg JM, Rogers MA, King DS, Holloszy JO (1993) Effect of endurance training on plasma free fatty acid turnover and oxidation during exercise. Am J Phys 265:E708–E714
Martin B, Robinson S, Robertshaw D (1978) Influence of diet on leg uptake of glucose during heavy exercise. Am J Clin Nutr 31:62–67
Maruyama K (1991) The discovery of adenosine triphosphate and the establishment of its structure. J Hist Biol 24:145154
Masoro EJ, Rowell LB, McDonald RM, Steiert B (1966) Skeletal muscle lipids. II. Nonutilization of intracellular lipid esters as an energy source for contractile activity. J Biol Chem 241:2626–2634
McNelly WC (1936) Some effects of training on the respiratory response to exercise. Am J Phys 116:100–101
Mendenhall LA, Swanson SC, Habash DL, Coggan AR (1994) Ten days of exercise training reduces glucose production and utilization during moderate-intensity exercise. Am J Phys 266:E136–E143
Meyerhof O, Suranyi J (1927) Über die Wärmetönungen der chemische Reaktionsphasen im Muskel. Biochem Z 191:106–124
Mittendorfer B, Klein S (2001) Effect of aging on glucose and lipid metabolism during endurance exercise. Int J Sports Nutr Exerc Metab 11(Suppl):S86–S91
Molé PA, Holloszy JO (1971) Adaptation of muscle to exercise. Increase in the levels of palmityl CoA synthetase, carnitine palmityltransferase, and palmityl CoA dehydrogenase, and in the capacity to oxidize fatty acids. J Clin Invest 50:2323–2330
Morgan TE, Cobb LA, Short FA, Ross R, Gunn DR (1971) Effect of long-term exercise on human muscle mitochondria. In: Pernow B, Saltin B (eds) Muscle metabolism during exercise. Plenum, New York, pp 87–95
Mulder GJ (1838) Sur la composition de quelques substances animales. Bulletin des Sciences Physiques et Naturelles en Néerlande 104-119
Neptune EM, Sudduth HC, Foreman DR, Fash FJ (1960) Phospholipid and triglyceride metabolism of excised rat diaphragm and the role of these lipids in fatty acid uptake and oxidation. J Lipid Res 1:229–235
Neufeld AH, Ross WD (1943) Blood ketone bodies in relation to carbohydrate metabolism in muscular exercise. Am J Phys 138:747–752
Paul P, Issekutz B Jr (1967) Role of extramuscular energy sources in the metabolism of exercising dogs. J Appl Physiol 22:615–622
Pette D, Vrbová G (2017) The contribution of neuromuscular stimulation in elucidating muscle plasticity revisited. Eur J Transl Myol 27:6368
Phillips SM, Atkinson SA, Tarnopolsky MA, MacDougall JD (1993) Gender differences in leucine kinetics and nitrogen balance in endurance athletes. J Appl Physiol 75:2134–2141

Powers SK, Riley W, Howley ET (1980) Comparison of fat metabolism between trained men and women during prolonged aerobic work. Res Q Exerc Sport 51:427–431

Robinson S (1938) Experimental studies of physical fitness in relation to age. Arbeit 10:251–323

Röckl KSC, Witczak CA, Goodyear LJ (2008) Signaling mechanisms in skeletal muscle: acute responses and chronic adaptations to exercise. IUBMB Life 60:145–153

Rowell LB, Masoro EJ, Spencer MJ (1965) Splanchnic metabolism in exercising man. J Appl Physiol 20:1032–1037

Saltin B, Nazar K, Costill DL, Stein E, Jansson E, Essén B, Gollnick PD (1976) The nature of the training response; peripheral and central adaptations to one-legged exercise. Acta Physiol Scand 96:289–305

Sial S, Coggan AR, Carroll R, Goodwin JS, Klein S (1996) Fat and carbohydrate metabolism during exercise in elderly and young subjects. Am J Phys 271:E983–E989

Smith E (1862) On the elimination of urea and urinary water. Philos Trans R Soc London 151:747–834

Steinhaus AH (1941) Exercise Ann Rev Physiol 3:695–716

Talanian JL, Holloway GP, Snook LA, Heigenhauser GJ, Bonen A, Spriet L (2010) Exercise training increases sarcolemmal and mitochondrial fatty acid transport proteins in human skeletal muscle. Am J Phys 299:E180–E188

Tarnopolsky MA, Atkinson SA, Phillips SM, MacDougall JD (1995) Carbohydrate loading and metabolism during exercise in men and women. J Appl Physiol 78:1360–1368

Tarnopolsky MA, Bosman M, MacDonald JR, Vandeputte D, Martin J, Roy BD (1997) Post-exercise protein-carbohydrate and carbohydrate supplements increase muscle glycogen in men and women. J Appl Physiol 83:1877–1883

Tarnopolsky LJ, MacDougall JD, Atkinson SA, Tarnopolsky MA, Sutton JR (1990) Gender differences in substrate for endurance exercise. J Appl Physiol 68:302–308

Tarnopolsky MA, Ruby BC (2001) Sex differences in carbohydrate metabolism. Curr Opin Clin Nutr Metab Care 4:521–526

Thyfault JP, Kraus RM, Hickner RC, Howell AW, Wolfe RR, Dohm GL (2004) Impaired plasma fatty acid oxidation in extremely obese women. Am J Phys 287:E1076–E1081

Varnauskas E, Björntorp P, Fahlén M, Přerovský I, Stenberg J (1970) Effects of physical training on exercise blood flow and enzymatic activity in skeletal muscle. Cardiovasc Res 4:418–422

Volk ME, Millington RH, Weinhouse S (1952) Oxidation of endogenous fatty acids of rat tissues in vitro. J Biol Chem 195:493–501

von Helmoltz H (1845) Über den Stoffverbrauch bei der Muskelaction. Müller's Arch Anat Physiol:72–83

von Liebig J (1842) Chimie organique appliquée à la physiologie animale et à la pathologie. Chez Fortin, Masson et Cie, Paris

von Pettenkofer M, Voit C (1866) Untersuchungen über den Stoffwechselverbrauch des normalen Menschen. Z Biol 2:459–573

Waclawik AJ, Lanska DJ (2019) Antecedents, development, adoption, and application of Duchenne's trocar for histopathologic studies of neuromuscular disorders in the nineteenth century. J Hist Neurosci 28:176–194

Wahren J, Felig P, Ahlborg G, Jorfeldt L (1971) Glucose metabolism during leg exercise in man. J Clin Invest 50:2715–2725

Watt MJ, Heigenhauser GJF, Spriet LL (2002) Intramuscular triacylglycerol utilization in human skeletal muscle during exercise: is there a controversy? J Appl Physiol 93:1185–1195

Weiss S (1871) Verbunden von Glycogen zu Muskelaktivität. Wien Acad Bericht 64:284–291

Zierler K (1976) Fatty acids as substrates for heart and skeletal muscle. Circ Res 38:459–463
Zuntz N, Hagermann 0 (1898) Untersuchungen über den Stoffwechsel des Pferdes bei Ruhe und Arbeit. Paul Parey, Berlin
Zuntz VN (1911) Umsatz der nahrstoffe. XI. Betrachtungen uber die beziehungen zwischen nahrstoffen und leistungen des korpers. In: Oppenheimer K (ed) Handbuch der Biochemie des Menschen und der Tierre. Fischer, Jena FRG, pp 826–855
Zuntz N, Schumberg WAEF (1901) Studien zu einer Physiologie des Marsches. Hirschwald, Berlin

Chapter 2
Overview of Exercise Metabolism

Mark Hargreaves

Abstract The supply of ATP is critical for ongoing skeletal muscle contractile activity during exercise. The metabolic pathways in muscle that ensure continual ATP supply are PCr degradation, glycolysis, and the oxidative metabolism of CHO (muscle glycogen and blood glucose) and fat (muscle triglyceride and plasma fatty acids). The relative contributions of these metabolic pathways are primarily determined by exercise intensity and duration but also influenced by training status, preceding diet, sex, and age. Various interventions designed to enhance sporting performance target the availability and utilization of metabolic substrates. In addition, metabolic perturbations and metabolic communication during exercise play key roles in the acute and adaptive responses to exercise.

Keywords ATP · Phosphocreatine · Carbohydrate · Fat · Metabolism

During exercise, the maintenance of adenosine triphosphate (ATP) levels is critical for maintaining sarcolemmal excitability, sarcoplasmic calcium (Ca^{2+}) release and uptake, and myofibrillar force production, thereby ensuring ongoing contractile activity. Fatigue during varying types of exercise is often associated with metabolic substrate (PCr (phosphocreatine), muscle glycogen, blood glucose) depletion and/or accumulation of by-products of metabolism (ADP, Pi, H^+, ROS (reactive oxygen species), heat). Various interventions designed to enhance exercise performance are targeted at metabolism (see Hargreaves and Spriet 2020 for review).

Since the intramuscular stores of ATP are relatively small (~5 $mmol.kg^{-1}$ wet muscle), other metabolic pathways must be activated to sustain rates of ATP utilization that may vary from 0.4 $mmol.kg^{-1}.s^{-1}$ during exercise at 200 W (~75% maximal oxygen uptake, VO_2 max) to 3.7 $mmol.kg^{-1}.s^{-1}$ at 900 W (~300% VO_2 max). These metabolic pathways are summarized in Fig. 2.1.

Phosphocreatine (PCr) serves as an immediate buffer for muscle [ATP] and is broken down rapidly during all-out, high-intensity exercise. The breakdown of

---

M. Hargreaves (✉)
Department of Anatomy & Physiology, University of Melbourne, Melbourne, Australia
e-mail: m.hargreaves@unimelb.edu.au

© The Author(s), under exclusive license to Springer Nature Switzerland AG 2022
G. McConell (ed.), *Exercise Metabolism*, Physiology in Health and Disease,
https://doi.org/10.1007/978-3-030-94305-9_2

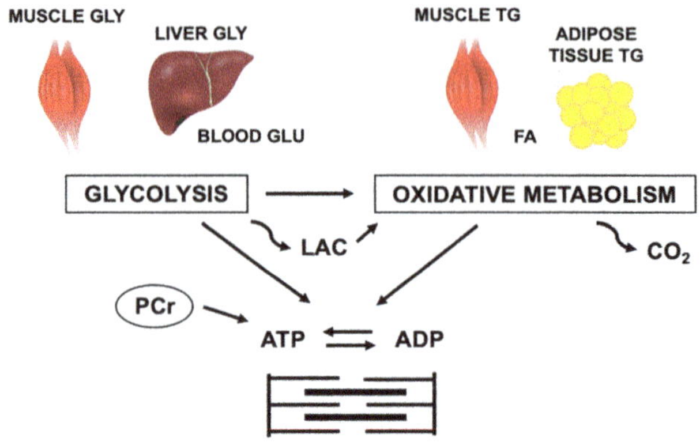

Fig. 2.1 Overview of metabolic pathways within skeletal muscle responsible for ATP generation during exercise. Abbreviations: *ATP* adenosine triphosphate; *ADP* adenosine diphosphate; CO_2 carbon dioxide; *LAC* lactate; *PCr* phosphocreatine. Original figure with permission from author

glucosyl units, derived from muscle glycogen and blood glucose, in glycolysis results in the production of pyruvate that is then converted to either lactate or acetyl CoA, depending on the balance between glycolysis and oxidative metabolism (Spriet et al. 2000). PCr degradation and glycolysis produce ATP via substrate-level phosphorylation that can occur in the absence of oxygen; hence, they are often described as the "anaerobic" energy system (see Chap. 4). The carbohydrate (CHO) fuel sources, muscle glycogen, and blood glucose, along with fatty acids derived from muscle and adipose triglyceride stores, produce ATP through a combination of substrate-level and oxidative phosphorylation in the tricarboxylic acid cycle and electron transport chain (see Chap. 3). This is critically dependent on mitochondrial oxygen availability ("aerobic" energy system), which in turn is determined by the respiratory and cardiovascular systems that facilitate the convective and diffusive transport of oxygen to contracting skeletal muscle during exercise (Hawley et al. 2014). The power (rate of ATP generation) is greater for the anaerobic system (PCr > glycolysis) than the aerobic system (CHO > fat), while the order is opposite for capacity (total amount of ATP generated), with that of the aerobic system (fat > CHO) being higher than that of the anaerobic system (glycolysis > PCr).

During high-intensity exercise of short duration (<1 min), ATP is primarily derived from PCr hydrolysis and glycolysis (Fig. 2.2; Medbø and Tabata 1993; Parolin et al. 1999), with muscle glycogen the dominant CHO source. They are also crucial for ATP generation during the transition from rest to steady-state exercise due to the lag in oxidative phosphorylation (Hughson et al. 2001); however, once exercise extends beyond ~1 min, oxidative phosphorylation is the primary pathway for ATP generation (Medbø and Tabata 1989). Anaerobic metabolism during exercise will be discussed in more detail in Chap. 4.

During exercise lasting from several minutes to hours, the oxidative metabolism of CHO and fat provides the vast majority of ATP for contracting skeletal muscle. Although amino acids from protein can be oxidized by the skeletal muscle during exercise (Rennie et al. 1981), their contribution to overall energy metabolism is small, except perhaps under conditions of reduced availability of other substrates such as muscle glycogen (Howarth et al. 2010). Of greater significance are the

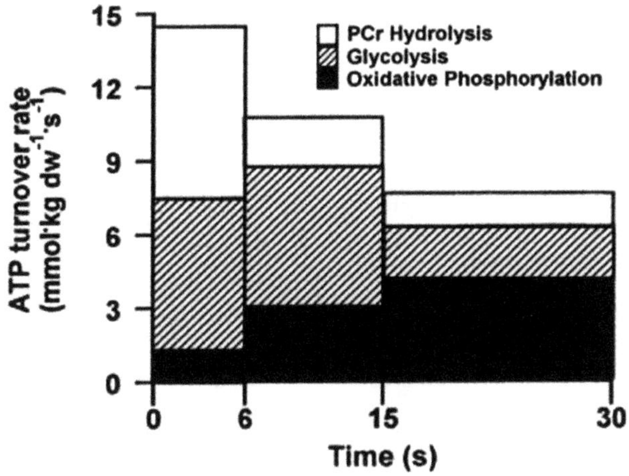

Fig. 2.2 Relative contributions of PCr, glycolysis, and oxidative phosphorylation to ATP generation during 30s of maximal exercise. From Parolin et al. (1999) with permission from the American Physiological Society

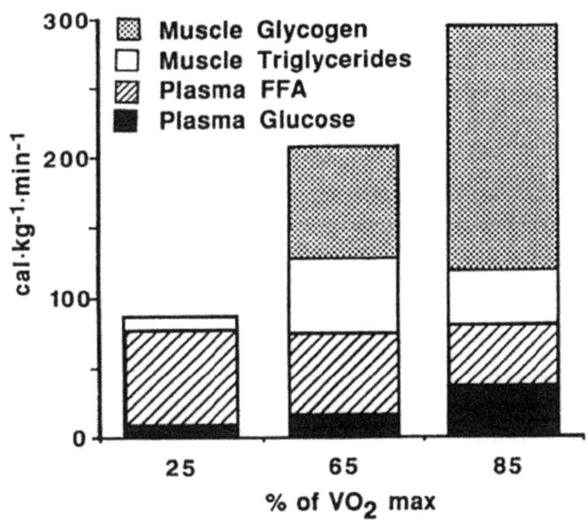

Fig. 2.3 Relative contributions of CHO and fat fuel sources to energy turnover during exercise of increasing intensity. From Romijn et al. (1993) with permission from the American Physiological Society

post-exercise increases in synthesis rates of myofibrillar, mitochondrial, and other proteins that underpin the adaptive responses to various modes of exercise training, including increased skeletal muscle mass, mitochondrial volume, and buffering capacity (Wilkinson et al. 2008; Chap. 9).

The intra- and extramuscular sources of CHO and fat are summarized in Fig. 2.1—muscle glycogen and blood glucose (derived from liver glycogenolysis and gluconeogenesis and the gut when CHO is ingested) and fatty acids from intramuscular (IMTG) and adipose triglyceride stores. The primary determinants of the relative contributions of CHO and fat to oxidative metabolism during exercise are exercise intensity and duration (Figs. 2.3 and 2.4; Romijn et al. 1993; van Loon et al. 2001; Watt et al. 2002). Other factors influencing exercise metabolism include

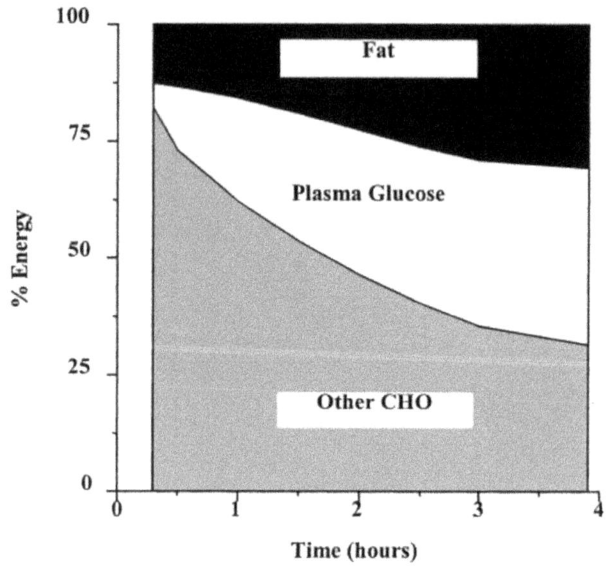

Fig. 2.4 Relative contributions of CHO and fat fuel sources to energy turnover during prolonged cycling exercise at~68% VO_2 max in trained men. "Other CHO" (mainly muscle glycogen) oxidation estimated as the difference between total CHO oxidation assessed from indirect calorimetry and tracer-determined glucose uptake. From Angus et al. (2002) with permission from the American Physiological Society

training status (Chap. 10), substrate availability as influenced by preceding diet, age (Chap. 14), sex (Chap. 15), and environmental factors (Febbraio 2000; Murray 2016).

Carbohydrate oxidation, notably from muscle glycogen, dominates at higher exercise intensities, while maximal fat oxidation occurs at ~60–65% VO_2 max (Romijn et al. 1993; van Loon et al. 2001). The reduction in fat oxidation at higher exercise intensities is due to both lower plasma fatty acid availability and intramuscular factors that limit mitochondrial fatty acid uptake and oxidation (Romijn et al. 1993). Given the dominance of CHO use during exercise at the intensities commonly observed in competitive athletic events (Hawley and Leckey 2015), the relatively finite endogenous CHO reserves, and the observation that fatigue is often associated with muscle glycogen depletion and hypoglycemia (see Chap. 18), there has been considerable interest in the use of CHOs as ergogenic aids in sports (Cermak and van Loon 2013). Ingestion of CHO during prolonged strenuous exercise maintains blood glucose levels, muscle glucose uptake and oxidation, and the ability to sustain exercise intensities above ~70% VO_2 max for longer periods (Angus et al. 2002; Coggan and Coyle 1987, 1988; Coyle et al. 1983). That said, while CHO ingestion delays fatigue, it does not prevent it, suggesting factors other than CHO availability contribute to fatigue, although it is also possible that muscle glycogen levels become depleted at intramuscular locations critical for the processes involved in excitation-contraction coupling (Ørtenblad et al. 2013).

The utilization of muscle glycogen (Chap. 5) and fatty acids derived from IMTG (Chap. 8) is highest during the early part of prolonged exercise and declines progressively, with concomitant increases in blood glucose and fatty acid uptake and oxidation by the contracting skeletal muscle (van Loon et al. 2005; Watt et al.

2002). The increase in muscle glucose uptake (Chap. 6) is accompanied by an increase in liver glucose output from both glycogenolysis and gluconeogenesis (Chap. 11; Ahlborg et al. 1974; Wasserman 2009). During prolonged strenuous exercise, liver glucose output falls below muscle glucose uptake resulting in hypoglycemia which ultimately limits both muscle and cerebral glucose uptake. The maintenance of blood glucose levels by CHO ingestion results in improved endurance exercise performance (Coyle et al. 1983; Coggan and Coyle 1987). Increased adipose tissue lipolysis during exercise results in a progressive increase in plasma fatty acid levels (Chap. 7; Horowitz and Klein 2000, supporting greater muscle fatty acid uptake and oxidation (Chap. 8; Turcotte 2000; Watt et al. 2002). Inhibition of adipose tissue lipolysis reduces muscle fatty acid uptake and results in greater reliance on muscle glycogen and IMTG, with no effect on muscle glucose uptake (van Loon et al. 2005). There has been some debate on IMTG utilization during exercise (Chap. 8; Kiens 2006), but it does appear to be an important substrate for contracting skeletal muscle (Stellingwerff et al. 2007). Despite activation of oxidative metabolism, there is production of lactate and its accumulation within muscle and blood during exercise (Spriet et al. 2000). Although for many years lactate was considered simply a by-product of metabolism, it is now recognized as an important substrate for oxidation, gluconeogenesis, and muscle glycogenesis and as a signalling molecule mediating inter-organ communication and exercise adaptations (Brooks 2020).

The regulation of metabolism during exercise involves the interplay of various local and systemic factors, as exemplified by the so-called dual control of muscle glycogenolysis by contractions and epinephrine (Richter et al. 1982). Exercise-induced changes in substrate, hormone, and electrolyte levels, whole body and local blood flows, and body temperature are all implicated in some way in the regulation of metabolism during exercise (Hawley et al. 2014). The effectiveness of metabolic regulation during exercise is demonstrated by the observation that ATP concentrations in contracting skeletal muscle are maintained at or close to resting levels, except perhaps at fatigue during maximal, sprint (Greenhaff et al. 1994) and prolonged, endurance exercise (Sahlin et al. 1990). Alterations in the intramuscular concentrations of Ca^{2+}, ADP, AMP, Pi, PCr, and glycogen, together with changes in cyclic AMP, nitric oxide (NO), ROS, and muscle tension and temperature, have both direct effects on metabolic pathways via allosteric and covalent regulation of the activities of key enzymes and translocation of substrate transporters such as GLUT4 and FAT CD36 and indirect effects via activation of various kinases and signalling pathways. Most attention in the skeletal muscle has focused on those responsive to changes in Ca^{2+} and energy status (Chen et al. 2000; Rose and Hargreaves 2003; Stephens et al. 2002; Wojtaszewski et al. 2000; Yu et al. 2003). It is teleologically appealing that AMP-activated protein kinase (AMPK), a sensor of muscle energy charge and glycogen levels, is implicated in the regulation of muscle metabolism during exercise, but this appears not to be the case (McConell 2020). Rather, AMPK may be more important in mediating post-exercise metabolism and insulin sensitivity and muscle adaptations to exercise such as increased mitochondrial biogenesis and GLUT4 expression (Flores-Opazo et al. 2020). In addition to effects on muscle function and metabolism during exercise, changes in sarcoplasmic Ca^{2+} levels also mediate key adaptive responses to exercise (Chin 2010).

It is likely that there are complex spatial, temporal, and redundant interactions between multiple signalling pathways in skeletal muscle during exercise. Indeed, a recent study employing phosphoproteomic analyses of human skeletal muscle samples before and after intense exercise identified >1000 phosphosites on >550 proteins, many of currently unknown functional relevance (Hoffman et al. 2015). Subsequent in vitro studies in myoblasts demonstrated that the simultaneous administration of Ca^{2+} and β-adrenergic agonists recapitulated much of the exercise phosphoproteomic signature, emphasizing again the potential importance of "dual control" (Needham et al. 2019). Alterations in muscle high energy phosphate (ATP, PCr) levels have been implicated in muscle adaptive responses to exercise (Ren et al. 1993; Yaspelkis et al. 1999), as have variations in muscle glycogen availability (Hawley et al. 2018), potentially via activation of AMPK and other signalling pathways. There has been considerable interest in the molecular regulation of the adaptive responses to acute and chronic exercise, with a focus on transcriptional and translational regulation and the roles of phosphorylation, methylation, and acetylation of key enzymes and transcriptional regulators (Egan and Zierath 2013; McGee and Hargreaves 2019). An emerging and future area of interest is the potential interaction between exercise metabolism and these molecular responses mediating muscle adaptation (Seaborne and Sharples 2020). Finally, over the last two decades, the important role of metabolic communication and inter-organ cross talk in mediating the acute responses and chronic adaptations to exercise has been recognized (Murphy et al. 2020).

In summary, the supply of ATP is critical for ongoing skeletal muscle contractile activity during exercise. The metabolic pathways in muscle that ensure this are PCr degradation, glycolysis, and the oxidative metabolism of CHO and fat. The relative contributions of these pathways are primarily determined by exercise intensity and duration but also influenced by training status, preceding diet, sex, and age. In addition, metabolic perturbations and metabolic communication during exercise play key roles in the acute and chronic responses to exercise. The subsequent chapters in this volume provide more detailed summaries of our contemporary understanding of exercise metabolism.

References

Ahlborg G, Felig P, Hagenfeldt L et al (1974) Substrate turnover during prolonged exercise in man. J Clin Invest 53:1080–1090

Angus DJ, Febbraio MA, Hargreaves M (2002) Plasma glucose kinetics during prolonged exercise in trained humans when fed carbohydrate. Am J Phys 283:E573–E577

Brooks GA (2020) Lactate as a fulcrum of metabolism. Redox Biol 35:101454

Cermak NM, van Loon LJ (2013) The use of carbohydrate during exercise as an ergogenic aid. Sports Med 43:1139–1155

Chen Z-P, McConell GK, Michell BJ et al (2000) AMPK signalling in contracting human skeletal muscle: acetyl-CoA carboxylase and NO synthase phosphorylation. Am J Phys 279:E1202–E1206

Chin ER (2010) Intracellular Ca^{2+} signaling in skeletal muscle: decoding a complex message. Exerc Sport Sci Rev 38:76–85

Coggan AR, Coyle EF (1987) Reversal of fatigue during prolonged exercise by carbohydrate infusion or ingestion. J Appl Physiol 63:2388–2395

Coggan AR, Coyle EF (1988) Effect of carbohydrate feedings during high-intensity exercise. J Appl Physiol 65:1703–1709

Coyle EF, Hagberg JM, Hurley BF et al (1983) Carbohydrate feeding during prolonged strenuous exercise can delay fatigue. J Appl Physiol 55:230–235

Egan B, Zierath JR (2013) Exercise metabolism and the molecular regulation of skeletal muscle adaptation. Cell Metab 17:162–184

Febbraio MA (2000) Does muscle function and metabolism affect exercise performance in the heat? Exerc Sport Sci Rev 28:171–176

Flores-Opazo M, McGee SL, Hargreaves M (2020) Exercise and GLUT4. Exerc Sport Sci Rev 48: 110–118

Greenhaff PL, Nevill ME, Söderlund K et al (1994) The metabolic responses of human type I and II muscle fibres during maximal treadmill sprinting. J Physiol 478:149–155

Hargreaves M, Spriet LL (2020) Skeletal muscle energy metabolism during exercise. Nat Metab 2: 817–828

Hawley JA, Hargreaves M, Joyner MJ et al (2014) Integrative biology of exercise. Cell 159:738–749

Hawley JA, Leckey JJ (2015) Carbohydrate dependence during prolonged, intense endurance exercise. Sports Med 45:S5–S12

Hawley JA, Lundby C, Cotter JD et al (2018) Maximising cellular adaptation to endurance exercise in skeletal muscle. Cell Metab 27:962–976

Hoffman NJ, Parker BL, Chaudhuri R et al (2015) Global phosphoproteomic analyses of human skeletal muscle reveals a network of exercise-regulated kinases and AMPK substrates. Cell Metab 22:922–935

Horowitz JF, Klein S (2000) Lipid metabolism during endurance exercise. Am J Clin Nutr 72: S558–S563

Howarth KR, Phillips SM, MacDonald MJ et al (2010) Effect of glycogen availability of human skeletal muscle turnover during exercise and recovery. J Appl Physiol 109:431–438

Hughson RL, Tschakovsky ME, Houston ME (2001) Regulation of oxygen consumption at the onset of exercise. Exerc Sci Sport Rev 29:129–133

Kiens B (2006) Skeletal muscle lipid metabolism in exercise and insulin resistance. Physiol Rev 86: 205–243

Medbø JI, Tabata I (1989) Relative importance of aerobic and anaerobic energy release during short-lasting exhausting bicycle exercise. J Appl Physiol 67:1881–1886

Medbø JI, Tabata I (1993) Anaerobic energy release in working muscle during 30s to 3 min of exhausting bicycling. J Appl Physiol 75:1654–1660

McConell GK (2020) It's well and truly time to stop stating that AMPK regulates glucose uptake and fat oxidation *during* exercise. Am J Phys 318:E564–E567

McGee SL, Hargreaves M (2019) Epigenetics and exercise. Trends Endocrinol Metab 30:636–645

Murphy RM, Watt MJ, Febbraio MA (2020) Metabolic communication during exercise. Nat Metab 2:805–816

Murray AJ (2016) Energy metabolism and the high-altitude environment. Exp Physiol 101:23–27

Needham EJ, Humphrey SJ, Cooke KC et al (2019) Phosphoproteomics of acute cell stressors targeting exercise signalling networks reveal drug interactions regulating protein secretion. Cell Rep 29:1524–1536

Ørtenblad N, Westerblad H, Nielsen J (2013) Muscle glycogen stores and fatigue. J Physiol 591: 4405–4413

Parolin ML, Chesley A, Matsos MP et al (1999) Regulation of skeletal muscle glycogen phosphorylase and PDH during maximal intermittent exercise. Am J Phys 277:E890–E900
Ren J-M, Semenkovich CF, Holloszy JO (1993) Adaptation of muscle to creatine depletion: effect on GLUT-4 glucose transporter expression. Am J Phys 264:C146–C150
Rennie MJ, Edwards RH, Krywawych S et al (1981) Effect of exercise on protein turnover in man. Clin Sci 61:627–639
Richter EA, Ruderman NB, Gavras H et al (1982) Muscle glycogenolysis during exercise: dual control by epinephrine and contractions. Am J Phys 242:E25–E32
Romijn JA, Coyle EF, Sidossis LS et al (1993) Regulation of endogenous fat and carbohydrate metabolism in relation to exercise intensity and duration. Am J Phys 265:E380–E391
Rose AJ, Hargreaves M (2003) Exercise increases Ca^{2+}-calmodulin-dependent protein kinase II activity in human skeletal muscle. J Physiol 553:303–309
Sahlin K, Katz A, Broberg S (1990) Tricarboxylic acid cycle intermediates in human muscle during prolonged exercise. Am J Phys 259:C834–C841
Seaborne RA, Sharples AP (2020) The interplay between exercise metabolism, epigenetics and skeletal muscle remodeling. Exerc Sport Sci Rev 48:188–200
Spriet LL, Howlett RA, Heigenhauser GJF (2000) An enzymatic approach to lactate production in human skeletal muscle during exercise. Med Sci Sports Exerc 32:756–763
Stellingwerff T, Boon H, Jonkers RA et al (2007) Significant intramyocellular lipid use during prolonged cycling in endurance trained males as assessed by three different methodologies. Am J Phys 292:E1715–E1723
Stephens TJ, Chen Z-P, Canny BJ et al (2002) Progressive increase in human skeletal muscle AMPKα2 activity and ACC phosphorylation during exercise. Am J Phys 282:E688–E694
Turcotte LP (2000) Muscle fatty acid uptake during exercise: possible mechanisms. Exerc Sport Sci Rev 28:4–9
van Loon LJ, Greenhaff PL, Constantin-Teodosiu D et al (2001) The effects of increasing exercise intensity on muscle fuel utilisation in humans. J Physiol 536:275–304
van Loon LJ, Thomason-Hughes M, Constantin-Teodosiu D et al (2005) Inhibition of adipose tissue lipolysis increases intramuscular lipid and glycogen use in vivo in humans. Am J Phys 289:E482–E493
Wasserman DH (2009) Four grams of glucose. Am J Phys 296:E11–E21
Watt MJ, Heigenhauser GJF, Dyck DJ et al (2002) Intramuscular triacylglycerol, glycogen and acetyl group metabolism during 4h of moderate exercise in man. J Physiol 541:969–978
Wilkinson SB, Phillips SM, Atherton PJ et al (2008) Differential effects of resistance and endurance exercise in the fed state on signalling molecule phosphorylation and protein synthesis in human muscle. J Physiol 586:3701–3717
Wojtaszewski JF, Nielsen P, Hansen BF et al (2000) Isoform-specific and exercise intensity-dependent activation of 5'-AMP-activated protein kinase in human skeletal muscle. J Physiol 528:221–226
Yaspelkis BB, Castle AL, Ding Z (1999) Attenuating the decline in ATP arrests the exercise training-induced increases in muscle GLUT4 protein and citrate synthase activity. Acta Physiol 165:71–79
Yu M, Stepto NK, Chibalin AV et al (2003) Metabolic and mitogenic signal transduction in human skeletal muscle after intense cycling exercise. J Physiol 546:327–335

Chapter 3
Exercise: Thermodynamic and Bioenergetic Principles

Jens Frey Halling, Anders Gudiksen, Henriette Pilegaard, and P. Darrell Neufer

Abstract The ATP energy charge needed to generate and sustain life is derived from the degradation of organic substrates with higher free energy to products with lower free energy. This difference in free energy is most efficiently harnessed by the mitochondrial oxidative phosphorylation (OXPHOS) system, which uses the complete oxidation of products to drive a series of thermodynamically based energy transformation steps to maximize the synthesis of ATP energy charge. Physical activity can dramatically increase the rate at which this ATP free energy charge is dissipated and therefore must be met by an equivalent increase in ATP production rate to sustain the activity. OXPHOS efficiency is obviously important to physical performance. As presented in this review, the efficiency of the OXPHOS system can be influenced by many factors that either optimize or at least partially decouple one or more energy transformation steps. Although much remains to be learned regarding how such processes are regulated, it is clear that modulating OXPHOS efficiency can have profound implications for exercise performance as well as overall health.

Keywords Mitochondria · Oxidative phosphorylation · Bioenergetic efficiency · Exercise performance

---

"Jens Frey Halling" and "Anders Gudiksen" are co-first authors. "Henriette Pilegaard" and "P. Darrell Neufer" are co-corresponding authors.

J. F. Halling · A. Gudiksen · H. Pilegaard (✉)
Section for Cell Biology and Physiology, Department of Biology, University of Copenhagen, Copenhagen, Denmark
e-mail: hpilegaard@bio.ku.dk

P. D. Neufer (✉)
East Carolina Diabetes & Obesity Institute, Departments of Physiology, Biochemistry & Molecular Biology, and Kinesiology, East Carolina University, North Carolina, USA
e-mail: NEUFERP@ecu.edu

© The Author(s), under exclusive license to Springer Nature Switzerland AG 2022
G. McConell (ed.), *Exercise Metabolism*, Physiology in Health and Disease,
https://doi.org/10.1007/978-3-030-94305-9_3

3.1 Introduction

Why should someone interested in exercise physiology care about thermodynamics and mitochondrial bioenergetics? The answer should be obvious—the transition from rest to high-intensity aerobic exercise imposes an energetic demand on the skeletal muscle that is unique compared with almost all other tissues. ATP is known as the energy currency of the cell, and yet the concentration of ATP rarely changes in the muscle except under the most intense, exhaustive exercise conditions. This ability to hold ATP concentrations steady reflects the remarkable ability of an energy transfer system that is exquisitely poised to match ATP production to rates of ATP utilization. The bioenergetics behind ensuring that the energy currency of the cell remains constant is based on the thermodynamic principles that govern transformations among all forms of energy in engineering as well as the physical and biological sciences. The goal of this chapter is to provide an overview of those thermodynamic and bioenergetic principles that are foundational to life, and by extension to energy metabolism during exercise, followed by consideration of various mechanisms by which mitochondrial bioenergetics may be regulated to affect metabolic efficiency and therefore exercise performance.

3.2 Thermodynamic Principles

Living cells and organisms must constantly perform chemical and physical work to sustain life, to grow, and to reproduce. These functions are dependent on the ability to harness and transform energy, which must follow the laws of thermodynamics.

The first law is the principle of conservation of energy and states that for any chemical or physical process, the total amount of energy within the universe remains constant; energy can be transformed among different forms (i.e., thermal, kinetic, chemical, electrical, etc.), but can be neither created nor destroyed. The second law of thermodynamics is the principle of irreversibility and states that for any natural process, the entropy (i.e., the degree of disorder or randomness) of the universe increases. When first encountered, the first two laws of thermodynamics seemingly present a conundrum: if the total amount of energy within the universe is constant and continuously becoming more disordered, then how is it possible that biological energy transformations are able to drive the assembly of molecules into highly organized living systems?

To understand how life itself does not violate the second law of thermodynamics requires defining "universe" in the context of biological systems and surroundings. The term "universe" can apply to any chemical or physical process within a biological system, be it a living organism, a cell, or a single reaction, and the surroundings in which that process is occurring. The *celestial universe*—the matter, energy, and even time in all of space—is an isolated system, that is, a system in which no exchange of energy (i.e., heat) or material can occur with the surroundings.

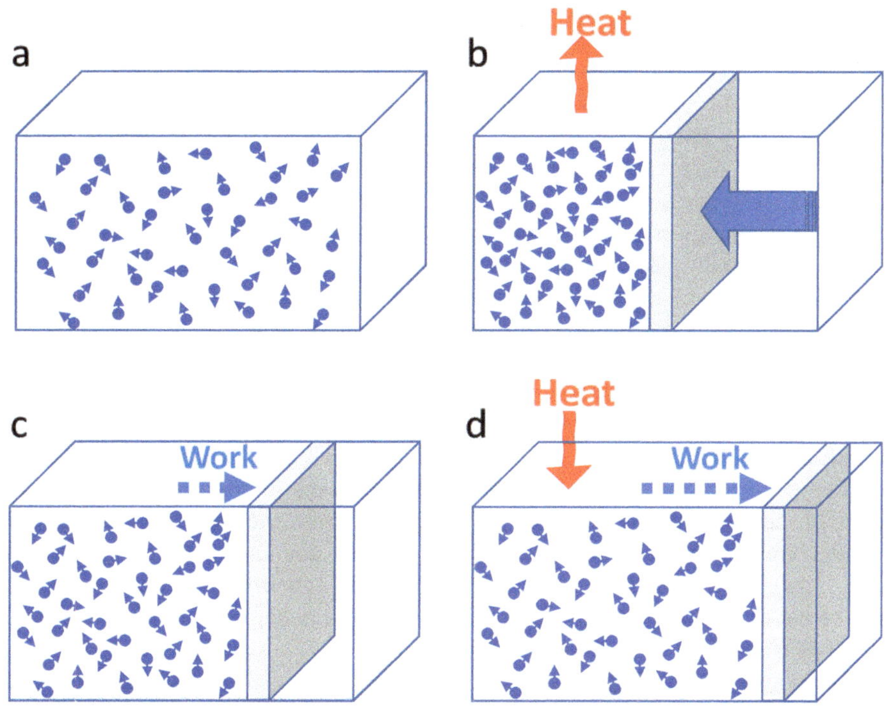

Fig. 3.1 Schematic diagram illustrating the principles of the first and second laws of thermodynamics. Refer to text for details

Conceptually, it is easier to think of an *isolated* system as a perfectly insulated box containing molecules of an ideal gas (Fig. 3.1a). With no exchange of energy or material, and an unrestricted degree of randomness, the energy within a simple isolated system is a function of the internal temperature as reflected by the Brownian motion of the gas particles, as long as the temperature is greater than absolute zero.

A *closed* system on the other hand is one in which heat, but no material, can exchange with the surroundings. Closed systems are helpful to conceptually illustrate how different parameters affect the energy within a system and its surroundings. Imagine a box (i.e., the system) that can exchange heat with the surroundings with one interior side divider that can be moved inward and fixed to condense the space occupied by the gas molecules (Fig. 3.1b). The change in available energy, known as Gibbs free energy (ΔG), of the gas molecules as a result of instantaneously condensing the size of the box is given by:

$$\Delta G = \Delta H - T\Delta S$$

where ΔH is the change in enthalpy (i.e., total heat content of a system), T is the absolute temperature, and ΔS is the change in entropy. Condensing the gas molecules initially raises the thermal energy within the system (i.e., +ΔH), but that heat is lost as the system re-equilibrates with the surroundings. ΔS is negative relative to the original state given the gas molecules occupy a smaller space and are therefore less random. Assuming T is unchanged due to the size of the surroundings relative to the system, the net change in ΔG relative to its original state is positive due to the loss of entropy. In other words, a net input of energy is required to condense the space occupied by the gas molecules. A caveat to this example is that the potential energy associated with the loss of entropy is available only if the divider is at least partially permeable to the gas molecules; otherwise, the gas molecules have no option to fill the voided space. If the divider is permeable, the gas molecules will spontaneously redistribute throughout the system, representing a negative ΔG due to the gain in entropy.

What would the scenario be if the divider is not fixed but capable of sliding within the box with some level of friction (Fig. 3.1c)? When the divider is moved from right to left condensing the space available to the gas molecules, a portion of the +ΔG created within the system can be used to perform work on the divider—to move it back to the right. If heat is added to the system from the surroundings, then the divider can be moved forward and/or a greater amount of friction overcome (Fig. 3.1d). This is another example of the first law of thermodynamics; the increase in internal energy of a system is equal to the energy added minus the amount lost as a result of work done by the system.

The point of this didactic exercise is to recognize that cells and organisms depend on and utilize the laws of thermodynamics to harness energy from food and transform it into the energy necessary to generate and sustain life (i.e., the energy for work = energy from food—energy lost to system). A simple example is an ion gradient across a membrane, which represents chemical and electrical potential energy based on ΔS. When coupled to another process (e.g., ATP synthesis), that potential energy can then be utilized to drive an otherwise unfavorable process. Living cells and organisms are *open* systems, meaning that they exchange both material and energy with their surroundings. How then can a process that represents more order, like the growth of cells, be possible without violating the second law of thermodynamics; i.e., the combined entropy of a system and its surroundings must increase during all chemical physical processes? The answer lies in the question; the entropy increase does not need to occur within the system itself. The order produced by the growth of cells is more than compensated for by the disorder imparted on the surroundings. In other words, living cells and organisms create and sustain internal order by taking free energy harnessed from nutrients or the sun, and returning to their surroundings a greater amount of energy as heat and entropy. A perfect example is physical exercise, which requires the uptake of complex carbon molecules and oxygen from the environment into the actively contracting muscle, seemingly creating more order within the system. However, the release of heat and carbon dioxide molecules to the environment results in a net increase in entropy. Thus, the energy in the complex carbon molecules is dissipated more efficiently by the system (i.e., the contracting muscle) to perform work (Fig. 3.2). There are numerous examples in nature where the dissipation of energy itself drives the formation of a

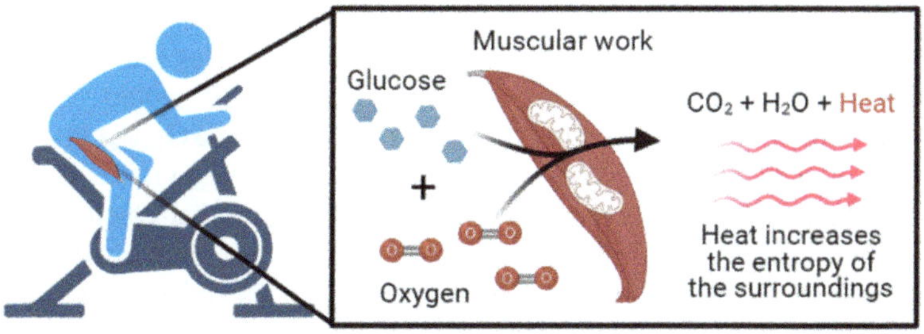

Fig. 3.2 The uptake of glucose and diffusion of oxygen into muscles during exercise are processes that reduce the entropy of the muscle, when viewed as a closed system, which is seemingly in conflict with the second law of thermodynamics. However, during muscular work, the metabolism of nutrients and the release of heat to the surrounding environment cause a net increase in entropy. Thus, when viewed as an open dissipative system, muscular work during exercise serves to increase the entropy of the universe

more orderly system to dispel energy more effectively (e.g., the formation of hurricanes/cyclones to dissipate the difference in heat energy between the hot sea surface and cold tropopause). This has led to the intriguing idea that evolution favors order out of disorder to achieve a greater efficiency of energy dissipation (Koch and Britton 2018; Prigogine 1978; Yun et al. 2006). Compared with inanimate objects, living systems are the embodiment of that principle.

3.3 Nonequilibrium Thermodynamics: Using Natural Forces to Establish Displacement from Equilibrium

From the previous section, the obvious question arises as to what natural source of energy can serve as a driving force sufficient to create and sustain the displacement from equilibrium that constitutes life for cells and organisms? For plants and some organisms, sunlight serves this function by powering the conversion of CO_2 and H_2O into O_2 and chemical energy via photosynthesis. For all other living organisms, the reduction of O_2 serves this function, albeit in a less direct way than sunlight. The oxygen atom is the second most electronegative element in the universe (fluorine is the most electronegative) due to two unpaired electrons in its outer orbital. The drive to acquire electrons extends to the oxygen molecule; in fact, the reduction of O_2 to H_2O has one of the highest standard reduction potentials ($E^{o\prime}$) in nature (+816 mV). Put simply, O_2 is an extremely powerful molecular electron magnet.

But is the reduction potential of O_2 sufficient to power the energy transformations necessary for life? Consider the catabolism of glucose through glycolysis, the pyruvate dehydrogenase complex and the citric acid cycle, which produces six molecules of CO_2, ten molecules of NADH, two molecules of $FADH_2$, and two molecules of ATP. The NAD^+/NADH and FAD/$FADH_2$ redox couples have standard reduction potentials of −320 mV and − 220 mV, respectively. Given the extremely high $E^{o\prime}$ of O_2, the metabolism of glucose could easily be completed by O_2 converting NADH and $FADH_2$ back to their oxidized forms (NAD^+, FAD) with two molecules of H_2O as the final reaction product. This would complete the chemistry but fail to capitalize on the >1 volt of free energy (ΔG_{redox}) that is present between the two electron carrier redox couples and the O_2/H_2O redox couple.

The key to metabolism, and perhaps life itself, is how that electrical potential energy is harnessed, transformed, and repurposed by the mitochondrial electron transport system (ETS). With redox pull by oxygen established as the driving force (i.e., analogous to gravity pulling water over a waterfall), electrons are drawn through a series of redox couples embedded within the multiprotein complexes that comprise the ETS. Beginning with oxygen ($E^{o\prime}$ = ~816 mV) drawing electrons from cytochrome a_3 ($E^{o\prime}$ = ~600 mV) in complex IV, each redox couple draws electrons from the redox couple with a lower $E^{o\prime}$ just upstream, continuing through cytochrome c, complex III, the coenzyme Q pool, and complex I to NADH. Electrons are also drawn into the Q pool from other redox proteins, including succinate dehydrogenase (complex II), the electron transfer flavoprotein, mitochondrial glycerol-3-phosphate dehydrogenase, as well as other dehydrogenases. It is useful to keep in mind that the flow of electrons, as in any electrical current, will be wide open as long as both the electron source and acceptor are available and no regulators are present, emphasizing that O_2 imposes a powerful and constant draw on electrons.

Dispersing ΔG_{redox} in smaller increments however does not accomplish anything. Three additional and quite remarkable features of nature contribute to the initial energy transformation. First, the entire ETS is embedded within the inner membrane of the mitochondria. Second, electron flux through complexes I, III, and IV is directly tied to proton translocation (i.e., pumping) from the inner to outer surface of the inner mitochondrial membrane, which generates an electrical ($\Delta\Psi_{m;}$ membrane potential) and chemical (H^+ concentration) gradient across the membrane collectively known as the proton motive force (pmf). The biophysical mechanism (s) by which electron flux is coupled to proton translocation is still unknown, but it is clear that the three largest changes in $E^{o\prime}$ between specific redox couples occur in complexes I, III, and IV, implying that the free energy released at each of these steps is used to power proton translocation. Again, it is difficult to understate how much energy ΔG_{pmf} represents at the molecular level. $\Delta\Psi_m$ is typically −170 to −200 mV in a resting cell, but when factoring in the high specific capacitance of lipid bilayers (ability to separate charge), the specific charge is estimated to be in excess of 300,000 V/cm (Nicholls and Ferguson 2013; Sperelakis 2021), which, at the molecular level, is approximately the energy contained in a lightning bolt. The third critical feature is that for each of the three complexes, electron flux and proton pumping are co-dependent—one cannot occur without the other. The fourth feature is an extension of the third—the buildup of ΔG_{pmf} on the outer surface of the inner membrane represents a backpressure, or brake, on the proton pumps and thus electron flux through the ETS.

The net effect is that the natural potential energy in ΔG_{redox} is partially transformed to a different form of potential energy, ΔG_{pmf} (analogous to

Fig. 3.1c). Importantly, the ETS is converted from a wide-open circuit to one that is limited by ΔG_{pmf}. In fact, by default, the system appears set to use ΔG_{redox} to drive ΔG_{pmf} until the backpressure on the pumps (i.e., ΔG_{pmf}) completely counterbalances the driving force of the pumps (i.e., ΔG_{redox}), at which point electron flux, proton pumping, and oxygen consumption would theoretically cease. Practically however, mitochondria never achieve such a "static head" because protons continuously conduct or "leak" at low rates back into the mitochondrial matrix, mediated primarily by membrane-bound proteins (Divakaruni and Brand 2011). ΔG_{pmf} is therefore less than maximum—i.e., the brake on the ETS is not fully engaged—and as such, electron flux and O_2 consumption proceed at a rate set by the basal rate of proton conductance. From an engineering perspective, the loss of some of the free energy associated with ΔG_{pmf} may seem unnecessarily wasteful. However, in the absence of basal proton conductance, the redox couples within the ETS are maximally reduced (i.e., oxidation potential is maximal) which will dramatically accelerate electron transfer directly to oxygen to form the highly reactive superoxide free radical (Murphy 2009). The basal proton conductance rate characteristic of mitochondria appears, in fact, to be sufficient to lower ΔG_{pmf} just enough to minimize superoxide production (Korshunov et al. 1997).

With the stage set, mitochondria are poised to complete a second energy transformation by using ΔG_{pmf} to drive ATP synthesis. It is important to recognize that the natural equilibrium of the ATP hydrolysis reaction under standard conditions heavily favors the products of the reaction with an apparent equilibrium constant of 10^5 M. In cells, however, the mitochondrial ΔG_{pmf} is sufficient to drive displacement of the reaction a full ten orders of magnitude away from equilibrium (i.e., to a mass action ratio of 10^{-5} M) (Davies et al. 1982). Thus, the electrochemical free energy of ΔG_{pmf} is transformed to a chemical free energy of ATP hydrolysis (i.e., ΔG_{ATP}). The extent to which ΔG_{pmf} can drive this displacement is not infinite but, similar to the backpressure of ΔG_{pmf} on ΔG_{redox}, is defined by how far the reaction can be displaced before the free energy of ATP hydrolysis (i.e., ΔG_{ATP}) counterbalances ΔG_{pmf}. At that point, the backpressure of ΔG_{ATP} is fully applied (i.e., second brake applied), ATP synthesis stops, and the energy charge is fully established. Conceptually, it is useful to think of ATP synthase as a reversible ATP hydrolyzing proton pump, meaning that in the absence of a membrane potential, the enzyme is an ATPase that pumps protons. The direction of proton movement through ATP synthase is thus a function of the balance between ΔG_{pmf} and ΔG_{ATP}.

ΔG_{ATP} is the source of free energy, the energetic "currency," for the rest of the cell. By coupling to the hydrolysis of ATP, hundreds if not thousands of reactions and processes (e.g., ion gradients, biosynthetic reactions, the cross-bridge cycling of actin/myosin, etc.) are displaced away from their default equilibrium state, with the extent of displacement being a direct function of the magnitude of ΔG_{ATP} available. It should be apparent that once all non-equilibrium thermodynamic states are established and held—that is, once the cell has been brought to life by ΔG_{ATP}—the subsequent rate of ATP utilization for the entire life of the cell is determined by the energy required to maintain or reestablish (if/when partially dissipated) those thermodynamic displacements from equilibrium. In other words, the rate of ATP production is governed by and continuously adjusting to the rate of ATP utilization.

3.4 How Are Bioenergetic Systems Engaged by Exercise?

Heavy exercise can increase the utilization rate of ATP in skeletal muscle of well-trained cyclists by 50- to 100-fold, equivalent to ~1.5–3.0 μmol ATP · g^{-1} muscle · s^{-1}. ATP concentration in muscle is only ~5 μmol · g^{-1} muscle, sufficient to last only a few seconds. Even during submaximal cycling exercise (75% VO_2 max) that can be sustained for >1 hour at an ATP utilization rate ~ 0.4 μmol ATP · g^{-1} muscle · s^{-1}, ATP concentration is sufficient to last only ~15 s (Hargreaves and Spriet 2020; Meyer and Wiseman 2012). Despite this limited supply, muscle ATP concentration does not decline during exercise except under the heaviest workloads at or near exhaustion. It should be apparent from the preceding section that this remarkable capacity to maintain energy charge reflects the massive free energy pent up in the OXPHOS system (i.e., the interplay between ΔG_{redox}, $\Delta G_{\Delta\Psi}$, and ΔG_{ATP}). Analogous to the controlled release of energy by a dam, the OXPHOS system is poised to respond instantly to an increase in ATP utilization rate over a wide range. This is exactly what is observed experimentally in isolated mitochondria and permeabilized fiber bundles; the addition of ADP (modeling an increase in ATP utilization) elicits a nearly instantaneous proportional steady-state increase in proton conductance, which is mediated (sequentially) through the ATP synthase, decrease in $\Delta\Psi$, increase in electron flux and oxygen consumption rate, increase in reducing equivalent oxidation rate, and therefore an increase in substrate oxidation rate.

In vivo however, the time required for muscle cells to reach a new steady-state rate of oxygen consumption when transitioning from one workload to another is not instantaneous but can require 60–120 seconds or more. This is due to the complexity of cells and the spatial challenge of coupling sites of ATP utilization to sites of ATP production. Total mitochondrial volume comprises only between 3 and 12% of cellular volume in skeletal muscle cells (depending on fiber type and imaging technique) (Dahl et al. 2015; Morgan et al. 1971; Palade 1952). Moreover, the diffusion rates of adenine nucleotides, particularly ADP, are orders of magnitude lower than required to support high rates of ATP turnover in muscle fibers (Kammermeier 1987; Yoshizaki et al. 1990). To overcome this problem, nature evolved to employ high-energy but inert phosphagen compounds, such as phosphocreatine (PCr), to serve in an energy transfer system. Creatine kinase (CK) catalyzes the reversible reaction $PCr^{2-} + MgADP^{-} + H^{+} \leftrightarrow MgATP^{2-} + Cr$ and thus can either utilize PCr to generate ATP or store energy from ATP-producing sites. Key features of the system include the large cytosolic pool of total creatine (~45 μmol·g^{-1}), the buildup of PCr within the pool (2:1 PCr:Cr) driven by OXPHOS ATP, the presence of specific CK isoforms localized to both sites of ATP utilization (i.e., myofibrillar, sarcoplasmic reticulum, sarcolemmal membrane) and ATP production (i.e., mitochondria and cytosol via glycolysis), and the high standard

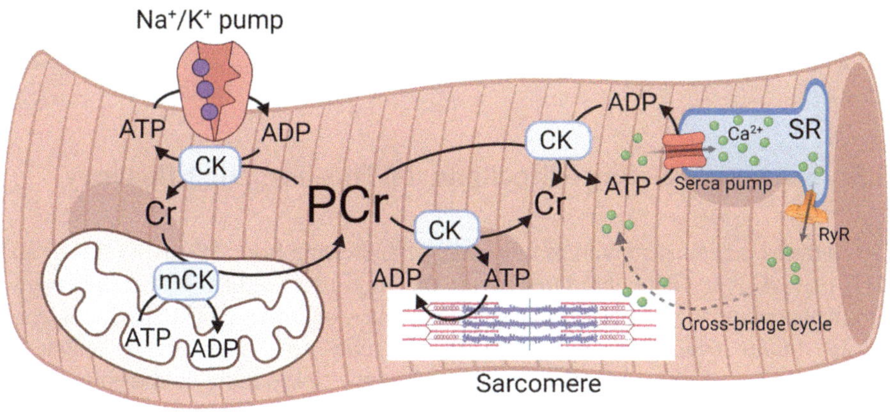

Fig. 3.3 The creatine kinase (CK)/phosphocreatine (PCr) shuttle system serves to increase the speed of energy transfer throughout the cell. Mitochondrial CK (mCK) is located in the mitochondrial intermembrane space and uses the ATP/ADP gradient to catalyze the phosphorylation of creatine (Cr) into PCr. Several different cellular compartments, such as the sarcoplasmic reticulum, the sarcolemmal Na^+/K^+ pump, and the sarcomeres, contain local CK isoforms that catalyze the reconversion of PCr to ATP to fuel ATPase activity at specific cellular compartments based on energetic demand. The net speed of energy transfer between sites of utilization and production is facilitated by flux through the cytosolic CK circuit and the higher diffusion coefficient of PCr compared with ATP

equilibrium (~150; i.e., heavily favoring ATP) (Golding et al. 1995) and rate constants (approximately tenfold greater than ATP synthase) of CK. As put forward by Wallimann et al. (Wallimann et al. 1992), the creatine shuttle system functions not only as a temporal buffer in localized microcompartments of ATPase activity but critically as a spatial energy transfer circuit that connects sites of ATP utilization to subcellular sites of ATP production. Consequently, any stepwise increase in steady-state ATPase activity will have the net effect of instantly decreasing PCr concentrations within the local microenvironment, which is then effectively transmitted through the cytosolic PCr/CK pool increasing ADP concentration at the mitochondria (via CK_{mito}) and thus ATP production, the energy from which is transmitted back through the same circuit to the site of ATPase activity. The net effect is a simultaneous decrease in [PCr] and increase in oxygen consumption rate over an equal but opposite exponential time course until a new steady-state energy transfer balance through the shuttle is reached (Fig. 3.3).

Maintaining ΔG_{ATP} over a wide range of ATP utilization rates is essential to preserving life for the cell—to keep exercise from inducing energetic suicide. In fact, during progressive exercise to exhaustion in humans, muscle ΔG_{ATP} only decreases from ~64 to 50 kJ · mole^{-1} (Jeneson and Bruggeman 2004). If ΔG_{ATP} declined by an additional 2 kJ · mole^{-1}, sufficient free energy would not be available for the sarcoplasmic reticulum Ca^{2+} ATPase to re-sequester calcium against the remaining concentration gradient, or for the Na^+/K^+-ATPase pump to reestablish the Na^+ and

K^+ gradients across the plasma membrane (Chen et al. 1998; Meyer and Wiseman 2012).

3.5 Factors Affecting Mitochondrial Bioenergetic Efficiency

Given the tremendous amount of free energy transferring from ΔG_{redox} to $\Delta G_{\Delta\Psi}$ and $\Delta G_{\Delta\Psi}$ to ΔG_{ATP}, it is not hard to imagine the importance of OXPHOS efficiency to physical performance capacity. In a sense however, optimizing OXPHOS efficiency is also in the eye of the beholder—advantageous to the endurance athlete to maximize performance, but disadvantageous to the obese individual trying to lose weight. Thermodynamic efficiency is defined as the ratio between the output and input powers of two coupled reactions, with power defined as the product of the flux and force for each reaction. Thus, conceptually, efficiency may be impacted or regulated at any of the energy transfer steps, i.e., H^+ pumping relative to electron transfer within each of the individual complexes (Di and Venditti 2001) or ATP production relative to H^+ translocation by ATP synthase. Net efficiency across the entire OXPHOS system can also be influenced by factors that decouple energy transfer between reactions or within the system overall (i.e., electron leak, non-ATP synthase-mediated proton conductance, proton "slipping," etc.), which in turn may be triggered by allosteric and/or structural changes within protein complexes, supercomplexes, and/or inner membrane phospholipids. This section will discuss factors that can influence mitochondrial bioenergetic efficiency, both in the context of exercise physiology where mitochondrial efficiency may have a huge impact on the economy of work and in relation to chronic energy overload (Fig. 3.4).

3.6 ROS Production as a Determinant of Bioenergetic Efficiency

ROS production from electron leak represents an uncoupling of electron transfer from proton translocation and thus a decrease in the conversion efficiency between ΔG_{redox} and $\Delta G_{\Delta\Psi}$. Mitochondria are known to produce and emit ROS, although the most generous estimates from the earliest studies of mitochondrial ROS production suggest that only 2–5% of the total oxygen consumption can be reduced to superoxide (Boveris and Chance 1973; Loschen et al. 1974). Later studies have suggested that the maximal fraction of oxygen consumption that can be used to form ROS in muscle mitochondria is 0.15% (St-Pierre et al. 2002).

It has often been proposed that mitochondria are a major source of ROS during exercise (Di and Venditti 2001; Kanter 1994; Malin and Braun 2016; Scheele et al. 2009; Urso and Clarkson 2003). This widespread assumption seems to have arisen from the fact that it was shown in the 1980s that contracting skeletal muscle

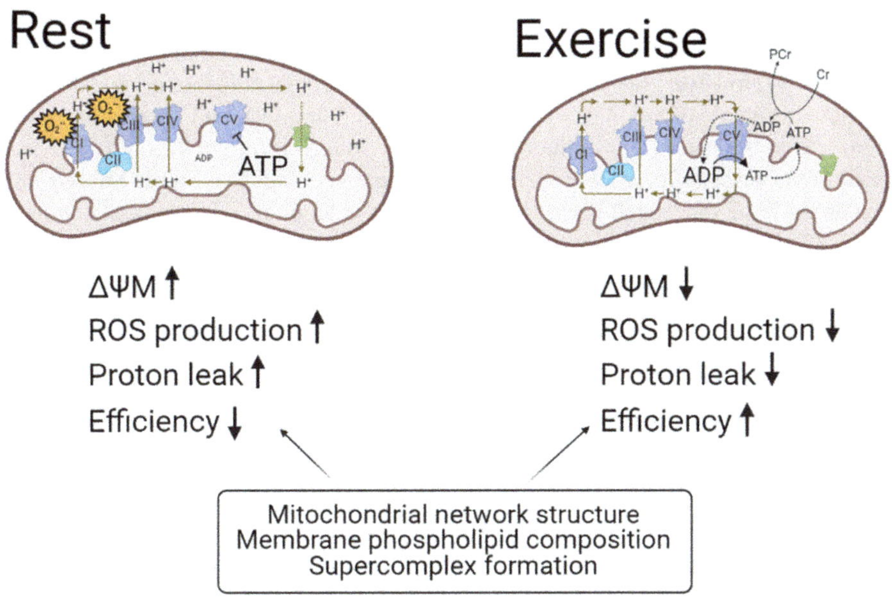

Fig. 3.4 Factors affecting mitochondrial efficiency during exercise. At rest and/or at times of nutritional overload, the high mitochondrial ATP concentration inhibits proton reentry through the ATP synthase. This increases mitochondrial membrane potential (ΔΨM) and creates a backpressure on the electron transport system resulting in higher proton leak, higher ROS production, and lower bioenergetic efficiency. During exercise, the high demand for ATP removes the brakes on proton conductance and electron flow, which automatically maximizes bioenergetic efficiency. Factors such as mitochondrial network structure, membrane phospholipid composition, and supercomplex formation may influence mitochondrial efficiency both at rest and during exercise

produces ROS (Davies et al. 1982; Jackson et al. 1985)—a finding that came relatively shortly after the discovery that mitochondria are important sources of ROS (Boveris and Chance 1973). However, if mitochondrial ROS production was proportional to the increased oxygen consumption during exercise, then mitochondrial ROS production should increase 50–100-fold during conditions resembling maximal aerobic contractions (i.e., maximal ADP-stimulated respiration). On the contrary, when mimicking the substrate and pH conditions observed in the skeletal muscle during mild and intense aerobic exercise, the rate of mitochondrial H_2O_2 production drops by 15–20% compared with "resting" conditions, despite a three- to tenfold increase in oxygen consumption during simulated "exercise" (Goncalves et al. 2015). Therefore, the apparent reduction in mitochondrial ROS production during exercise can be considered a mechanism to increase mitochondrial efficiency, albeit this can likely only explain a minor portion of the four- to fivefold increase in mitochondrial efficiency that occurs when going from submaximal to maximal ADP-stimulated respiration (Lark et al. 2016).

Non-mitochondrial sources of ROS have been shown to contribute substantially more than mitochondria to exercise-induced ROS production (Powers and Jackson

2008), which could theoretically contribute to reduce bioenergetic efficiency at the whole-body level by "stealing" oxygen that would otherwise be available for ATP production. However, the studies showing increased non-mitochondrial ROS production during exercise have primarily employed indirect and qualitative ROS measurements, such as difluorofluorescein (Sakellariou et al. 2013) and dichlorofluorescein fluorescence (Henriquez-Olguin et al. 2019). To our knowledge, no studies have succeeded in quantifying the molar amounts of oxygen that is used for ROS production during exercise. Therefore, improved methods for quantifying non-mitochondrial exercise-induced ROS will be necessary to determine the degree to which this contributes to altering metabolic efficiency.

3.7 Protein-Mediated Proton Leak

3.7.1 Uncoupling Proteins (UCPs)

The main physiological mechanism of regulating mitochondrial efficiency is thought to be proton leak through transporters in the inner mitochondrial membrane. Thus, a number of proteins in the solute carrier (SLC)25 family have been identified as "uncoupling proteins" meaning that they can mediate the leakage of protons from the intermembrane space to the matrix and thereby uncouple the proton motive force from ATP production. Most prominent is perhaps uncoupling protein 1 (UCP1), or SLC25A7, which mediates the thermogenic function of brown adipose in rodents (for a review, see Nicholls & Rial, (Nicholls and Rial 1999)). In that regard, it would be shameful to miss the opportunity to mention that UCP1 was discovered by Gillian *Heaton* and colleagues (Heaton et al. 1978). However, although UCP1 is vital for the ability of mice to tolerate cold exposure without prior acclimation (Enerback et al. 1997), adult humans have very limited brown adipose depots (Leitner et al., PNAS, 2017). Humans are also generally not exposed to prolonged cold in daily life, suggesting that UCP1 may be of minor relevance in modern human physiology. On the other hand, it has been proposed that exercise training may cause a "browning" of white adipose tissue. This has fueled an ongoing hypothesis that exercise-induced stress mediates a UCP1-dependent "browning" of white adipose, which may lower metabolic efficiency at the whole-body level (Aldiss et al. 2018).

Adipose tissue only consumes 5–10% of the basal metabolic rate in humans (Nookaew et al. 2013), whereas 40–50% is consumed in the skeletal muscle, which does not express UCP1 (Zurlo et al. 1990). Therefore, mechanisms other than UCP1-mediated uncoupling likely contribute to the regulation of bioenergetic efficiency at the whole-body level. Skeletal muscle cells express the UCP1 homologs UCP2 (SLC25A8) and UCP3 (SLC25A9), which share ~60% sequence similarity with UCP1, and can catalyze proton leak when activated by peroxidation products and/or fatty acids (Brand and Esteves 2005). However, mice lacking UCP2 or UCP3 show very little phenotypic alterations, including having normal cold tolerance (Vidal-Puig et al. 2000; Zhang et al. 2001). Thus, the exact physiological functions of UCP2 and UCP3, including their role in regulating mitochondrial efficiency, are not clear. Interestingly, genetic association studies have shown effects of polymorphisms in the UCP2 gene on metabolic functions, including a higher exercise efficiency in bearers of an Ala/Val55 substitution (Buemann et al. 2001), indicating that uncoupling proteins in skeletal muscle may play a role in regulating mitochondrial efficiency during exercise.

3.7.2 ANT1

The adenine nucleotide transporter (ANT)1, or SLC25A4, exchanges cytosolic ADP for mitochondrial ATP across the inner mitochondrial membrane. In addition, it has been shown that the basal proton leak of muscle mitochondria is chiefly mediated by ANT1 (Brand et al. 2005). This was recently substantiated by electrophysiology studies showing that proton transport is an inherent function of ANT1 (Bertholet et al. 2019). The proton flux through ANT1 is partly inhibited by ADP/ATP exchange (Bertholet et al. 2019), suggesting that ANT1 can function in two distinct modes: Either to lower mitochondrial efficiency by leaking protons or to increase efficiency by exchanging ATP for ADP. Therefore, changes in mitochondrial efficiency during increased ATP demand, such as during exercise, may be regulated at the level of ANT1. Interestingly, UCP- and ANT1-mediated mitochondrial proton leak is strongly activated by the lipid peroxidation product 4-hydroxynonenal and by fatty acids (Echtay et al. 2003; Klingenberg and Winkler 1985). Aerobic exercise can change the muscle levels of both 4-HNE (Parker et al. 2016) and fatty acids (Havel et al. 1967), raising the possibility that these metabolites may influence mitochondrial efficiency during exercise although this remains to be addressed. In addition, ANT1 has been shown to be acetylated in vivo in the human skeletal muscle, which was predicted through molecular dynamics modeling to regulate ADP affinity (Mielke et al. 2014; Perry et al. 2012). However, it remains to be addressed whether posttranslational modifications contribute to regulating ADP/ATP-exchange during exercise.

3.7.3 Other SLC25 Family Proteins

Several other members of the SLC25 family of mitochondrial carriers can potentially contribute to altering mitochondrial efficiency during exercise. For example, the mitochondrial phosphate carrier (PIC), or SLC25A3, imports phosphate in symport with protons, and both SLC25A12 and SLC25A22 cotransport glutamate and protons (Ruprecht and Kunji 2020). However, their potential relevance in regulating mitochondrial efficiency during exercise and in response to exercise training is still unknown.

3.8 Posttranslational Regulation of Mitochondrial Function

Although a number of proteins have been found to contribute to proton leak in the skeletal muscle, it is not fully understood how exercise regulates these proteins acutely. ANT1 and UCP3 mRNA levels have been shown to increase in the skeletal muscle with exercise training ((Sparks et al. 2016; Tsuboyama-Kasaoka et al. 1998) as a part of the general mitochondrial biogenic signal in response to exercise. However, it is not known whether posttranslational modifications of mitochondrial proteins are involved in the four- to fivefold increase in mitochondrial efficiency that occurs when going from submaximal to maximal ADP-stimulated respiration (Lark et al. 2016).

Hundreds of phospho-sites have been identified in rodent and human mitochondria (Bak et al. 2013; Zhao et al. 2014), suggesting that protein phosphorylation is a posttranslational mechanism for regulating mitochondrial function. The A-kinase anchor protein 1 (AKAP1) has been identified as a mitochondrial target of the 5' AMP-activated protein kinase (AMPK), which is strongly stimulated by exercise (Hoffman et al. 2015). Specifically, it has been shown that AKAP1 mitochondrial localization and AMPK-mediated AKAP1 Ser-103 phosphorylation were enriched after acute exercise and that AKAP1 Ser-103 silencing strongly reduced mitochondrial oxidative function (Hoffman et al. 2015).

Other posttranslational mechanisms include mitochondrial protein acetylation, which has been shown to be altered in rat strains bred for high vs. low running capacity (Overmyer et al. 2015). In addition, mitochondrial proteins have been shown to be deacetylated in rat skeletal muscle in response to an acute running bout, which was suggested to facilitate more efficient oxidation of fatty acids and branched-chain amino acids during exercise (Overmyer et al. 2015). Exercise-induced deacetylation of mitochondrial proteins may be mediated by the mitochondrial deacetylase sirtuin 3, which has been shown to increase in response to contractile activity in rat skeletal muscle (Gurd et al. 2012) and to deacetylate the ATP synthase in mouse skeletal muscle during exercise (Vassilopoulos et al. 2014). While these studies provide circumstantial evidence of regulation of mitochondrial function by acetylation, more recent studies have been unable to confirm a direct impact of acetylation on respiratory function in either skeletal muscle or heart (Davidson et al. 2020; Williams et al. 2020; Bertholet et al. 2019; Divakaruni and Brand 2011).

3.9 The Mitochondrial Network

Mitochondrial network structure is a factor with potential crucial impact on the efficiency of oxidative phosphorylation in skeletal muscle. Experiments using three-dimensional focused ion beam scanning electron microscopy revealed four different mitochondrial network morphologies in mitochondria-rich muscle fibers (Glancy et al. 2015). Moreover, the observations suggested that the mitochondrial reticulum allowed efficient membrane potential conductance through the muscle fiber with generation of the proton motive force in mitochondrial pools near capillaries and use of the proton motive force to produce ATP in mitochondrial pools near ATPases. The authors suggest that such a mechanism of electrical conduction is important for skeletal muscle energy distribution enabling the support of an immediate increase in energy demand throughout the muscle fiber during (intense) muscle contractions (Glancy et al. 2015).

Mitochondrial morphology has been shown to be muscle fiber type specific with oxidative fibers characterized by (more) elongated and interconnected mitochondria (Glancy et al. 2015), while glycolytic fibers have punctuate mitochondria (Mishra et al. 2015). Four weeks of exercise training resulted in parallel (myosin heavy chain, MHC) fiber type switching and development of an elongated mitochondrial morphology in mouse muscle indicating that mitochondrial structure can respond to the metabolic activity of the individual fiber. In addition, incubation of glycolytic muscle ex vivo in oxidative media (acetoacetate) increased oxygen consumption, mitochondrial membrane potential, and induced development of interconnected, highly tubular mitochondria (Mishra et al. 2015). In accordance, lifelong exercise training has been shown to prevent, and 7 weeks of exercise training at old age has been shown to reverse, an age associated fragmentation of the mitochondrial network in mouse skeletal muscle (Halling et al. 2019; Halling et al. 2017). Moreover, the changes in mitochondrial network structure with age and exercise training at old age were associated with changes in maximal and submaximal mitochondrial respiration in permeabilized mouse muscle fibers measured ex vivo (Halling et al. 2019), supporting a functional importance of the mitochondrial morphology. Together, this indicates a highly dynamic regulation of mitochondrial structure in skeletal muscle linked to the metabolic activity of the muscle.

The mitochondrial network in skeletal muscle is dynamically regulated by fission and fusion processes. This involves dynamin-related protein (DRP)1 and fission protein (FIS)1 as regulators of fission and mitofusin (MFN)1, MFN2, and optic atrophy (OPA)1 as regulators of fusion (Malka et al. 2005). The importance of these factors for mitochondrial network structure is evident by an observed mitochondrial fragmentation both with tissue-specific knockout of MFN2 (Sebastian et al. 2012) and overexpression of FIS1 and DRP1 (Romanello et al. 2010). Moreover, exercise training has been reported to increase the protein content of both fusion and fission proteins in human skeletal muscle (Konopka et al. 2014; Perry et al. 2010) suggesting an enhanced fusion and fission capacity in skeletal muscle with exercise training. On the other hand, an exercise training-induced prevention of an age-associated mitochondrial fragmentation in mouse skeletal muscle seemed to involve reduction in fission proteins rather than an increase in fusion proteins (Halling et al. 2017). In addition, using an in vivo approach, fusion was shown to be a highly dynamic process with higher rates in oxidative fibers than glycolytic fibers (Mishra et al. 2015). In addition, because the OMM also exclusively features the anchored proteins mediating fusion and fission processes as well as inter-organelle tethering, vis-à-vis contact site formation (Detmer and Chan 2007; Koshiba et al. 2004; Scorrano et al. 2019), it allows coupling of responsive

Chapter 4
Anaerobic Metabolism During Exercise

Lawrence L. Spriet

Abstract A constant supply of adenosine triphosphate (ATP) is essential for function in all cells and especially so in skeletal muscle cells to power the contractions needed to enable the many forms of movement required in our daily lives and for exercise and sporting events. The muscle stores of ATP are small, and metabolic pathways must maintain the required rates of ATP resynthesis when the demand for ATP is high. Oxidative ("aerobic") phosphorylation uses reducing equivalents from the metabolism of carbohydrate and fat to produce ATP and is the default energy system in skeletal muscle. Substrate-level phosphorylation or "anaerobic metabolism" also plays a very important role in supplementing or buffering ATP production when aerobic ATP production cannot meet the needs of an activity. These situations include the transitions from rest to exercise and from one power output to a higher one, exercise that demands ATP provision rates above what can be provided aerobically, and in situations of suboptimal oxygen supply. Anaerobic energy is provided from phosphocreatine and muscle glycogen breakdown (anaerobic glycolysis). These systems can provide energy very quickly and at very high rates but are limited to short periods of time during high intensity exercise due to substrate depletion and increasing muscle acidosis. In most exercise and sporting situations, energy provision is maintained by contributions from both the aerobic and anaerobic sources to ensure that ATP resynthesis closely matches the exercise ATP demand.

Keywords Anaerobic metabolism · Skeletal muscle · Phosphocreatine · Anaerobic glycolysis · Glycogen · Carbohydrate · Lactate · ATP provision · Acidosis · High-intensity exercise

L. L. Spriet (✉)
Department of Human Health & Nutritional Sciences, College of Biological Science, University of Guelph, Guelph, ON, Canada
e-mail: lspriet@uoguelph.ca

© The Author(s), under exclusive license to Springer Nature Switzerland AG 2022
G. McConell (ed.), *Exercise Metabolism*, Physiology in Health and Disease,
https://doi.org/10.1007/978-3-030-94305-9_4

4.1 Introduction

The ability to move in our environment is important for daily functioning and survival. Skeletal muscles receive signals from the nervous system and contract with the necessary force, frequency, and sequence to permit the many forms of movement required in our daily lives. The energy that powers these movements is adenosine triphosphate (ATP) and is provided by a number of metabolic pathways that ultimately combine inorganic phosphate (Pi) and adenosine diphosphate (ADP) to provide a steady source of newly generated ATP (Table 4.1). The requirement for ATP during movement or exercise can vary from the low demands of slow movements to the extremely high demands of ballistic and sprint-like movements. Aerobic metabolism is the predominant or default source of ATP generation in most movement and exercise situations. It is defined as the energy that is produced in the mitochondria of cells using oxygen (O_2), ADP, and Pi, and reducing equivalents from food in a process called oxidative phosphorylation (Table 4.1). As powerful as this form of ATP production is in skeletal muscles, it is limited in terms of the rate at which it can be activated at the onset of exercise and also by the maximal rate of ATP production that can be provided.

Anaerobic ATP generation plays an important role in providing ATP in situations where the aerobic system cannot provide all the energy needed for specific movements. Anaerobic ATP provision is defined as the ability of metabolic pathways to generate ATP without the immediate use of O_2, and these processes are termed substrate phosphorylation. Anaerobic energy is provided from phosphocreatine and muscle glycogen breakdown (anaerobic glycolysis) (Fig. 4.1). Anaerobic ATP is generally needed in the following situations: (1) the transition from rest to exercise in the so-called aerobic domain (power outputs that elicit 100% maximal O_2 uptake (VO_2max) or less), where it may take 60–90 s for the aerobic system to fully activate

Table 4.1 Energy producing pathways in skeletal muscle

ATP utilization
ATP + H_2O → ADP + Pi + H^+ + energy
ATP resynthesis
Substrate phosporylation
PCr + ADP + H^+ ↔ ATP + creatine
Glycogen + 3 ADP + 2Pi → 2 lactate + $2H^+$ + 3 ATP
2 ADP ↔ ATP + AMP
Oxidative phosphorylation
Glucose + $6O_2$ + 36 – 38 ADP → $6CO_2$ + $6H_2O$ + 36 - 8 ATP
Palmitate + $23O_2$ + 136 - 8 ADP → $16CO_2$ + $16H_2O$ + 136 - 8 ATP

ADP adenosine diphosphate, ATP adenosine triphosphate, AMP adenosine monophosphate, Pi inorganic phosphate, PCr phosphocreatine

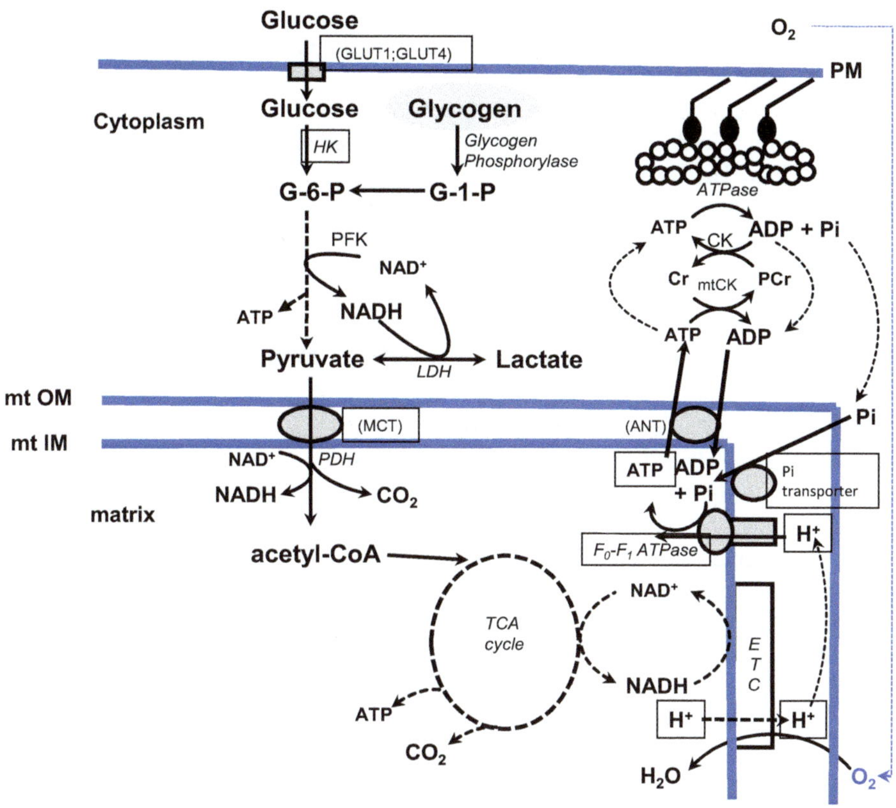

Fig. 4.1 A schematic of the anaerobic energy producing pathways in skeletal muscle. Key metabolic pathways in contracting skeletal muscle during exercise and the major sites of regulation. *PM* plasma membrane; *O_2* oxygen; *GLUT1, 4* glucose transport proteins 1 and 4; *HK* hexokinase; *PFK* phosphofructokinase; *ATP* adenosine triphosphate; *NAD and NADH* unreduced and reduced nicotinamide adenine dinucleotide; *LDH* lactate dehydrogenase; *Pi* inorganic phosphate; *G-6-P* glucose 6-phosphate; *G-1-P* glucose 1-phosphate; *ADP* adenosine diphosphate; *Cr* creatine; *PCr* phosphocreatine; *CK* creatine kinase; *mCK* matrix, mitochondrial matrix; mitochondrial creatine kinase; *mt OM and mt IM* outer and inner mitochondrial membranes; *MCT* monocarboxylase transporter; *ANT* adenine transport protein; *PDH* pyruvate dehydrogenase; *TCA* tricarboxylic acid; *ETC* electron transport chain; *H^+* hydrogen ion; *CO_2* carbon dioxide; *H_2O* water; *F_o-Fi ATPase* catalytic portion (Fi) and a proton channel (Fo) ATPase

and provide the needed ATP; (2) the transition from one power output to a higher power output where again there is a lag in the response of the aerobic system to respond to the new higher demand for ATP; (3) situations where the demand for ATP is greater than the maximal rate of aerobic ATP provision, such as exercise at power outputs above those required to elicit 100% VO_2max and during ballistic and sprint movements where the power output may be several fold higher than the VO_2max power output; and (4) situations where the delivery of O_2 is compromised or less than optimal and limits aerobic ATP production, including exercising at altitude with a lower barometric pressure and swimming where optimal breathing may be limited.

The purpose of this chapter is to examine the importance of anaerobic metabolism in skeletal muscle during exercise with an emphasis on information derived from adult humans. The reader may be interested in several previous review articles on this topic (Hermansen 1969; Gollnick and Hermansen 1973; Saltin 1990; Spriet 1992, 2006) and related topics (Sahlin, 2014; Hargreaves and Spriet 2020).

4.2 Aerobic and Anaerobic Metabolism Work Together

The production of anaerobic ATP occurs through the conversion of phosphate from phosphocreatine (PCr) to ADP in a single metabolic reaction with very high activity and at two sites in the glycolytic pathway where substrate phosphorylation produces ATP and where lactate is an end product (Fig. 4.1). These sites of ATP production are in the cytoplasm of the cell and close to the sites where ATP is consumed by the actin-myosin, calcium (Ca^{2+}), and sodium (Na^+)-potassium (K^+) ATPases.

The most common misconception when discussing skeletal muscle aerobic and anaerobic metabolism with students is the belief that they work in isolation—when exercising at a steady-state power output where aerobic metabolism produces most of the energy, anaerobic metabolism is not being used, and when sprinting all of the energy production is anaerobic. This is not the case as the signals related to ATP use in the muscle during exercise activate all the metabolic pathways in proportion for the need for ATP and ultimately determine how active each pathway will be. This belief may stem from the fact that when a person is riding a cycle ergometer at a very constant power output in the laboratory and O_2 uptake is measured after 60–90 s when a "steady state" is reached, the energy required is provided by aerobic metabolism. However, riding a bicycle in the real world will not be as "steady state" like, and the rider will need to speed up now and then, climb a grade, etc., and anaerobic energy will help provide the extra energy needed for these transitions and increases in power output.

An example at the other end of the spectrum is an all-out 30 s sprint like the classic "Wingate test." Large increases in ADP, adenosine monophosphate (AMP), Pi, and Ca^{2+} in the cell as well as increases in epinephrine in the blood activate all the major pathways in the muscle once exercise commences. The pathways associated with aerobic ATP production are maximally stimulated but take some time for this source of ATP to be significant, and therefore anaerobic ATP must provide the majority of the energy early on. Measurements have shown that on average a person will reach ~75% VO_2max by 30 s, and some people reach VO_2max at this time (Kowalchuk et al. 1988). However, even when VO_2max is reached, not enough aerobic ATP is being generated, and some contribution from anaerobic energy is still needed. In terms of total energy provided during a 30 s all-out sprint, ~70–80% is provided by anaerobic sources and only ~20–30% by aerobic sources (Jacobs et al.

1982, 1983; Parolin et al. 1999). However, during the 25–30 s time period, aerobic metabolism provides ~50% of the total energy (Parolin et al. 1999).

4.3 Brief History of Anaerobic Metabolism Investigations

The study of muscle metabolism began following the discovery of O_2 in the late eighteenth century and the subsequent development of methods to measure the amount and nature of substances taken up and given off by the body (see Asmussen 1971). The French scientist Lavoisier is generally believed to be the first "work or exercise scientist" as he and his colleagues performed experiments on plants and animals and had human subjects do light exercise while measuring an increase in the use of "vital air" (O_2) compared to the resting condition.

Over the course of the nineteenth century, much work focused on attempts to determine the nature of the foodstuffs responsible for the energy utilized at rest and during exercise (Asmussen 1971). Zuntz and colleagues in 1894 believed that work was performed through the combustion of both fat and carbohydrate (CHO) fuels, while Chavreau and collaborators in 1896 argued that CHO was the sole source of fuel for muscular contractions. Both groups based their postulations on estimations of the respiratory exchange ratios as direct measurements of the complete combustion of CHO and fat resulted in carbon dioxide (CO_2) produced/O_2 consumed ratios of 1.0 and 0.7, respectively (Asmussen 1971). A few years later in the early twentieth century, the classic work of Fletcher and Hopkins (1907) using direct metabolic measurements in amphibian muscles was published. They demonstrated that lactic acid existed in surviving resting muscle and that lactic acid was produced in contracting muscle. They also reported that (1) an insufficient O_2 level in resting or contracting muscle increased lactic acid production, (2) administering O_2 caused the lactic acid to disappear, (3) muscle fatigue was greatest in the presence of lactic acid and least in its absence, and lastly, (4) CHO in the form of muscle glycogen was the proposed precursor of muscle lactic acid formation.

Later in the twentieth century, the study of muscle physiology and metabolism was dominated by Otto Meyerhoff and AV Hill who shared the 1922 Nobel prize for their work with frog muscles (Asmussen 1971). Meyerhoff integrated his chemical findings with the thermodynamic and mechanical findings of the time and formulated a hypothesis explaining the physiology of muscular contractions (Sacks and Sacks 1933). Hill combined his myothermic findings with the biochemical findings of others and published his explanation of muscular exertion (Hill and Lupton 1923; Hill et al. 1924). Their combined work led to the "Hill-Meyerhoff theory of muscular metabolism" which appeared to explain most of the mechanical and metabolic changes produced by muscular contractions. Essentially, the primary event in contracting muscle was the anaerobic breakdown of glycogen to lactic acid. During exercise, CO_2 was driven off, heat release was proportional to the lactic acid production, hydrogen ion (H^+) concentration of the muscle increased, and O_2 was consumed to oxidize lactic acid during exercise and in the recovery period after exercise. So, while the researchers were aware that O_2 was used in the combustion of foodstuffs to supply the energy required for body functions, they felt that it was not used in the primary breakdown processes for contraction, only in the recovery processes (Hill and Lupton 1923).

Over the following years, evidence accumulated to suggest that the Hill-Meyerhoff theory was incorrect and too simplistic. Central to these arguments was the discovery of the presence of phosphate compounds in muscle by two independent groups of investigators that published their results in the same year. Eggleton and Eggleton (1927) improved the methods for extracting and measuring Pi and organic phosphate compounds in muscle, and their experiments demonstrated that contractions producing rapid muscular fatigue decreased organic phosphate and increased Pi levels. Working with heart, skeletal, and smooth muscle suggested that phosphagen levels correlated with the muscle's ability to respond to sudden demands for violent activity. Fiske and Subbarow (1927) also improved methods for measuring phosphate compounds and reported that fatiguing contractions in frog muscle decreased organic phosphates, with total depletion occurring when muscle blood flow was occluded. The compound in question was a derivative of creatine and labelled creatine phosphate (CP). Following contractions this compound was quickly resynthesized in the presence of O_2. In 1930, Lundsgaard, working in Meyerhoff's laboratory, poisoned the glycolytic pathway with iodoacetic acid and showed that the muscles were still able to perform a series of anaerobic contractions with no lactic acid formation and that the development of mechanical tension was proportional to the breakdown of CP to creatine (Cr) and Pi (Sacks and Sacks 1933). It was consequently suggested that PCr was the immediate source of energy for muscular contractions and that lactic acid production served to resynthesize CP stores.

However, at about the same time, Lohmann's group in 1929 (Bessman and Geiger 1981) and Fiske and Subbarow, also in Fiske and Subbarow 1929, both discovered the presence of another high-energy phosphate compound in muscle, ATP. In 1934, Lohmann reported that PCr breakdown only occurred when ADP was present to accept the phosphate from CP, producing ATP and Cr (Bessman and Geiger 1981). These results suggested that CP was used to replenish ATP levels which then transferred energy directly to the contractile mechanism, although an earlier study reported no decrease in ATP levels with electrical stimulation of muscles resulting in tetani (Sacks and Sacks 1933). A few years later, Engelhardt and Ljubimowa (1939) also reported that myosin, a contractile constituent of muscle fibrils, possessed an ATPase enzyme which hydrolyzed ATP to ADP and Pi (Asmussen 1971).

Finally, 1962, Cain and Davies conclusively established that ATP was the immediate source of energy for muscle contractions by chemically inhibiting the breakdown of CP and demonstrating a decreased ATP content following a single contraction. This finding brought into focus the realization that both CP and anaerobic glycolysis could replenish the ATP that was being used during intense muscular contractions and that the rates of ATP provision from CP degradation and glycolytic activity were very high during intense muscular contractions. And of course, if the exercise lasted some seconds, ATP produced in the mitochondria would begin to contribute significantly to the resynthesis of ATP.

Margaria et al. (1964, 1969) examined repeated bouts of heavy exercise lasting 10–15 s and believed that only PCr (more commonly used than CP) or the "alactic" component of anaerobic metabolism was involved in these short exercise bouts. They reported no lactate accumulation in the blood as long as recovery periods were sufficient to allow the replenishment of PCr stores. However, with the reintroduction of the muscle needle biopsy technique for sampling human skeletal muscle and the refinement of analytical techniques to measure muscle high energy phosphates, more precise information was subsequently provided (Bergstrom and Hultman 1967; Hultman et al. 1967). Muscle biopsy studies in the 1970s, 1980s, and early 1990s from several independent laboratories demonstrated the simultaneous breakdown of PCr and production of lactate during intense exercise of ~30 s in males on a cycle ergometer (Karlsson and Saltin 1970; Jacobs et al. 1983; Jones et al. 1985; McCartney et al. 1986; Withers et al. 1991; Gaitanos et al. 1993) and sprinting on a treadmill (Cheetham et al. 1986). Jacobs et al. (1982) also reported similar results for females during 30 s of all-out cycling.

4.4 Regulation of Anaerobic Energy Provision During High-Intensity, Short-Term Exercise

The store of ATP in skeletal muscle is low and would be used in a few seconds during high-intensity exercise if no resynthesis of ATP occurred. However, the [ATP] in skeletal muscle is well defended. While this may not be surprising during aerobic exercise, it has been reported from several laboratories that ATP levels decreased by only ~20–30% during one or more bouts of volitional sprint exercise (Karlsson and Saltin 1970; Bogdanis et al. 1996; Hargreaves et al. 1998; Jones et al. 1985; Spriet et al. 1989). This suggests that the high rate of ATP resynthesis from anaerobic energy sources defends the muscle [ATP]. An interesting experiment with the aim of exhausting the capacity of the anaerobic energy provision systems and potentially driving [ATP] lower bypassed the central nervous system and electrically stimulated the vastus lateralis muscles of male volunteers (Spriet et al. 1987a). Blood flow to the legs was also occluded, and the leg muscles were maximally stimulated for 100 s with biopsies taken before and at 25 s intervals during the stimulation. Remarkably, while the [ATP] decreased at each sampling time point, it only decreased to 57% of the resting level at 100 s. It appeared that once the ability to provide anaerobic energy was exhausted, by-products of the high glycolytic activity (e.g., H^+) inhibited the contractile events in the muscle before the entire ATP store was used (Spriet et al. 1987b). Interestingly, in other animals such as fish, the entire ATP store was used during intense sprinting, presumably in an attempt to survive a predator (Pearson et al. 1990). These animals then hide and rest for many hours while their ATP stores are replenished. In summary, ATP levels are reasonably well

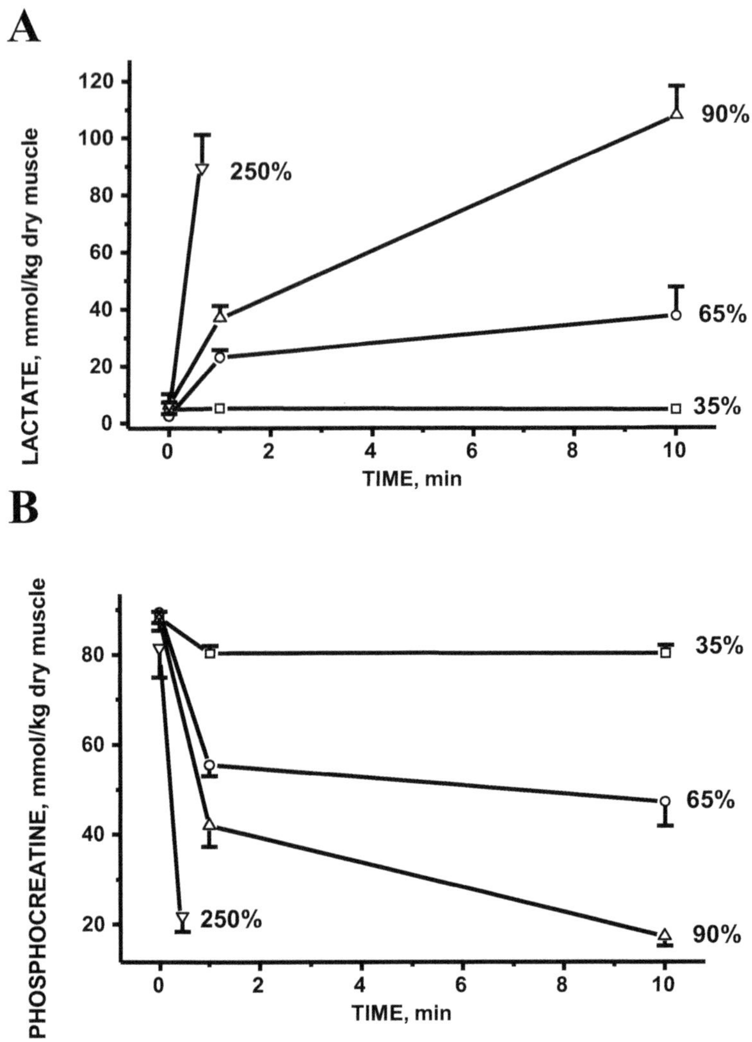

Fig. 4.2 Lactate accumulation and phosphocreatine use in human skeletal muscle during 10 min of exercise at 35, 65, and 90% VO_2max and 30 s at ~250% VO_2 max (Reproduced from Howlett et al. 1998 and Parolin et al. 1999)

maintained in human skeletal muscle even during very intense exercise requiring maximal rates of anaerobic energy provision.

Direct measures of muscle PCr and lactate in whole muscle before and after sprint exercise bouts (~250% VO_2 max) revealed significant decreases in PCr and increases in lactate (Parolin et al. 1999; Howlett et al. 1998; Gaitanos et al. 1993; Medbø and Tabata, 1993) (Figs. 4.2 and 4.3). Similar results were obtained when examining the response of single fibers to sprint exercise. At rest, type I fibers had

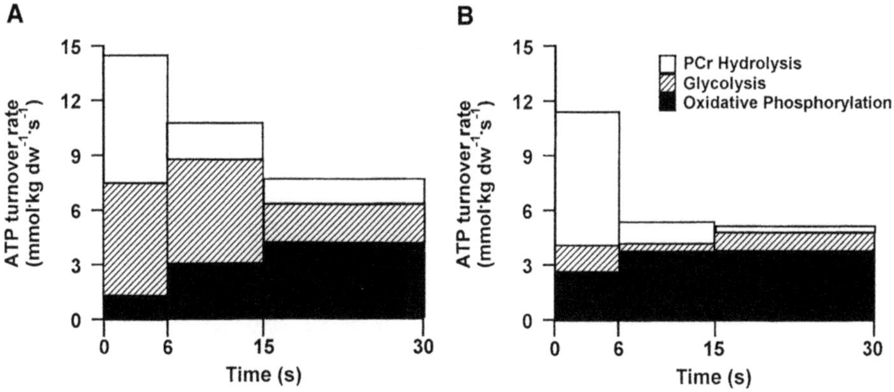

Fig. 4.3 Contribution of phosphocreatine (PCr), anaerobic glycolysis, and oxidative phosphorylation to energy (ATP) provision in human skeletal muscle during the first 30 s all-out sprint at ~250% VO_2 max (A) and the third sprint (B) with 4 min rest between the first, second, and third sprints (Reproduced from Parolin et al. 1999)

10% less PCr and 20% less muscle glycogen than type II fibers (Greenhaff et al. 1994). When subjects were asked to sprint for 30 s on a treadmill, PCr was largely depleted in both fiber types, but type I fibers contributed 20% less ATP from PCr than type II fibers and ~ 40% less ATP from anaerobic glycolysis (Greenhaff et al. 1994).

PCr is a remarkable fuel source as only one metabolic reaction is required to provide ATP (Table 4.1). The enzyme that catalyzes this reaction (creatine phosphokinase, CPK) is in high abundance and only regulated by the concentration of its substrates and products—a so-called near-equilibrium enzyme. As soon as contractions begin and ATP is degraded and the ADP concentration increases, this reaction will move from left to right, and ATP is regenerated in a few msec (Table 4.1). Direct muscle measurements have shown that 60–75% of the PCr store can be used in 6–15 s of all-out sprinting (Gaitanos et al. 1993; Parolin et al. 1999).

At the same time, cellular Ca^{2+} (and to some extent epinephrine from outside the cell) activates phosphorylase kinase to move glycogen phosphorylase from its less active "b" form to the more active "a" form (covalent regulation) (Chasiotis et al. 1982). Increases in ADP and AMP also activate phosphorylase a directly (allosteric regulation) to degrade glycogen and combine with Pi to produce glucose 1-phosphate, glucose 6-phosphate (G-6-P), and fructose 6-phosphate (F-6-P) in the glycolytic pathway (Ren and Hultman 1989). Phosphorylase is considered a "non-equilibrium enzyme" as it is controlled by external factors and not just substrates and products. This combination of covalent and allosteric regulation explains how the flux through phosphorylase can increase from very low rates at rest to very high rates during sprint exercise in only a few msec (Fig. 4.1). The increases in the allosteric regulators ADP, AMP, and Pi (which are by-products of ATP breakdown) and the accumulating substrate F-6-P also activate the regulatory enzyme phosphofructokinase, and flux through the reactions of the glycolytic pathway continues with a net

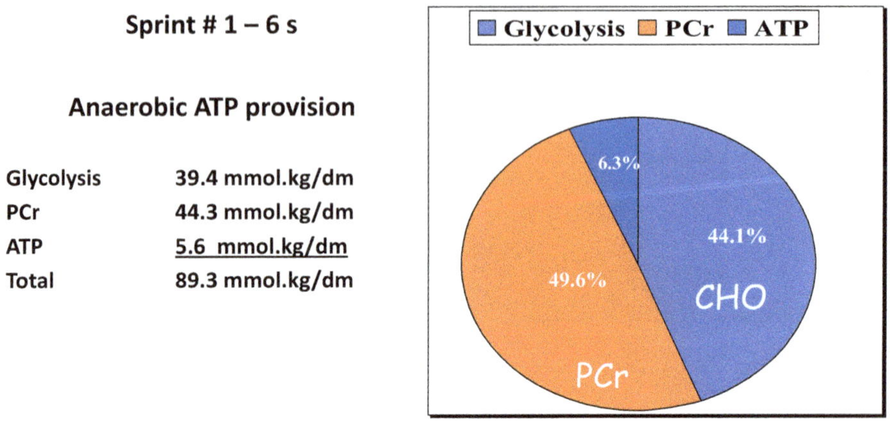

Fig. 4.4 Anaerobic energy provision from phosphocreatine (PCr), anaerobic glycolysis (CHO), and the ATP store in human skeletal muscle during one all-out 6 s cycle sprint (Reproduced from Gaitanos et al. 1993)

production of 3 ATP and lactate formation for each glucose moiety stored as muscle glycogen (Table 4.1). Large amounts of lactate are produced in the reaction catalyzed by lactate dehydrogenase which also regenerates NAD for use higher up in the glycolytic pathway to help maintain the high flux and ATP production in the pathway (Spriet et al. 2000) (Fig. 4.1). Even though there are more reactions involved in producing ATP in the glycolytic pathway, compared to PCr, the production of ATP through anaerobic glycolysis is also activated over a msec time course! Lactate accumulation has been measured in the muscle after a 1 s contraction (Hultman and Sjoholm 1983), and the contribution of anaerobic energy from PCr and anaerobic glycolysis is essentially equivalent after 6–10 s of intense exercise (Figs. 4.3 and 4.4) (Gaitanos et al. 1993; Parolin et al. 1999). When intense exercise is sustained and the demand for ATP is great, a small amount of energy can be also produced in the near-equilibrium myokinase reaction where 2 ADP generate ATP and AMP (Table 4.1), but the contribution from this source is quantitatively small.

The capacity of the PCr energy store is a function of its resting content (~75 mmol/kg dm) and as mentioned can be mostly used up in 10–15 s of all out exercise. The capacity of the anaerobic glycolytic system is about threefold higher (~225 mmol/kg dm) over exercise lasting about 2 min and is limited not by glycogen substrate but by the increasing acidity (Bangsbo et al. 1990; Medbø and Tabata, 1993). The increases in ATP utilization, glycolysis, and strong ion fluxes during sprint exercise result in metabolic acidosis (Kowalchuk et al. 1988; Spriet et al. 1989). The decline in power output during single and repeated bouts of maximal exercise is associated with PCr depletion and the accumulation of metabolic by-products (H^+, ADP, AMP, Pi, K^+) that decrease the excitation-contraction coupling processes within skeletal muscle (Casey et al. 1996; Hargreaves et al. 1998; Medbø and Tabata, 1993; Kowalchuk et al. 1988; Spriet et al. 1989). The above information was generated during all-out sprints where the power output in the initial

5–10 s was ~1000 W and decreased to ~400–500 W by 30 s. Sprinting efforts at lower power outputs and for short periods of time (2–5 s) require lower rates of anaerobic energy provision. Shorter sprints are common in stop-and-go sports, although there will be many sprints in an entire training session or game (Krustrup et al. 2006; Vigh-Larsen et al. 2020).

Another important aspect of high-intensity short-term energy production relates to the ability to rapidly resynthesize PCr when the exercise intensity falls to low levels or the athlete rests. This is common in stop-and-go sports where short sprints are interspersed with rest periods where continued aerobic ATP production fuels the regeneration of PCr, such that the store can be recovered to 65–90% of resting levels in 60–120 s (Harris et al. 1976; Sahlin et al. 1979; Hultman et al. 1967; Bogdanis et al. 1995). This is extremely important for the ability to repeatedly sprint in stop-and-go or intermittent sports as the recovery of the glycolytic system from prolonged sprinting (20–120 s) and the associated muscle acidity takes minutes, not seconds, and can limit performance (Gaitanos et al. 1993; Parolin et al. 1999). The ability to buffer the produced acid is also paramount for success in one-off sprints and in stop-and-go sports, and buffering capacities are generally very high in these athletes due to a combination of genetic endowment and adaptation to sprint training (Bishop et al. 2004). In most stop-and-go sports, sprints are usually kept short such that increasing acidity is minimized and the PCr store is not completely exhausted.

It should not be forgotten that the production of aerobic ATP also turns on during very intense exercise and 70–100% of the VO_2 max can be reached in an all-out 30 s sprint (Kowalchuk et al. 1988). However, the time for aerobic ATP contribution is short, and while little is provided in the first 5–10 s, ~50% of the energy contribution in the last 5 s of a 30 s sprint is aerobic (Parolin et al. 1999). If the exercise task lasts beyond about 1 min, oxidative phosphorylation becomes the major ATP-generating pathway (Medbø and Tabata, 1993). During the transition from rest to intense exercise, the substrate for the increasing aerobic ATP production is from muscle glycogen as a small amount of the produced pyruvate is transported into the mitochondria to produce acetyl-CoA and the reducing equivalent NADH in the pyruvate dehydrogenase reaction (Fig. 4.1). This enzyme is also under covalent control existing in an inactive form at rest and moved to a fully active form by Ca^{2+} during exercise. The power of Ca^{2+}, with assistance from pyruvate, keeps the enzyme in the active form, despite increases in acetyl-CoA that would normally inactivate the enzyme at rest (Howlett et al. (1998).

4.5 Intermittent High-Intensity Exercise

A number of laboratories have directly examined anaerobic energy provision during repeated bouts of high-intensity exercise lasting 6–30 s (Trump et al. 1996; Spriet et al. 1989; Gaitanos et al. 1993; Casey et al. 1996; Bogdanis et al. 1996; Parolin et al. 1999). In one study, subjects performed ten maximal cycling sprints lasting 6 s and separated by 30 s of rest (Gaitanos et al. 1993). Mean power output decreased by

Sprint # 10 - 6 s

Anaerobic ATP provision

Glycolysis	**5.1 mmol.kg/dm**
PCr	**25.3 mmol.kg/dm**
ATP	**1.2 mmol.kg/dm**
Total	**31.6 mmol.kg/dm**

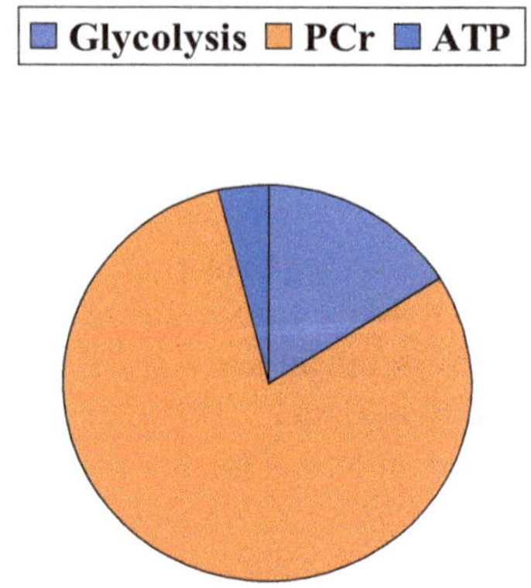

Fig. 4.5 Anaerobic energy provision from phosphocreatine (PCr), anaerobic glycolysis, and the ATP store in human skeletal muscle during the tenth all-out 6 s cycle sprint, where sprints were separated by 30 s of recovery (Reproduced from Gaitanos et al. 1993)

27% from sprint 1 to 10 (Figs. 4.4 and 4.5). Anaerobic energy production decreased by 65% from bout 1 to 10, suggesting that aerobic ATP production played a larger role in the later sprints (O_2 uptake was not measured in this study). The contribution from anaerobic glycolysis decreased by ~90%, and the PCr contribution decreased by 55% of total energy production from bouts 1 to 10 (Figs. 4.4 and 4.5). The short 30 s recovery time was not enough time for the glycolytic pathway of PCr to recover from the previous bout. In fact, it appeared that the glycolytic pathway could not be reactivated in the final bout possibly due to the muscle acidosis, while PCr was partially resynthesized in 30s, but more time was needed for full resynthesis (Bogdanis et al. 1995).

A second study had subjects complete 3–30 s all-out cycling sprints with 4 min of passive (recovery) between bouts (Parolin et al. 1999). Subjects cycled for 6, 15, or 30 s in bouts 1 and 3 with muscle biopsies at these time points over a number of days to complete the study. Breath by breath O_2 uptake measures were also made during exercise. In the first 6 s of bout 1, equal contributions of anaerobic energy came from PCr degradation and glycolysis (Fig. 4.3). In the 6–15 s period, the PCr contribution waned, while the glycolytic contribution was maintained (and became the dominant energy source), and the aerobic contribution increased. In the final 15 s of the first sprint, PCr contribution was again low, the glycolytic contribution decreased, and the aerobic contribution was ~50% of the total energy provision (Fig. 4.3). Muscle PCr decreased from 88 to 8 mmol/kg dm, glycogen decreased from 480 to 400 mmol/kg dm, lactate increased from 5 to 60 mmol/kg dm, and pH decreased from 7.05 to 6.74. Following 4 min of rest, a second 30 s sprint, and 4 additional min (of) rest, PCr had recovered to 70 mmol/kg dm, glycogen was further reduced to 370 mmol/min, lactate remained very high at 95 mmol/kg dm, and muscle pH remained low at 6.66. While the muscle acidosis severely slowed recovery of the glycolytic processes, PCr was able to almost fully resynthesize in the 4 min between bouts.

In the third bout, energy provision and power output from 0–6 s were ~ 70% of the first bout values, largely on the back of PCr energy provision, as the glycolytic contribution was very small and the aerobic contribution was also small but larger than the first bout (Parolin et al. 1999). Force and anaerobic energy provision dropped precipitously in the 6–15 and 15–30 s periods as the PCr store was depleted and glycolysis simply could not be reactivated (Fig. 4.3). Aerobic energy provision contributed most of the energy although force production in these time periods was much lower than in bout 1. It appeared that the accumulating and extreme acidosis that occurred during these sequential all-out 30 s sprints made it impossible to reactivate the glycolytic pathway through inhibition of the key enzymes glycogen phosphorylase and phosphofructokinase. On the other hand, PCr had time to resynthesize during the 4 min rest periods and could contribute ATP again during the subsequent sprints. Aerobic metabolism became more important as the sprints progressed in large part because O_2 uptake did not fully recover to resting values in 4 min and was therefore partially activated at the beginning of the next bout.

The findings from these repeated sprint studies have some practical applications. First, it seems important to keep all-out sprints short (<6 s) such that muscle acidosis is minimized. Second, sufficient time for the recovery between bouts would be needed (~60–90 s) to ensure near or full resynthesis of PCr, and third, a high VO_2max is desirable in repeated sprint situations to contribute more aerobic energy during the sprints, increasing amounts of energy in successive sprints, and provide ATP for the resynthesis of PCr in the rest periods (and possibly contribute to the oxidation of lactate if the rest period is long enough).

4.6 Other Methods to Estimate Anaerobic Energy Contributions

Several other methods have been used over the years to estimate the importance of anaerobic energy provision. While the results from these approaches will not be discussed in detail in this chapter, the reader is directed to the studies listed below. Nuclear magnetic resonance (NMR) or magnetic resonance spectroscopy (MRS) can measure the content of ^{31}P in human skeletal muscle, and estimates of PCr, the three phosphates of ATP, and Pi can be obtained at rest and during exercise when subjects are encased in a magnet (Burt et al. 1976; Miller et al. 1988; Wilson et al. 1988). Studies have examined the kinetics of PCr pre- and post-training (Kent-Braun et al. 1990), the differences between sprinters and long-distance runners (McCully 1993), and in single fibers during recovery from contractions (Walter et al. 1997). Simultaneous MRS and biochemical measurements of PCr and Pi have been comparable, but MRS estimates of ATP, H^+, and lactate concentrations during exercise have overestimated the biochemical findings (Bangsbo et al. 1993; Constantin-Teodosiu et al. 1997).

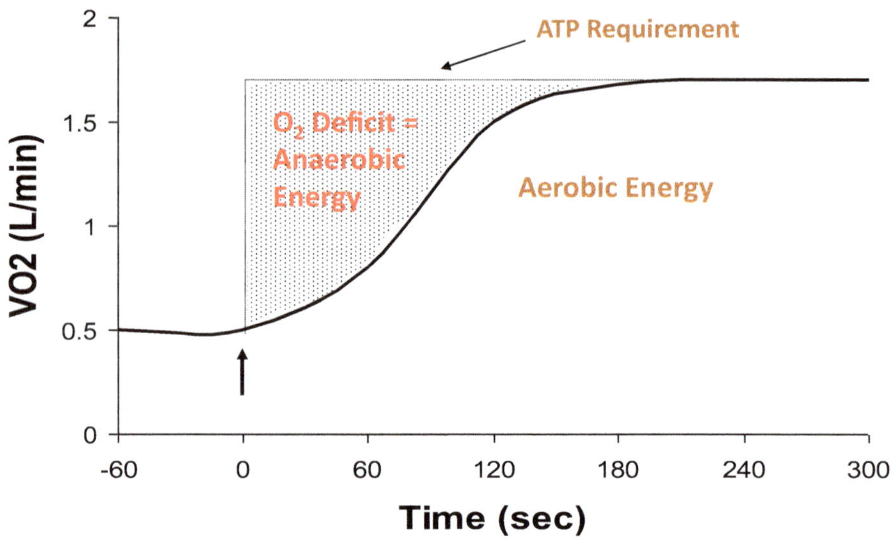

Fig. 4.6 Schematic diagram showing the oxygen (O_2) uptake kinetics (VO_2) during the transition from rest to a submaximal power output and the requirement for anaerobic provision during this transition (O_2 deficit)

Medbø et al. (1988) described a method to indirectly estimate the contribution of anaerobic energy at the onset of exercise by essentially determining how much whole-body O_2-derived energy was missing to account for the work done—the "maximum accumulated oxygen deficit (MAOD)." This requires the ability to measure O_2 uptake in a breath-by breath manner to accurately follow the O_2 kinetics during the transition from rest to a steady-state level during submaximal exercise or VO_2max in supramaximal exercise. The amount of energy that cannot be provided through aerobic means at the onset of exercise is called the O_2 deficit (Fig. 4.6). The O_2 deficit is larger as the power output increases and the contracting muscles require larger amounts of anaerobic energy from PCr and anaerobic glycolysis (Fig. 4.2). The study of O_2 kinetics measured at the mouth and what limits the rate of O_2 uptake at the onset of exercise is a major field of study on its own and will not be examined in detail here. However, studies have shown that speeding of the exercise onset O_2 kinetics occurs with training in men and women (Murias et al. 2011; Grey et al. 1985), which means the O_2 deficit decreases as does the need for anaerobic energy. Discussion of the best way to collect and interpret the O_2 kinetics data also continues (Benson et al. 2017). The relationships between measurements of O_2 kinetics, O_2 delivery to the contracting muscles, and actual O_2 utilization continue to be an active research area (Grassi et al. 2021; Korzeniewski and Rossiter 2021).

Medbø and Tabata (1989) estimated that the anaerobic energy contribution to the total energy production during exercise designed to exhaust the subjects in ~30 s, 60 s, and 2–3 min was ~60, 50, and 35%, respectively. They also reported that the anaerobic capacity was 30% higher in male sprinters compared to male distance runners or untrained subjects and that 6 weeks of training increased the anaerobic capacity by 10% in a group of men and women (Medbø and Burgers 1990). Lastly, they reported that 2 min of exhausting exercise was needed to maximize anaerobic energy release and, importantly, that there was a close relationship between the estimated whole body $\dot{O}_2$ deficit and direct muscle biopsy measurements of anaerobic energy release (Medbø and Tabata 1993). The MAOD technique continues to be used to estimate the anaerobic contribution to total energy production in in many sporting situations (Spencer and Gastin 2001; Andrade et al. 2021; Campos et al. 2017).

Other indirect attempts to estimate anaerobic power and capacity and the contributions of PCr and anaerobic glycolysis to the total anaerobic energy release in human subjects have been made during sprint tests (Serresse et al. 1988; Smith and Hill 1991). These assessments require several assumptions which in many cases have been shown to be incorrect—for example, the suggestion that only PCr contributes anaerobic energy in the first 10 s of an all-out sprint test (Gaitanos et al. 1993; Parolin et al. 1999).

4.7 Sprint Training and Creatine Supplementation

Not all the Cr in skeletal muscle is phosphorylated as the total Cr content is ~120–130 mmol/kg dm, and ~ 60–70% (of) the Cr is phosphorylated at rest. A major unanswered question in physiology and sports science is what determines the amount of total creatine (TCr), free Cr, and PCr that is stored in human skeletal muscle? In most cases, we want to know why it is not higher and how can we augment it to hopefully enhance high-intensity exercise performance. There is natural variation in the muscle content between people ranging from ~100 to 150 mmol/kg dm, possibly predisposing people with high levels to greater success in sports requiring high-energy provision in sprint-like activities. Remarkably, the contents of Cr and PCr in muscle do not increase following short- or long-term aerobic (Chesley et al. 1996; Talanian et al. 2007; Perry et al. 2008), sprint training (Cheetham et al. 1986; Nevill et al. 1989), or resistance training (MacDougall et al. 1977) in men or women. This (contrasts) with a 10–20% increase in the glycolytic capacity to produce anaerobic energy following sprint training (Cheetham et al. 1986; Nevill et al. 1989). Cr does turnover in the muscle, and ~ half of the Cr is provided from the diet and half from Cr synthesis in the body (Negro et al. 2019). In a landmark study by Harris et al. in 1992, the authors suggested that the delivery of Cr to the muscle may limit the ability of Cr transporters in the cell membrane to take up additional Cr and increase the muscle Cr store. They demonstrated that ingesting a large quantity of Cr (~5 g) could increase the plasma [Cr] several fold and maintain these increases for 4–6 hours. This resulted in ~20% increases in the total Cr content in muscle and smaller but significant increases in PCr content (Harris et al. 1992). Of course, this amount of Cr could not be ingested in actual food, but single and multiple ingestions of this large amount of Cr presented the body with an unphysiological increase in plasma Cr for many hours. They also reported that the increases in muscle Cr with supplementation were largest in people with the lowest muscle Cr content and that people who already had high total Cr muscle contents of ~140–150 mmol/kg dm responded very little or not at all.

Additional work by the same group identified what are now termed the "rapid" and "slow" Cr loading regimens (Hultman et al. 1996). The "rapid" loading regimen involves ingesting 5 g of creatine monohydrate every 4–5 hours coinciding with breakfast, lunch, dinner, and an evening snack. This ensures that the plasma [Cr] will be high for ~16 hours/day or more, and in this situation, Cr is taken up by the muscles with most occurring in the first 2–3 days, based on the excretion of Cr in the urine (Harris et al. 1992). The authors also showed that ingesting 2 g/day following the rapid loading phase maintained the new higher total Cr content. If no additional Cr was ingested beyond the rapid loading phase muscle [Cr] returned to baseline in about 28 days. The "slow" loading regimen simply required the ingestion of 3 g Cr/day for ~30 days and the same increase in muscle [Cr] as the rapid loading phase was achieved (Hultman et al. 1996).

The ultimate goal of Cr supplementation is to increase the resting [PCr] and therefore the capacity of this system to provide ATP during intense sprint-like activities. Many studies have since examined whether increased performance results from Cr supplementation in appropriate forms of exercise, and many studies, but not all, have confirmed this (Antonio et al. 2021). Numerous research papers and review articles now exist summarizing the methods to enhance Cr uptake by the muscles, the performance effects of supplementation, and the safety concerns related to long-term supplementation, among many other topics (Green et al. 1996; Steenge et al. 2000; Negro et al. 2019; Antonio et al. 2021).

4.8 Summary

Aerobic metabolism is the dominant pathway for ATP production in human skeletal muscle in many movement, exercise, and sport situations. However, anaerobic ATP production also plays an important role to supplement or buffer the need for ATP when aerobic metabolism cannot meet the ATP demands. These situations include the transitions from rest to exercise and from one power output to a higher one, high-intensity exercise that demands ATP provision rates above what can be provided aerobically, and in situations of suboptimal oxygen supply. Anaerobic energy is provided from phosphocreatine and muscle glycogen breakdown (anaerobic glycolysis). These systems can provide energy very quickly and at very high rates but are limited to short periods of time during high-intensity exercise due to substrate depletion and increasing muscle acidosis. This is especially important as the amount of energy stored in muscle is low and would be consumed in just a few seconds of high-intensity exercise. The capacity of the PCr store is about 3 times the amount of energy stored in the muscle as ATP and can be depleted in ~10–15 s of intense exercise. The capacity of glycolysis to produce anaerobic energy is about 3 times

that of PCr but needs ~90–120 s for this contribution. After long sprints, PCr can be rapidly resynthesized in a few minutes, whereas anaerobic glycolysis needs 30–-60 minutes to remove the accumulated acidosis. Sprint training can increase the capacity of anaerobic glycolysis by 10–20% while the PCr store in muscle is unchanged. Supplementation with creatine can increase the skeletal muscle total creatine and PCr stores in most individuals. Importantly, energy provision is maintained by contributions from both the aerobic and anaerobic sources to ensure that ATP resynthesis closely matches the exercise ATP demand in most exercise and sporting situations. Anaerobic ATP provision makes its mark during very intense bursts of activity requiring ballistic and powerful sprint-like movements.

References

Andrade VL, Zagatto AM, Kalva CA et al (2021) Running-based anaerobic sprint test as a procedure to evaluate anaerobic power. Int J Sports Med 36:1156–1162

Antonio J, Candow DG, Forbes SC et al (2021) Common questions and misconceptions about creatine supplementation: what does the scientific evidence really show? J Int Soc Sports Nutr 18:13

Asmussen E. Muscle metabolism during exercise in man(1971) A historical survey. In: Muscle metabolism during exercise. Plenum Press, New York, pp 1–11

Bangsbo J, Gollnick PB, Graham TE et al (1990) Anaerobic energy production and 0_2 deficit–debt relationship during exhaustive exercise in humans. J Physiol 422:539–559

Bangsbo J, Johansen B, Quistorff B et al (1993) NMR and analytic biochemical evaluation of PCr nucleotides in the human calf during muscle contraction. J Appl Physiol 74:2034–2039

Benson AP, Bowen TS, Ferguson C et al (2017) Data collection, handling, and fitting strategies to optimize accuracy and precision of oxygen uptake kinetics estimation from breath-by-breath measurements. J Appl Physiol 23:227–242

Bergstrom J, Hultman E (1967) A study of glycogen metabolism during exercise in man. Scand J Clin Lab Invest 19:218–228

Bessman SP, Geiger PJ (1981) Transport of energy in muscle: the phosphorylcreatine shuttle. Science 211:448–452

Bishop D, Edge J, Goodman C (2004) Muscle buffer capacity and aerobic fitness are associated with repeated-sprint ability in women. Eur J Appl Physiol 92:540–547

Bogdanis GC, Nevill ME, Boobis LH et al (1995) Recovery of power output and muscle metabolites following 30 s of maximal sprint cycling in man. J Physiol 482:467–480

Bogdanis GC, Nevill ME, Boobis LH et al (1996) Contribution of phosphocreatine and aerobic metabolism to energy supply during repeated sprint exercise. J Appl Physiol 80:876–884

Burt CT, Glonek T, Barany M (1976) Analysis of phosphate metabolites, the intracellular pH, and the state of adenosine triphosphate in intact muscle by phosphorous nuclear magnetic resonance. J Biol Chem 251:2584–2591

Cain DF, Davies RE (1962) Breakdown of adenosine triphosphate during a single contraction of working muscle. Biochem Biophys Res Commun 8:361–367

Campos EZ, Kalva-Filho CA, Gobbi RB et al (2017) Anaerobic contribution determined in swimming distances: relation with performance. Front Physiol 8:755

Casey A, Constantin-Teodosiu D, Howell S et al (1996) Metabolic responses of type I and II muscle fibres during repeated bouts of maximal exercise in humans. Am J Physiol Endocrinol Metab 271:E38–E43

Chasiotis D, Sahlin K, Hultman E (1982) Regulation of glycogenolysis in human muscle at rest and during exercise. J Appl Physiol 53:708–715

Cheetham ME, Boobis LH, Brooks S et al (1986) Human muscle metabolism during sprint running. J Appl Physiol 61:54–60

Chesley A, Heigenhauser GJF, Spriet LL (1996) Regulation of muscle glycogen phosphorylase activity following short-term endurance training. Am J Physiol Endocrinol Metab 270:E328–E335

Constantin-Teodosiu D, Greenhaff PL, McIntyre B et al (1997) Anaerobic energy production in human skeletal muscle in intense contraction: a comparison of 31P magnetic resonance spectroscopy and biochemical techniques. Exp Physiol 82:593–601

Eggleton P, Eggleton MG (1927) The significance of phosphorous in muscular contractions. Nature 119:194–195

Engelhardt WA, Ljubimowa MN (1939) Myosin and adenosine triphosphate. Nature 144:668–669

Fiske CH, Subbarow Y (1927) The nature of the "inorganic phosphate" in voluntary muscle. Science 65:401–403

Fiske CH, Subbarow Y (1929) Phosphorous compounds of muscle and liver. Science 70:381–382

Fletcher WM, Hopkins FG (1907) Lactic acid in amphibian muscles. J Physiol 35:247–309

Gaitanos GC, Williams C, Boobis LH et al (1993) Human muscle metabolism during intermittent maximal exercise. J Appl Physiol 75:712–719

Gollnick PD, Hermansen L (1973) Biochemical adaptation to exercise: anaerobic metabolism. In: Wilmore JH (ed) Exercise and sport sciences review. Academic Press, New York, pp 1–43

Grassi B, Hogan MC, Gladden LB (2021) Microvascular O_2 delivery and O_2 utilization during metabolic transitions in skeletal muscle. One-hundred years after the pioneering work by August Krogh. Comp Biochem Physiol A Mol Integr Physiol 252:110842

Green AL, Hultman E, Macdonald IA et al (1996) Carbohydrate ingestion augments skeletal muscle creatine accumulation during creatine supplementation in humans. Am J Phys 271:E821–E826

Greenhaff PL, Nevill ME, Soderlund K et al (1994) The metabolic responses of human type I and II muscle fibres during maximal treadmill sprinting. J Physiol 478:149–155

Grey TM, Spencer MD, Belfry GR et al (1985) Effects of age and long-term endurance training on VO2 kinetics. Med Sci Sports Exerc 47:289–298

Hargreaves M, McKenna MJ, Jenkins DG et al (1998) Muscle metabolites and performance during high intensity, intermittent exercise. J Appl Physiol 84:1687–1691

Hargreaves M, Spriet LL (2020) Skeletal muscle energy metabolism during exercise. Nat Metab 2: 817–828

Harris RC, Edwards RH, Hultman E et al (1976) The time course of phosphorylcreatine resynthesis during recovery of the quadriceps muscle in man. Pflugers Arch 367:137–142

Harris RC, Soderlund K, Hultman E (1992) Elevation of creatine in resting and exercised muscle of normal subjects by creatine supplementation. Clin Sci 83:367–374

Hermansen L (1969) Anaerobic energy release. Med Sci Sports 1:32–38

Hill AV, Lupton H (1923) Muscular exercise, lactic acid, and the supply and utilization of oxygen. Quart J Med 16:135–171

Hill AV, Long CNH, Lupton H (1924) Muscular exercise, lactic acid, and the supply and utilization of oxygen. Proc Roy Soc Lond Ser B 96:438–475

Howlett RA, Parolin ML, Dyck DJ et al (1998) Regulation of skeletal muscle glycogen phosphorylase and PDH at varying exercise power outputs. Am J Physiol Regul Integr Comp Physiol 275:R418–R425

Hultman E, Bergstrsm J, McLennan-Anderson N (1967) Breakdown and resynthesis of phosphorylcreatine and adenosine triphosphate in connection with muscular work in man. Scand J Clin Lab Invest 19:56–66

Hultman E, Sjoholm H (1983) Energy metabolism and contraction force of human skeletal muscle in situ during electrical stimulation. J Physiol 345:525–532

Hultman E, Soderlund K, Timmons JA (1996) Muscle creatine loading in men. J Appl Physiol 81: 232–237

Jacobs I, Bar-Or O, Karlsson J et al (1982) Changes in muscle metabolites in females with 30-s exhaustive exercise. Med Sci Sports Exerc 14:457–460

Jacobs I, Tesch P, Bar-Or O et al (1983) Lactate in human skeletal muscle after 10 and 30 s of supramaximal exercise. J Appl Physiol 55:365–367

Jones NL, McCartney N, Graham T et al (1985) Muscle performance and metabolism in maximal isokinetic cycling at slow and fast speeds. J Appl Physiol 59:132–136

Karlsson J, Saltin B (1970) Lactate, ATP and CP in working muscle during exhaustive exercise in men. J Appl Physiol 29:598–602

Kent-Braun JA, McCully KK, Chance B (1990) Metabolic effects of training in humans: a ^{31}P-MRS study. J Appl Physiol 69:1165–1170

Korzeniewski B, Rossiter HB (2021) Factors determining training-induced changes in $\dot{V}O_{2max}$, critical power, and $\dot{V}O_2$ on-kinetics in skeletal muscle. J Appl Physiol 130:498–507

Kowalchuk JM, Heigenhauser GJ, Lindinger MI (1988) Factors influencing hydrogen ion concentration in muscle after intense exercise. J Appl Physiol 65:2080–2089

Krustrup P, Mohr M, Steensburg A et al (2006) Muscle and blood metabolites during a soccer game: implications for sprint performance. Med Sci Sports Exerc 38:1165–1174

MacDougall JD, Ward GR, Sale DG et al (1977) Biochemical adaptation of human skeletal muscle to heavy resistance training and immobilization. J Appl Physiol 43:700–703

Margaria R, Cerretelli P, Mangili E (1964) Balance and kinetics of anaerobic energy release during strenuous exercise in man. J Appl Physiol 19:623–628

Margaria R, Oliva D, DiPrampero PE et al (1969) Energy utilization in intermittent exercise of supramaximal intensity. J Appl Physiol 26:752–756

McCartney N, Spriet LL, Heigenhauser GJF et al (1986) Muscle power and metabolism in maximal intermittent exercise. J Appl Physiol 60:1164–1169

McCully KK (1993) ^{31}P-MRS of quadriceps reveals quantitative differences between sprinters and long-distance runners. Med Sports Sci Exerc 25:1299–1300

Medbø JI, Mohn A-C, Tabata I (1988) Anaerobic capacity determined by maximal accumulated O_2 deficit. J Appl Physiol 64:50–60

Medbø JI, Tabata I (1989) Relative importance of aerobic and anaerobic energy release during short-lasting exhausting bicycle exercise. J Appl Physiol 67:1881–1886

Medbø JI, Burgers S (1990) Effect of training on the anaerobic capacity. Med Sci Sports Exerc 22: 501–507

Medbø JI, Tabata I (1993) Anaerobic energy release in working muscle during 30 s to 3 min of exhausting bicycling. J Appl Physiol 75:1654–1660

Miller RG, Boska MD, Moussavi RS et al (1988) ^{31}P nuclear magnetic resonance studies of high energy phosphates and pH in human muscle fatigue. J Clin Invest 81:1190–1196

Murias JM, Kowalchuk JM, Paterson DH (2011) Speeding of VO2 kinetics in response to endurance-training in older and young women. Eur J Appl Physiol 111:235–243

Negro M, Avanzato I, D'Antona G (2019) Creatine in skeletal muscle physiology. In: Nonvitamin and nonmineral nutritional supplements. Academic Press

Nevill ME, Boobis LH, Brooks S et al (1989) Effect of training on muscle metabolism during treadmill sprinting. J Appl Physiol 67:2376–2382

Parolin ML, Chesley A, Matsos MP et al (1999) Regulation of skeletal muscle glycogen phosphorylase and PDH during maximal intermittent exercise. Am J Physiol Endocrinol Metab 277: E890–E900

Pearson MP, Spriet LL, Stevens ED (1990) Effects of sprint training on swim performance and white muscle metabolism during exercise and recovery in rainbow trout (Salmo gairdnerii Richardson). J Exp Biol 149:45–60

Perry CGR, Heigenhauser GJF, Bonen A et al (2008) High-intensity aerobic interval training increases fat and carbohydrate metabolic capacities in human skeletal muscle. Appl Physiol Nutr Metab 33:1112–1123

Ren JM, Hultman E (1989) Regulation of glycogenolysis in human skeletal muscle. J Appl Physiol 67:2243–2248

Sacks J, Sacks WC (1933) The fundamental chemical changes in contracting mammalian muscle. Am J Phys 105:151–161

Sahlin K, Harris RC, Hultman E (1979) Resynthesis of creatine phosphate in human muscle after exercise in relation to intramuscular pH and availability of oxygen. Scand J Clin Lab Invest 79: 551–558

Sahlin K (2014) Muscle energetics during explosive activities and potential effects of nutrition and training. Sports Med 44(Suppl 2):S167–S173

Saltin B (1990) Anaerobic capacity: past, present and prospective. In: Taylor AW, Gollnick PD, Green HJ et al (eds) Biochemistry of exercise VII. Human Kinetics, Champaign, pp 387–412

Serresse O, Lortie G, Bouchard C (1988) Estimation of the contribution of the various energy systems during maximal work of short duration. Int J Sports Med 9:456–360

Smith JC, Hill DW (1991) Contribution of energy systems during a Wingate power test. Br J Sports Med 25:196–199

Spencer MR, Gastin PB (2001) Energy contribution during 200-to1500 m running in highly trained athletes. Med Sci Sports Exerc 33:157–162

Spriet LL (1992) Anaerobic metabolism in human skeletal muscle during short-term, intense activity. Can J Physiol Pharm 70:157–165

Spriet LL (2006) Anaerobic metabolism during exercise. In: Hargreaves M, Spriet LL (eds) Exercise metabolism, 2nd edn. Human Kinetics, Windsor, pp 7–28

Spriet LL, Soderlund K, Bergstrom M et al (1987a) Anaerobic energy release in skeletal muscle during electrical stimulation in man. J Appl Physiol 62:611–615

Spriet LL, Soderlund K, Bergstrom M et al (1987b) Skeletal muscle glycogenolysis, glycolysis and pH during electrical stimulation in man. J Appl Physiol 62:616–621

Spriet LL, Lindinger MI, McKelvie RS et al (1989) Muscle glycogenolysis and H^+ concentration during maximal intermittent cycling. J Appl Physiol 66:8–13

Spriet LL, Howlett RA, Heigenhauser GJF (2000) An enzymatic approach to lactate production in human skeletal muscle during exercise. Med Sci Sports Exerc 32:756–763

Steenge GR, Simpson EJ, Greenhaff PL (2000) Protein- and carbohydrate-induced augmentation of whole body creatine retention in humans. J Appl Physiol 89:1165–1171

Talanian JL, Galloway SDR, Heigenhauser GJF et al (2007) Two weeks of high-intensity aerobic interval training increases the capacity for fat oxidation in women. J Appl Physiol 102:1439–1447

Trump ME, Heigenhauser GJF, Putman CT (1996) Importance of phosphocreatine during intermittent maximal cycling. J Appl Physiol 80:1574–1580

Vigh-Larsen JF, Ermidis G, Rago V et al (2020) Muscle metabolism and fatigue during simulated ice hockey match-play in elite players. Med Sci Sports Exerc 52:2162–2171

Walter G, Vandenborne K, McCully KK (1997) Noninvasive measurement of phosphocreatine recovery kinetics in single human muscles. Am J Phys 272:C525–C534

Wilson JR, McCully KK, Mancini DM et al (1988) Relationship of muscular fatigue to pH and deprotonated pi in humans: a ^{31}P-NMR study. J Appl Physiol 64:2333–2339

Withers RT, Sherman WM, Clark DG et al (1991) Muscle metabolism during 30, 60 and 90 s of maximal cycling on an air-braked ergometer. Eur J Appl Physiol Occup Physiol 63:354–362

Chapter 5
Exercise and Muscle Glycogen Metabolism

Niels Ørtenblad, Joachim Nielsen, James P. Morton, and José L. Areta

Abstract Muscle glycogen is an important fuel source for contracting skeletal muscle, and it is well documented that exercise performance is impaired when the muscle's stores of glycogen are exhausted. The role of carbohydrate (CHO) availability on exercise performance has been known for more than a century, while the specific role of muscle glycogen for muscle function has been known for half a century. Nonetheless, the precise cellular and molecular mechanisms by which glycogen availability regulates cell function and contractile-induced fatigue are unresolved. Alterations of pre-exercise muscle glycogen reserves by dietary and exercise manipulations or modifying the degree of dependency on endogenous glycogen during exercise have collectively established a close relationship between muscle glycogen and the resistance to fatigue. It is also apparent that glycogen availability regulates rates of muscle glycogenolysis and resynthesis, muscle glucose uptake, key steps in excitation-contraction coupling, and exercise-induced cell signaling regulating training adaptation. The present review provides both a historical and contemporary overview of the effects of exercise on muscle glycogen metabolism, addressing factors affecting glycogen use during exercise as well as the evolving concepts of how glycogen and glycolysis are integrated with cell function, skeletal muscle fatigue, and training adaptation.

Keywords Glycogenolysis, glycogen particle · Diet · Exercise · E-C coupling, fatigue, performance

N. Ørtenblad (✉) · J. Nielsen
Department of Sports Science and Clinical Biomechanics, University of Southern Denmark, Odense, Denmark
e-mail: nortenblad@health.sdu.dk

J. P. Morton · J. L. Areta
Research Institute for Sport and Exercise Sciences, School of Sport and Exercise Sciences, Liverpool John Moores University, Liverpool, UK

© The Author(s), under exclusive license to Springer Nature Switzerland AG 2022
G. McConell (ed.), *Exercise Metabolism*, Physiology in Health and Disease,
https://doi.org/10.1007/978-3-030-94305-9_5

5.1 Introduction

It is remarkable how skeletal muscle fibers can adapt acutely to provide the necessary production of energy during exercise, where a several-fold elevated energy turnover can be sustained for hours or a more than a hundred-fold increase can be executed for minutes. This ability to balance the energy turnover during various types of exercise is achieved by an integration of different energy pathways and by an efficient regulatory system, ensuring that ATP resynthesis is closely matched to the ATP demand of exercise. By the end of the 1930s, it had already been established that both fat and carbohydrate (CHO) could be used as fuel sources for aerobic metabolism during exercise and that fuel used during exercise can be modified by dietary manipulation (Zuntz 1896; Frentzel and Reach 1901; Krogh and Lindhard 1920; Edwards et al. 1934; Christensen and Hansen 1939). It was also understood that CHO was the predominant metabolic substrate when exercising at high intensities and that a relation existed between increasing CHO utilization with increased intensity. Although these early studies had documented that CHO is a major substrate during exercise and that the diet plays an important role in endurance capacity, it was the introduction of the needle biopsy technique in the 1960s that initially demonstrated that work time to exhaustion is highly correlated with muscle glycogen concentration (Bergström et al. 1967). Furthermore, it was also established that muscle glycogen content can be easily altered by isocaloric diets with varied CHO content and that at exhaustion there was a near depletion of muscle glycogen. These seminal studies on the important role of muscle glycogen on prolonged submaximal exercise performance have since been confirmed and extended numerous times. It is also well established that endurance training increases the basal stores of muscle glycogen, as demonstrated in rodent muscle already in the 1930s (Palladin 1945) and later in humans (Taylor et al. 1992; Gollnick et al. 1974). Additionally, trained humans have higher muscle glycogen content and a lower carbohydrate utilization for a given absolute submaximal exercise intensity when compared to untrained subjects (Karlsson et al. 1974; Saltin et al. 1976).

Based on the fundamental findings that carbohydrate combustion is more efficient than fat combustion (~6.2 ATP per O_2 and ~ 5.6 ATP per O_2, respectively) (Krogh and Lindhard 1920; McGilvery 1975) and that the degradation of glycogen to lactate or CO_2 and H_2O in muscles provides a more rapid energy production than that provided by the utilization of fatty acids (~1, 0.5 and 0.24 μmoles ATP g^{-1} s^{-1}, respectively) (Margaria et al. 1964), a large body of studies have been undertaken from around the mid-1960s to today to investigate which factors influence glycogen utilization during exercise and how this affects the function of skeletal muscle fibers and athletic performance. As discussed above, a fundamental observation is that exercise performance is impaired when the muscle's stores of glycogen are exhausted. During exercise, glycogen is utilized and can be depleted to very low levels often reaching one-fifth to one-sixth of the pre-exercise level (Gollnick et al. 1974). This is observed in humans, who are unable to withstand exercise at or above moderate intensity for a prolonged time when the stores are depleted to very low levels (<150 mmol kg^{-1} dw) compared to pre-exercise levels of 500–900 mmol kg^{-1} dw (Bergström et al. 1967; Hermansen et al. 1967) even with carbohydrate supplementation during the exercise (Coyle et al. 1986; Rauch et al. 1995). Therefore, the understanding of factors affecting the graded utilization of glycogen during exercise is key to avoid unforeseen glycogen-dependent muscle fatigue. Indeed, the precise mechanisms underpinning the role of glycogen in muscle function and performance are far from understood.

With this in mind, the present chapter provides both a historical and contemporary overview of the regulation of muscle glycogen metabolism during exercise in humans. After an initial discussion of glycogen storage and regulatory processes of glycogenolysis, a critical review of the factors that modulate glycogen utilization during exercise (i.e., intensity, duration, training status, modality, fiber type, sex, subcellular location, and environmental factors) is then presented. Subsequently, the evolving concepts of how glycogen and glycolysis are integrated with cell function and skeletal muscle fatigue are discussed from both a biochemical and physiological context. Finally, we close by outlining how glycogen utilization may serve as a signal to regulate cell signaling processes associated with modulating the endurance phenotype (i.e., training adaptations).

5.2 Glycogen Storage and Regulation

5.2.1 Biochemistry of the Glycogen Particle and Its Turnover

Glycogen is a unique molecule among several glucose polymers found in nature with structural and energy storage functions. Polymers of glucose with structural function include chitin (polymer of n-acetylglucosamine, a derivative of glucose), predominantly in arthropods and fungi, and cellulose in plants, algae, and oomycetes, which represents the most abundant polymer on earth (Klemm et al. 2005). For energy storage, the main polymers of glucose are starch in plants and glycogen which is by far the most widespread form of storage, found in archaea, bacteria, and eukaryotes (from protozoa and fungi to mammals) (Ball et al. 2011). Synthesis of glycogen through digestion of other glucose polymers is possible in humans for which starch represents a digestible form, while other mammals (e.g., ruminants and ungulates) can also digest cellulose by endosymbiont bacteria in the gut. In this context, two facts make glycogen standout: (1) it is a highly evolutionarily conserved molecule in prokaryotes and from unicellular eukaryotes to mammals, and (2) it represents the largest storage form of energy for high metabolic power output processes in mammalian cells, for which different mechanisms have evolved to maximize its synthesis and storage. These facts attest its key role in energy storage and place it in the centerstage of cellular energy production in skeletal muscle for exercise.

The glycogen polysaccharide is termed as a glycogen molecule, granule, or particle (Fig. 5.1). Due to its physical association with several proteins, it has also

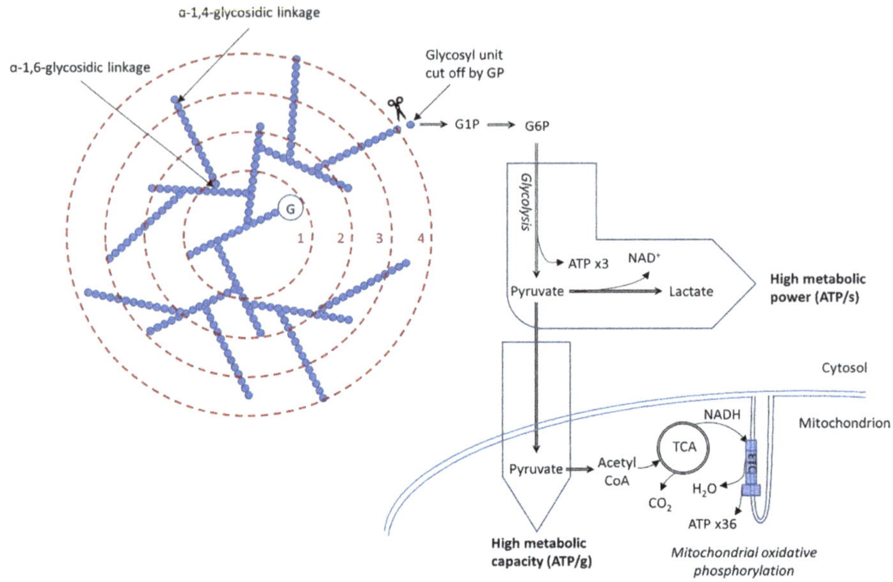

Fig. 5.1 Schematic diagram of a glycogen particle (with four concentric tiers of chains of glycosyl units) and the turnover of glycosyl units to lactate generating a high metabolic power or to CO_2 and H_2O ensuring a high metabolic capacity. *G* glycogenin, *GP* glycogen phosphorylase, *G1P* glucose-1-phosphate, *G6P* glucose-6-phosphate, *TCA* tricarboxylic acid cycle, *ETC* electron transport chain

been termed glycosome to indicate its more organelle-like structure (Rybicka 1996). When glucose enters the cell, it is phosphorylated to glucose 6-phosphate by hexokinase I and can either be metabolized in the glycolytic pathway or be added to existing glycogen particles by the action of phosphoglucomutase, UDP-glucose pyrophosphorylase, glycogen synthase, and branching enzyme (Roach et al. 2012). While the first two steps activate the glucose, the glycogen synthase catalyzes the reaction, where a α-1,4-glycosidic linkage connects the new glycosyl units with the nonreducing end of a chain of glycosyl units. The chain grows to a size of 11–13 glycosyl units. The branching enzyme transfers a set of glycosyl units from a chain to another position of a chain by creating α-1,6-glycosidic linkages. Since each chain branches out two times with new chains organized in concentric tiers, the number of glycosyl units in the glycogen particle increases exponentially from around 200 in a small particle of 4 tiers (diameter of 11 nm) to around 55,000 in the largest particle of 12 tiers (diameter of 42 nm). In resting skeletal muscles, the typical (around 80% of all particles observed) diameter of glycogen particles is 20–32 nm (Marchand et al. 2002; Nielsen et al. 2010; Hokken et al. 2020), corresponding to 1000 to 9000 glycosyl units per particle (Melendez-Hevia et al. 1993). This structure of the glycogen particle is probably optimized by evolution to increase solubility and decrease the osmotic effect (Melendez-Hevia et al. 1993). Despite this, the relatively high numerical density of glycogen particles equals an osmotic effect of around 3 grams of water per 1 gram of glycogen (Olsson and Saltin 1970), which can accumulate up to 2 kilograms of glycogen-dependent stored water in humans during glycogen-loaded conditions (Olsson and Saltin 1970; Shiose et al. 2016).

Although the true numerical density of glycogen particles cannot be investigated with the currently available techniques, measures of the size of the observable glycogen particles indicate that the utilization of glycogen during exercise is mainly attributable to a decrease in the average particle size (Marchand et al. 2007; Gejl et al. 2017a, b; Hokken et al. 2020); on the other hand, with training and diet interventions, the increase in resting glycogen content is ascribed to an increase in particle number (Nielsen et al. 2010; Jensen et al. 2021). Thus, in resting muscles, there may be a preferable size of particles, which could be the result of a trade-off between storage efficiency (larger particles store more glycosyl units per volume) and metabolic power (smaller particles possess a higher glycogenolytic rate) as also suggested by Shearer and Graham (2004). An increase in glycogen particle number has been recognized to require a de novo synthesis of the self-glycosylating protein backbone glycogenin (Roach et al. 2012), but recent studies have suggested an increase in glycogen particle number based on a mechanism, which is independent on glycogenin (Testoni et al. 2017; Visuttijai et al. 2020).

The degradation of glycogen is conducted by the action of glycogen phosphorylase, which cuts the α-1,4-glycosidic linkages and liberates the glycosyl units to the glycolytic pathway. Interestingly, given the branched structure of the glycogen particle, around one third of the glycosyl units are in the outermost tier (Melendez-Hevia et al. 1993) and readily available for glycogen phosphorylase securing a fast mobilization of energy. A more pronounced degradation of the particle requires a coordinated action of glycogen debranching enzyme cutting the α-1,6-glycosidic linkages. As discussed, the breakdown of muscle glycogen to produce glucose 1-phosphate is thus under the control of glycogen phosphorylase, and this reaction requires both glycogen and P_i as substrates. Phosphorylase, in turn, exists as a more active *a* form (which is under the control of phosphorylation by phosphorylase kinase) and also as *a* more inactive *b* form (which exists in a dephosphorylated form due to the action of protein phosphatase 1). Phosphorylase can be transformed via covalent modification (i.e., phosphorylation by phosphorylase kinase) as mediated through epinephrine (Roach et al. 2012). Additionally, Ca^{2+} is a potent positive allosteric regulator of phosphorylase kinase through binding to the calmodulin subunit (Jensen and Richter 2012), though glycolytic flux is not controlled by Ca^{2+} directly, but by factors related to energy state (Ørtenblad et al. 2009). During contractile activity, the increased accumulation of Pi as a result of increased ATP hydrolysis can increase the rate of glycogenolysis as it provides increased substrate required for the reaction. Furthermore, greater accumulations of free ADP and AMP can also subsequently fine-tune the activity of phosphorylase a through allosteric regulation (Howlett et al. 1998). When taken collectively, the regulation of glycogen phosphorylase is dependent on hormonal control, substrate availability, and local allosteric regulation, the precise contribution of which is dependent on the specific exercise challenge.

5.2.2 Measurement of Muscle Glycogen

The golden standard for determination of muscle glycogen is the biochemical technique where the glycosyl units are liberated by acid hydrolysis and the amount determined spectrophotometrically (Passonneau and Lowry 1993). This is performed in skeletal muscle biopsies in small specimens of ~10 mg (wet weight), and with small adjustments can also be applied to single fibers (Hintz et al. 1982). The latter approach makes it possible to directly combine glycogen content with other biochemical or physiological measures and to distinguish between different fiber types. Noninvasively, nuclear magnetic resonance (NMR) has been optimized to detect muscle glycogen signals (Taylor et al. 1992) and can serve as an alternative to the biochemical assessment when biopsies are unsuitable for the target group.

To investigate fiber-type-specific glycogen, a semi-quantitative histochemical periodic acid-Schiff (PAS) staining has been employed in a large body of studies. Here, the glycogen content is assessed by staining intensity from empty to full using a scale of 4–5 steps and compared with the myosin ATPase characteristics (Pearse 1961). The PAS staining is based on a reaction of periodic acid with all sugars in the muscle, and it is therefore not specific for glycogen. It has been suggested to be replaced by an antibody-based technique (Nakamura-Tsuruta et al. 2012; Skurat et al. 2017), which can also be used by dot blotting (Albers et al. 2015).

With the binding properties of glycogen particles to a reduced form of osmium, a protocol for staining glycogen in transmission electron microscopy has also been developed (de Bruijn 1973; Marchand et al. 2002). With this approach, the subcellular distribution of glycogen can be envisaged along with information on fiber types (based on Z-discs and M-band appearances) and other ultrastructural parameters (Sjöström et al. 1982b).

5.2.3 Inter-fiber Variability and Subcellular Differences

While most studies on glycogen content are conducted using homogenates of small pieces of muscle, a string of studies have shown large inter-fiber variability ranging from around 100 to 1000 mmol kg^{-1} dw (Essén and Henriksson 1974). With discrimination between fiber types based on myosin ATPase activity, it has repeatedly been shown that type 2 fibers contain about 10–30% more glycogen than type 1 fibers in human skeletal muscles (e.g., Essén and Henriksson 1974; Ball-Burnett et al. 1991; Greenhaff et al. 1993). At the subcellular level, glycogen particles are dispersed heterogeneously throughout the myoplasm with one large pool located in the intermyofibrillar space close to sarcoplasmic reticulum and mitochondria and two small pools located in the intramyofibrillar and subsarcolemmal spaces, respectively (Sjöström et al. 1982a; Fridén et al. 1985, 1989; Marchand et al. 2002, see Fig. 5.4, Sect. 5.3). Although all the pools are utilized during exercise taxing the endogenous glycogen stores (see later), their content seems not to be related in resting muscles (Nielsen et al. 2010), suggesting that local independent factors are involved in their regulation.

5.3 Utilization of Glycogen During Exercise

As alluded to earlier, a series of studies from Scandinavian researchers during the late 1960s and early 1970s collectively demonstrated the role of skeletal muscle glycogen as a key fuel source for prolonged exercise capacity, particularly when completed at higher exercise intensities (Bergström and Hultman 1966a, b; Bergström et al. 1967; Saltin and Karlsson 1971). For example, pioneering work from Bergström et al. (1967) highlighted the role of dietary carbohydrate in elevating muscle glycogen reserves and subsequently demonstrated that high pre-exercise muscle glycogen concentration (~890 mmol.kg^{-1} dw) extends exercise capacity at 75% of VO_{2max} by a remarkable 320% (from 59 to 189 min) when compared to low (~170 mmol.kg^{-1} dw) muscle glycogen stores. It was later demonstrated that increasing exercise intensity results in an exponential increase in the rate of muscle glycogen utilization (Saltin and Karlsson 1971).

Further association between starting skeletal muscle glycogen and prolonged aerobic exercise performance (Karlsson and Saltin 1971) solidified the central role of skeletal muscle glycogen as key substrate for intense exercise of prolonged duration. These seminal studies paved the way for the next 50 years of research examining the dietary and exercise-related factors that can subsequently affect the pattern of glycogen utilization during exercise. Such factors have been the subject of a recent meta-analysis (Areta and Hopkins 2018) and are portrayed conceptually in Fig. 5.2. The text to follow provides a critical overview of factors that can affect glycogen utilization during exercise.

5.3.1 Duration and Intensity

Although the duration of exercise that can be maintained is inevitably dependent on exercise intensity, the effect of intensity and duration on muscle glycogen utilization can be analyzed independently. Glycogen utilization increases in accordance with increased exercise duration though it is noteworthy that the rates of utilization vary at different stages during exercise. For example, during the first ~20 min of exercise, the rates of glycogen utilization seem to be the highest, a factor that is likely related to higher activity of glycogenolytic enzymes and lower availability of other metabolic substrates (Green et al. 1995; Chesley et al. 1996; Dyck et al. 1996; Graham et al. 2001), over and above the fact that higher glycogen concentration per se is associated with increased glycogen utilization (see below substrate availability and glycogen utilization). Thereafter, rates of utilization during prolonged exercise seem to be rather constant until they reach low levels (~200 mmol· kg^{-1} dw) at which

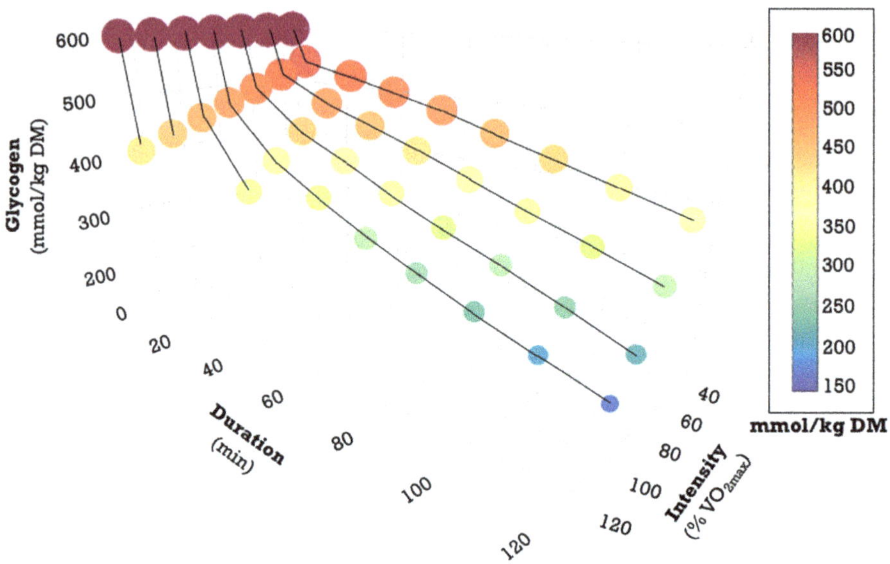

Fig. 5.2 Conceptual representation of skeletal muscle glycogen utilization dynamics at different intensities and durations, based on mathematical modelling of data extracted from a large sample of published literature (Areta and Hopkins 2018). Note that the starting muscle glycogen has been set constant at 600 mmol· kg^{-1} dw for exercise of intensities ranging from equivalent to 40 to 120% of VO_{2max}, ranging from 5 to 120 min duration

point the rates are then reduced significantly (Coyle et al. 1986; Bosch et al. 1993, 1996). Fatigue during prolonged submaximal exercise (i.e., 60–70% of VO_{2max}) typically occurs after a duration of 2 h with glycogen at exhaustion within the range of ~100 to 200 mmol·kg^{-1} dw (Areta and Hopkins 2018). Indeed, early studies on muscle glycogen and athletic performance suggested that the inability to maintain high rates of work would coincide with theoretical muscle glycogen content of 70–120 mmol·kg^{-1} dw during an endurance competition of ~135 min (Karlsson and Saltin 1971). With this in mind, it is most likely that only competitive events lasting >90 min would benefit from CHO loading strategies that aim to super compensate pre-exercise muscle glycogen stores, where performance has been suggested to improve by ~2–3% (Hawley et al. 1997a, b).

During intense exercise, ATP provision is achieved principally by the oxidation of carbohydrate, and muscle glycogen utilization increases exponentially with exercise intensity. At intensities ranging from 75% VO_{2max} to near maximal workloads, glycogen is the main energy substrate (Saltin 1973; Gaitanos et al. 1993; Hultman and Greenhaff 1999). Thus, even a single 6 s or 30 s all-out sprint can reduce muscle glycogen by15 or 20–30%, respectively (Gaitanos et al. 1993; Bogdanis et al. 1996; Parolin et al. 1999). Such high glycogen utilization is achieved by estimated glycogenolytic rates of around 4.5 mmol glucosyl units·kg^{-1}·dw·s^{-1} (Gaitanos et al. 1993; Parolin et al. 1999). During longer durations of high-intensity exercise, the glycogenolytic rate is known to decreases though it is noteworthy that muscle glycogen stores in both arms and legs are reduced by 20–25% during 4 min of high-intensity all-out cross-country sprint skiing (Gejl et al. 2014).

The first study to systematically investigate the effect of intensity demonstrated a clear exponential increase in muscle glycogen use (Fig. 5.3), with a disproportionate increase in rates at intensities close to VO_{2max} and above (Saltin and Karlsson 1971). More recent studies addressing the contribution of different energy substrates using metabolic tracers show a clear disproportionate increase in reliance on skeletal muscle glycogen with increasing intensities when compared to all other fuel sources including intracellular lipids and plasma lipids and glucose (Romijn et al. 1993; van Loon et al. 2001). Corroborating these findings, a meta-analytic evaluation of the effect of exercise intensity on muscle glycogen use demonstrates moderate and large effects with increasing exercise intensity, albeit with substantial variation between studies (Areta and Hopkins 2018). The variation in glycogen use observed in different studies and individuals at the same intensity may be related to the fact that % VO_{2max} is typically used as the "default" intensity parameter, and this parameter may not necessarily match the metabolic and substrate demands of the effort at the muscular level. Indeed, muscular oxidative capacity can vary greatly between individuals with the same VO_{2max} (Holloszy 1973; Holloszy and Coyle 1984), and lactate threshold represents a more suitable method for which to match intensity within and between studies (Coyle et al. 1988; Poole et al. 2020).

Given that phosphorylase can be transformed via covalent modification (i.e., phosphorylation by phosphorylase kinase) mediated through epinephrine, it would be reasonable to expect that greater phosphorylase transformation from *b* to *a* may be one mechanism to explain increased glycogenolysis that is evident with increasing exercise intensity. This would also be logical given that sarcoplasmic Ca^{2+} levels would be increased with high-intensity exercise (given the need for more rapid cross-bridge cycling) and that Ca^{2+} is a potent positive allosteric regulator of phosphorylase kinase through binding to the calmodulin subunit. However, the percentage of phosphorylase in the more active a form does not appear to be increased with exercise intensity and in actual fact is decreased after only 10 min of high-intensity exercise, which may be related to the reduced pH associated with intense exercise (Howlett et al. 1998). Whereas this mechanism of transformation (mediated by Ca^{2+} signaling) may be in operation within seconds of the onset of contraction (Parolin et al. 1999), it appears that post-transformational mechanisms are in operation during more prolonged periods of high-intensity exercise given that glycogenolysis still occurs despite reduced transformation. In this regard, vital signals related to the energy status of the cell play a more prominent role. Indeed, as exercise intensity progresses from moderate- to high-intensity exercise, the rate of ATP hydrolysis increases so much so that there is a greater accumulation of ADP, AMP, and Pi, thus providing allosteric and substrate level control (Howlett et al. 1998).

One final consideration is that the rapid decrease in muscle glycogen at intensities close to and above VO_{2max} most likely represents mere glycogenolysis, rather than oxidation of its glucose units. For example, all-out exercise of ~2 min at an intensity well above the VO_{2max} decreases muscle glycogen by about 25% (Medbø 1993; Medbø et al. 2006), but the net oxidation of the glycogen utilized is estimated to be

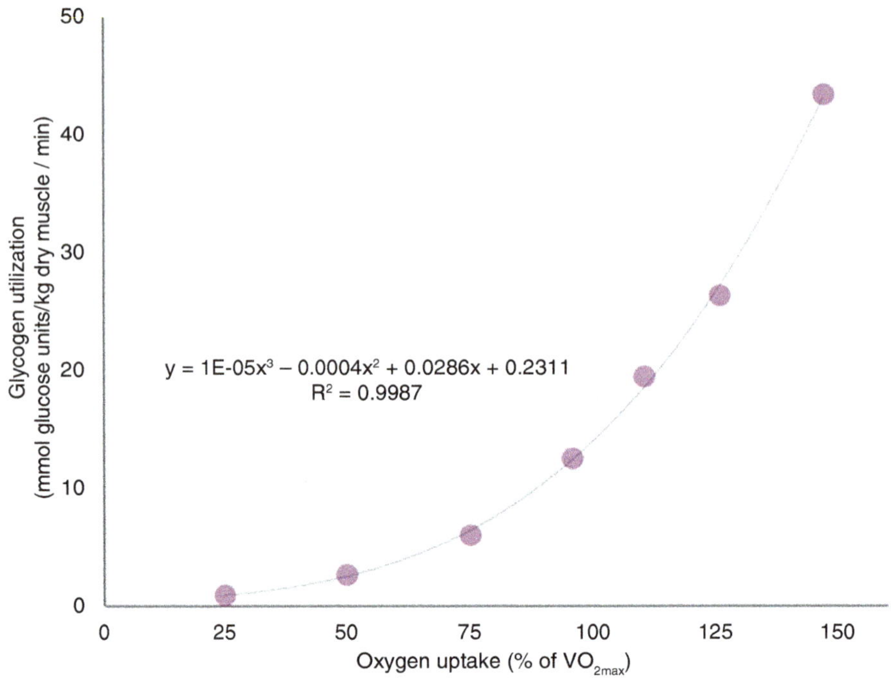

Fig. 5.3 Relationship between exercise intensity and rates of skeletal muscle glycogen utilization. Linear increases in exercise intensity (as % of VO_{2max} equivalent) result in exponential increases in skeletal muscle glycogen utilization, as shown by early research (Figure redrawn from Saltin and Karlsson 1971)

~4–13% (Medbø 1993). In this scenario, most of it ends as lactate, of which half is released to the bloodstream and the other half is utilized for rapid muscle glycogen resynthesis within the muscle (Medbø et al. 2006). Bangsbo et al. (1991) estimated that between 13 and 27% of lactate released from high-intensity exercise was converted back to glycogen (Bangsbo et al. 1991). From a practical perspective, it is worth noting that despite the disproportionate increase of skeletal muscle glycogen with increased intensity, prolonged steady-state exercise results in increased absolute muscle glycogen utilization when compared with shorter high-intensity intervals (Impey et al. 2020).

5.3.2 *Substrate Availability*

Although exercise intensity and duration are key drivers of glycogen use, high or low glycogen availability can result in higher or lower glycogenolysis, respectively, despite divergent exercise intensities (Arkinstall et al. 2004). Due to the significant effects of nutrient availability on glycogen utilization dynamics and exercise performance, manipulation of substrate availability and skeletal muscle glycogen has therefore been a major area of focus in research. Indeed, given the relationship between high muscle glycogen and increased work capacity, a number of dietary practices manipulating substrate availability have been tested to maximize muscle glycogen stores or to minimize its use during exercise in order to further increase work capacity. The most noteworthy practices manipulating substrate availability are the increase of muscle glycogen itself, the increase in availability and oxidative capacity of lipids (through diet and intravenous provision of lipids or lipolytic agents), and the ingestion of glucose during exercise.

5.3.2.1 Increased Muscle Glycogen Stores and Glycogen Use

Skeletal muscle glycogen content is directly related to dietary carbohydrate as shown by seminal studies (Bergström et al. 1967; Gollnick et al. 1972), a finding that has systematically been corroborated in the literature (Areta and Hopkins 2018). Depleting muscle glycogen through exhaustive exercise followed by high carbohydrate intake has been shown to subsequently overshoot skeletal muscle glycogen reserves above normal resting values in the previously contracted muscle (Bergström and Hultman 1966a, b), a phenomenon that was termed as "glycogen supercompensation." The precise molecular mechanisms underpinning the supercompensation effect remain an active area of research (Hingst et al. 2018). Since the original work documenting this effect, it is now accepted that prior depletion of muscle glycogen is not necessary for individuals who regularly practice aerobic-type exercise and that a taper of training load in conjunction with elevated dietary carbohydrate intake of ~10–12 g/kg of body mass/day can maximize skeletal muscle glycogen concentration within 24–48 h (Burke et al. 2017). Regardless, higher resting muscle glycogen leads to higher rates of muscle glycogen use during exercise, likely a reflection of substrate regulation of glycogen phosphorylase (Hargreaves et al. 1995). Indeed, when intensity and duration of exercise are kept constant, commencing exercise with high muscle glycogen (e.g., > 600 mmol·kg^{-1} dw) leads to higher muscle glycogen use compared to normal (e.g., ~450 mmol·kg^{-1} dw) or low (e.g., <300 mmol·kg^{-1} dw) starting muscle glycogen (Galbo et al. 1979; Hargreaves et al. 1995; Areta and Hopkins 2018). Accordingly, when examining performance testing data of prolonged endurance exercise of a set distance and workload, exercise is often finished with similar muscle glycogen levels despite significant differences in starting muscle glycogen (Karlsson and Saltin 1971; Sherman et al. 1981; Rauch et al. 1995; Tomcik et al. 2018).

5.3.2.2 Increased Lipid Availability and Glycogen Sparing

Given the comparatively unlimited storage of energy in the form of fat, increasing the use of lipids as a fuel source during exercise to "spare" muscle glycogen with the goal of improving physical performance has been a topic of intense research. Intravenous lipid infusion, lipolytic agents, high-fat meals, and high-fat diets have been all been used to test this hypothesis. Interventions using intralipid infusion and heparin have been consistently shown to spare muscle glycogen between ~15 and 45% at moderate and high intensities lasting 15–60 min (Vukovich et al. 1993; Dyck et al. 1993, 1996; Odland et al. 1998; Hawley 2002). The effects of a high-fat meal consumed prior to exercise, however, have been less consistent with studies ranging from 40% glycogen sparing to no glycogen sparing at all (Costill et al. 1977; Vukovich et al. 1993; Hawley 2002;). The possibility of sparing skeletal muscle glycogen via dietary fat intake has been expanded through the use of high-fat diets during days and weeks.

The use of high-fat diets consumed over durations greater than 3 days has been termed "fat adaptation" and is effective in increasing circulating free fatty acids (FFA), FFA uptake by muscle, and increased fat oxidation during exercise (Stellingwerff et al. 2006). The increased rates of fat oxidation, however, occur in parallel with decreased skeletal muscle glycogen due to reduced dietary carbohydrate intake (Hammond et al. 2019; Areta et al. 2020) and hence, as detailed in the previous section, lower muscle glycogen results in lower glycogen utilization. Nonetheless, 24 h of restoration of carbohydrate intake after days of fat adaptation permits skeletal muscle glycogen storage while still retaining the increased capacity to utilize higher amounts of fat during exercise and spare glycogen to some extent (Burke et al. 2000; Stellingwerff et al. 2006). However, fat adaptation protocols also decrease pyruvate dehydrogenase (PDH) activity (Stellingwerff et al. 2006), the enzyme that represents a rate-limiting step in the flux of glucose into mitochondrial ATP, and thereby this condition impairs the capacity of generating metabolic power through the oxidation of glycogen. Therefore, in the majority of the studies to date, fat adaptation has not been able to enhance physical performance in endurance exercise, likely due to impairment of the capacity for high-intensity exercise (Burke 2006). However, over prolonged exercise (>5 h), fat adaptation with carbohydrate restoration is potentially beneficial for performance (Rowlands and Hopkins 2002). In conclusion, increasing fat availability acutely and chronically has the capacity to reduce muscle glycogen use during exercise, likely due to the independent and combined effects of increasing fat utilization and decreasing the capacity to oxidize glucose. Nonetheless, these strategies do not seem conducive to promoting exercise performance.

5.3.2.3 CHO Supplementation During Exercise and Glycogen Sparing

The fact that carbohydrate consumption during exercise increases carbohydrate oxidation led to the idea that muscle glycogen would be spared as a fuel source at its expense (Coyle et al. 1986). This intuitive idea, however, has not been supported by research findings. When considering whole skeletal muscle glycogen, a meta-analytic evaluation of CHO supplementation during exercise ($n = 24$ studies) determined there was no glycogen-sparing effect (Areta and Hopkins 2018). It is important to highlight that this meta-analysis only evaluated the presence or absence of CHO consumption. Indeed, although the effects of consuming 0, 15, 30, and 60 g/h of CHO showed no differences in glycogen use during exercise (Smith et al. 2010), a more nuanced analysis should perhaps determine whether low, medium, and high amounts of carbohydrate (e.g., <30, ~60, and ≥ 90 g/h respectively) across different exercise durations can affect glycogen utilization. No unequivocal evidence exists to support sparing with specific fibers, with reports of no sparing in either fiber type (Coyle et al. 1986; Mitchell et al. 1989), sparing in type 2a fibers (De Bock et al. 2006), and sparing in type 1 fibers (Yaspelkis et al. 1993; Tsintzas et al. 1995, 1996; Tsintzas et al. 2001). Such differences between studies may be related to differences in CHO availability prior to exercise, the CHO feeding strategy, and contrasting exercise protocols related to duration, modality, and intensity. In a recent study, Fell et al. (2021) demonstrated that consuming 45 or 90 g/h of CHO (in the form of solids, gels, and fluids) does not spare glycogen use in type 1 or 2 fibers when compared with no CHO ingestion, as assessed during 3 h of steady-state cycling conducted at lactate threshold. Importantly, exercise was commenced after a 36 h CHO loading protocol of 12 g/kg as well as consumption of a pre-exercise meal of 2 g/kg, thus replicating nutritional practices of elite road cyclists. Despite the apparent consensus that CHO feeding does not exert glycogen sparing, future studies should also examine the effects of CHO feeding on utilization within the subcellular glycogen storage pools of both type 1 and 2 fibers.

5.3.3 *Training Status*

For a given absolute exercise intensity, longitudinal studies demonstrate that glycogen utilization is reduced with exercise training (Karlsson et al. 1974), an effect that is confined locally to the actual muscles that were trained (Saltin et al. 1976). The reduced glycogenolysis observed after training is not due to any change in phosphorylation transformation but rather allosteric mechanisms (Chesley et al. 1996). Indeed, exercise in the trained state is associated with reduced content of ADP, AMP, and Pi thereby providing a mechanism leading to reduced phosphorylase activity. Importantly, the reduced rates of glycogenolysis that are evident with training are also apparent despite the fact that training induces elevations in resting glycogen stores and that higher basal glycogen is normally associated with increased glycogen utilization. As such, local allosteric control exerts a more pronounced regulatory role than substrate-level regulation. On this basis, it is now accepted that glycogen utilization during exercise is inversely related to training status and that exercise in the trained state, at same absolute intensity, requires comparatively less muscle glycogen. However, despite this well-documented finding, a meta-analytic evaluation of the effects of training status on glycogen utilization has shown only a trivial or small effect in relation to increases of VO_{2max} of 10 ml/kg/BM (Areta and Hopkins 2018).

This latter finding may be attributable to the fact that VO_{2max} is not an ideal parameter of training status when assessing changes in local muscle metabolism, despite it being considered as the main parameter for maximal aerobic capacity. Rather, a parameter of training status (and indeed exercise intensity) that is more reflective of skeletal muscle oxidative capacity is likely a more suitable approach to assessing the changes in glycogen utilization associated with training. For example, as little as 3 days of prolonged (2 h) aerobic-type training induces glycogen sparing by ~60% when assessed during exercise undertaken at 65% VO_{2peak} in healthy individuals (Green et al. 1995). As such, parameters of training status that are more indicative of the oxidative capacity of skeletal muscle (i.e., lactate threshold) are likely a better parameter for classification of status when determining the effects of training on skeletal muscle glycogen utilization. Indeed, individuals with comparable VO_{2max} but who possess a "high" or "low" lactate threshold present with distinct differences in glycogen utilization when cycling for 30 min at 80% VO_{2max}, i.e., subjects with a low threshold utilize more than twice as much muscle glycogen during 30 min exercise (Coyle et al. 1988).

In summary, substantial evidence demonstrate that a more trained skeletal muscle (i.e., with higher oxidative capacity) is less reliant on skeletal muscle glycogen use during exercise. However, there is a large variation in rates of muscle glycogen use in individuals when exercising at a percentage of their VO_{2max}. The use of a parameter of intensity and training status representative of skeletal muscle oxidative capacity will be important to further characterize the estimation of muscle glycogen use during exercise.

5.3.4 Exercise Mode

5.3.4.1 Running vs. Cycling

When comparing glycogen utilization within the vastus lateralis muscle, it has been consistently demonstrated that cycling induces greater absolute utilization than when running at a matched relative exercise intensity and duration. For example, Arkinstall et al. (2004) compared glycogen utilization in the vastus lateralis of moderately trained males during 60 minutes of cycling and running at lactate threshold and observed an absolute utilization of approximately 220 and 120 $mmol·kg^{-1}$ dw, respectively. Accordingly, a recent meta-analysis demonstrated that absolute glycogen utilization is "small but very likely" reduced during matched protocols of running versus cycling (i.e., relative exercise intensity and duration), where the expected reduction in running equates to 70 $mmol·kg^{-1}$ dw (Areta and Hopkins 2018). Such differences between modalities are, of course, reflective of greater recruitment of the vastus lateralis muscle during cycling when compared with running. The role of muscle recruitment in modulating glycogen utilization is also evident during running where both the absolute and rate of glycogen utilization are higher in the gastrocnemius muscle when compared with the vastus lateralis (Costill et al. 1974). This pattern of greater glycogen utilization within the gastrocnemius versus vastus lateralis when running is evident in both males and females during moderate- and high-intensity exercise protocols (Impey et al. 2020).

5.3.4.2 Resistance Exercise

Resistance exercise sessions typically consist of several sets (interspaced by pauses of 1–4 min) of repeated near maximal force productions. Here, the short duration and the high intensity of each force production necessitate a predominant use of the rapid energy systems taxing glycogen and creatine phosphate as substrates. During recovery periods between sets, however, other substrates such as intramuscular triglycerides can also contribute to the overall energy provision (Koopman et al. 2006). The most important factor for the utilization of glycogen during resistance exercise seems to be the total work performed rather than the exercise intensity (Robergs et al. 1991). Typical high-volume resistance exercise sessions decrease skeletal muscle glycogen by about 20–40% (100–250 mmol·kg^{-1} dw) (Tesch et al. 1986; Essén-Gustavsson and Tesch 1990; Robergs et al. 1991; Pascoe et al. 1993; MacDougal et al. 1999; Haff et al. 2003; Harber et al. 2010; Samuelson et al. 2016), but, in contrast to endurance exercise, the decrease is larger in type 2 fibers than in type 1 fibers (Tesch et al. 1998; Koopman et al. 2006; Morton et al. 2019).

5.3.5 Lower vs Upper Body

Vastus lateralis of the quadriceps has been the muscle predominantly sampled in research on skeletal muscle glycogen and used as a model that is normally extrapolated to all other muscles. However, several lines of evidence point to different intrinsic regulations of metabolism in the upper and lower body. These differences include a higher relative fat oxidation in the leg compared with arm exercise (Helge et al. 2008; Larsen et al. 2009), and a higher mitochondrial oxygen flux is present in the vastus lateralis compared to the arm (deltoid) muscle measured by high-resolution respirometry (Larsen et al. 2009). Further, it is well recognized that lactate release is higher during arm compared with leg exercise of comparable intensity (Jensen-Urstad and Ahlborg 1992; Jensen-Urstad et al. 1993). Thus, Ahlborg and Jensen-Urstad (1991) had two groups performing arm cranking and leg cycling, respectively, and demonstrated a higher relative carbohydrate utilization during arm cranking due to both a higher muscle glycogenolysis and a higher glucose uptake, while arms also had a higher lactate release (Ahlborg and Jensen-Urstad 1991). In a later study, the glucose uptake was not different between limbs when expressed per muscle volume, whereas the lactate release was noticeably higher in the arm than in the leg (Jensen-Urstad et al. 1993). Overall, this suggests that exercising arm muscle displays a different metabolic response compared with leg muscle, with a higher glycogen utilization and lactate release compared to leg muscle during same relative exercise intensity, implying a possible qualitative difference between muscles from the arm vs. the leg. However, another explanation for the differences described above may simply be that the upper body musculature, including the arm muscle, resides in less trained musculature than that of the lower body, as in the leg muscles (see Sect. 5.3.3). Still, when arm-trained athletes are exercising with either arm or leg exercise at the same relative intensity, there was a markedly higher lactate release during arm exercise (Jensen-Urstad and Ahlborg 1992). Also, in highly trained cross-country skiers, a net lactate release from arms and uptake by legs has been demonstrated during exercise involving both the upper and lower body (van Hall et al. 2003). In line with this, direct comparisons of the highly trained arm and leg muscles of elite cross-country skiers reveal that despite same mitochondrial volume percentage and citrate synthase activity in the legs and arms, the muscles exhibited clear difference in their enzyme-linked ability to oxidize fatty acids (HAD capacity) and a fourfold higher intramyocellular lipid volume contents in leg muscles (Ørtenblad et al. 2018; Koh et al. 2018). These data point to a clear limb difference in metabolism between the leg and the arm, which cannot be explained by training status or different fiber-type distributions. Taken together, the current data suggests that when exercising at the same relative intensity, arm muscle has a higher muscle glycogen use and lactate release as compared to the leg. This is also apparent in subjects with trained both upper and lower body.

5.3.6 Temperature

Exercise in the heat results in exaggerated fatigue concomitant with major alterations in several physiological and metabolic factors. The majority of the research conducted on the effects of heat stress on energy metabolism during exercise has demonstrated a shift toward increased carbohydrate use. During exercise in the heat, the rate of muscle glycogen degradation is significantly increased (Fink et al. 1975; Febbraio et al. 1994) with an increase in both carbohydrate oxidation and lactate accumulation at a given exercise intensity, while the muscle glucose uptake and utilization appear to be unaltered during exercise in the heat, despite hyperglycemia and an augmented liver glucose output. Thus, the increase in carbohydrate utilization is largely explained by an increased muscle glycogenolysis observed via both aerobic and anaerobic energy turnover. The mechanisms thought to be responsible for the enhanced muscle glycogenolysis likely are due to increased sympatho-adrenal response and increased muscle temperature (Hargreaves et al. 1996). Although exercise in the heat increases the intramuscular glycogen utilization (Fink et al. 1975; Febbraio et al. 1994; Hargreaves et al. 1996), depletion seems not to be the cause of fatigue during exercise in the heat. Thus, a general observation is that muscle glycogen stores are not critically low at fatigue and exhaustion, i.e., <250–300 $mmol \cdot kg^{-1}$ dw, suggesting that exercise in the heat is terminated before available glycogen stores have been limiting. Further, the total amount of carbohydrate oxidized during exercise in the heat is relatively low as exercise time is shorter than exercising in moderate or low temperatures (Fink et al. 1975; Febbraio et al. 1994; Galloway and Maughan 1997). Still, diet-induced increase in muscle glycogen before exercise is associated with enhanced exercise capacity in the heat, and carbohydrate ingestion during exercise increases exercise capacity in the heat (Pitsiladis and Maughan 1999; Carter et al. 2003). These outcomes cannot be explained by direct effects on either hyperthermia or substrate depletion but may exert an ergogenic effect related to factors peripheral to the muscle. Taken together, exercising in the heat results in an increase in intramuscular glycogen utilization; however, fatigue in these circumstances appears to be related to factors other than muscle glycogen per se.

5.3.7 Altitude

As the contribution of carbohydrate is determined by the exercise intensity, the hypoxic exposure experienced at altitude causes, in most studies, an increase in relative carbohydrate utilization, when exercising at the same absolute intensity as at sea level (Young et al. 1991; Brooks et al. 1991; Brooks 1992; Roberts et al. 1996a, b; Lundby et al. 2004; Katayama et al. 2009), but not all (Braun et al. 2000; O'Hara et al. 2017; Matu et al. 2018). A more direct comparison of possible effects on substrate utilization during exercise at hypoxic conditions can be gained by comparing exercise at sea level and at altitude, matched for the same relative intensities. A meta-analysis of the effects of exposure to hypoxia during exercise matched for relative intensities, compared with normoxia, demonstrated no consistent change in the relative contribution of carbohydrate to the total energy yield (Griffiths et al. 2019). This has been evidently demonstrated in a study where subjects exercised at both the same absolute and relative intensity as at sea level and under acute hypoxia, before and after 4 weeks exposure to 4100 m altitude (Lundby and Van Hall 2002). Submaximal substrate utilization was unchanged with acute and chronic hypoxia when exercising at the same relative intensity, while the carbohydrate utilization was increased when exercising at the same absolute intensity. Further, 4 weeks of acclimatization to altitude did not affect substrate utilization, also confirming data from rodents (McClelland et al. 1998). Overall, these studies demonstrate that relative work intensity is the main factor determining fuel selection during exercise and prolonged hypoxia does not cause a significant shift in fuel selection. However, little is known about the relative contributions of muscle glycogen during hypoxia.

5.3.8 Sex Differences

Sex-based differences in substrate metabolism during endurance exercise are well documented in that it is now recognized that females have reduced reliance on whole-body CHO metabolism (typically reflective of reduced liver glycogen metabolism) to support energy production (Devries 2016). Although not always consistent, there is also some evidence that females use less glycogen than males. For example, when assessed during the luteal phase, females use less glycogen in the vastus lateralis muscle (25%) compared with males, as assessed during 90 minutes of cycling at 65% VO_{2peak} (Devries et al. 2006). When tested in the mid-follicular phase, it has also been reported that females use less glycogen (25%) in the vastus lateralis muscle than males when running a set distance of 15.5 km on a treadmill at 65% VO_{2peak} (Tarnopolsky et al. 1990). When completing a 16 km road run undertaken at lactate threshold in the mid-follicular phase, we also recently reported that females use less glycogen in both the vastus lateralis (30%) and gastrocnemius muscles (20%) when compared with males (Impey et al. 2020). It is, of course, difficult to offer definitive mechanisms underpinning such differences in local muscle metabolism owing to the challenge of matching resting glycogen concentration, training status, total work done, or distance covered between sexes (i.e., thus reflective of exercise intensity and duration). Additionally, variations in phase of the menstrual cycle studied as well as the use of contraceptives also make it difficult to compare between studies. On the basis that studies examining the impacts of sex hormones (e.g., estrogen and progesterone) have generally indicated they exert minimal regulatory effects on muscle glycogen utilization (Devries 2016), it is possible that the aforementioned factors may indeed play a more subtle but influential role. A recent meta-analysis demonstrated that sex-based differences in glycogen utilization are indicative of a "likely small" reduction on absolute glycogen utilization of approximately 30 $mmol{\cdot}kg^{-1}$ dw (Areta and Hopkins 2018). From a practical perspective, we therefore suggest that such magnitudes of differences in glycogen utilization are unlikely to require sex-specific nutritional strategies and that both males and females should simply ensure they commence their training or competitive scenario with sufficient glycogen stores to meet the subsequent metabolic demand.

5.3.9 *Subcellular Compartmentalization*

In the above sections, the utilization of glycogen has been described based on mixed muscle homogenates or histochemical-defined specific fiber types. This view assumes a uniform utilization of glycogen within the muscle fibers, i.e., with no spatial compartmentalization of metabolic reactions and no existence of local gradients of ions and metabolites. A wealth of studies have shown that this assumption is not valid and, in contrast, demonstrated that the muscle fibers' interior is extremely crowded (high concentrations of, e.g., proteins, metabolites, and ions) with limited free diffusion, but with a highly developed compartmentalized network of enzymatic reactions (e.g., Srere 1967; Sear 2019). Within this crowded interior, glycogen particles are dispersed in distinct compartments. The definition of these compartments is man-made and may be limited by the typical 2D portray of a muscle fiber.

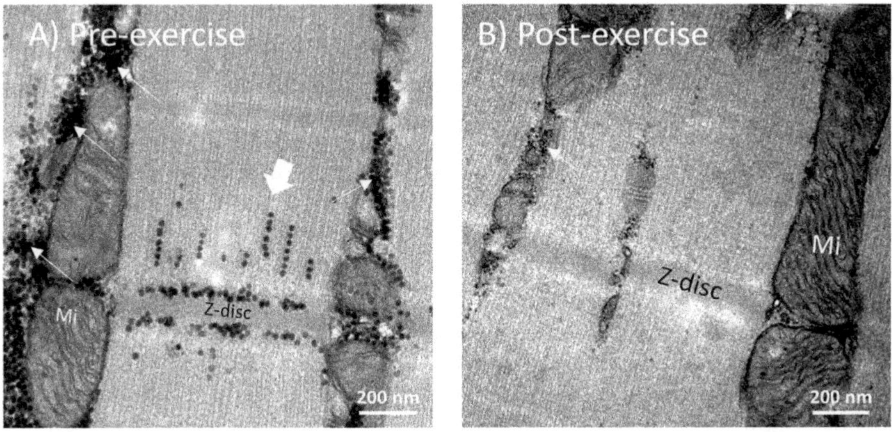

Fig. 5.4 Representative transmission electron micrographs showing the typical pattern of glycogen storage (glycogen particles are the black dots) at pre- (**a**) and post-prolonged exhaustive exercise (**b**). Mi, mitochondria. The glycogen particles located within the sarcomere often close to the Z-disc and as strings between the filaments are termed intramyofibrillar glycogen (thick arrow). The particles located close to mitochondria (Mi) between the myofibrils are termed intermyofibrillar glycogen (thin arrows). The images are collected as a part of the project described in Nielsen et al. 2011

However, the studies from independent research groups (Sjöström et al. 1982a; Marchand et al. 2002; Nielsen et al. 2011) have all defined three clearly separated compartments (Fig. 5.4): (1) the intermyofibrillar space (between the myofibrils); (2) the intramyofibrillar space (within the myofibrils); and (3) the subsarcolemmal space (just beneath the surface membrane). In addition, some studies have addressed glycogen particles in physical interaction with the sarcoplasmic reticulum (Wanson and Drochmans 1968; Goldstein et al. 1985; Tammineni et al. 2020) or have discriminated between intramyofibrillar glycogen located either in the I- or A-band of the sarcomere (Fridén et al. 1985, 1989). In the subsarcolemmal space, glycogen particles have been described as perinuclear (Caulfield and Klionsky 1959), nuclear (Sun et al. 2019), or lysosomal (Viragh et al. 1982). Since quantitative data is available only for the three clearly separated compartments, this will form the basis for the discussion below.

Of the three compartments, intramyofibrillar glycogen are utilized relatively most during various types of exercise (Marchand et al. 2007; Nielsen et al. 2011; Gejl et al. 2017a, b, c; Jensen et al. 2020b; Hokken et al. 2020, Fig. 5.4). In a recent study where participants cycled at 75% of VO_2max, exhaustion was associated with mixed muscle homogenate glycogen concentration well above zero (120 mmol kg dw), but with intramyofibrillar glycogen levels close to zero in about 60% of the type 1 fibers (Jensen et al. 2020b). In comparison, subsarcolemmal and intermyofibrillar glycogen levels were close to zero in only 40 and 10% of the type 1 fibers, respectively, clearly suggesting a link between the depletion of intramyofibrillar glycogen and exhaustion. However, in a few exceptions, intramyofibrillar glycogen was not found to be preferentially utilized, which forms some basis for the understanding of compartmentalized glycogen utilization. Firstly, with repeated high-intensity exercise (4 × 4 min sprint skiing), only intermyofibrillar glycogen was utilized during the fourth exercise bout (Gejl et al. 2017a, b, c), indicating that the energy production by degradation of intramyofibrillar glycogen can be replaced by another energy source if the exercise is repeated. This could be from intermyofibrillar glycogen degradation, which in absolute terms was utilized more during the fourth exercise bout, or from phosphocreatine, which is also localized within the myofibrils (Wallimann and Eppenberger 1985) and known to super-compensate in response to repeated exercise (Sahlin et al. 1997). Secondly, if subsarcolemmal glycogen is super-compensated, its utilization rate is increased concomitant with a reduced utilization of intramyofibrillar glycogen (Jensen et al. 2020b). This implies the existence of a mechanism linking subsarcolemmal glycogen with intramyofibrillar glycogen utilization. Thirdly, after resistance exercise, intermyofibrillar glycogen was preferentially utilized in type 1 fibers, which contrasts to type 2 fibers, where glycogen was utilized from all compartments (Hokken et al. 2020). The resistance exercise was characterized by work periods interspaced by 2–4 min pauses, which may facilitate some resynthesis of glycogen. With a low rate of glycogen utilization in type 1 fibers, a small preferential resynthesis of intramyofibrillar glycogen during rest periods may almost equal the degradation during work, which underscores that net results (pre-post) from intermittent exercises should be carefully interpreted.

5.4 Glycogen Depletion and Fatigue

Most of the studies investigating the role of glycogen in muscle fatigue are based on associations or correlational findings, but the causative effect and mechanisms explaining glycogen depletion-induced fatigue are not clear. To definitively address this experimentally, these human studies with correlative findings should be combined with mechanistic in vitro studies, but while glycogen can be removed enzymatically by amylase, it cannot be instantly added exogenously. Due to these inherent limitations in research designs, our understanding of the role of glycogen in muscle fatigue must include careful interpretations of the available data.

5.4.1 Correlations with Performance

Pioneering research by A.V. Hill suggested that glycogen was the sole fuel for muscle work and that lactate was necessary to activate the muscle (e.g., Hill 1913). Today, it is widely known that ATP generated through several metabolic pathways is the fuel for muscle work and that Ca^{2+} ions ultimately mediate switching off the brake on actin filaments and facilitate cross-bridge formation, which is the main feat of ATP consumption of contracting skeletal muscle. Although Hill unintendedly exaggerated the role of glycogen in skeletal muscle, much evidence still points toward glycogen as a key fuel source mediating these processes. In the following sections, we review the literature examining the role of glycogen for performance during prolonged, short-term, and resistance exercise.

5.4.1.1 Prolonged Exercise (60–180 Min)

When performance is evaluated as maximal work conducted at a fixed protocol of working intensity (time to exhaustion), most (Bergström et al. 1967; Galbo et al. 1979; Lamb et al. 1994; Walker et al. 2000; McInerney et al. 2005; Duhamel et al. 2006c; Alghannam et al. 2016; Jensen et al. 2020b), but not all (Madsen et al. 1990), studies demonstrate improved performance with high pre-exercise glycogen stores. Closed-end tests (e.g., time trials) show small effects (Karlsson and Saltin 1971; Widrick et al. 1993; Rauch et al. 1995) or no effect (Sherman et al. 1981; Hawley et al. 1997a, b; Burke et al. 2000; Tomcik et al. 2018) of high pre-exercise glycogen stores on performance. Independent of test protocol, the characteristics of the studies showing a positive effect of elevated glycogen levels are that the participants either worked for 90–180 min and/or had large differences (>50%) in pre-exercise glycogen levels between the experimental conditions.

5.4.1.2 Short-Term Exercise (<15 Min)

If the pre-exercise glycogen level is well above the utilization level, the current consensus is that there are no effects of above-normal levels of glycogen on short-term exercise performance. This is based on tests of time to exhaustion at 125% (Vandenberghe et al. 1995) and 85% (Lambert et al. 1994) of VO_{2max} as well as a 75 sec all-out time trial (Hargreaves et al. 1997). However, if the pre-exercise glycogen level is very low, both repeated 6 sec all-out sprint (Balsom et al. 1999) and repeated one-legged intense exercise (Bangsbo et al. 1992) performance are impaired. Collectively, muscle glycogen seems to be important for both continuous high-intensity exercise tolerance (>60 s duration) and single or repeated sprint performance (<60 s duration), only if a substantial degree of depletion is achieved, whereas loading the stores above normal levels imposes no consistent additional benefit (Vigh-Larsen et al. 2021).

5.4.1.3 Resistance Exercise

Several studies have found that a carbohydrate-restricted diet is not associated with reduced strength and power output (i.e., Mitchell et al. 1997; Sawyer et al. 2013), suggesting that resistance exercise performance is not related to the muscle glycogen levels. This is in accordance with one study, where a very low pre-exercise glycogen level did not affect maximal voluntary isometric contraction or peak force during

50 repetitions (Symons and Jacobs 1989). While this is corroborated by the finding that only around 40% of the glycogen stores in the active muscles are utilized during typical resistance exercise protocols (Tesch et al. 1986; Essén-Gustavsson and Tesch 1990; Robergs et al. 1991; MacDougal et al. 1999), a recent study demonstrated, however, that a local pool (intramyofibrillar) of glycogen in type 2 fibers is very low after resistance exercise (Hokken et al. 2020), suggesting that fatigue development of some type 2 fibers may limit resistance exercise performance.

5.4.2 A Causal Link to Fatigue?

The causal link between glycogen depletion and impaired performance is most likely multifactorial, but studies on isolated muscles from rodents (Chin and Allen 1997) and amphibians (Stephenson et al. 1999) clearly suggest a local factor within the muscles. To our knowledge, till date, only one paper has questioned this finding. Here, mice were lacking glycogen due to a whole-body *gys1* (glycogen synthase) deletion and showed surprisingly normal exercise capacity (Pederson et al. 2005). However, only around 10% of the mice survived birth suggesting a survival phenotype, which may not be comparable to a wild-type phenotype, and a later study using inducible skeletal muscle-specific *gys1* deletion showed considerable reduced exercise capacity (Xirouchaki et al. 2016). Thus, the lack of glycogen inevitable leads to diminished muscle function. Nonetheless, the mechanisms explaining why and how glycogen and glycolytic rate are integrated with cell function are far from understood.

According to one's intuition and the recognized "energy crisis" theory, the association between low muscle glycogen levels and impaired contractile function is that low glycogen causes a slowed glycogenolytic and glycolytic flux, compromising the required rate of ATP regeneration for the sustained muscle function during a given intensity (Green 1991; Sahlin et al. 1998). Consequently, the muscle is unable to maintain an adequate ATP supply to one or more of the processes involved in E-C coupling, leading to impaired muscle function, i.e., fatigue. This is supported by observations of whole muscle cell PCr decreases along with an increase in free ADP and IMP following prolonged glycogen depleting exercise (Norman et al. 1988; Sahlin et al. 1997). However, this theory is challenged by both in vitro and in vivo studies demonstrating a strong association between low glycogen and decreased muscle function even after recovery periods, where ATP levels would be normal (Bangsbo et al. 1992; Chin and Allen 1997). Moreover, muscular fatigue is also observed even when glycogen is far from depleted in different conditions where glycogen is decreased prior to the start of exercise (Duhamel et al. 2006a, b; Ørtenblad et al. 2011). Also, low glycogen affects muscle function under experimental circumstances in vitro where global ATP and PCr levels can be maintained at near resting levels (Kabbara et al. 2000; Helander et al. 2002; Nielsen et al. 2009). Together, these series of experiments do not provide experimental support for the energy crisis hypothesis.

It is noteworthy, however, that the ATP concentration inside cells may not be uniform at a subcellular level (Jones 1986). The highly organized muscle cell forms many compartments and hence microenvironments with high ATPase activity, and restricted diffusional access of metabolites and observations on whole muscle experiments or fiber preparations does not rule out a role of glycogen in maintaining a subcellular compartment energy status. Such a functional compartmentalization of glycolytic metabolism is known in a variety of tissues, including skeletal muscle. In this way, the model that has evolved is that glycolytic-derived ATP regulates key steps in the muscle excitation-contraction (E-C) coupling by delivering ATP in the microenvironment of the triad junction (Han et al. 1992; Korge and Campbell 1995). This is particularly significant in the muscle triad junction between the transverse tubular system and the sarcoplasmic reticulum (SR), with a diffusional restricted space around 12 nm wide and with a high metabolic activity (Dulhunty 1984). Consistent with the notion of compartmentalized glycolysis, most of the glycolytic enzymes are associated with membranes of intracellular compartments such as the SR (Wanson and Drochmans 1972; Entman et al. 1980; Xu and Becker 1998). Furthermore, glycogen is stored in particles located in specific regions of the muscle fiber, and variable utilization of these depots occurs during exercise, with the depot localized near the triad region being repeatedly associated with muscle function and whole-body exhaustion (see Sect. 5.3.9). Physiologically, this organization places muscular energy stores in close proximity to their site of utilization and provides support for the emerging concept for functionally compartmentalized energetic networks, ensuring an efficient energy transfer and signal transduction between energy production and utilization in different cellular compartments (Korge and Campbell 1995; Saks et al. 2008; Nielsen and Ørtenblad 2013). The following sections will focus on the experimental evidence of how glycogenolytically/glycolytically derived products preferentially regulate key steps in the muscle E-C coupling, i.e., SR Ca^{2+} regulation and muscle excitability, thus providing an explanation of the observed association between muscle glycogen contents and fatigue.

5.4.2.1 SR Ca^{2+} Regulation

Contraction of skeletal muscle is governed by the series of events in the E-C coupling, in which the Ca^{2+} release and uptake from the sarcoplasmic reticulum (SR) are an integral part through initiation and termination of the cross-bridge cycling. The SR Ca^{2+} release is triggered through an action potential (AP) activation of the voltage-sensor molecules in the t-system membrane, which open the SR Ca^{2+} release channels (RyR), leading to a rise in intracellular free concentration ($[Ca^{2+}]_i$) and generation of force by the contractile apparatus (Melzer et al. 1995; Stephenson 1996).

Both direct and indirect evidence point to a modulating role of glycogen availability on SR Ca^{2+} handling, as demonstrated in animal (Chin and Allen 1997; Stephenson et al. 1999; Kabbara et al. 2000; Barnes et al. 2001; Helander et al. 2002; Nielsen et al. 2009) and human models (Gejl et al. 2014; Duhamel et al. 2006b;

Ørtenblad et al. 2011). Using both single fibers and muscle bundles, Chin and Allen (1997) elegantly demonstrated that muscle force and $[Ca^{2+}]_i$ are associated with muscle glycogen content. Thus, through the manipulation of glucose availability in the recovery phase after fatiguing contractions, it was shown that a reduced resting level of glycogen was associated with a faster decrease in tetanic $[Ca^{2+}]_i$ and force during subsequent contractions. These results have subsequently been confirmed (Kabbara et al. 2000; Helander et al. 2002; Nielsen et al. 2014), and together, experiments on rodent muscle suggest that the change in tetanic $[Ca^{2+}]_i$ associated with fatigue and recovery has a component that is glycogen dependent.

The mechanisms linking low muscle glycogen with decreased $[Ca^{2+}]_i$ has further been elucidated by direct measures of SR vesicle Ca^{2+} release rate (Duhamel et al. 2006a, b, c; Ørtenblad et al. 2011). These studies on SR vesicles from the human muscle, where glycogen levels have been modulated by the diet either before or during the recovery phase after exercise, demonstrate a clear association between the SR vesicle Ca^{2+} release rate and muscle glycogen levels (Duhamel et al. 2006a, b; Ørtenblad et al. 2011; Gejl et al. 2014; Watanabe and Wada 2019). Importantly, there seems to be a critical level of muscle glycogen at around 250–300 $mmol \cdot kg^{-1}$ dw below which the SR Ca^{2+} release rate is impaired (Duhamel et al. 2006a; Ørtenblad et al. 2011; Gejl et al. 2014). Such data explain why minor decreases in muscle glycogen do not cause significant impairments in SR Ca^{2+} release rate and why exhaustive exercise starting with a low or high muscle glycogen store decreases or improves the endurance performance, respectively, possibly affected by SR Ca^{2+} regulation (Ørtenblad and Nielsen 2015; Ørtenblad et al. 2013).

The use of the mechanically skinned fiber preparation has provided a unique means to investigate the possible interactions between glycogen and basic muscle function (Lamb and Stephenson 2018). Indeed, the selective removal of the sarcolemma allows for the study of muscle function and the effects of glycogen content per se while maintaining the cellular architecture and control of the intracellular milieu, i.e., keeping PCr and ATP high and constant. Studies using the mechanically skinned fiber have provided experimental evidence that low glycogen content in the muscle fiber is associated with force depression during repeated voltage sensor activated contractions in most studies (Stephenson et al. 1999; Barnes et al. 2001; Watanabe and Wada 2019), but not all (Goodman et al. 2005), as well as AP-induced contractions (Nielsen et al. 2009: Jensen et al. 2020a).

Taken together, emerging evidence suggests that low muscle glycogen affects the SR Ca^{2+} release and in turn $[Ca^{2+}]_i$ and muscle function despite global ATP being held constant. This conclusion supports the concept of a functionally compartmentalized energetic network regulating key steps in the muscle E-C coupling. In support hereof are the observations (from both mechanically skinned fiber preparation, intact mouse fibers, and human SR vesicles) that the specific pool of intramyofibrillar glycogen within the myofibrils is associated with SR Ca^{2+} release and $[Ca^{2+}]_i$ (Nielsen et al. 2009; Ørtenblad et al. 2011; Nielsen et al. 2014). At present, little is known about the precise mechanism(s) which links glycogen levels in the muscle with SR Ca^{2+} release rate.

In skeletal muscle, Ca^{2+} is released from the SR Ca^{2+} stores via the specific Ca^{2+} channels (RyR1), located at the junctional SR of the triad, which ensures efficient Ca^{2+} release to the contractile proteins. In relation to the role of glycogen affecting the SR Ca^{2+} release, low glycogen may in the compartmentalized cell lead to changes in metabolic status, especially in triad region with the RyR localization. This may lead to increase in free [Mg^{2+}] and decrease in free [ATP], which are strong regulators of the RyR1, also in the physiological range of changes (Lamb and Stephenson 1994; Blazev and Lamb 1999). Furthermore, glycolytic intermediates as fructose 1,6-bisphosphate have been demonstrated to increase the open probability of the RyR channels (Han et al. 1992). However, low glycogen has also been demonstrated to modulate the Ca^{2+} release rate in isolated vesicles without restricted metabolic space and during resting metabolic conditions. This may indicate a crucial role of the metabolic machinery associated with the SR. Also, the RyR protein contains numerous phosphorylation sites, which may be affected by PKA and Ca^{2+}-calmodulin-dependent kinase II in fast twitch fibers (Fill and Copello 2002). However, the impact of phosphorylation and/or dephosphorylation on single RyR channel behavior and the role of glycogen and energy status are at present not fully unraveled.

5.4.2.2 Muscle Excitability and Na, K-Pump

With repeated intense activation, a change in the electrochemical gradients for K^+ can cause a substantial membrane depolarization leading to failure of excitation and SR Ca^{2+} release and an ensuing decrease in force responses (Sejersted and Sjøgaard 2000; Clausen and Nielsen 2007), although several mechanisms interact during exercise to counteract the depressive effects of elevated extracellular levels of K^+ (Pedersen et al. 2003; de Paoli et al. 2007, 2010). There is a reasonably well-established relationship between glycolytic-derived ATP and Na, K-pump activity, and evidence exists supporting that glycolysis and the Na, K-pump are functionally coupled. This seems to be an evolutionary conserved metabolic coupling and has been observed in several tissue types, including the mammalian erythrocytes (Schrier 1966; Mercer and Dunham 1981; Kennedy et al. 1986), axons (Caldwell et al. 1960), cardiac myocytes (Philipson and Nishimoto 1983; Hasin and Barry 1984; MacLeod 1989), and skeletal muscle (Clausen 1965; James et al. 1999; Okamoto et al. 2001; Jensen et al. 2020a). In line with this, many tissues generate pyruvate and lactate under aerobic conditions (aerobic glycolysis) in a process linking glycolytic ATP supply to the activity of the Na, K-pump (Brooks 1986; Dhar-Chowdhury et al. 2007). Indeed, aerobic glycolysis and glycogenolysis occur in resting, well-oxygenated skeletal muscles and is closely linked to stimulation of active membrane Na, K-pump transport due to epinephrine release (James et al. 1996, 1999a, b; Bundgaard et al. 2003; Levy et al. 2005). The role of glycolysis and oxidative phosphorylation in providing fuel to the Na, K-pump in the skeletal muscle was investigated using the Na, K-pump inhibitor ouabain in resting rat extensor digitorum longus muscles, demonstrating that Na, K-pump activity is only impaired when the glycolysis is inhibited (Okamoto et al. 2001). A tight coupling between the glycogenolytic rate and Na, K-pump activity is further demonstrated by the observation that intracellular Na^+ decreases if glycogen breakdown is stimulated with epinephrine at rest, while ouabain significantly attenuates glycogen utilization (James et al. 1999b). The data suggest that in the skeletal muscle, glycolysis is the predominant source of the fuel for the Na, K-pump. Further, a direct link between muscle glycogenolysis and Na, K-pump activity was demonstrated by the observation of decreased glycogen utilization in resting muscle when the muscle Na, K-pump activity was blocked with ouabain (Clausen 1965). Moreover, lactic acid production was increased in proportion to activation of the Na, K-pump. Taken together, these data indicate a clear association of glycogenolytic/glycolytically derived ATP on active cation transport across the muscle membranes. Indeed, a direct link between energy state and excitability of the muscle was demonstrated by blocking cross-bridge cycling and SR Ca^{2+} release with the cross-bridge cycling blocker BTS (N-benzyl-p-toluene sulphonamide) and dantrolene, respectively, thereby conserving energy during repeated electrical stimulations, which in turn improved the ability of muscles to maintain excitability during high-frequency stimulation (Macdonald et al. 2007).

The effects of glycogenolytically derived ATP on muscle excitability are further substantiated in mechanically skinned fibers with a high and constant global ATP, demonstrating that glycogenolytically derived energy is associated with fiber contractile endurance irrespective of global ATP levels. Thus, enzymatically lowering glycogen by 70% led to a reduction in both voltage sensors-activated and AP-induced forces in the skinned fibers, with a larger decrease in the AP force by the glycogen lowering. These data suggest that low glycogen and glycogenolytic rate affects t-system polarization and excitability, as the voltage sensor inactivation is displaced to markedly more positive *V*m values compared with the AP inactivation (Ørtenblad and Stephenson 2003; Nielsen et al. 2004) and thereby force production will be less affected by voltage sensor compared to AP activation. For a more direct estimate of the Na, K-pump function, one can quantify the membrane ability to respond to two closely spaced AP. The repriming time until the second pulse generates an AP is in turn dependent on the Na, K-pump function. A depolarization of the t-system increases the repriming time as expected; however, the addition of phosphoenolpyruvate, which increases glycolytic ATP resynthesis, decreases t-system repriming period (Dutka and Lamb 2007a, b). Also, when glycogen is enzymatically lowered by glucoamylase treatment, the repriming period increases (Watanabe and Wada 2019; Jensen et al. 2020a). This was confirmed by the use of glycogen phosphorylase inhibitors and glycogen lowering treatment in mechanically skinned fibers, which invariably prolonged repriming time, strongly indicating an attenuated Na, K-pump activity (Jensen et al. 2020a). Collectively, substantial evidence indicates a tight coupling between glycogenolytic-glycolytic-derived ATP production on Na-K-pump function, with low glycogen or inhibited GP in turn leading to an attenuated t-system excitability and ultimately force production.

5.4.2.3 Insights from McArdle Patients

Insights into the role and importance of muscle glycogen can be gained from McArdle patients (glycogen storage disease type V) who lack glycogen phosphorylase and thereby not able to catalyze the breakdown of glycogen into glucose-1-phosphate in muscle fibers (McArdle 1951; Santalla et al. 2014). McArdle patients' have low peak work capacities and exercise intolerance and a suppressed plasma lactate concentration during exercise (De Stefano et al. 1996; Ørngreen et al. 2015). The exercise intolerance is especially evident during the onset of exercise, during high-intensity exercise and high force output contractions with smaller muscle groups (Lucia et al. 2008; Santalla et al. 2014). These patients may provide a unique paradigm to gain insight to the role of muscle glycogen per se in muscle function and exercise tolerance. Interestingly, a higher surface electromyography signal is measured during submaximal contractions in McArdle patients, which is indicative of a need to activate larger muscle mass for a given force output, suggesting reduced muscle excitability (Santalla et al. 2014). Interestingly, McArdle patients have less Na^+-K^+-ATPases as compared to controls (Haller et al. 1998). Taken together, insights from McArdle patients are in line with the idea of glycolytic/glycogenolytic-derived ATP as required for muscle E-C coupling, possibly by a tight coupling between glycogenolytic/glycolytic rate and Na^+-K^+-ATPase activity.

5.5 Muscle Glycogen as a Regulator of Skeletal Muscle Training Adaptations

In addition to the well-documented role as a metabolic substrate for ATP provision, it is becoming increasingly accepted that the glycogen granule can also act as a regulator of training adaptations (Impey et al. 2018; Impey et al. 2016). Accordingly, the concept of deliberately training with reduced pre-exercise muscle glycogen availability in an attempt to enhance the activation of the molecular signaling pathways that regulate mitochondrial biogenesis (the so-called train low paradigm) has received significant research in the previous decade (Burke et al. 2018). In this regard, findings from acute exercise studies demonstrate that commencing exercise with "reduced" pre-exercise muscle glycogen concentration upregulates cell signaling pathways with putative roles in the regulation of both the nuclear and mitochondrial genomes (Pilegaard et al. 2002; Wojtaszewski et al. 2003; Yeo et al. 2010; Bartlett et al. 2013; Psilander et al. 2013). Furthermore, repeated bouts of train-low exercise can subsequently augment many hallmark muscle adaptations inherent to the endurance phenotype (as reviewed in Impey et al. 2018). For example, the strategic periodization of dietary CHO in order to commence exercise with low muscle glycogen (during 3–10 weeks of training) enhances mitochondrial enzyme activity and protein content (Hansen et al. 2005; Morton et al. 2009; Yeo et al. 2008) and whole body and intra-muscular lipid metabolism (Hulston et al. 2010) and in some instances improves exercise capacity (Hansen et al. 2005) and performance (Marquet et al. 2016a, b), though performance enhancing effects are not always evident (Yeo et al. 2008; Hulston et al. 2010; Burke et al. 2017; Gejl et al. 2017a, b; 2018). As such, the train-low paradigm and wider CHO periodization strategies have subsequently gained increased recognition among athletic populations (Stellingwerff 2012; Burke et al. 2018; Impey et al. 2018). It should also be noted that some of the enhanced adaptations associated with "train-low" (at least in the twice per day training model) may be due to performing two consecutive training sessions in close proximity to one another, as opposed to the effects of low pre-exercise muscle glycogen per se (Andrade-Souza et al. 2020).

Skeletal muscle glycogen may exert its regulatory effects upon training adaptation through modulation of the AMP-activated protein kinase (AMPK)-peroxisome proliferator-activated receptor γ coactivator 1α (PGC- 1α) signaling axis. For example, exercise-induced AMPKα2 activity (Wojtaszewski et al. 2003), phosphorylation (Yeo et al. 2010), and nuclear abundance (Steinberg et al. 2006) are all augmented under conditions of reduced pre-exercise muscle glycogen. These effects may be partly mediated through the glycogen-binding domain present on the β-subunit of AMPK (McBride and Hardie 2009; McBride et al. 2009). Commencing acute exercise with reduced muscle glycogen also potentiates the phosphorylation of the tumor suppressor protein, p53 (Bartlett et al. 2013), that in turn may coordinate regulation of the mitochondrial genome, through modulation of mitochondrial transcription factor A (Tfam). It is noteworthy that training with low muscle glycogen also increases epinephrine, stimulates lipolysis, increases circulating free fatty acids (FFAs), and therefore augments both whole-body and intramuscular lipid oxidation (Hearris et al. 2018, 2020; Maunder et al. 2021). In this way, it is possible that FFAs may also regulate the enhanced adaptations associated with training low through acting as stimulatory signaling molecules for PPARδ signaling.

Given that the enhanced training response associated with train-low is potentially mediated by muscle glycogen availability, it is pertinent to consider the absolute glycogen concentrations required to facilitate the response. In relation to research design, high glycogen trials are commonly commenced with muscle glycogen concentrations between 400 and 600 mmol$\cdot$kg^{-1} dw and remain above 300 mmol$\cdot$kg^{-1} dw after exercise (Wojtaszewski et al. 2003; Roepstorff et al. 2005; Bartlett et al. 2013). In such instances, these researchers observed attenuated (Wojtaszewski et al. 2003) or abolished (Roepstorff et al. 2005; Bartlett et al. 2013) activation of cell signaling pathways. For example, we have previously observed that both AMPK and p53-related signaling is reduced when exercise is commenced and finished with high (reducing from ~500 to ~300 mmol.kg^{-1} dw, respectively) versus low (reducing from ~150 to ~50 mmol.kg^{-1} dw, respectively) muscle glycogen (Bartlett et al. 2013). A detailed examination of available train low studies (see Impey et al. 2018) indeed demonstrates that adaptations associated with CHO restriction are particularly evident when the absolute pre-exercise muscle glycogen concentration permits exercise-induced depletion to post-exercise concentrations that are $\leq$200–300 mmol$\cdot$kg^{-1} dw. Accordingly, we also observed that comparable cell signaling responses occur if exhaustive (Hearris et al. 2019) and non-exhaustive

(Hearris et al. 2020) exercise is finished with similar absolute post-exercise muscle glycogen concentrations (i.e., 100–300 mmol·kg^{-1} dw), despite commencing exercise with graded levels of muscle glycogen concentration (i.e., 300–600 mmol·kg^{-1} dw). On this basis, it was suggested that the absolute post-exercise muscle glycogen concentration may be a more influential factor than pre-exercise muscle glycogen concentration in relation to modulation of exercise-induced cell signaling. In keeping with this hypothesis, CHO feeding during exercise attenuates AMPK-mediated signaling but only when glycogen sparing occurs (Akerstrom et al. 2006). Furthermore, restoring post-exercise glycogen levels to >500 mmol·kg^{-1} dw attenuates exercise-induced changes in gene expression (Pilegaard et al. 2005). From a practical perspective, such data collectively suggest that distinct differences in post-exercise muscle glycogen concentration would likely be required between low (e.g., <200 mmol·kg^{-1} dw) and high (> 300 mmol·kg^{-1} dw) glycogen conditions in order to achieve any potential physiological advantage of reduced muscle glycogen availability in relation to augmenting training adaptation. While the optimal approach to CHO periodization remains to be determined, further research is also required to examine the role of the specific subcellular storage pools of glycogen as a regulator of skeletal muscle cell signaling pathways. Moreover, despite the theoretical rationale for carefully scheduled periods of "training-low," it is noteworthy that definitive evidence supporting that this approach to training induces superior improvements in performance is currently limiting. In this regard, there is also a requirement to utilize experimental testing protocols (e.g., exercise durations of 3–6 h) in which the physiological adaptations associated with training low (i.e., increased oxidative capacity and lipid oxidation) may actually manifest as improved exercise performance.

5.6 Concluding Remarks and Future Directions

Glycogen is an important fuel source for contracting skeletal muscle to sustain high metabolic power output during exercise, and modern research has also shown it to be far more than just an inert energy source. Early research determined a clear direct association between the capacity to withstand intense exercise for prolonged periods and starting muscle glycogen, showing that the onset of fatigue coincides with the depletion of glycogen to very low levels, even with exogenous carbohydrate supplementation. In addition to exercise intensity and duration, baseline muscle glycogen and training status have also been identified as important factors in determining its rate of use, with other factors such as sex substrate availability and environment also having some impact. Glycogen use has shown to be divergent in different fiber types, and more recent research has identified specific subcellular loci of glycogen (namely, intramyofibrillar, intermyofibrillar, and subsarcolemmal), with possible diverse utilization and metabolic function depending on their localization. The mechanisms explaining how glycogen and glycolytic rate are integrated with muscle function are far from understood; however, emerging evidence link glycogen availability with key steps in the excitation-contraction coupling. In addition, it is also apparent that glycogen availability can exert important effects on metabolic regulation and exercise-induced cell signaling, regulating training adaptation. Thus, glycogen has a far more diverse function than just being an energy storage. Despite over 100 years of research in carbohydrate metabolism, fundamental aspects of muscle glycogen metabolism and regulation continue to challenge the scientific community and remain an evergreen area of exciting research. At several levels, we appear to have a complete understanding of the physiology of muscle glycogen, and yet at others, we know very little. Indeed, many aspects of muscle glycogen metabolism in exercise and cell function warrant further knowledge. Table 5.1 lists

Table 5.1 Possible future directions in the study of muscle glycogen metabolism

Topic	Comment
Factors affecting glycogen utilization	
Skeletal muscle glycogen loading and performance	Currently unclear effect of placebo in events $\geq$90 min or with >25% differences in glycogen content
Identification of glycogen utilization at intensities relative to "lactate threshold"	Glycogen utilization typically characterized relative to VO_{2max}, provides large inter-individual variation on prediction of glycogenolysis
Repeated glycogen supercompensation and resynthesis and performance	Unclear current capacity to resynthesize depleted glycogen stores within 24 h and enhance performance repetitively
Glycogen metabolism in female's vs males	Currently gender effect is unclear
Effect of different muscles on glycogen utilization	Majority of current evidence based on vastus lateralis
Skeletal glycogen sparing with increasing doses of carbohydrate ingestion during exercise	Unclear effect on mixed muscle, fiber types, and subcellular localization
Exercise task dependency of glycogen pool metabolism	Little is known about the glycogen pool-specific utilization with exercise intensity and duration and during intermittent exercise
Effects of glycogen on training adaptations	
Ketogenic high-fat diets and glycogen dynamics	Unclear effect of "keto-adapted" state on use and resynthesis of glycogen
Effect of chronically training with low skeletal muscle glycogen on selected session on performance	Currently early adaptive responses show promising results, but longer interventions show mostly no improvement in performance
Role and effects of glycogen on muscle function and fatigue	
Fiber-type-specific glycogen-dependent fatigue mechanisms during exercise	If a mechanism of fatigue sets in in only one specific fiber type, it can be difficult to detect and probably completely overlooked in a mixed muscle sample
Role of specific glycogen pools on muscle E-C coupling and fatigue	It is not known if pools are used by specific ATPases and/or ion channels

(continued)

Table 5.1 (continued)

Topic	Comment
Mechanism(s) linking low muscle glycogen levels with impaired SR Ca^{2+} release	Precise mechanism by which low (intramyofibrillar) glycogen levels regulates the SR RyR channel
Role of muscle glycogen on SR Ca^{2+} uptake	It is currently unclear if glycogen levels affect SR Ca^{2+} uptake mechanisms (SERCA activity and/or SR leak)

possible future studies further generating our understanding of muscle glycogen metabolism and regulation, as well as the causal link underlying the association between glycogen availability in the etiology of muscle fatigue and key steps in muscle function and regulator of training adaptations.

References

Ahlborg G, Jensen-Urstad M (1991) Metabolism in exercising arm vs. leg muscle. Clin Physiol 11: 459–468

Akerstrom TC, Birk JB, Klein DK, Erikstrup C, Plomgaard P, Pedersen BK, Wojtaszewski JF (2006) Oral glucose ingestion attenuates exercise-induced activation of 5′-AMP-activated protein kinase in human skeletal muscle. Biochem Biophys Res Commun 342:949–955

Albers PH, Pedersen AJT, Birk JB, Kristensen DE, Vind BF, Baba O, Nøhr J, Højlund K, Wojtaszewski J (2015) Human muscle fiber type-specific insulin signaling: impact of obesity and type 2 diabetes. Diabetes 64:485–497

Alghannam AF, Jedrzejewski D, Tweddle MG, Gribble H, Bilzon J, Thompson D, Tsintzas K, Betts JA (2016) Impact of muscle glycogen availability on the capacity for repeated exercise in man. Med Sci Sports Exerc 48:123–131

Andrade-Souza VM, Ghiarone T, Sansonio A, Santos Silva KA, Tomazini F, Enrico Perri F, Saner N, Kuang K, Bertuzzi R, Leandro CG, Bishop DJ, Lima-Silva AE (2020) Exercise twice-a-day potentiates markers of mitochondrial biogenesis in men. FASEB J 34:1602–1609

Areta JL, Hopkins WG (2018) Skeletal muscle glycogen content at rest and during endurance exercise in humans a meta-analysis. Sports Med 48:2091–2102

Areta JL, Iraki J, Owens DJ, Joanisse S, Philp A, Morton JP, Hallén J (2020) Achieving energy balance with a high-fat meal does not enhance skeletal muscle adaptation and impairs glycemic response in a sleep-low training model. Exp Physiol:EP088795

Arkinstall MJ, Bruce CR, Clark SA, Rickards CA, Burke LM, Hawley JA (2004) Regulation of fuel metabolism by preexercise muscle glycogen content and exercise intensity. J Appl Physiol 97(6):2275–2283

Ball S, Colleoni C, Cenci U, Raj JN, Tirtiaux C (2011) The evolution of glycogen and starch metabolism in eukaryotes gives molecular clues to understand the establishment of plastid endosymbiosis. J Exp Bot 62(6):1775–1801

Ball-Burnett M, Green HJ, Houston ME (1991) Energy metabolism in human slow and fast twitch fibres during prolonged cycle exercise. J Physiol 437:257–267

Balsom PD, Gaitanos GC, Söderlund K, Ekblom B (1999) High-intensity exercise and muscle glycogen availability in humans. Acta Physiol Scand 165:337–345

Bangsbo J, Gollnick PD, Graham TE, Saltin B (1991) Substrates for muscle glycogen synthesis in recovery from intense exercise in man. J Physiol 434:423–440

Bangsbo J, Graham TE, Kiens B, Saltin B (1992) Elevated muscle glycogen and anaerobic energy production during exhaustive exercise in man. J Physiol 451:205–227

Barnes M, Gibson LM, Stephenson DG (2001) Increased muscle glycogen content is associated with increased capacity to respond to T-system depolarisation in mechanically skinned skeletal muscle fibres from the rat. Pflugers Arch 442(1):101–106

Bartlett, J. D., Louhelainen, J., Iqbal, Z., Cochran, A. J., Gibala, M. J., Gregson, W., . . . Morton, J. P. (2013). Reduced carbohydrate availability enhances exercise-induced p53 signaling in human skeletal muscle: implications for mitochondrial biogenesis. American journal of physiology. Regulatory, Integrative and Comparative Physiology, 304, R450–R458

Bergström J, Hultman E (1966a) The effect of exercise on muscle glycogen and electrolytes in normals. Scand J Clin Lab Invest 18(1):16–20

Bergström J, Hultman E (1966b) Muscle glycogen synthesis after exercise: an enhancing factor localized to the muscle cells in man. Nature 210:309–310

Bergström J, Hermansen L, Hultman E, Saltin B (1967) Diet, muscle glycogen and physical performance. Acta Physiol Scand 71:140–150

Blazev R, Lamb GD (1999) Low [ATP] and elevated [Mg2+] reduce depolarization-induced Ca^{2+} release in rat skinned skeletal muscle fibres. J Physiol 520(Pt 1):203–215

Bogdanis GC, Nevill ME, Boobis LH, Lakomy HK (1996) Contribution of phosphocreatine and aerobic metabolism to energy supply during repeated sprint exercise. J Appl Physiol 80(3):876–884

Bosch AN, Dennis SC, Noakes TD (1993) Influence of carbohydrate loading on fuel substrate turnover and oxidation during prolonged exercise. J Appl Physiol 74(4):1921–1927

Bosch AN, Weltan SM, Dennis SC, Noakes TD (1996) Fuel substrate kinetics of carbohydrate loading differs from that of carbohydrate ingestion during prolonged exercise. Metabolism 45(4):415–423

Braun B, Mawson JT, Muza SR et al (2000) Women at altitude: carbohydrate utilization during exercise at 4,300 m. J Appl Physiol 88:246–256

Brooks GA (1986) Lactate production under fully aerobic conditions: the lactate shuttle during rest and exercise. Fed Proc 45(13):2924–2929

Brooks GA (1992) Increased glucose dependency in circulatory compensated hypoxia. In: Houston C, Coates J (eds) Hypoxia and mountain medicine. Queen City, Burlington, VT, pp 213–226

Brooks GA, Butterfield GE, Wolfel RR et al (1991) Increased dependence on blood glucose after acclimatization to 4,300 m. J Appl Physiol 70:919–927

Bundgaard H, Kjeldsen K, Suarez Krabbe K, van Hall G, Simonsen L, Qvist J, Hansen CM, Moller K, Fonsmark L, Lav Madsen P, Klarlund Pedersen B (2003) Endotoxemia stimulates skeletal muscle Na+-K+-ATPase and raises blood lactate under aerobic conditions in humans. Am J Physiol Heart Circ Physiol 284(3):H1028–H1034

Burke LM (2006) "Fat adaptation" for athletic performance: the nail in the coffin? J Appl Physiol 100(1):7–8

Burke LM, Angus DJ, Cox GR, Cummings NK, Febbraio MA, Gawthorn K, Hawley JA, Minehan M, Martin DT, Hargreaves M (2000) Effect of fat adaptation and carbohydrate restoration on metabolism and performance during prolonged cycling. J Appl Physiol 89(6): 2413–2421

Burke LM, Loon LJC, Hawley JA (2017) Postexercise muscle glycogen resynthesis in humans. J Appl Physiol 122(5):1055–1067

Burke LM, Hawley JA, Jeukendrup A, Morton JP, Stellingwerff T, Maughan RJ (2018) Toward a common understanding of diet–exercise strategies to manipulate fuel availability for training and competition preparation in endurance sport. Int J Sport Nutr Exerc Metab 28:451–463

Caldwell PC, Hodgkin AL, Keynes RD, Shaw TL (1960) The effects of injecting energy-rich phosphate compounds on the active transport of ions in the giant axons of Loligo. J Physiol 152: 561–590

Carter J, Jeukendrup AE, Mundel T et al (2003) Carbohydrate supplementation improves moderate and high-intensity exercise in the heat. Pflugers Arch 446:211–219
Caulfield J, Klionsky B (1959) Myocardial ischemia and early infarction: an electron microscopy study. Amer J Path 35:489–523
Chesley A, Heigenhauser GJ, Spriet LL (1996) Regulation of muscle glycogen phosphorylase activity following short-term endurance training. Am J Phys 270(2 Pt 1):E328–E335
Chin ER, Allen DG (1997) Effects of reduced muscle glycogen concentration on force, Ca^{2+} release and contractile protein function in intact mouse skeletal muscle. J Physiol 498:17–29
Clausen T (1965) The relationship between the transport of glucose and cations across cell membranes in isolated tissues. I. Stimulation of glycogen deposition and inhibition of lactic acid production in diaphragm, induced by ouabain. Biochim Biophys Acta 109:164–171
Clausen T, Nielsen OB (2007) Potassium, Na+,K+-pumps and fatigue in rat muscle. J Physiol 584 (Pt 1):295–304
Costill DL, Jansson E, Gollnick PD, Saltin B (1974) Glycogen utilization in leg muscles of men during level and uphill running. Acta Physiol Scand 91:475–481
Costill DL, Coyle E, Dalsky G, Evans W, Fink W, Hoopes D (1977) Effects of elevated plasma FFA and insulin on muscle glycogen usage during exercise. J Appl Physiol Respir Environ Exerc Physiol 43(4):695–699
Coyle EF, Coggan AR, Hemmert MK, Ivy JL (1986) Muscle glycogen utilization during prolonged strenuous exercise when fed carbohydrate. J Appl Physiol 61(1):165–172
Coyle EF, Coggan AR, Hopper MK, Walters TJ (1988) Determinants of endurance in well-trained cyclists. J Appl Physiol. (Bethesda, Md.: 1985), 64(6):2622–2630
De Bock K, Derave W, Ramaekers M, Richter EA, Hespel P (2006) Fiber type-specific muscle glycogen sparing due to carbohydrate intake before and during exercise. J Appl Physiol 102(1): 183–188
de Bruijn WC (1973) Glycogen, its chemistry and morphologic appearance in the electron microscope. I. a modified OsO 4 fixative which selectively contrasts glycogen. J Ultrastruct Res 42:29–50
de Paoli FV, Overgaard K, Pedersen TH, Nielsen OB (2007) Additive protective effects of the addition of lactic acid and adrenaline on excitability and force in isolated rat skeletal muscle depressed by elevated extracellular K^{+}. J Physiol 581:829–839
de Paoli FV, Ørtenblad N, Pedersen TH, Jørgensen R, Nielsen OB (2010) Lactate per se improves the excitability of depolarized rat skeletal muscle by reducing the cl- conductance. J Physiol 588 (Pt 23):4785–4794
De Stefano N, Argov Z, Matthews PM et al (1996) Impairment of muscle mitochondrial oxidative metabolism in McArdles's disease. Muscle Nerve 19(6):764–769
Devries MC (2016) Sex-based differences in endurance exercise muscle metabolism: impact on exercise and nutritional strategies to optimize health and performance in woman. Exp Physiol 101:243–249
Devries MC, Hamadeh MJ, Phillips SM, Tarnopolsky MA (2006) Menstrual cycle phase and sex influence muscle glycogen utilization and glucose turnover during moderate-intensity endurance exercise. Am J Physiol Regul Intergr Comp Physiol 291:1120–1128
Dhar-Chowdhury P, Malester B, Rajacic P, Coetzee WA (2007) The regulation of ion channels and transporters by glycolytically derived ATP. Cell Mol Life Sci 64:3069–3083
Duhamel TA, Green HJ, Perco JG, Ouyang J (2006a) Comparative effects of a low-carbohydrate diet and exercise plus a low-carbohydrate diet on muscle sarcoplasmic reticulum responses in males. Am J Physiol Cell Physiol 291:C607–C617
Duhamel TA, Green HJ, Perco JG, Ouyang J (2006b) Effects of prior exercise and a low-carbohydrate diet on muscle sarcoplasmic reticulum function during cycling in women. J Appl Physiol 101:695–706
Duhamel TA, Perco JG, Green HJ (2006c) Manipulation of dietary carbohydrates after prolonged effort modifies muscle sarcoplasmic reticulum responses in exercising males. Am J Physiol Regul Integr Comp Physiol 291:R1100–R1110

Dulhunty AF (1984) Heterogeneity of T-tubule geometry in vertebrate skeletal muscle fibres. J Muscle Res Cell Motil 5:333–347
Dutka TL, Lamb GD (2007a) Na+-K+ pumps in the transverse tubular system of skeletal muscle fibers preferentially use ATP from glycolysis. Am J Physiol-Cell Physiol 293:C967–C977
Dutka TL, Lamb GD (2007b) Transverse tubular system depolarization reduces tetanic force in rat skeletal muscle fibers by impairing action potential repriming. Am J Physiol Cell Physiol 292: C2112–C2121
Dyck DJ, Putman CT, Heigenhauser GJ, Hultman E, Spriet LL (1993) Regulation of fat-carbohydrate interaction in skeletal muscle during intense aerobic cycling. Am J Phys 265(6 Pt 1):E852–E859
Dyck DJ, Peters SJ, Wendling PS, Chesley A, Hultman E, Spriet LL (1996) Regulation of muscle glycogen phosphorylase activity during intense aerobic cycling with elevated FFA. Am J Phys 270(1 Pt 1):E116–E125
Edwards HT, Margaria R, Dill DB (1934) Metabolic rate, blood sugar, and the utilization of carbohydrate. Am J Phys 58:203–209
Entman ML, Keslensky SS, Chu A, Van Winkle B (1980) The sarcoplasmic reticulum-glycogenolytic complex in mammalian fast twitch skeletal muscle. J Biol Chem 255:6245–6252
Essén B, Henriksson J (1974) Glycogen content of individual muscle fibres in man. Acta Physiol Scand 90:645–647
Essén-Gustavsson B, Tesch PA (1990) Glycogen and triglyceride utilization in relation to muscle metabolic characteristics in men performing heavy-resistance exercise. Eur J Appl Physiol 61: 5–10
Febbraio MA, Snow RJ, Hargreaves M et al (1994) Muscle metabolism during exercise and heat stress in trained men: effect of acclimation. J Appl Physiol 76:589–597
Fell JM, Hearris MA, Ellis DG, Moran J, Jevons EFP, Owens DJ, Strauss JA, Cocks MC, Louis JB, Shepherd SO, Morton JP (2021) Carbohydrate intake improves exercise capacity but does not affect subcellular lipid droplet morphology, AMPK and p53 signalling in human skeletal muscle. J Physiol 599:2823–2849
Fill M, Copello JA (2002) Ryanodine receptor calcium release channels. Physiol Rev 82(4): 893–922
Fink WJ, Costill DL, Van Handel PJ (1975) Leg muscle metabolism during exercise in the heat and cold. Eur J Appl Physiol 34:183–190
Frentzel J, Reach F (1901) Untersuchungen zur Frage nach der Quelle der Muskelkraft. Pflugers Arch 83:477–508
Fridén J, Seger J, Ekblom B (1989) Topographical localization of muscle glycogen: an ultrahistochemical study in the human vastus lateralis. Acta Physiol Scand 135:381–391
Fridén J, Seger J, Ekblom B (1985) Implementation of periodic acid-thiosemicarbazide-silver proteinate staining for ultrastructural assessment of muscle glycogen utilization during exercise. Cell Tissue Res 242:229–232
Gaitanos GC, Williams C, Boobis LH, Brooks S (1993) Human muscle metabolism during intermittent maximal exercise. J Appl Physiol 75(2):712–719
Galbo H, Holst JJ, Christensen NJ (1979) The effect of different diets and of insulin on the hormonal response to prolonged exercise. Acta Physiol Scand 107:19–32
Galloway SDR, Maughan RJ (1997) Effects of ambient temperature on the capacity to perform prolonged cycle exercise in man. Med Sci Sports Exerc 29:1240–1249
Gejl K, Hvid LG, Ulrik Frandsen U, Jensen K, Sahlin K, Ørtenblad N (2014) Muscle glycogen content modifies SR Ca2+ release rate in elite endurance athletes. Med Sci Sports Ex 46(3): 496–505
Gejl KD, Ørtenbad N, Andersson E, Plomgaard P, Holmberg HC, Nielsen J (2017a) Local depletion of glycogen with supra-maximal exercise in human skeletal muscle fibres. J Physiol 595:2809–2821

Gejl KD, Thams L, Hansen M, Rokkedal-Lausch T, Plomgaard P, Nybo L, Larsen FJ, Cardinale DA, Jensen K, Holmberg HC, Vissing K, Ørtenblad N (2017b) No superior adaptations to carbohydrate periodization in elite endurance athletes. Med Sci Sports Exerc 49(12):2486–2497

Gejl KD, Ørtenblad N, Andersson E, Plomgaard P, Holmberg HC, Nielsen J (2017c) Local depletion of glycogen with supra-maximal exercise in human skeletal muscle fibres. J Physiol 595(9):2809–2821

Gejl KD, Vissing K, Hansen M, Thams L, Rokkedal-Lausch T, Plomgaard P, Meinild Lundby AK, Nybo L, Jensen K, Holmberg HC, Ørtenblad N (2018) Changes in metabolism but not myocellular signaling by training with CHO-restriction in endurance athletes. Physiol Rep 6(17):e13847

Goldstein MA, Murphy DL, Van Winkle WB, Entman ML (1985) Cytochemical studies of a glycogen-sarcoplasmic reticulum complex. J Muscle Res Cell Motility 6:177–187

Gollnick PD, Piehl K, Saubert CW, Armstrong RB, Saltin B (1972) Diet, exercise, and glycogen changes in human muscle fibers. J Appl Physiol 33(4):421–425

Gollnick PD, Piehl K, Saltin B (1974) Selective glycogen depletion pattern in human muscle fibres after exercise of varying intensity and at varying pedalling rates. J Physiol 241:45–57

Goodman C, Blazev R, Stephenson G (2005 Sep) Glycogen content and contractile responsiveness to T-system depolarization in skinned muscle fibres of the rat. Clin Exp Pharmacol Physiol 32(9):749–756

Graham TE, Adamo KB, Shearer J, Marchand I, Saltin B (2001) Pro- and macroglycogenolysis: relationship with exercise intensity and duration. J Appl Physiol 90(3):873–879

Green HJ (1991) How important is endogenous muscle glycogen to fatigue in prolonged exercise? Can J Physiol Pharmacol 69(2):290–297

Green HJ, Ball-Burnett M, Jamieson G, Cadefau J, Cussó R (1995) Metabolic adaptations to short-term training are expressed early in submaximal exercise. Can J Physiol Pharmacol 73(4): 474–482

Greenhaff PL, Söderlund K, Ren JM, Hultman E (1993) Energy metabolism insingle human muscle fibres during intermittent contraction with occluded circulation. J Physiol 460:443–453

Griffiths A, Shannon OM, Matu J, King R, Deighton K, O'Hara JP (2019) The effects of environmental hypoxia on substrate utilisation during exercise: a meta-analysis. J Int Soc Sports Nutr 16(1):10

Haff GG, Lehmkuhl MJ, McCoy LB, Stone MH (2003) Carbohydrate supplementation and resistance training. Strength Cond Res 17(1):187–196

Haller RG, Clausen T, Vissing J (1998) Reduced levels of skeletal muscle Na+K+ -ATPase in McArdle disease. Neurology 50:37–40

Hammond KM, Sale C, Fraser W, Tang J, Shepherd SO, Strauss JA, Close GL, Cocks M, Louis J, Pugh J, Stewart C, Sharples AP, Morton JP (2019) Post-exercise carbohydrate and energy availability induce independent effects on skeletal muscle cell signalling and bone turnover: implications for training adaptation. J Physiol 597(18):4779–4796

Han JW, Thieleczek R, Varsanyi M, Heilmeyer LM Jr (1992) Compartmentalized ATP synthesis in skeletal muscle triads. Biochemistry 31:377–384

Hansen AK, Fischer CP, Plomgaard P, Andersen JL, Saltin B, Pedersen BK (2005) Skeletal muscle adaptation: training twice every second day vs. training once daily. J Appl Physiol 98(1):93–99

Harber MP, Crane JD, Douglass MD, Weindel KD, Trappe TA, Trappe SW, Fink WF (2010) Resistance exercise reduces muscular substrates in women. Int J Sports Med 29(9):719–725

Hargreaves M, McConell G, Proietto J (1995) Influence of muscle glycogen on glycogenolysis and glucose uptake during exercise in humans. J Appl Physiol 78(1):288–292

Hargreaves M, Angus D, Howlett K et al (1996) Effect of heat stress on glucose kinetics during exercise. J Appl Physiol 81:1594–1597

Hargreaves M, Finn JP, Withers RT, Halbert JA, Scroop GC, Mackay M, Snow RJ, Carey MF (1997) Effect of muscle glycogen availability on maximal exercise performance. Eur J Appl Physiol 75:188–192

Hasin Y, Barry WH (1984) Myocardial metabolic inhibition and membrane potential, contraction, and potassium uptake. Am J Phys 247:H322–H329

Hawley JA (2002) Effect of increased fat availability on metabolism and exercise capacity. Med Sci Sports Exerc 34(9):1485–1491

Hawley JA, Palmer GS, Noakes TD (1997a) Effects of 3 days of carbohydrate supplementation on muscle glycogen content and utilisation during a 1-h cycling performance. Eur J Appl Physiol Occup Physiol 75:407–412

Hawley JA, Schabort EJ, Noakes TD, Dennis SC (1997b) Carbohydrate-loading and exercise performance. Sports Med 24(2):73–81

Hearris MA, Hammond KM, Fell JM, Morton JP (2018) Regulation of muscle glycogen metabolism during exercise: implications for endurance performance and training adaptations. Nutrients 10(3):298

Hearris MA, Hammond KM, Seaborne RA, Stocks B, Shepherd SO, Philp A, Sharples AP, Morton JP, Louis JB (2019) Graded reductions in preexercise muscle glycogen impair exercise capacity but do not augment skeletal muscle cell signaling: implications for CHO periodization. J Appl Physiol 126(6):1587–1597

Hearris MA, Owens DJ, Strauss JA, Shepherd SO, Sharples AP, Morton JP, Louis JB (2020) Graded reductions in pre-exercise glycogen concentration do not augment exercise-induced nuclear AMPK and PGC-1α protein content in human muscle. Exp Physiol 105(11):1882–1894

Helander I, Westerblad H, Katz A (2002 Jun) Effects of glucose on contractile function, [Ca2+]i, and glycogen in isolated mouse skeletal muscle. Am J Physiol Cell Physiol 282(6):C1306–C1312

Helge JW, Damsgaard R, Overgaard K, Andersen JL, Donsmark M, Dyrskog SE, Hermansen K, Saltin B, Daugaard JR (2008) Low-intensity training dissociates metabolic from aerobic fitness. Scand J Med Sci Sports 18:86–94

Hermansen L, Hultman E, Saltin B (1967) Muscle glycogen during prolonged severe exercise. Acta Physiol 71(2–3):129–139

Hill AV (1913) The energy degraded in the recovery processes of stimulated muscles. J Physiol 46 (1):28–80

Hingst JR, Bruhn L, Hansen MB, Rosschou MF, Birk JB, Fentz J, Foretz M, Viollet B, Sakamoto K, Færgeman NJ, Havelund JF, Parker BL, James DE, Kiens B, Richter EA, Jensen J, Wojtaszewski JFP (2018) Exercise-induced molecular mechanisms promoting glycogen supercompensation in human skeletal muscle. Molecular Metabolism

Hintz CS, Chi MM-Y, Fell RD, Ivy JL, Kaiser KK, Lowry CV, Lowry OH (1982) Metabolite changes in individual rat muscle fibers during stimulation. Am J Physiol Cell Physiol 242: C218–C228

Hokken R, Laugesen S, Aagaard P, Suetta C, Frandsen U, Ørtenblad N, Nielsen J (2020) Subcellular localization- and fibre type-dependent utilization of muscle glycogen during heavy resistance exercise in elite power and Olympic weightlifters. Acta Physiol (Oxf) e13561

Holloszy JO (1973) Biochemical adaptations to exercise: aerobic metabolism. Exerc Sport Sci Rev 1:45–71

Holloszy JO, Coyle EF (1984) Adaptations of skeletal muscle to endurance exercise and their metabolic consequences. J Appl Physiol Respir Environ Exerc Physiol 56(4):831–838

Howlett RA, Parolin ML, Dyck DJ, Jones NL, Heigenhauser GJ, Spriet LL (1998) Regulation of skeletal muscle glycogen phosphorylase and PDH at varying exercise power outputs. Am J Phys 275:R418–R425

Hulston CJ, Venables MC, Mann CH, Martin C, Philp A, Baar K, Jeukendrup AE (2010) Training with low muscle glycogen enhances fat metabolism in well-trained cyclists. Med Sci Sports Exerc 42:2046–2055

Hultman, Greenhaff (1999) Role of submaximal exercise in promoting creatine and glycogen accumulation in human skeletal muscle. J Appl Physiol 87(2):598–604

Impey SG, Hammond KM, Shepherd SO, Sharples AP, Stewart C, Limb M et al (2016) Fuel for the work required: a practical approach to amalgamating train-low paradigms for endurance athletes. Physiol Rep 4:e12803

Impey SG, Hearris MA, Hammond KM, Bartlett JD, Louis J, Close GL, Morton JP (2018) Fuel for the work required: a theoretical framework for carbohydrate periodization and the glycogen threshold hypothesis. Sports Med 48:1031–1048

Impey SG, Jevons E, Mees G, Cocks M, Strauss J, Chester N, Laurie I, Target D, Hodgson A, Shepherd SO, Morton JP (2020) Glycogen utilization during running: intensity, sex, and muscle-specific responses. Med Sci Sports Exerc 52(9):1966–1975

James JH, Fang CH, Schrantz SJ, Hasselgren PO, Paul RJ, Fischer JE (1996) Linkage of aerobic glycolysis to sodium-potassium transport in rat skeletal muscle. Implications for increased muscle lactate production in sepsis. J Clin Invest 98(10):2388–2397

James JH, Luchette FA, McCarter FD, Fischer JE (1999a) Lactate is an unreliable indicator of tissue hypoxia in injury or sepsis. Lancet 354(9177):505–508

James JH, Wagner KR, King J-K, Leffler RE, Upputuri RK, Balasubramanian A, Friend LA, Shelly DA, Paul RJ, Fischer JE (1999b) Stimulation of both aerobic glycolysis and Na^+-K^+-ATPase activity in skeletal muscle by epinephrine or amylin. Am J Physiol Endocrinol Metab 277: E176–E186

Jensen TE, Richter EA (2012) Regulation of glucose and glycogen metabolism during and after exercise. J Physiol 590(5):1069–1076

Jensen R, Nielsen J, Ørtenblad N (2020a) Inhibition of glycogenolysis prolongs action potential repriming period and impairs muscle function in rat skeletal muscle. J Physiol 598(4):789–803

Jensen R, Ørtenblad N, Stausholm MH, Skjærbæk MC, Larsen DN, Hansen M, Holmberg HC, Plomgaard P, Nielsen J (2020b) Heterogeneity in subcellular muscle glycogen utilisation during exercise impacts endurance capacity in men. J Physiol 598:4271–4292

Jensen R, Ørtenblad N, Stausholm MH, Skjærbæk MC, Larsen DN, Hansen M, Holmberg HC, Plomgaard P, Nielsen J (2021) Glycogen supercompensation is due to increased number, not size, of glycogen particles in human skeletal muscle. Exp Physiol 106:1272–1284

Jensen-Urstad M, Ahlborg G (1992) Is the high lactate release during arm exercise due to a low training status? Clin Physiol 12:487–496

Jensen-Urstad M, Ahlborg G, Sahlin K (1993) High lactate and NH3 release during arm vs. leg exercise is not due to beta-adrenoceptor stimulation. J Appl Physiol 74:2860–2867

Jones DP (1986) Intracellular diffusion gradients of O2 and ATP. Am J Phys 250:C663–C675

Kabbara AA, Nguyen LT, Stephenson GM, Allen DG (2000) (2000). Intracellular calcium during fatigue of cane toad skeletal muscle in the absence of glucose. J Muscle Res Cell Motil 21(5): 481–489

Karlsson J, Saltin B (1971) Diet, muscle glycogen, and endurance performance. J Appl Physiol 31(2):203–206

Karlsson J, Nordesjö LO, Saltin B (1974) Muscle glycogen utilization during exercise after physical training. Acta Physiol 90:210–217

Katayama K, Goto K, Ishida K, Ogita F (2009) Substrate utilization during exercise and recovery at moderate altitude. Metabolism 59(7):959–966

Kennedy BG, Lunn G, Hoffman JF (1986 Jan) Effects of altering the ATP/ADP ratio on pump-mediated Na/K and Na/Na exchanges in resealed human red blood cell ghosts. J Gen Physiol 87(1):47–72

Klemm D, Heublein B, Fink H-P, Bohn A (2005) Cellulose: fascinating biopolymer and sustainable raw material. Angew Chem Int Ed Engl 44(22):3358–3393

Koh HE, Ørtenblad N, Winding KM, Hellsten Y, Mortensen SP, Nielsen J (2018) High-intensity interval, but not endurance training induces muscle fiber type-specific subsarcolemmal lipid droplet size reduction in type 2 diabetic patients. Am J Physiol Endocrinol Metab 315(5):E872–E884

Koopman R, Manders RJ, Jonkers RA, Hul GB, Kuipers H, van Loon LJ (2006) Intramyocellular lipid and glycogen content are reduced following resistance exercise in untrained healthy males. Eur J Appl Physiol 96:525–534
Korge P, Campbell KB (1995) The importance of ATPase microenvironment in muscle fatigue: a hypothesis. Int J Sports Med 16:172–179
Krogh A, Lindhard J (1920) The relative value of fat and carbohydrate as sources of muscular energy: with appendices on the correlation between standard metabolism and the respiratory quotient during rest and work. Biochem J 14:290–363
Lamb GD, Stephenson DG (1994) Effects of intracellular pH and [Mg2+] on excitation-contraction coupling in skeletal muscle fibres of the rat. J Physiol 478(Pt 2):331–339
Lamb GD, Stephenson DG (2018) Measurement of force and calcium release using mechanically skinned fibers from mammalian skeletal muscle. J Appl Physiol 125(4):1105–1127
Lambert EV, Speechly DP, Dennis SC, Noakes TD (1994) Enhanced endurance in trained cyclists during moderate intensity exercise following 2 weeks adaptation to a high fat diet. Eur J Appl Physiol Occup Physiol 69:287–293
Larsen S, Ara I, Rabol R, Andersen JL, Boushel R, Dela F, Helge JW (2009) Are substrate use during exercise and mitochondrial respiratory capacity decreased in arm and leg muscle in type 2 diabetes? Diabetologia 52:1400–1408
Levy B, Gibot S, Franck P, Cravoisy A, Bollaert PE (2005) Relation between muscle Na+K+ ATPase activity and raised lactate concentrations in septic shock: a prospective study. Lancet 365(9462):871–875
Lucia A, Nogales-Gadea G, Perez M et al (2008) McArdle disease: what do neurologists need to know? Nat Clin Pract Neurol 4(10):568–577
Lundby C, Van Hall G (2002) Substrate utilization in sea level residents during exercise in acute hypoxia and after 4 weeks of acclimatization to 4100 m. Acta Physiol Scand 176(3):195–201
Lundby C, Calbet JAL, van Hall G, Saltin B, Sander M (2004) Pulmonary gas exchange at maximal exercise in Danish lowlanders during 8 wk of acclimatization to 4100 m and in high-altitude Aymara natives. Am J Phys Regul Integr Comp Phys 287(5):R1202–R1208
Macdonald WA, Ørtenblad N, Nielsen OB (2007) Energy conservation attenuates the loss of skeletal muscle excitability during intense contractions American. Journal of Physiology (Endocrin and Metab) 292:E771–E778
MacDougal JD, Ray S, Slae DG, McCartney N, Lee P, Garner S (1999) Muscle substrate utilization and lactate production during weightlifting. Can J Appl Physiol 24(3):209–215
MacLeod KT (1989) Effects of hypoxia and metabolic inhibition on the intracellular sodium activity of mammalian ventricular muscle. J Physiol 416:455–468
Madsen K, Pedersen PK, Rose P, Richter EA (1990) Carbohydrate supercompensation and muscle glycogen utilization during exhaustive running in highly trained athletes. Eur J Appl Physiol Occup Physiol 61:467–472
Marchand I, Chorneyko K, Tarnopolsky M, Hamilton S, Shearer J, Potvin J, Graham TE (2002) Quantification of subcellular glycogen in resting human muscle: granule size, number, and location. J Appl Physiol 93:1598–1607
Marchand I, Tarnopolsky M, Adamo KB, Bourgeois JM, Chorneyko K, Graham TE (2007) Quantitative assessment of human muscle glycogen granules size and number in subcellular locations during recovery from prolonged exercise. J Physiol 580(2):617–628
Margaria R, Cerretelli P, Mangili F (1964) Balance and kinetics of anaerobic energy release during strenuous exercise in man. J Appl Physiol 19:623–628
Marquet LA, Brisswalter J, Louis J, Tiollier E, Burke LM, Hawley JA, Hausswirth C (2016a) Enhanced endurance performance by periodization of carbohydrate intake: "sleep low" strategy. Med Sci Sports Exerc 48:663–672
Marquet LA, Hausswirth C, Molle O, Hawley JA, Burke LM, Tiollier E, Brisswalter J (2016b) Periodization of carbohydrate intake: short- term effect on performance. Nutrients 8:755

Matu J, Gonzalez JT, Ispoglou T, Duckworth L, Deighton K (2018) The effects of hypoxia on hunger perceptions, appetite-related hormone concentrations and energy intake: a systematic review and meta-analysis. Appetite 125:98–108

Maunder E, Bradley HE, Deane CS, Hodgson AB, Jones M, Joanisse S, Turner AM, Breen L, Philp A, Wallis GA (2021) Effects of short term graded dietary carbohydrate intake on intramuscular and whole body metabolism during moderate intensity exercise. J Appl Physiol 131(1):376–387

McArdle B (1951) Myopathy due to a defect in muscle glycogen breakdown. Clin Sci 10:20

McBride A, Hardie DG (2009) AMP-activated protein kinase–a sensor of glycogen as well as AMP and ATP? Acta Physiol 196:99–113

McBride A, Ghilagaber S, Nikolaev A, Hardie DG (2009) The glycogen-binding domain on the AMPK β subunit allows the kinase to act as a glycogen sensor. Cell Metab 9:23–34

McClelland GB, Hochachka PW, Weber JM (1998) Carbohydrate utilization during exercise after high-altitude acclimation: a new perspective. Proc Natl Acad Sci 95:10288–10293

McGilvery RW (1975) The use of fuels for muscular work. In: Howald H et al (eds) Metabolic adaptation to prolonged physical exercise. Springer, Basel

McInerney P, Lessard SJ, Burke LM, Coffey VG, Lo Giudice SL, Southgate RJ, Hawley JA (2005) Failure to repeatedly supercompensate muscle glycogen stores in highly trained men. Med Sci Sports Exerc 37(3):404–411

Medbø JI (1993) Glycogen breakdown and lactate accumulation during high-intensity cycling. Acta Physiol Scand 149(1):85–89

Medbø JI, Jebens E, Noddeland H, Hanem S, Toska K (2006) Lactate elimination and glycogen resynthesis after intense bicycling. Scand J Clin Lab Invest 66(3):211–226

Melendez-Hevia E, Wadell TG, Shelton ED (1993) Optimization of molecular design in the evolution of metabolism: the glycogen molecule. Biochem J 295:477–483

Melzer W, Herrmann-Frank A, Lüttgau HC (1995) The role of Ca2+ ions in excitation-contraction coupling of skeletal muscle fibres. Biochim Biophys Acta 1241(1):59–116

Mercer RW, Dunham PB (1981 Nov) Membrane-bound ATP fuels the Na/K pump. Studies on membrane-bound glycolytic enzymes on inside-out vesicles from human red cell membranes. J Gen Physiol 78(5):547–568

Mitchell JB, Costill DL, Houmard JA, Fink WJ, Pascoe DD, Pearson DR (1989) Influence of carbohydrate dosage on exercise performance and glycogen metabolism. J Appl Physiol 67(5): 1843–1849

Mitchell JB, DiLauro PC, Pizza FX, Cavender DL (1997) The effect of preexercise carbohydrate status on resistance exercise performance. Int J Sport Nutr 7(3):185–196

Morton JP, Croft L, Bartlett JD, MacLaren DPM, Reilly T, Evans L et al (2009) Reduced carbohydrate availability does not modulate training-induced heat shock protein adaptations but does upregulate oxidative enzyme activity in human skeletal muscle. J Appl Physiol 106: 1513–1521

Morton RW, Sonne MW, Zuniga AF, Mohammad IYZ, Jones A, McGlory C, Keir PJ, Potvin JR, Philips SM (2019) Muscle fibre activation is unaffected by load and repetition duration when resistance exercise is performed to task failure. J Physiol 597:4601–4613

Nakamura-Tsuruta S, Yasuda M, Nakamura T, Shinoda E, Furuyashiki T, Kakutani R, Takata H, Kato Y, Ashida H (2012) Comparative analysis of carbohydrate-binding specificities of two anti-glycogen monoclonal antibodies using ELISA and surface plasmon resonance. Carbohydr Res 350:49–54

Nielsen J, Ørtenblad N (2013) Physiological aspects of the subcellular localization of glycogen in skeletal muscle (review). Appl Physiol Nutr Metab 38(2):91–99

Nielsen OB, Ørtenblad N, Lamb GD, Stephenson DG (2004) Excitability of the T-tubular system in rat skeletal muscle: roles of K+ and Na+ gradients and Na+-K+ pump activity. J Physiol 557: 133–146

Nielsen J, Schrøder HD, Rix CG, Ørtenblad N (2009) Distinct effects of subcellular glycogen localization on tetanic relaxation time and endurance in mechanically skinned rat skeletal muscle fibres. J Physiol 587:3679–3690

Nielsen J, Mogensen M, Vind BF, Sahlin K, Højlund K, Schrøder HD, Ørtenblad N (2010) Increased subsarcolemmal lipids in type 2 diabetes: effect of training on localization of lipids, mitochondria, and glycogen in sedentary human skeletal muscle. Am J Physiol Endocrinol Metab 298:E706–E713

Nielsen J, Holmberg HC, Schrøder HD, Saltin B, Ørtenbald N (2011) Human skeletal muscle glycogen utilization in exhaustive exercise: role of subcellular localization and fibre type. J Physiol 589(11):2871–2885

Nielsen J, Cheng AJ, Ørtenblad N, Westerblad H (2014) Subcellular distribution of glycogen and decreased tetanic Ca2+ in fatigued single intact mouse muscle fibres. J Physiol 592(9): 2003–2012

Norman B, Sollevi A, Jansson E (1988) Increased IMP content in glycogen-depleted muscle fibres during submaximal exercise in man. Acta Physiol Scand 133(1):97–100

Odland LM, Heigenhauser GJF, Wong D, Hollidge-Horvat MG, Spriet LL (1998) Effects of increased fat availability on fat-carbohydrate interaction during prolonged exercise in men. Am J Phys Regul Integr Comp Phys 274(4):R894–R902

O'Hara JP, Woods DR, Mellor A, Boos C, Gallagher L, Tsakirides C, Arjomandkhah NC, Holdsworth DA, Cooke CB, Morrison DJ, Preston T, King RF (2017) A comparison of substrate oxidation during prolonged exercise in men at terrestrial altitude and normobaric normoxia following the coingestion of 13C glucose and 13C fructose. Physiol Rep 5(1):e13101

Okamoto K, Wang W, Rounds J, Chambers EA, Jacobs DO (2001) ATP from glycolysis is required for normal sodium homeostasis in resting fast-twitch rodent skeletal muscle. Am J Physiol Endocrinol Metab 281:E479–E488

Olsson K-E, Saltin B (1970) Variation in total body water with muscle glycogen changes in man. Acta Physiol Scand 80:11–18

Ørngreen MC, Jeppesen TD, Taivassalo T, Hauerslev S, Preisler N, Heinicke K, Haller RG, Vissing J, van Hall G (2015) Lactate and energy metabolism during exercise in patients with blocked Glycogenolysis (McArdle disease). J Clin Endocrinol Metab 100(8):E1096–E1104

Ørtenblad N, Stephenson GD (2003) A novel signalling pathway originating in mitochondria modulates muscle membrane excitability. J Physiol 548:139–145

Ørtenblad N, Nielsen J (2015) Muscle glycogen and cell function–location, location, location. Scand J Med Sci Sports Suppl 4:34–40

Ørtenblad N, Macdonald WA, Sahlin K (2009) Glycolysis in contracting rat skeletal muscle is controlled by factors related to energy state. Biochem J 420:161–168

Ørtenblad N, Nielsen J, Saltin B, Holmberg HC (2011) Role of glycogen availability in sarcoplasmic reticulum Ca2+ kinetics in human skeletal muscle. J Physiol 589(3):711–725

Ørtenblad N, Westerblad H, Nielsen J (2013) Muscle glycogen stores and fatigue (review). J Physiol 591(18):4405–4413

Ørtenblad N, Nielsen J, Boushel R, Söderlund K, Saltin B, Holmberg H-C (2018) The muscle fiber profiles, mitochondrial content and enzyme activities of the exceptionally well-trained arm and leg muscles of elite cross-country skiers. Front Physiol 9:1031

Palladin AV (1945) The biochemistry of muscle training. Science 102:576–578

Parolin ML, Chesley A, Matsos MP, Spriet LL, Jones NL, Heigenhauser GJF (1999) Regulation of skeletal muscle glycogen phosphorylase and PDH during maximal intermittent exercise. Am J Phys 277:E890–E900

Pascoe DD, Costill DL, Fink WJ, Robergs RA, Zachwieja JJ (1993) Glycogen resynthesis in skeletal muscle following resistive exercise. Med Sci Sports Exerc 25(3):349–354

Passonneau JV, Lowry OH (1993) Enzymatic analysis, vol 2. The Humana Press Inc., New Jersey, USA

Pearse AGE (1961) Carbohydrates. In: Histochemistry, theoretical and applied, Boston, Little, Brown and Company, pp 228–280

Pedersen TH, Clausen T, Nielsen OB (2003) Loss of force induced by high extracellular [K^+] in rat muscle: effect of temperature, lactic acid and beta2-agonist. J Physiol 551:277–286
Pederson BA, Schroeder JM, Parker GE, Smith MW, DePaoli-Roach AA, Roach PJ (2005) Glucose metabolism in mice lacking muscle glycogen synthase. Diabetes 54:3466–3473
Philipson KD, Nishimoto AY (1983) ATP-dependent Na+ transport in cardiac sarcolemmal vesicles. Biochim Biophys Acta 733:133–141
Pilegaard H, Keller C, Steensberg A, Helge JW, Pedersen BK, Saltin B, Darrell Neufer P (2002) Influence of pre-exercise muscle glycogen content on exercise-induced transcriptional regulation of metabolic genes. J Physiol 541:261–271
Pilegaard H, Osada T, Andersen LT, Helge JW, Saltin B, Neufer PD (2005) Substrate availability and transcriptional regulation of metabolic genes in human skeletal muscle during recovery from exercise. Metabolism 54:1048–1055
Pitsiladis YP, Maughan RJ (1999) The effects of exercise and diet manipulation on the capacity to perform prolonged exercise in the heat and in the cold in trained humans. J Physiol 517:919–930
Poole DC, Rossiter HB, Brooks GA, Gladden LB (2020) The anaerobic threshold: 50+ years of controversy. J Physiol JP279963
Psilander N, Frank P, Flockhart M, Sahlin K (2013) Exercise with low glycogen increases PGC-1α gene expression in human skeletal muscle. Eur J Appl Physiol 113:951–963
Rauch LH, Rodger I, Wilson GR, Belonje JD, Dennis SC, Noakes TD, Hawley JA (1995) The effects of carbohydrate loading on muscle glycogen content and cycling performance. Int J Sport Nutr 5(1):25–36
Roach PJ, DePaoli-Roach AA, Hurley TD, Tagliabracci VS (2012) Glycogen and its metabolism: some new developments and old themes. Biochem J 441:763–787
Robergs RA, Pearson DR, Costill DD, Fink WJ, Pascoe DD, Benedict MA, Lambert CP, Zachweija JJ (1991) Muscle glycogenolysis during different intensities of weight-resistance exercise. J Appl Physiol 70(4):1700–1706
Roberts AC, Butterfield GE, Cymerman A, Reeves JT, Wolfel EE, Brooks GA (1996a) Acclimatization to 4,300-m altitude decreases reliance on fat as a substrate. J Appl Physiol 81:1762–1771
Roberts AC, Reeves JT, Butterfield GE et al (1996b) Altitude and b-blockade augments glucose utilization during submaximal exercise. J Appl Physiol 80:605–615
Roepstorff C, Halberg N, Hillig T, Saha AK, Ruderman NB, Woj- taszewski JF, Richter EA, Kiens B. (2005) Malonyl-CoA and carnitine in regulation of fat oxidation in human skeletal muscle during exercise. Am J Physiol Endocrinol Metab 288:E133–E142
Romijn JA, Coyle EF, Sidossis LS, Gastaldelli A, Horowitz JF, Endert E, Wolfe RR (1993) Regulation of endogenous fat and carbohydrate metabolism in relation to exercise intensity and duration. Am J Phys 265(3 Pt 1):E380–E391
Rowlands DS, Hopkins WG (2002) Effects of high-fat and high-carbohydrate diets on metabolism and performance in cycling. Metab Clin Exp 51(6):678–690
Rybicka KK (1996) Glycosomes--the organelles of glycogen metabolism. Tissue Cell 28:253–265
Sahlin K, Söderlund K, Tonkonogi M, Hirakoba K (1997) Phosphocreatine content in single fibers of human muscle after sustained submaximal exercise. Am J Phys 273(1 Pt 1):C172–C178
Sahlin K, Tonkonogi M, Söderlund K (1998) Energy supply and muscle fatigue in humans. Acta Physiol Scand 162(3):261–266
Saks V, Beraud N, Wallimann T (2008) Metabolic compartmentation–a system level property of muscle cells. Int J Mol Sci 9:751–767
Saltin B (1973) Metabolic fundamentals in exercise. Med Sci Sports 5(3):137–146
Saltin B, Karlsson J (1971) Muscle glycogen utilization during work of different intensities. In: Pernow B, Saltin B (eds) Muscle metabolism during exercise, vol 11. Springer, US, pp 289–299
Saltin B, Nazar K, Costill DL, Stein E, Jansson E, Essen B, Gollnick PD (1976) The nature of the training response: peripheral and central adaptations of one-legged exercise. Acta Physiol Scand 96:289–305

Samuelson H, Moberg M, Apró W, Ekblom B, Blomstrand E (2016) Intake of branched-chain or essential amino acids attenuates the elevation in muscle levels of PGC-1α4 mRNA caused by resistance exercise. Am J Physiol Endocrinol Metab 311(1):246–251
Santalla A, Nogales-Gadea G, Ørtenblad N, Brull A, de Luna N, Pinós T, Lucia A (2014) McArdle disease: a unique study model in sports medicine. (review). Sports Med 44(11):1531–1544
Sawyer JC, Wood RJ, Davidson PW, Collins SM, Matthews TD, Gregory SM, Paolone VJ (2013) Effects of a short-term carbohydrate restricted diet on strength and power performance. J Strength Cond Res 27:2255–2262
Schrier SL (1966 Jan) Organization of enzymes in human erythrocyte membranes. Am J Phys 210(1):139–145
Sear RP (2019) Diffusionphoresis in cells: a general nonequilibrium, nonmotor mechanism for the metabolism-dependent transport of particles in cells. Phys Rev let 122:128101
Sejersted OM, Sjøgaard G (2000) Dynamics and consequences of potassium shifts in skeletal muscle and heart during exercise. Physiol Rev 80:1411–1481
Sherman WM, Costill DL, Fink WJ, Miller JM (1981) Effect of exercise-diet manipulation on muscle glycogen and its subsequent utilization during performance. Int J Sports Med 2:114–118
Shearer J, Graham T (2004) Novel aspects of skeletal muscle glycogen and its regulation during rest and exercise. Exerc Sport Sci Rev 32:120–126
Shiose K, Yamada Y, Motonaga K, Sagayama H, Higaki Y, Tanaka H, Takahashi H (2016) Segmental extracellular and intracellular water distribution and muscle glycogen after 72-h carbohydrate loading using spectroscopic techniques. J Appl Physiol 121(1):205–211
Sjöström M, Fridén J, Ekblom B (1982a) Fine structural details of human muscle fibres after fibre type specific glycogen depletion. Histochemistry 76:425–438
Sjöström M, Kidman S, Larsén KH, Ängquist KA (1982b) Z- and M-band appearance in different Histochemically defined types of human skeletal muscle fibers. J Histochem Cytochem 30(1): 1–11
Skurat AV, Segvich DM, DePaoli-Roach AA, Roach PJ (2017) Novel method for detection of glycogen in cells. Glycobiology 27(5):416–424
Smith JW, Zachwieja JJ, Péronnet F, Passe DH, Massicotte D, Lavoie C, Pascoe DD (2010) Fuel selection and cycling endurance performance with ingestion of [13C]glucose: evidence for a carbohydrate dose response. J Appl Physiol 108(6):1520–1529
Srere PA (1967) Enzyme concentrations in tissues. Science 158(3803):936–937
Steinberg GR, Watt MJ, McGee SL, Chan S, Hargreaves M, Febbraio MA et al (2006) Reduced glycogen availability is associated with increased AMPKα2 activity, nuclear AMPKα2 protein abundance, and GLUT4 mRNA expression in contracting human skeletal muscle. Appl Physiol Nutr Metab 31:302–312
Stellingwerff T (2012) Case study: nutrition and training periodization in three elite marathon runners. Int J Sport Nutr Exer Metab 22(5):392–400
Stellingwerff T, Spriet LL, Watt MJ, Kimber NE, Hargreaves M, Hawley JA, Burke LM (2006) Decreased PDH activation and glycogenolysis during exercise following fat adaptation with carbohydrate restoration. American Journal of Physiology-Endocrinology and Metabolism 290(2):E380–E388
Stephenson DG (1996) Molecular cogs in machina carnis. Clin Exp Pharmacol Physiol 23(10–11): 898–907
Stephenson DG, Nguyen LT, Stephenson GM (1999) Glycogen content and excitation-contraction coupling in mechanically skinned muscle fibres of the cane toad. J Physiol 519:177–187
Sun RC, Dukhande VV, Zhou Z, Young LEA, Emanuelle S, Brainson CF, Gentry MS (2019) Nuclear glycogenolysis modulates histone acetylation in human non-small cell lung cancers. Cell Metab 30(5):903–916.e7
Symons JD, Jacobs I (1989) High-intensity exercise performance is not impaired by low intramuscular glycogen. Med Sci Sports Exerc 21(5):550–557

Tammineni ER, Kraeva N, Figueros L, Manno C, Ibarra CA, Klip A, Riazi S, Rio... Intracellular calcium leak lowers glucose storage in human muscle, promoting hyperglyc... and diabetes. eLife 9:e53999

Tarnopolsky LJ, MacDougall JD, Atkinson SA, Tarnopolsky MA, Sutton JR (1990) Gender differences in substrate for endurance exercise. J Appl Physiol 68:302–308

Taylor R, Price TB, Rothman DL, Shulman RG, Shulman GI (1992) Validation of 13C NMR measurement of human skeletal muscle glycogen by direct biochemical assay of needle biopsy samples. Magn Reson Med 27(1):13–20

Tesch PA, Colliander EB, Kaiser P (1986) Muscle metabolism during intense, heavy resistance exercise. Eur J Appl Physiol 55:362–366

Tesch PA, Ploutz-Snyder LL, Yström L, Castro MJ, Dudley GA (1998) Skeletal muscle glycogen loss evoked by resistance exercise. J Strength Cond Res 12(2):67–73

Testoni G, Duran J, García-Rocha M, Vilaplana F, Serrano AL, Sebastián D, López-Soldado I, Sullivan MA, Slebe F, Vilaseca M, Munoz-Cánoves P, Guinovart JJ (2017) Lack of glycogenin causes glycogen accumulation and muscle function impairment. Cell Metab 26:256–266

Tomcik KA, Camera DM, Bone JL, Ross ML, Jeacocke NA, Tachtsis B, Senden J, LJC VANL, Hawley JA and Burke LM. (2018) Effects of Creatine and carbohydrate loading on cycling time trial performance. Med Sci Sports Exerc 50:141–150

Tsintzas OK, Williams C, Boobis L, Greenhaff P (1995) Carbohydrate ingestion and glycogen utilization in different muscle fibre types in man. J Physiol 489(Pt 1):243–250

Tsintzas OK, Williams C, Boobis L, Greenhaff P (1996) Carbohydrate ingestion and single muscle fiber glycogen metabolism during prolonged running in men. J Appl Physiol 81(2):801–809

Tsintzas K, Williams C, Constantin-Teodosiu D, Hultman E, Boobis L, Clarys P, Greenhaff P (2001) Phosphocreatine degradation in type I and type II muscle fibres during submaximal exercise in man: effect of carbohydrate ingestion. J Physiol 537(1):305–311

Van Hall G, Jensen-Urstad M, Rosdahl H, Holmberg HC, Saltin B, Calbet JA (2003) Leg and arm lactate and substrate kinetics during exercise. Am J Physiol Endocrinol Metab 284(1):E193–E205

van Loon LJ, Greenhaff PL, Constantin-Teodosiu D, Saris WH, Wagenmakers AJ (2001) The effects of increasing exercise intensity on muscle fuel utilisation in humans. J Physiol 536(Pt 1): 295

Vandenberghe K, Hespel P, Vanden Eynde B, Lyssens R, Richter EA (1995) No effect of glycogen level on glycogen metabolism during high intensity exercise. Med Sci Sports Exerc 27:1278–1283

Vigh-Larsen JF, Ørtenblad N, Spriet LL, Overgaard K, Mohr M (2021) Muscle glycogen metabolism and high-intensity exercise performance–a narrative review. Sports Med 51(9):1855–1874

Virágh S, Szabó E, Challice CE (1982) Glycogen-containing lysosomes and glycogen loss in the cardiomyocytes of embryonic and neonatal mice. Adv Myocardiol 3:553–561

Visuttijai K, Hedberg-Oldfors C, Thomsen C, Glamuzina E, Kornblum C, Tasca G, Hernandez-Lain A, Sandstedt J, Dellgren G, Roach P, Oldfors A (2020) Glycogenin is deispensable for glycogen synthesis in human muscle, and glycogen deficiency causes polyglucosan storage. J Clin Endocrinol Metab 105:557–566

Vukovich MD, Costill DL, Hickey MS, Trappe SW, Cole KJ, Fink WJ (1993) Effect of fat emulsion infusion and fat feeding on muscle glycogen utilization during cycle exercise. J Appl Physiol 75(4):1513–1518

Walker JL, Heigenhauser GJ, Hultman E, Spriet LL (2000) Dietary carbohydrate, muscle glycogen content, and endurance performance in well-trained women. J Appl Physiol (Bethesda, Md: 1985) 88:2151–2158

Wallimann T, Eppenberger HM (1985) Localization and function of M-line-bound creatine kinase. J Muscle Res Cell Motility 6:239–285

Wanson JC, Drochmans P (1968) Rabbit skeletal muscle glycogen. A morphological and biochemical study of glycogen betaparticles isolated by the precipitation-centrifugation method. J Cell Biol 38:130–150
Wanson JC, Drochmans P (1972) Role of the sarcoplasmic reticulum in glycogen metabolism. J Cell Biol 54:206–224
Watanabe D, Wada M (2019) Effects of reduced muscle glycogen on excitation–contraction coupling in rat fast-twitch muscle: a glycogen removal study. J Muscle Res Cell Motil
Widrick JJ, Costill DL, Fink WJ, Hickey MS, McConell GK, Tanaka H (1993) Carbohydrate feedings and exercise performance: effect of initial muscle glycogen concentration. J Appl Physiol (Bethesda, Md: 1985) 74:2998–3005
Wojtaszewski JFP, MacDonald C, Nielsen JN, Hellsten Y, Hardie DG, Kemp BE et al (2003) Regulation of 5′AMP- activated protein kinase activity and substrate utilization in exercising human skeletal muscle. Am J Physiol Endocrinol Metab 284:E813–E822
Xirouchaki CE, Mangiafico SP, Bate K, Ruan Z, Huang AM, Tedjosiswoyo BW, Lamont B, Pong W, Favaloro J, Blair AR, Zajac JD, Proietto J, Andrikopoulos S (2016) Impaired glucose metabolism and exercise capacity with muscle-specific glycogen synthase 1 (gys1) deletion in adult mice. Mol Metab 5:221–232
Xu KY, Becker LC (1998) Ultrastructural localization of glycolytic enzymes on sarcoplasmic reticulum vesticles. J Histochem Cytochem 46:419–427
Yaspelkis BB, Patterson JG, Anderla PA, Ding Z, Ivy JL (1993) Carbohydrate supplementation spares muscle glycogen during variable-intensity exercise. J Appl Physiol 75(4):1477–1485
Yeo WK, Paton CD, Garnham AP, Burke LM, Carey AL, Hawley JA (2008) Skeletal muscle adaptation and performance responses to once a day versus twice every second day endurance training regimens. J Appl Physiol 105:1462–1470
Yeo WK, McGee SL, Carey AL, Paton CD, Garnham AP, Hargreaves M, Hawley JA (2010) Acute signalling responses to intense endurance training commenced with low or normal muscle glycogen. Exp Physiol 95:351–358
Young AJ, Young PM, McCullough RE, Moore LG, Cymerman A, Reeves JT (1991) Effect of betaadrenergic blockade on plasma lactate concentration during exercise at high altitude. Eur J Appl Physiol Occup Physiol 63:315–322
Zuntz N (1896) Über die Rolle des Zuckers im tierischen Stoffwechsel. Arch Physiol:538–542

Chapter 6
Exercise-Regulated Skeletal Muscle Glucose Uptake

Thomas E. Jensen, Jonas R. Knudsen, Carlos Henriquez-Olguin, Lykke Sylow, Glenn McConell, and Erik A. Richter

Abstract Muscle glucose uptake during exercise is regulated by a coordinated increase in glucose delivery (via increased blood flow and glucose moving out of the capillaries into the interstitial space), by facilitated glucose transport into the myocytes and by intramyocellular metabolism. The facilitative glucose transporter GLUT4 is translocated to the sarcolemma and the t-tubules, and GLUT4 translocation is essential for glucose transport into the myocytes during exercise. Several molecular mechanisms have been proposed to regulate insulin-independent GLUT4 translocation during in vivo conditions, but the regulation of both exercise-stimulated GLUT4 translocation and the integrative glucose uptake process by exercise remains incompletely understood. GLUT4 intrinsic transporter activity may also be regulated during exercise although there is no firm evidence for this. The multitude of mechanisms involved in muscle glucose uptake stimulation during exercise ensure the delivery of easily combustible fuel to the working muscles.

Keywords Exercise · Glucose uptake · Human · Skeletal muscle · GLUT-4 · Metabolism

---

"Thomas E. Jensen" and "Erik A. Richter" are co-corresponding authors.

T. E. Jensen (✉) · J. R. Knudsen · C. Henriquez-Olguin · L. Sylow · E. A. Richter (✉)
August Krogh Section for Molecular Physiology, Department of Nutrition, Exercise and Sports, Faculty of Science, University of Copenhagen, Copenhagen, Denmark
e-mail: TEJensen@nexs.ku.dk; ERichter@nexs.ku.dk

G. McConell
Institute for Health and Sport, Victoria University, Melbourne, VIC, Australia

© The Author(s), under exclusive license to Springer Nature Switzerland AG 2022
G. McConell (ed.), *Exercise Metabolism*, Physiology in Health and Disease,
https://doi.org/10.1007/978-3-030-94305-9_6

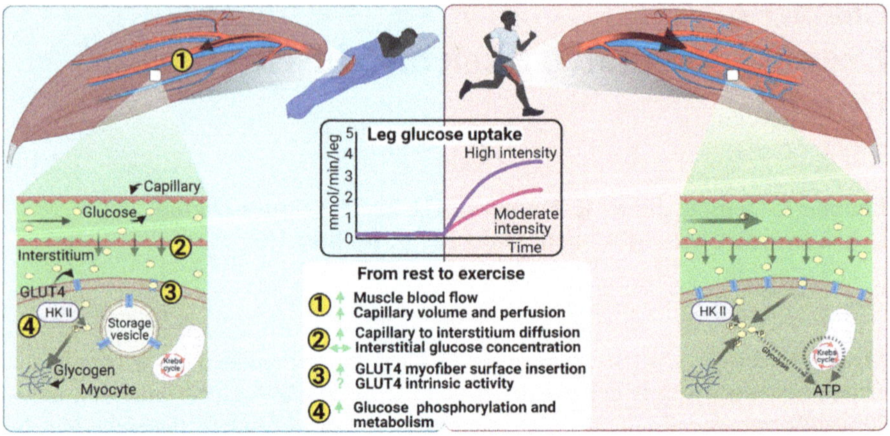

Fig. 6.1 Human glucose uptake regulation during exercise. Skeletal muscle glucose uptake results from four tightly coupled steps including (1) glucose delivery to the capillaries, (2) diffusion from the capillary to the interstitium, (3) transport across the muscle fiber surface (sarcolemma and t-tubules), and (4) intramyocellular phosphorylation of glucose and subsequent metabolism or storage. At rest (left half), the muscle glucose uptake is low due to low leg blood flow rates (0.3 to 0.4 l/min in overnight fasted humans), a low amount of muscle-fiber surface inserted GLUT4, and a net conversion of glucose to glycogen. Contractile activity (right half) increases the muscle glucose uptake in proportion to the exercise intensity and duration. This increase results from increased bulk leg blood flow (up to 7–10 l/min), increased capillary recruitment, and a greater amount of muscle fiber surface inserted GLUT4. During exercise, a switch occurs from net glycogen synthesis to glycogen breakdown. These processes provide substrate for ATP production by glycolysis and mitochondrial oxidative phosphorylation

6.1 Muscle Glucose Uptake During Exercise

Exercise is a powerful stimulator of glucose uptake in the working muscles. In humans, physiological experiments in the 1960s and 1970s showed that, as a general rule, muscle glucose uptake during exercise is a function of exercise duration and intensity (Fig. 6.1). This was demonstrated in studies of muscle glucose uptake in humans utilizing radiolabelled glucose tracers or arteriovenous glucose difference and blood flow measurements across active forearm and leg muscles (Ahlborg et al. 1974; Reichard et al. 1961; Sanders et al. 1964; Wahren et al. 1971; Whichelow et al. 1968; Havel et al. 1967). These studies were performed in males, but subsequent studies comparing females and males have shown no major sex difference in leg glucose uptake during exercise at the same relative intensity (Lundsgaard and Kiens 2014).

How big a contribution to energy expenditure during exercise does glucose uptake make? Studies measuring substrate turnover in endurance trained cyclists demonstrated that glucose utilization contributed between 10 and 18% of whole-body energy turnover during increasing exercise intensities (Romijn et al. 1993; Van Loon et al. 2001). However, in prolonged exercise, blood glucose could account for up to 40% of oxidative metabolism once muscle glycogen was depleted (Ahlborg et al. 1974; Wahren et al. 1971; Coyle et al. 1983). In contrast, during short-term high-intensity exercise, utilization of blood glucose is quantitatively unimportant, and glycogen is the major fuel (Katz et al. 1986a).

Mechanistically, glucose uptake in muscle during exercise depends upon at least four factors: Glucose delivery to the muscle capillaries, glucose transport out of the capillaries into the interstitial space, glucose transport across the sarcolemma and t-tubules into the cytoplasm, and glucose phosphorylation and subsequent further metabolism in the cell (Richter and Hargreaves 2013). It is noteworthy that three of these factors do not depend on GLUT4 translocation.

Glucose delivery, glucose transport out of the capillaries, transport across the muscle cell membrane, and glucose phosphorylation and subsequent further metabolism are increased markedly from rest to exercise, and therefore it is often difficult to determine the rate-limiting step for glucose utilization during any given exercise. The combined increase in these regulatory factors results in huge increases in muscle glucose uptake. In an overnight fasted individual, leg glucose uptake is very low at rest as reflected by the small femoral arterio-venous (A-V) glucose difference in the order of 0.05–0.1 mmol/l and the leg blood flow around 300–400 ml/min (Capaldo et al. 1999; DeFronzo et al. 1981; Kjaer et al. 1991; Kristiansen et al. 1996c; Roepstorff et al. 2005; Sjøberg et al. 2017). This comes to an uptake of 15–40 μmol/min/leg. Leg blood flow can reach 7–10 l/min per leg (Katz et al. 1986a; Calbet et al. 2007; Mortensen et al. 2005), and A-V difference may increase to 0.3 to 0.4 mmol/l during intense exercise (Katz et al. 1986a). Thus, leg glucose uptake may increase to 2100–4000 μmol/min/leg, in other words, up to an approximate 100-fold increase from resting values. During moderate exercise of 50–60% of maximal oxygen uptake, leg blood flow is around 5–6 l/min and the A-V difference for glucose 0.2 to 0.3 mM (Katz et al. 1986a; Roepstorff et al. 2005; Katz et al. 1991; Roepstorff et al. 2002). In this situation, the increase in leg glucose uptake is 25–50-fold, still a substantial increase. Leg glucose uptake has also been measured by positron emission tomography scanning. In one such experiment, leg glucose uptake was found to increase by approximately 20-fold from rest to exercise at 75% of VO2 max (Kemppainen et al. 2002). However, measurements were performed after exercise, and exercise intensity was not maximal. Therefore, since it is well-known that muscle glucose uptake decreases rapidly upon exercise cessation (Bangsbo et al. 1997), a 20-fold increase in leg glucose uptake after submaximal exercise does seem to confirm the large fold increase in leg glucose uptake during intense exercise.

6.2 Glucose Delivery

Glucose delivery to the muscle is increased by the marked increase in muscle blood flow during exercise (Joyner and Casey 2015). In addition, increased hepatic glucose production serves to replace the glucose taken up by the muscle and thereby maintains plasma glucose concentration. During exercise, the large increase in leg blood flow leads to increased capillary perfusion. In humans, this has been demonstrated by experiments with contrast-enhanced ultrasound which directly have shown large increases in muscle capillary volume and perfusion during exercise compared to rest (Sjøberg et al. 2011; Dawson et al. 2002). This increase in capillary perfusion is well matched to the increase in uptake of glucose by the muscle cells because it has been demonstrated that the interstitial muscle glucose concentration is maintained close to the plasma glucose concentration during exercise (MacLean et al. 1999) indicating that the rate of delivery of glucose into the interstitial space matches the rate of uptake into the myocytes. Direct evidence of the importance of muscle perfusion has been obtained in isolated perfused rat hindlimbs where glucose uptake during muscle contractions was dependent upon the perfusion rate (Schultz et al. 1977; Hespel et al. 1995).

Another aspect of glucose delivery is the plasma glucose concentration. Muscle glucose uptake follows Michaelis-Menten saturation kinetics with Km values reported to be 5 mM in dogs (Zinker et al. 1993) and 11 mM in humans (Richter 1996). Since plasma glucose concentration in heathy individuals is usually around 5 mM, it follows that changes in plasma glucose concentration will almost proportionally lead to changes in glucose uptake in turn indicating that at a fixed exercise intensity glucose supply is an important rate-limiting step in glucose uptake during exercise. This has been directly demonstrated in experiments when glucose was ingested during exercise resulting in higher plasma glucose concentration and higher glucose disposal (McConell et al. 1994). However, since plasma insulin concentrations are also increased with glucose ingestion, this will also tend to add to the effect of the increase in plasma glucose concentrations per se. Therefore, experiments with glucose ingestion show the combined effect of increasing plasma glucose and insulin. Furthermore, increased plasma insulin will also decrease plasma fatty acid concentrations which also tends to increase glucose uptake during exercise (Hargreaves et al. 1991).

6.3 Glucose Transport Out of the Capillaries

Glucose transport out of the muscle capillaries is thought to occur through the pores or slits between the endothelial cells without involvement of transport molecules. Experiments in the 1950s by Pappenheimer and co-workers indicated that escape from the capillaries is dependent upon the size of the molecules (Pappenheimer et al. 1951). Small molecules like glucose or urea can easily escape through the capillary pores, while large molecules like albumin cannot (Pappenheimer et al. 1951). This is different from the brain where the GLUT1 transporter is important for transcapillary glucose transport (Yazdani et al. 2019). GLUT1 is also expressed as the dominant endothelial glucose transporter in muscle capillaries (Rohlenova et al. 2018), and it is possible that these glucose transporters could contribute to transcapillary glucose transport also in the muscle. However, this remains to be studied. During exercise, the enlargement of capillary volume and perfusion markedly increases the surface available for glucose escape from the capillary into the interstitial space. Measurements of interstitial glucose concentration during exercise indicate that the interstitial glucose concentration is maintained at values similar to or even higher than at rest (MacLean et al. 1999) which in turn indicates that transport of glucose out of the capillaries is unlikely to be a major limitation for glucose uptake in muscle during exercise.

6.4 Glucose Transport into the Muscle Fibers

Once glucose has entered the interstitial space, glucose entry into the myocytes during exercise is dependent upon translocation of GLUT4 glucose transporters from an intracellular storage site to the sarcolemma and t-tubular membrane. This is supported by studies in GLUT4 knockout mice where exercise and insulin do not lead to increased glucose uptake in muscle (Fueger et al. 2007; Howlett et al. 2013; Ryder et al. 1999a; Zisman et al. 2000). In humans, attempts to quantify GLUT4 translocation in muscle during exercise have used membrane fractionation techniques or purified sarcolemmal vesicles (Kennedy et al. 1999; Kristiansen et al. 1996a, 1997). These studies have generally shown a twofold increase in sarcolemmal content of GLUT4 from rest to strenuous submaximal exercise which is far less than the 25–50-fold and up to 100-fold increase in glucose uptake during submaximal and maximal exercise, respectively (Roepstorff et al. 2002; Katz et al. 1986b; Katz et al. 1991). It would therefore seem that either the methodology to assess GLUT4 translocation in human muscle is vastly underestimating the true translocation or that there is a marked increase in GLUT4 intrinsic activity, as discussed recently (Richter 2020). Surface labelling to measure the presence of GLUT4 in the sarcolemmal surface membrane in intact muscle showed a fourfold increase in ex vivo incubated rat soleus muscle following maximal electrical stimulation which fully accounted for the similar fourfold increase in muscle glucose transport (Lund et al. 1995). Furthermore, using transmission electron microscopy in the rat muscle, ninefold and 23–29-fold increases in sarcolemmal and t-tubule GLUT4 translocation were demonstrated in response to in situ contraction (Ploug et al. 1998). Thus, some rodent studies showed a significantly higher GLUT4 translocation with electrically stimulated muscle contractions than studies in human muscle although only a doubling of GLUT4 at the sarcolemmal surface membrane was found in mouse muscle after submaximal but strenuous treadmill running (Sylow et al. 2016). Nevertheless, the translocation and perhaps increase in the ability of each GLUT4 transporter to transport glucose (intrinsic activity of GLUT4) result in a marked increase in muscle surface membrane permeability to glucose. This is in fact possible to estimate using published data of the interstitial muscle glucose concentration. Given that the interstitial glucose concentration does not change appreciably from rest to graded exercise in man (MacLean et al. 1999), it follows that influx and efflux to and from the interstitial space are of equal size. Since glucose efflux from the interstitial space during exercise is transported into the myocytes, it follows that the average permeability of the muscle membrane to glucose increases to a similar extent as the leg glucose uptake. Whether changes in GLUT4 intrinsic transporter activity contribute to increased glucose transport is far less studied compared to the number of studies of GLUT4 translocation. Still, there are studies that indicate that GLUT4 transport activity can be altered, for instance, by unmasking a GLUT4 epitope (Wang et al. 1996) or by binding of enzymes (Zaid et al. 2009) or glucose analogues (Shamni et al. 2017) or by vanadate (Kristiansen et al. 1996b). However, at this point, there is no clear concept of how and when GLUT4 transporter activity may change with insulin or with exercise/muscle contraction.

Another factor that might influence intrinsic activity of the GLUT4 transporter is temperature. Muscle temperature at rest depends on the external temperature and is typically 2–3 degrees centigrade below core temperature so often 34–35 degrees centigrade (González-Alonso et al. 1999a, b). During exercise, muscle temperature increases and may increase to approximately 40–41 degrees during intense exercise (González-Alonso et al. 1999a, b). This is a potential increase in muscle temperature of up to 6–7 degrees centigrade. It is likely that such a temperature increase may increase the transporter activity, but detailed investigations of temperature effects on GLUT4 transporter activity have not been performed. In addition, during muscle contraction and relaxation, the mechanical deformation of the muscle tissue may cause increased GLUT4 transporter activity. In fact, passive leg movement increases muscle glucose uptake (Mortensen et al. 2012), and stretch of mouse muscle also increases glucose uptake (Jensen et al. 2014a; Sylow et al. 2015). Together, these two stimuli may increase GLUT4 intrinsic activity during exercise.

6.5 Glucose Metabolism

Glucose metabolism is the fourth step in glucose utilization during exercise. Obviously, glucose metabolism is accelerated during the transition from rest to exercise, and as previously noted, the regulated steps in glucose utilization during exercise are mostly matched closely. This is shown by measurements of free glucose and glucose-6-phosphate in the muscle, both of which do not change appreciably from rest to moderate exercise (Katz et al. 1991), indicating that glucose metabolic flux is keeping up with glucose influx to the myocytes in moderate exercise. However, during intense exercise (Katz et al. 1986b) or when pre-exercise, muscle glycogen levels are high (Hespel and Richter 1990), glucose-6-phosphate concentrations may increase, and the ensuing inhibition of hexokinase II may lead to accumulation of free glucose inside the cell (Hespel and Richter 1990). This decrease in glucose gradient across the muscle cell then leads to decreased glucose uptake as shown in perfused rat muscle (Hespel and Richter 1990).

6.6 Evidence Linking Glucose Transport to GLUT4 in Transgenic Mice

In human skeletal muscle, GLUT4 mRNA and protein are highly expressed (Gaster et al. 2004; Gaster et al. 2000; Huang and Czech 2007). The observation that cytochalasin B, a non-specific direct inhibitor of GLUT transporter function, abolishes glucose uptake in skeletal muscle demonstrates the necessity of GLUTs for glucose uptake into muscle (Ryder et al. 1999a; Lawrence et al. 1992).

GLUT4 is considered the major GLUT isoform regulating glucose uptake into healthy adult skeletal muscle. The most compelling evidence for this comes from muscle-specific skeletal muscle GLUT4 KO mice (referred to as G4mKO#1 hereafter (Abel et al. 1999)). These mice display an ~80% reduced unstimulated glucose uptake into ex vivo incubated soleus and EDL muscles due to a ~ 95% reduction in GLUT4 content (Zisman et al. 2000). Furthermore, contraction-stimulated glucose uptake in both soleus and EDL muscles was nearly prevented in G4mKO#1 mice (Zisman et al. 2000). Similarly, whole-body GLUT4 KO mice exhibited ~20–40% lower basal glucose uptake and abrogated glucose uptake response to ex vivo electrically induced contraction (Katz et al. 1995), swimming exercise (Ryder et al. 1999a, b; Katz et al. 1995; Zierath et al. 1998; Stenbit et al. 1996), and treadmill running (Fueger et al. 2007). Additional evidence for GLUT4's essential role comes from another independently generated muscle-specific GLUT4 mouse model (referred to as G4mKO #2 hereafter (Kaczmarczyk et al. 2003)) with a 85–95% reduction in muscle GLUT4 protein displaying no increase in glucose uptake during moderate-high-intensity treadmill exercise (Howlett et al. 2013).

Whole-body GLUT4 heterozygous ($^{+/-}$) (Fueger et al. 2004b) mice, with 40–70% reduction in GLUT4 protein expression, have been used to investigate the relative role of GLUT4-dependent glucose transport vs. hexokinase II (HKII)-dependent phosphorylation in exercise-stimulated glucose uptake. HKII overexpression augmented treadmill exercise-stimulated glucose uptake in WT mice but not in whole-body GLUT4$^{+/-}$ mice (Fueger et al. 2004b), suggesting that exercise-stimulated muscle glucose uptake by GLUT4 is also dependent on glucose phosphorylation capacity in the working muscle. The glucose uptake response was even higher in mice overexpressing both GLUT4 and HKII but not in mice overexpressing GLUT4 alone (Fueger et al. 2004a), suggesting that GLUT4-dependent glucose transport is only limiting exercise-stimulated glucose uptake when the HKII-dependent glucose phosphorylation capacity is elevated.

Yet, GLUT4 deficient mouse models show a variable glucose uptake dependence on GLUT4 which might suggest compensation by other GLUTs (Katz et al. 1995; Stenbit et al. 1996; Kaczmarczyk et al. 2003; Fam et al. 2012; McMillin et al. 2017). For example, in G4mKO#1 mice (Abel et al. 1999), the stimulation of glucose uptake into plantaris muscle by 5 days of synergist-ablation overload was unaffected by the absence of GLUT4, while GLUT1, GLUT3, GLUT6, and GLUT10 protein levels increased (McMillin et al. 2017). Thus, other GLUT isoforms in skeletal muscle could facilitate glucose transport in the partial or complete absence of

GLUT4. Another example suggesting a potential contribution of other GLUT isoforms is the whole-body GLUT4 hypomorphic (Howlett et al. 2013) or heterozygous ($^{+/-}$) (Fueger et al. 2004b) mouse models that display 40–70% reduction in GLUT4 protein expression, yet show normal or even elevated treadmill-exercise-stimulated glucose uptake in most muscles (Howlett et al. 2013; Fueger et al. 2004b). However, the lack of glucose uptake phenotype with partial reductions in muscle GLUT4 content could also reflect a large spare capacity of GLUT4 under normal conditions. The extent to which other GLUTs contribute to physiological glucose uptake in the presence of GLUT4 is uncertain but is likely of quantitatively minor—if any—significance. Overall, these studies suggest a clear and major dependence of skeletal muscle glucose uptake on the GLUT4 isoform during contraction-stimulated conditions.

6.7 Exercise-Activated Signals Regulating Glucose Uptake

Although it is one of multiple processes contributing to in vivo glucose uptake, GLUT4 translocation is mechanistically the most studied process in relation to muscle glucose uptake in response to both insulin and exercise. Contracting muscle activates abundant and complex signaling cascades, many of which could potentially stimulate the redistribution of GLUT4 to the muscle cell surface to promote glucose uptake. The cell signaling is likely cell autonomous since contraction-stimulated glucose uptake is confined to the actively contracting musculature in vivo and can be potently increased in muscles isolated from the body, although a contribution from local auto/paracrine signaling has been suggested. The three stimuli that have historically received the most attention are Ca^{2+}, mechanical, and metabolic stress. These stimuli and some potential mediators are discussed below with a graphical summary of the discussed mechanisms in Fig. 6.2. For discussion of the regulation of exercise-stimulated muscle blood flow and metabolism, see (Sylow et al. 2017a). For more in-depth glucose uptake cell signaling reviews, see (Sylow et al. 2017a; Jensen et al. 2014; Sylow et al. 2014; Hong et al. 2014).

6.8 Ca^{2+} Signaling

Ca^{2+} regulates many aspects of vesicle trafficking in non-muscle cells and remains a candidate to regulate exercise/contraction-stimulated GLUT4 translocation. The most obvious Ca^{2+} source during contraction is ryanodine receptor (RyR)1 channel Ca^{2+} release from terminal cisternae of the sarcoplasmic reticulum (SR) during excitation-contraction coupling. However, dissociations have been reported in incubated mouse muscles between glucose transport stimulation by contractions, AMPK activators and/or passive stretch, and phosphorylation of the RyR1-dependent (dantrolene-blockable) Ca^{2+} readout eEF2 Thr57 (Jensen et al. 2014, c; Sylow

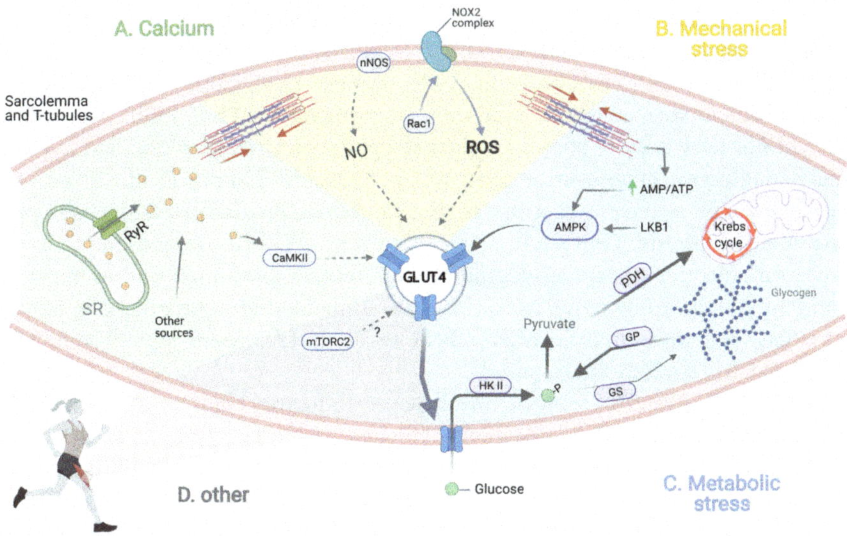

Fig. 6.2 Potential stimuli and mediators regulating exercise-stimulated glucose uptake in the skeletal muscle. (**a**) depolarization of the muscle fiber induces Ca^{2+} release from the sarcoplasmic reticulum to stimulate cross-bridge cycling and force production. Ca^{2+} may stem from other compartments as well. The Ca^{2+}/calmodulin-dependent kinase, CaMKII, in particular is activated by exercise in contracting muscle fibers and linked by multiple independent studies to glucose transport stimulation. CaMKII activation may be regulated by mechanisms other than Ca^{2+}, particularly NOX2-dependent ROS production. (**b**) Mechanical stress during muscle contraction activates stretch-responsive proteins including the small GTPase, Rac1, which is required for NOX2-dependent ROS generation and exercise-simulated GLUT4 translocation and glucose uptake. (**c**) Exercise-induced ATP hydrolysis increases the AMP/ATP and AMP/ADP ratio thereby activating the AMP-activated kinase, AMPK, to stimulate muscle glucose transport. (**d**) Rodent studies have implicated mTORC2 as a requirement for exercise-induced glucose uptake, but the underlying up- and downstream mechanisms are yet to be resolved. *SR* sarcoplasmic reticulum; *CaMKII* Ca^{2+}/calmodulin-dependent protein kinase II; *RYR* ryanodine receptor; *nNOS* nitric oxide synthase; *NO* nitric oxide; *NOX2* NADPH oxidase 2; *Rac1* p21-Rac1; *ROS* reactive oxygen species; *TBC1D1* TDC1 domain family member 1; *AMPK* AMP-activated protein kinase; *AMP* adenosine monophosphate; *ATP* adenosine triphosphate; *LKB1*, liver kinase B1; *GP* glycogen phosphorylase; *GS* glycogen synthase; *HK II* hexokinase II; *mTORC2* mammalian target of rapamycin complex 2

et al. 2017b), suggesting that RyR1 Ca^{2+} release is neither necessary nor sufficient for muscle glucose uptake. Consistent with these observations, a series of articles in incubated rat muscles systematically varied tetanic stimulation frequency and force development and concluded glucose uptake stimulation to correlate with events secondary to contraction (tension and metabolic perturbance) rather than electrical stimulation frequency (with each electrically induced depolarization eliciting SR Ca^{2+} release) (Ihlemann et al. 1999, 2000, 2001). However, there are other potential sources of Ca^{2+} during muscle contraction. Among these, Ca^{2+} entry from the extracellular space is a known requirement for vesicle exocytosis in many cell types and therefore might regulate GLUT4 exocytosis. Indeed, near-surface-membrane Ca^{2+} in adult mouse muscle increases with insulin, presumably via transient receptor potential channel (TRPC)3 Ca^{2+} channels, and TRPC3 co-localizes with GLUT4 along t-tubuli (Bruton et al. 1999; Lanner et al. 2006, 2009). However, pharmacological stimulation or inhibition of extracellular Ca^{2+} influx was neither sufficient nor necessary to stimulate glucose uptake or for contraction and hypoxia-stimulated glucose uptake in isolated rodent muscles (Lanner et al. 2006). This suggests that cell surface Ca^{2+} entry is specific to insulin-stimulated glucose uptake in the skeletal muscle. One study suggested cyclic ADP-ribose and nicotinic acid adenine dinucleotide phosphate-regulated Ca^{2+} release, presumably from the sarcoplasmic reticulum, to be required for exercise-stimulated glucose uptake in isolated mouse gastrocnemius muscle fibers (Park et al. 2015). More studies into these alternative Ca^{2+} sources are required.

Regardless of the Ca^{2+} source, the hetero-multimeric kinase Ca^{2+}/calmodulin-activated protein kinase II (CaMKII) is activated by human (Rose et al. 2006) and rodent adult muscle exercise/contraction (Rose et al. 2007) and by mild mechanical stress ex vivo and may, based on pharmacological inhibitor studies, be required for mouse adult muscle glucose uptake in response to contraction but not insulin (Park et al. 2015; Wright et al. 2004; Jensen et al. 2007; Witczak et al. 2010). However, mechanical stress could also increase Ca^{2+}-independent CaMKII activity via NOX2-dependent methionine oxidation (Wang et al. 2018; Erickson et al. 2008; Luczak and Anderson 2014). For discussion of other Ca^{2+} regulated proteins, see (Jensen et al. 2014; Sylow et al. 2017a). Overall, the role of various Ca^{2+} sources and potential mediators in stimulating muscle glucose uptake is not entirely clear and has not been studied during physiological exercise conditions.

6.9 Mechanical Stress

Mechanical stress elicited by passive stretching of incubated rodent muscles is sufficient to stimulate muscle glucose uptake (Jensen et al. 2014c; Ihlemann et al. 1999; Sakamoto et al. 2003; Chambers et al. 2009), and studies using myosin ATPase inhibitors that prevent cross-bridge cycling (Jensen et al. 2014c; Blair et al. 2009), or adjusting resting length to abolish force development during contraction (Ihlemann et al. 1999), suggest that mechanical stress is one of several necessary signals activated following contractile activity to stimulate glucose uptake.

The small rho family GTPase Rac1 is emerging as a mechanical stress-sensitive necessary signaling component regulating exercise/contraction-stimulated glucose uptake. Rac1 is activated by muscle contraction and exercise in both rodent and human skeletal muscle and by passive stretch in incubated mouse muscles (Sylow et al. 2013a, 2015). Studies in Rac1 knockout mice show that Rac1 is required for insulin, contraction, and passive stretch-stimulated glucose uptake ex vivo (Sylow et al. 2013a, b, 2015), in vivo insulin-stimulated GLUT4 translocation (Ueda et al. 2010) and glucose uptake (Raun et al. 2018), and in vivo treadmill exercise-stimulated GLUT4 translocation and glucose uptake (Sylow et al. 2017b, 2016). In L6 muscle cells, insulin stimulates GLUT4 translocation via Rac1-dependent cortical actin remodeling (Khayat et al. 2000). However, deletion of the two muscle isoforms of the Rac1-activated putative actin regulator p21-activated kinase (PAK)1 and 2 did not inhibit contraction-stimulated glucose transport (Møller et al. 2020), suggesting that PAK1 and PAK2 are not important downstream mediators of Rac1's stimulating effect on muscle glucose uptake during contractions.

Another recently proposed distinct mechanism involves Rac1-dependent activation of the superoxide-producing enzyme NADPH oxidase 2 (NOX2), found in the sarcolemma and t-tubules (Henríquez-Olguín et al. 2019). Combining redox histology and biosensors, NOX2 activation during moderate-intensity treadmill exercise was found to require Rac1 in mouse skeletal muscle (Henríquez-Olguin et al. 2019), consistent with NOX2 regulation in other cell types (refs in (Henríquez-Olguín et al. 2019)). Mice lacking another regulatory NOX2 subunit p47phox had no detectable increase in cytosolic H_2O_2 and displayed large reductions in exercise-stimulated sarcolemmal GLUT4 translocation and glucose uptake (Henríquez-Olguin et al. 2019) similar to Rac1 KO mice (Sylow et al. 2015, 2017b), suggesting that Rac1 regulates GLUT4 translocation via NOX2. Whether t-tubular GLUT4 translocation is equally affected by NOX2 inhibition is unknown. The potential downstream mediators of NOX2-stimulated GLUT4 translocation are unclear but presumably reside close to NOX2 at the cell surface.

Our NOX2-based working model seems to fit with previous observations that passive stretch and contraction-stimulated glucose uptake in incubated mouse muscles are sensitive to the exogenous antioxidants N-Acetylcysteine (NAC) and ebselen (Chambers et al. 2009; Merry et al. 2010b; Sandström et al. 2006). Worth noting, millimolar exogenous H_2O_2 stimulation of incubated rodent muscles is also sufficient to stimulate adult muscle glucose uptake (Cartee and Holloszy 1990) but unlike contraction appears to work via a PI3K-dependent mechanism (Higaki et al. 2008). In contrast, NAC infusion does not appear to affect exercise/contraction-stimulated glucose uptake in perfused rat hindlimb (Merry et al. 2010a) or bicycling humans (Merry et al. 2010d) except after blood flow-restricted exercise training that increased GLUT4 and NOS protein content in muscle (Christiansen et al. 2020). It remains unclear if NAC prevented the relevant localized redox signaling in vivo, and it seems reasonable to assume that the inhibition ex vivo was more complete. Thus, the interaction of general antioxidants with training mode and status in humans seems quite complex, making firm conclusions difficult regarding the role of exercise-stimulated cytosolic H_2O_2 production as a regulator of glucose uptake in humans.

6.10 Metabolic Stress and AMPK

Like tension development, the degree of metabolic stress correlated with muscle glucose uptake in incubated rodent muscles (Jensen et al. 2014c; Ihlemann et al. 1999, 2001; Blair et al. 2009). Furthermore, isolated metabolic stress induced by hypoxia (Mu et al. 2001) and mitochondrial poisoning (Fujii et al. 2005) is sufficient to increase sarcolemmal GLUT4 and glucose uptake into incubated mouse EDL muscle via a process fully dependent on AMP-activated protein kinase (AMPK) which has long been a candidate to mediate metabolic stress-stimulated glucose uptake during contraction.

As its name implies, AMPK is activated allosterically by increased ATP turnover increasing AMP/ATP and ADP/ATP ratios, which also promote its covalent activation by LKB1 and perhaps CaMKK-dependent phosphorylation and reduce its inactivation by phosphatases (Kjøbsted et al. 2018). Regardless of the mechanism, AMPK is intensely and time-dependently activated in skeletal muscle by human exercise (Wojtaszewski et al. 2000; Birk and Wojtaszewski 2006; Treebak et al. 2007; Chen et al. 2003). AMPK is a hetero-trimeric protein complex consisting of a catalytic α-subunit (α1 or 2) bound to regulatory β (β1 or 2) and γ-subunits (γ1, 2 or 3). In human skeletal muscle, there are three major complexes, α2β2γ1, α1β2γ1, and α2β2γ3, estimated to contribute ~ 65, 20%, and 15% of total AMPK complexes, among which α2β2γ3 is the only complex activated by short-term intense exercise, whereas the γ1 associated complexes are only activated during prolonged exercise (Birk and Wojtaszewski 2006).

Pharmacological AMPK activation is sufficient to stimulate muscle glucose uptake (Mu et al. 2001; Rhein et al. 2021; Jørgensen et al. 2021) and augments submaximal but not maximal contraction stimulated glucose uptake in incubated mouse EDL muscle (Bultot et al. 2016), arguing that AMPK activation mobilizes part of the contraction-regulated GLUT4 pool to increase glucose uptake. However, it is also increasingly clear that AMPK is not necessary during exercise/contraction to increase glucose uptake in AMPK-deficient mouse models (Sylow et al. 2017b; Maarbjerg et al. 2009; Kjøbsted et al. 2019; Hingst et al. 2020), except perhaps at the onset of contraction before reaching steady state (Abbott et al. 2011). Rather, AMPK is required to maintain elevated glucose uptake after cessation of exercise/contraction in mice (Kjøbsted et al. 2019) and to increase fat oxidation, thereby redirecting glucose away from oxidation and promoting glycogen synthesis in the recovery after exercise (Fritzen et al. 2015). Furthermore, human studies suggest that low-intensity exercise (Wojtaszewski et al. 2000; Chen et al. 2003; Jensen et al. 2012) or moderate-high-intensity exercise in exercise-trained individuals does not activate AMPK but does increase glucose uptake (Wahren et al. 1971; McConell et al. 2005; McConell et al. 2020). However, a limitation of these studies is that AMPK α2β2γ3 activity cannot be estimated from measuring AMPK Thr172 or α2 AMPK activity (Birk and Wojtaszewski 2006) and exercise/contraction-stimulated ACC2 phosphorylation is probably shared by multiple kinases (Dzamko et al. 2008), making the activity of the major human exercise-responsive AMPK trimer unknown. Still, as recently highlighted elsewhere (McConell 2020), the collective evidence in transgenic mice and humans clearly shows that AMPK is not necessary for steady-state exercise-stimulated glucose uptake. However, since pharmacological activation of AMPK is sufficient to stimulate muscle glucose uptake, AMPK probably contributes to human muscle glucose uptake when activated.

6.11 Nitric Oxide

There is some evidence, especially in humans, that nitric oxide (NO) may regulate skeletal muscle glucose uptake during contraction and exercise. Nitric oxide synthase (NOS) is expressed in the endothelium of skeletal muscle with the alternatively spliced isoform of nNOS, nNOSµ, being the primary source of skeletal muscle NO during contraction in mouse muscle (Lau et al. 2000) and in contracting muscle cells (Hirschfield et al. 2000; Patwell et al. 2004). Skeletal muscle NOS activity increases during ex vivo contractions in mouse muscle (Merry et al. 2010b), during in situ contractions in rats (Ross et al. 2007), and during in vivo exercise in humans (Linden et al. 2011).

In humans, local infusion of the NOS inhibitor L-NMMA into the femoral artery during cycling exercise substantially attenuated the increase in leg glucose uptake in healthy individuals (~30%) and in people with T2D (up to 75%) without affecting total leg blood flow (Bradley et al. 1999; Kingwell et al. 2002). In a study by Mortenson et al. (Mortensen et al. 2009) involving moderate one-legged knee extension exercise, it was possible to calculate that L-NMMA attenuated increases in leg glucose uptake during exercise using the presented leg blood flow and leg femoral arterial and venous blood glucose values. Another study found no effect of NOS inhibition on muscle glucose uptake during relatively low-intensity one-legged knee extension exercise (Heinonen et al. 2013). This may have been because the exercise intensity was insufficient to activate NOS (Silveira et al. 2003; Lee-Young et al. 2009) or because of an order effect given participants always performed a saline infusion trial before the NOS inhibition trial (Heinonen et al. 2013).

Rodent studies have yielded conflicting results with NOS inhibition attenuating skeletal muscle glucose uptake in some (Merry et al. 2010b, c; Ross et al. 2007; Roberts et al. 1997; Balon and Nadler 1997) but not all (Etgen et al. 1997; Higaki et al. 2001; Hong et al. 2015a, b, 2016) studies. In addition, lack of NOS (either nNOSµ KO or eNOS KO) does not attenuate glucose uptake during contraction or exercise (Hong et al. 2015b; Hong et al. 2016; Lee-Young et al. 2010). This is baffling given that studies in humans using NOS inhibitors are generally supportive of a role of NO in glucose uptake during exercise. It is possible that L-NMMA is having non-specific effects given that glucose uptake during ex vivo contraction is normal in nNOSµ KO muscles, but L-NMMA attenuates the increase in glucose uptake during contraction of these muscles (Hong et al. 2015b). Taken together, the overall contribution of NO and NOS enzymes to exercise-stimulated glucose uptake remains somewhat unclear.

6.12 Mechanistic Target of Rapamycin Complex 2

mTORC2 comprises the catalytic mTOR kinase in complex with the core regulatory subunit Rictor and a number of other proteins (Knudsen et al. 2020). mTORC2 has been suggested to reside in multiple subcellular compartments including most of the endomembrane system, the cytosol, nucleus, endoplasmic reticulum, and mitochondria (Knudsen et al. 2020), but mTORC2 localization has not been investigated in adult skeletal muscle. Based on the phosphorylation of the indirect mTORC2-substrate NDRG1 downstream of serum and glucocorticoid-responsive kinase 1 (SGK1), mTORC2 may be activated by exercise/contraction (Kleinert et al. 2017). However, given that Rictor muscle KO mice displayed increased resting NDRG1 phosphorylation (Kleinert et al. 2017) and SGK1 activation seems to require other cell signaling pathways and is also regulated by acute changes in SGK1 expression (Lang et al. 2009), the validity of NDRG1 phosphorylation as a mTORC2 activity measure is uncertain.

Regardless of the mTORC2 activation status, glucose uptake during treadmill exercise is reduced in mice lacking the Rictor subunit in skeletal muscle (Kleinert et al. 2017), suggesting that mTORC2 is required for this process. It is unclear which step of the muscle glucose uptake process is affected in the absence of Rictor. However, Rictor mKO likely affects GLUT4 translocation since exercise capacity, exercise-stimulated glycogen breakdown, lactate production, and expression of GLUT4 and hexokinase were not affected in the Rictor mKO mice compared to wild type (Kleinert et al. 2017).

In summary, exercise/contraction-stimulated GLUT4 translocation appears to be primarily stimulated by factors secondary to contraction itself, such as mechanical and metabolic stress, rather than to depolarization-related events such as SR Ca^{2+} release. In vivo, exercise-stimulated GLUT4 translocation may require input from NOX2 downstream of Rac1 and potentially also from mTORC2 activity and CaMKII. The requirement for nitric oxide is not clear based on the current evidence. AMPK is not necessary for exercise-stimulated steady-state GLUT4 translocation or glucose uptake, but since AMPK is a sufficient signal to increase glucose uptake, AMPK likely augments exercise-stimulated glucose uptake when activated.

6.13 Conclusion

The mechanisms regulating exercise/contraction-stimulated adult skeletal muscle glucose uptake remain somewhat unclear. The muscle glucose uptake response to exercise is determined by a coordinated increase in glucose delivery, facilitated glucose transport in the myocytes, and intramyocellular metabolism which may increase glucose uptake across the working leg by up to 100-fold during maximal exercise in humans. The lack of accumulation of interstitial and intramyocellular glucose in humans during exercise, except during very intense exercise, suggests that the increases in glucose transport and phosphorylation are well coordinated to the increase in glucose delivery and do not impose significant barriers to glucose uptake in vivo.

GLUT4 is the main transporter facilitating muscle glucose uptake during exercise in healthy adult skeletal muscle. Exercise-stimulated GLUT4 translocation does not require proximal insulin signaling and appears to be stimulated by distinct aspects of intracellular GLUT4 trafficking compared to insulin. Studies in isolated adult rodent muscles suggest that glucose transport is mainly stimulated by stress signals produced secondary to actual contraction including mechanical and metabolic stress. Stimulation of the transporter activity of individual GLUT4 proteins by environmental factors such as increased temperature, mechanical stress, or post-translational modifications remains a possibility that deserves further scrutiny.

Ultimately, progress in the field is limited by the fact that most mechanistic insight currently comes from rodent muscle studies which may not faithfully mirror the mechanistic "wiring" of exercising human muscle. A much cited example of this is clathrin heavy chain (CHC)22, a GLUT4 regulating protein in human skeletal muscle which is not expressed in mice (Vassilopoulos et al. 2009). To overcome this limitation, better models of human skeletal muscle must be developed, be it more metabolically mature human in vitro myotube models, isolated adult muscle strips or fibers from humans, or more human-like large animal models.

Funding This work was supported by grants from the Novo Nordisk foundation, the Independent Research Fund Denmark, and/or the Lundbeck foundation to LSH, TEJ, GKM, and EAR and the Danish Diabetes Academy funded by the Novo Nordisk Foundation to COL, JRK, and GKM.

References

Abbott MJ, Bogachus LD, Turcotte LP (2011) AMPKα2 deficiency uncovers time dependency in the regulation of contraction-induced palmitate and glucose uptake in mouse muscle. J Appl Physiol 111:125–134

Abel ED et al (1999) Cardiac hypertrophy with preserved contractile function after selective deletion of GLUT4 from the heart. J Clin Invest 104:1703–1714

Ahlborg G, Felig P, Hagenfeldt L, Hendler R, Wahren J (1974) Substrate turnover during prolonged exercise in man. Splanchnic and leg metabolism of glucose, free fatty acids, and amino acids. J Clin Invest 53:1080–1090

Balon TW, Nadler JL (1997) Evidence that nitric oxide increases glucose transport in skeletal muscle. J Appl Physiol 82:359–363

Bangsbo J, Madsen K, Kiens B, Richter EA (1997) Muscle glycogen synthesis in recovery from intense exercise in humans. Am J Physiol–Endocrinol Metab 273

Birk JB, Wojtaszewski JFP (2006) Predominant α2/β2/γ3 AMPK activation during exercise in human skeletal muscle. J Physiol 577:1021–1032

Blair DR, Funai K, Schweitzer GG, Cartee GD (2009) A myosin II ATPase inhibitor reduces force production, glucose transport, and phosphorylation of AMPK and TBC1D1 in electrically stimulated rat skeletal muscle. Am J Physiol–Endocrinol Metab 296

Bradley SJ, Kingwell BA, McConell GK (1999) Nitric oxide synthase inhibition reduces leg glucose uptake but not blood flow during dynamic exercise in humans. Diabetes 48:1815–1821

Bruton JD, Katz A, Westerblad H (1999) Insulin increases near-membrane but not global Ca2+ in isolated skeletal muscle. Proc Natl Acad Sci U S A 96
Bultot L et al (2016) Benzimidazole derivative small-molecule 991 enhances AMPK activity and glucose uptake induced by AICAR or contraction in skeletal muscle. Am J Physiol–Endocrinol Metab 311:E706–E719
Calbet JAL et al (2007) Cardiac output and leg and arm blood flow during incremental exercise to exhaustion on the cycle ergometer. J Appl Physiol 103:969–978
Capaldo B et al (1999) Splanchnic and leg substrate exchange after ingestion of a natural mixed meal in humans. Diabetes 48:958–966
Cartee GD, Holloszy JO (1990) Exercise increases susceptibility of muscle glucose transport to activation by various stimuli. Am J Physiol–Endocrinol Metab 258
Chambers MA, Moylan JS, Smith JD, Goodyear LJ, Reid MB (2009) Stretch-stimulated glucose uptake in skeletal muscle is mediated by reactive oxygen species and p38 MAP-kinase. J Physiol 587:3363–3373
Chen ZP et al (2003) Effect of exercise intensity on skeletal muscle AMPK signaling in humans. Diabetes 52:2205–2212
Christiansen D, Eibye K, Hostrup M, Bangsbo J (2020) Training with blood flow restriction increases femoral artery diameter and thigh oxygen delivery during knee-extensor exercise in recreationally trained men. J Physiol 598:2337–2353
Coyle EF et al (1983) Carbohydrate feeding during prolonged strenuous exercise can delay fatigue. J Appl Physiol Respir Environ Exerc Physiol 55:230–235
Dawson D et al (2002) Vascular recruitment in skeletal muscle during exercise and hyperinsulinemia assessed by contrast ultrasound. Am J Physiol–Endocrinol Metab 282
DeFronzo RA, Ferrannini E, Sato Y, Felig P, Wahren J (1981) Synergistic interaction between exercise and insulin on peripheral glucose uptake. J Clin Invest 68:1468–1474
Dzamko N et al (2008) AMPK-independent pathways regulate skeletal muscle fatty acid oxidation. J Physiol 586:5819–5831
Erickson JR et al (2008) A dynamic pathway for calcium-independent activation of CaMKII by methionine oxidation. Cell 133:462–474
Etgen GJ, Fryburg DA, Gibbs EM (1997) Nitric oxide stimulates skeletal muscle glucose transport through a calcium/contraction- and phosphatidylinositol-3-kinase-independent pathway. Diabetes 46:1915–1919
Fam BC et al (2012) Normal muscle glucose uptake in mice deficient in muscle GLUT4. J Endocrinol 214:313–327
Fritzen AM et al (2015) 5′-AMP activated protein kinase α2 controls substrate metabolism during post-exercise recovery via regulation of pyruvate dehydrogenase kinase 4. J Physiol 593:4765–4780
Fueger PT, Bracy DP, Malabanan CM, Pencek RR, Wasserman DH (2004a) Distributed control of glucose uptake by working muscles of conscious mice: roles of transport and phosphorylation. Am J Physiol–Endocrinol Metab 286
Fueger PT et al (2004b) Control of exercise-stimulated muscle glucose uptake by GLUT4 is dependent on glucose phosphorylation capacity in the conscious mouse. J Biol Chem 279: 50956–50961
Fueger PT et al (2007) Glucose kinetics and exercise tolerance in mice lacking the GLUT4 glucose transporter. J Physiol 582:801–812
Fujii N et al (2005) AMP-activated protein kinase α2 activity is not essential for contraction-and hyperosmolarity-induced glucose transport in skeletal muscle. J Biol Chem 280:39033–39041
Gaster M, Poulsen P, Handberg A, Schrøder HD, Beck-Nielsen H (2000) Direct evidence of fiber type-dependent GLUT-4 expression in human skeletal muscle. Am J Physiol–Endocrinol Metab 278
Gaster M et al (2004) GLUT11, but not GLUT8 or GLUT12, is expressed in human skeletal muscle in a fibre type-specific pattern. Pflugers Arch Eur J Physiol 448:105–113

González-Alonso J, Calbet JAL, Nielsen B (1999a) Metabolic and thermodynamic responses to dehydration-induced reductions in muscle blood flow in exercising humans. J Physiol 520:577–589

González-Alonso J et al (1999b) Influence of body temperature on the development of fatigue during prolonged exercise in the heat. J Appl Physiol 86:1032–1039

Hargreaves M, Kiens B, Richter EA (1991) Effect of increased plasma free fatty acid concentration on muscle metabolism in exercising men. J Appl Physiol 70:194–201

Havel RJ, Pernow B, Jones NL (1967) Uptake and release of free fatty acids and other metabolites in the legs of exercising men. J Appl Physiol 23:90–99

Heinonen I et al (2013) Effect of nitric oxide synthase inhibition on the exchange of glucose and fatty acids in human skeletal muscle. Nutr Metab (Lond) 10:43

Henríquez-Olguín C et al (2019) The emerging roles of nicotinamide adenine dinucleotide phosphate oxidase 2 in skeletal muscle redox signaling and metabolism. Antioxid Redox Signal 31: 1371–1410

Henríquez-Olguin C et al (2019) Cytosolic ROS production by NADPH oxidase 2 regulates muscle glucose uptake during exercise. Nat Commun 10

Hespel P, Richter EA (1990) Glucose uptake and transport in contracting, perfused rat muscle with different pre-contraction glycogen concentrations. J Physiol 427:347–359

Hespel P, Vergauwen L, Vandenberghe K, Richter EA (1995) Important role of insulin and flow in stimulating glucose uptake in contracting skeletal muscle. Diabetes 44:210–215

Higaki Y, Hirshman MF, Fujii N, Goodyear LJ (2001) Nitric oxide increases glucose uptake through a mechanism that is distinct from the insulin and contraction pathways in rat skeletal muscle. Diabetes 50:241–247

Higaki Y et al (2008) Oxidative stress stimulates skeletal muscle glucose uptake through a phosphatidylinositol 3-kinase-dependent pathway. Am J Physiol–Endocrinol Metab 294

Hingst JR et al (2020) Inducible deletion of skeletal muscle AMPKα reveals that AMPK is required for nucleotide balance but dispensable for muscle glucose uptake and fat oxidation during exercise. Mol. Metab. 40

Hirschfield W et al (2000) Nitric oxide release and contractile properties of skeletal muscles from mice deficient in type III NOS. Am J Physiol–Regul Integr Comp Physiol 278

Hong YH, Betik AC, Mcconell GK (2014) Role of nitric oxide in skeletal muscle glucose uptake during exercise. Exp Physiol 99:1569–1573

Hong YH, Yang C, Betik AC, Lee-Young RS, McConell GK (2016) Skeletal muscle glucose uptake during treadmill exercise in neuronal nitric oxide synthase-μ knockout mice. Am J Physiol–Endocrinol Metab 310:E838–E845

Hong YH et al (2015a) No effect of NOS inhibition on skeletal muscle glucose uptake during in situ hindlimb contraction in healthy and diabetic Sprague-Dawley rats. Am J Physiol–Regul Integr Comp Physiol 308:R862–R871

Hong YH et al (2015b) Glucose uptake during contraction in isolated skeletal muscles from neuronal nitric oxide synthase μ knockout mice. J Appl Physiol 118:1113–1121

Howlett KF, Andrikopoulos S, Proietto J, Hargreaves M (2013) Exercise-induced muscle glucose uptake in mice with graded, muscle-specific GLUT-4 deletion. Physiol Rep 1

Huang S, Czech MP (2007) The GLUT4 glucose transporter. Cell Metab 5

Ihlemann J, Ploug T, Galbo H (2001) Effect of force development on contraction induced glucose transport in fast twitch rat muscle. Acta Physiol Scand 171:439–444

Ihlemann J, Ploug T, Hellsten Y, Galbo H (1999) Effect of tension on contraction-induced glucose transport in rat skeletal muscle. Am J Physiol–Endocrinol Metab 277

Ihlemann J, Ploug T, Hellsten Y, Galbo H (2000) Effect of stimulation frequency on contraction-induced glucose transport in rat skeletal muscle. Am J Physiol–Endocrinol Metab 279

Jensen TE, Angin Y, Sylow L, Richter EA (2014) Is contraction-stimulated glucose transport feedforward regulated by Ca2+? Exp Physiol 99

Jensen TE et al (2007) Possible CaMKK-dependent regulation of AMPK phosphorylation and glucose uptake at the onset of mild tetanic skeletal muscle contraction. Am J Physiol–Endocrinol Metab 292

Jensen TE et al (2012) EMG-normalised kinase activation during exercise is higher in human gastrocnemius compared to soleus muscle. PLoS One 7

Jensen TE et al (2014a) Contraction-stimulated glucose transport in muscle is controlled by AMPK and mechanical stress but not sarcoplasmatic reticulum Ca2+ release. Mol Metab 3

Jensen TE et al (2014c) Contraction-stimulated glucose transport in muscle is controlled by AMPK and mechanical stress but not sarcoplasmic reticulum Ca2+ release. Mol Metab 3

Jørgensen NO et al (2021) Direct small molecule ADaM-site AMPK activators reveal an AMPKγ3-independent mechanism for blood glucose lowering. Mol Metab 51:101259

Joyner MJ, Casey DP (2015) Regulation of increased blood flow (hyperemia) to muscles during exercise: A hierarchy of competing physiological needs. Physiol Rev 95:549–601

Kaczmarczyk SJ et al (2003) Threshold effects of glucose transporter-4 (GLUT4) deficiency on cardiac glucose uptake and development of hypertrophy. J Mol Endocrinol 31:449–459

Katz A, Broberg S, Sahlin K, Wahren J (1986a) Muscle ammonia and amino acid metabolism during dynamic exercise in man. Clin Physiol 6:365–379

Katz A, Broberg S, Sahlin K, Wahren J (1986b) Leg glucose uptake during maximal dynamic exercise in humans. Am J Physiol–Endocrinol Metab 251

Katz A, Sahlin K, Broberg S (1991) Regulation of glucose utilization in human skeletal muscle during moderate dynamic exercise. Am J Physiol–Endocrinol Metab 260

Katz EB, Stenbit AE, Hatton K, Depinho R, Charron MJ (1995) Cardiac and adipose tissue abnormalities but not diabetes in mice deficient in GLUT4. Nature 377:151–155

Kemppainen J et al (2002) Myocardial and skeletal muscle glucose uptake during exercise in humans. J Physiol 542:403–412

Kennedy JW et al (1999) Acute exercise induces GLUT4 translocation in skeletal muscle of normal human subjects and subjects with type 2 diabetes. Diabetes 48:1192–1197

Khayat ZA, Tong P, Yaworsky K, Bloch RJ, Klip A (2000) Insulin-induced actin filament remodeling colocalizes actin with phosphatidylinositol 3-kinase and GLUT4 in L6 myotubes. J Cell Sci 113:279–290

Kingwell BA, Formosa M, Muhlmann M, Bradley SJ, McConell GK (2002) Nitric oxide synthase inhibition reduces glucose uptake during exercise in individuals with type 2 diabetes more than in control subjects. Diabetes 51:2572–2580

Kjaer M, Kiens B, Hargreaves M, Richter EA (1991) Influence of active muscle mass on glucose homeostasis during exercise in humans. J Appl Physiol 71:552–557

Kjøbsted R et al (2018) AMPK in skeletal muscle function and metabolism. FASEB J 32:1741–1777

Kjøbsted R et al (2019) AMPK and TBC1D1 regulate muscle glucose uptake after, but not during, exercise and contraction. Diabetes 68:1427–1440

Kleinert M et al (2017) Mammalian target of rapamycin complex 2 regulates muscle glucose uptake during exercise in mice. J Physiol 595:4845–4855

Knudsen JR et al (2020) Growth factor-dependent and -independent activation of mTORC2. Trends Endocrinol Metab 31:13–24

Kristiansen S, Hargreaves M, Richter EA (1996a) Exercise-induced increase in glucose transport, GLUT-4, and VAMP-2 in plasma membrane from human muscle. Am J Physiol–Endocrinol Metab 70(1):E197–E201

Kristiansen S, Youn J, Richter EA (1996b) Effect of vanadate on glucose transporter (GLUT4) intrinsic activity in skeletal muscle plasma membrane giant vesicles. Biochim Biophys Acta 1282:71–75

Kristiansen S, Asp S, Richter EA (1996c) Decreased muscle GLUT-4 and contraction-induced glucose transport after eccentric contractions. Am J Physiol–Regul Integr Comp Physiol 271

Kristiansen S, Hargreaves M, Richter EA (1997) Progressive increase in glucose transport and GLUT-4 in human sarcolemmal vesicles during moderate exercise. Am J Physiol–Endocrinol Metab 272

Lang F, Artunc F, Vallon V (2009) The physiological impact of the serum and glucocorticoid-inducible kinase SGK1. Curr Opin Nephrol Hypertens 18:439–448

Lanner JT et al (2006) The role of Ca2+ influx for insulin-mediated glucose uptake in skeletal muscle. Diabetes 55:2077–2083

Lanner JT et al (2009) Knockdown of TRPC3 with siRNA coupled to carbon nanotubes results in decreased insulin-mediated glucose uptake in adult skeletal muscle cells. FASEB J 23:1728–1738

Lau KS et al (2000) nNOS and eNOS modulate cGMP formation and vascular response in contracting fast-twitch skeletal muscle. Physiol Genomics 2000:21–27

Lawrence JC, Piper RC, Robinson LJ, James DE (1992) GLUT4 facilitates insulin stimulation and cAMP-mediated inhibition of glucose transport. Proc Natl Acad Sci U S A 89:3493–3497

Lee-Young RS et al (2009) Skeletal muscle AMP-activated protein kinase is essential for the metabolic response to exercise in vivo. J Biol Chem 284:23925–23934

Lee-Young RS et al (2010) Endothelial nitric oxide synthase is central to skeletal muscle metabolic regulation and enzymatic signaling during exercise in vivo. Am J Physiol–Regul Integr Comp Physiol 298

Linden KC, Wadley GD, Garnham AP, McConell GK (2011) Effect of l-arginine infusion on glucose disposal during exercise in humans. Med Sci Sports Exerc 43:1626–1634

Luczak ED, Anderson ME (2014) CaMKII oxidative activation and the pathogenesis of cardiac disease. J Mol Cell Cardiol 73:112–116

Lund S, Holman GD, Schmitz O, Pedersen O (1995) Contraction stimulates translocation of glucose transporter GLUT4 in skeletal muscle through a mechanism distinct from that of insulin. Proc Natl Acad Sci U S A 92:5817–5821

Lundsgaard AM, Kiens B (2014) Gender differences in skeletal muscle substrate metabolism–molecular mechanisms and insulin sensitivity. Front Endocrinol 5

Maarbjerg SJ et al (2009) Genetic impairment of AMPKα2 signaling does not reduce muscle glucose uptake during treadmill exercise in mice. Am J Physiol–Endocrinol Metab 297

MacLean DA, Bangsbo J, Saltin B (1999) Muscle interstitial glucose and lactate levels during dynamic exercise in humans determined by microdialysis. J Appl Physiol 87:1483–1490

McConell GK (2020) It's well and truly time to stop stating that AMPK regulates glucose uptake and fat oxidation during exercise. Am J Physiol Endocrinol Metab 318:E564–E567

McConell G, Fabris S, Proietto J, Hargreaves M (1994) Effect of carbohydrate ingestion on glucose kinetics during exercise. J Appl Physiol 77:1537–1541

McConell GK, Wadley GD, Le Plastrier K, Linden KC (2020) Skeletal muscle AMPK is not activated during 2 h of moderate intensity exercise at ~65% VO2peak in endurance trained men. J Physiol 598:3859–3870

McConell GK et al (2005) Short-term exercise training in humans reduces AMPK signalling during prolonged exercise independent of muscle glycogen. J Physiol 568:665–676

McMillin SL, Schmidt DL, Kahn BB, Witczak CA (2017) GLUT4 is not necessary for overload-induced glucose uptake or hypertrophic growth in mouse skeletal muscle. Diabetes 66:1491–1500. (American Diabetes Association Inc.

Merry TL, Dywer RM, Bradley EA, Rattigan S, McConell GK (2010a) Local hindlimb antioxidant infusion does not affect muscle glucose uptake during in situ contractions in rat. J Appl Physiol 108:1275–1283

Merry TL, Steinberg GR, Lynch GS, McConell GK (2010b) Skeletal muscle glucose uptake during contraction is regulated by nitric oxide and ROS independently of AMPK. Am J Physiol–Endocrinol Metab 298

Merry TL, Lynch GS, McConell GK (2010c) Downstream mechanisms of nitric oxide-mediated skeletal muscle glucose uptake during contraction. Am J Physiol–Regul Integr Comp Physiol 299

Merry TL et al (2010d) N-acetylcysteine infusion does not affect glucose disposal during prolonged moderate-intensity exercise in humans. J Physiol 588:1623–1634
Møller LLV et al (2020) The p21-activated kinase 2 (PAK2), but not PAK1, regulates contraction-stimulated skeletal muscle glucose transport. Physiol Rep 8
Mortensen SP, Askew CD, Walker M, Nyberg M, Hellsten Y (2012) The hyperaemic response to passive leg movement is dependent on nitric oxide: a new tool to evaluate endothelial nitric oxide function. J Physiol 590:4391–4400
Mortensen SP et al (2005) Limitations to systemic and locomotor limb muscle oxygen delivery and uptake during maximal exercise in humans. J Physiol 566:273–285
Mortensen SP et al (2009) ATP-induced vasodilation and purinergic receptors in the human leg: roles of nitric oxide, prostaglandins, and adenosine. Am J Physiol–Regul Integr Comp Physiol 296
Mu J, Brozinick JT, Valladares O, Bucan M, Birnbaum MJ (2001) A role for AMP-activated protein kinase in contraction- and hypoxia-regulated glucose transport in skeletal muscle. Mol Cell 7: 1085–1094
Pappenheimer JR, Renkin EM, Borrero LM (1951) Filtration, diffusion and molecular sieving through peripheral capillary membranes; a contribution to the pore theory of capillary permeability. Am J Phys 167:13–46
Park DR, Park KH, Kim BJ, Yoon CS, Kim UH (2015) Exercise ameliorates insulin resistance via Ca2+ signals distinct from those of insulin for GLUT4 translocation in skeletal muscles. Diabetes 64:1224–1234
Patwell DM, McArdle A, Morgan JE, Patridge TA, Jackson MJ (2004) Release of reactive oxygen and nitrogen species from contracting skeletal muscle cells. Free Radic Biol Med 37:1064–1072
Ploug T et al (1998) Analysis of GLUT4 distribution in whole skeletal muscle fibers: identification of distinct storage compartments that are recruited by insulin and muscle contractions. J Cell Biol 142. http://www.jcb.org
Raun SH et al (2018) Rac1 muscle knockout exacerbates the detrimental effect of high-fat diet on insulin-stimulated muscle glucose uptake independently of Akt. J Physiol 596
Reichard GA et al (1961) Blood glucose metabolism in man during muscular work. J Appl Physiol 16:1001–1005
Rhein P et al (2021) Compound- and fiber type-selective requirement of AMPKγ3 for insulin-independent glucose uptake in skeletal muscle. Mol Metab 51:101228
Richter EA (1996) Glucose utilization. Handb Physiol Sect 12 Exerc Regul Integr Mult Syst Ed by Rowell LB Shepherd JT New York Oxford Univ Press
Richter EA (2020) Is GLUT4 translocation the answer to exercise stimulated muscle glucose uptake? Am J Physiol Endocrinol Metab 320(2):E240–E243
Richter EA, Hargreaves M (2013) Exercise, GLUT4, and skeletal muscle glucose uptake. Physiol Rev 93:993–1017
Roberts CK, Barnard RJ, Scheck SH, Balon TW (1997) Exercise-stimulated glucose transport in skeletal muscle is nitric oxide dependent. Am J Physiol–Endocrinol Metab 273
Roepstorff C et al (2002) Gender differences in substrate utilization during submaximal exercise in endurance-trained subjects. Am J Physiol–Endocrinol Metab 282
Roepstorff C et al (2005) Malonyl-CoA and carnitine in regulation of fat oxidation in human skeletal muscle during exercise. Am J Physiol–Endocrinol Metab 288
Rohlenova K, Veys K, Miranda-Santos I, De Bock K, Carmeliet P (2018) Endothelial cell metabolism in health and disease. Trends Cell Biol 28:224–236
Romijn JA et al (1993) Regulation of endogenous fat and carbohydrate metabolism in relation to exercise intensity and duration. Am J Physiol Endocrinol Metab 265
Rose AJ, Alsted TJ, Kobberø JB, Richter EA (2007) Regulation and function of Ca2+−calmodulin-dependent protein kinase II of fast-twitch rat skeletal muscle. J Physiol 580:993–1005
Rose AJ, Kiens B, Richter EA (2006) Ca2+−calmodulin-dependent protein kinase expression and signalling in skeletal muscle during exercise. J Physiol 574:889–903

Ross RM, Wadley GD, Clark MG, Rattigan S, McConell GK (2007) Local nitric oxide synthase inhibition reduces skeletal muscle glucose uptake but not capillary blood flow during in situ muscle contraction in rats. Diabetes 56:2885–2892
Ryder JW et al (1999a) In vitro analysis of the glucose-transport system in GLUT4-null skeletal muscle. Biochem J 342:321–328
Ryder JW et al (1999b) Postexercise glucose uptake and glycogen synthesis in skeletal muscle from GLUT4-deficient mice. FASEB J 13:2246–2256
Sakamoto K, Aschenbach WG, Hirshman MF, Goodyear LJ (2003) Akt signaling in skeletal muscle: regulation by exercise and passive stretch. Am J Physiol–Endocrinol Metab 285
Sanders CA, Levinson GE, Abelmann WH, Freinkel N (1964) Effect of exercise on the peripheral utilization of glucose in man. N Engl J Med 271:220–225
Sandström ME et al (2006) Role of reactive oxygen species in contraction-mediated glucose transport in mouse skeletal muscle. J Physiol 575:251–262
Schultz TA, Lewis SB, Westbie DK (1977) Glucose delivery: a modulator of glucose uptake in contracting skeletal muscle. Am J Physiol Endocrinol Metab Gastrointest Physiol 2
Shamni O et al (2017) Supportive data on the regulation of GLUT4 activity by 3-O-methyl-D-glucose. Data Br 14:329–336
Silveira LR, Pereira-Da-Silva L, Juel C, Hellsten Y (2003) Formation of hydrogen peroxide and nitric oxide in rat skeletal muscle cells during contractions. Free Radic Biol Med 35:455–464
Sjøberg KA, Rattigan S, Hiscock N, Richter EA, Kiens B (2011) A new method to study changes in microvascular blood volume in muscle and adipose tissue: real-time imaging in humans and rat. Am J Physiol Heart Circ Physiol 301
Sjøberg KA et al (2017) Exercise increases human skeletal muscle insulin sensitivity via coordinated increases in microvascular perfusion and molecular signaling. Diabetes 66:1501–1510
Stenbit AE et al (1996) Diverse effects of Glut 4 ablation on glucose uptake and glycogen synthesis in red and white skeletal muscle. J Clin Invest 98:629–634
Sylow L, Kleinert M, Richter EA, Jensen TE (2017a) Exercise-stimulated glucose uptake-regulation and implications for glycaemic control. Nat Rev Endocrinol 13:133–148
Sylow L, Møller LLV, Kleinert M, Richter EA, Jensen TE (2014) Rac1–a novel regulator of contraction-stimulated glucose uptake in skeletal muscle. Exp Physiol 99:1574–1580
Sylow L, Møller LLV, Kleinert M, Richter EA, Jensen TE (2015) Stretch-stimulated glucose transport in skeletal muscle is regulated by Rac1. J Physiol 593
Sylow L et al (2013a) Rac1 is a novel regulator of contraction-stimulated glucose uptake in skeletal muscle. Diabetes 62:1139–1151
Sylow L et al (2013b) Rac1 signaling is required for insulin-stimulated glucose uptake and is dysregulated in insulin-resistant murine and human skeletal muscle. Diabetes 62(6):1865–1875
Sylow L et al (2016) Rac1 governs exercise-stimulated glucose uptake in skeletal muscle through regulation of GLUT4 translocation in mice. J Physiol 594:4997–5008
Sylow L et al (2017b) Rac1 and AMPK account for the majority of muscle glucose uptake stimulated by ex vivo contraction but not in vivo exercise. Diabetes 66
Treebak JT et al (2007) AS160 phosphorylation is associated with activation of α 2β2γ1- but not α 2β2γ3-AMPK trimeric complex in skeletal muscle during exercise in humans. Am J Physiol–Endocrinol Metab 292
Ueda S et al (2010) Crucial role of the small GTPase Rac1 in insulin-stimulated translocation of glucose transporter 4 to the mouse skeletal muscle sarcolemma. FASEB J 24(7):2254–2261
Van Loon LJC, Greenhaff PL, Constantin-Teodosiu D, Saris WHM, Wagenmakers AJM (2001) The effects of increasing exercise intensity on muscle fuel utilisation in humans. J Physiol 536: 295–304
Vassilopoulos S et al (2009) A role for the CHC22 clathrin heavy-chain isoform in human glucose metabolism. Science (80-) 324:1192–1196
Wahren J, Felig P, Ahlborg G, Jorfeldt L (1971) Glucose metabolism during leg exercise in man. J Clin Invest 50:2715–2725

Wang W, Hansen PA, Marshall BA, Holloszy JO, Mueckler M (1996) Insulin unmasks a COOH-terminal GLUT4 epitope and increases glucose transport across T-tubules in skeletal muscle. J Cell Biol 135:415–430
Wang Q et al (2018) Oxidized CaMKII (ca 2+ /calmodulin-dependent protein kinase II) is essential for ventricular arrhythmia in a mouse model of Duchenne muscular dystrophy. Circ Arrhythmia Electrophysiol 11
Whichelow MJ, Butterfield WJH, Abrams ME, Sterky G, Garratt CJ (1968) The effect of mild exercise on glucose uptake in human forearam tissues in the fasting state and after oral glucose administration. Metabolism 17:84–96
Witczak CA et al (2010) CaMKII regulates contraction- but not insulin-induced glucose uptake in mouse skeletal muscle. Am J Physiol–Endocrinol Metab 298
Wojtaszewski JFP, Nielsen P, Hansen BF, Richter EA, Kiens B (2000) Isoform-specific and exercise intensity-dependent activation of 5'-AMP-activated protein kinase in human skeletal muscle. J Physiol 528:221–226
Wright DC, Hucker KA, Holloszy JO, Han DH (2004) Ca2+ and AMPK both mediate stimulation of glucose transport by muscle contractions. Diabetes 53:330–335
Yazdani S, Jaldin-Fincati JR, Pereira RVS, Klip A (2019) Endothelial cell barriers: transport of molecules between blood and tissues. Traffic 20:390–403
Zaid H, Talior-Volodarsky I, Antonescu C, Liu Z, Klip A (2009) GAPDH binds GLUT4 reciprocally to hexokinase-II and regulates glucose transport activity. Biochem J 419:475–484
Zierath JR et al (1998) Restoration of hypoxia-stimulated glucose uptake in GLUT4-deficient muscles by muscle-specific GLUT4 transgenic complementation. J Biol Chem 273:20910–20915
Zinker BA, Lacy DB, Bracy DP, Wasserman DH (1993) Role of glucose and insulin loads to the exercising limb in increasing glucose uptake and metabolism. J Appl Physiol 74:2915–2921
Zisman A et al (2000) Targeted disruption of the glucose transporter 4 selectively in muscle causes insulin resistance and glucose intolerance. Nat Med 6:924–928

Chapter 7
Adipose Tissue Lipid Metabolism During Exercise

Jeffrey F. Horowitz

Abstract Adipose tissue-derived fatty acids are the primary source of energy during low-intensity exercise. Although the relative contribution of fat to energy production during exercise decreases with increasing exercise intensity, fatty acids still contribute meaningfully to energy production even during vigorous exercise. How exercise triggers the liberation of fatty acids from this "remote" energy storage site for subsequent oxidation in the exercising muscle is complex, with multiple integrated steps and some seemingly paradoxical regulation. Adipose tissue metabolic function and composition differ considerably in subcutaneous vs. visceral adipose tissue beds, and there are also sizable differences between subcutaneous adipose tissue in different regions of the human body (e.g., subcutaneous abdominal vs. subcutaneous femoral/gluteal), whereas most of the fatty acids used for energy during exercise are derived from triacylglycerol stored in subcutaneous abdominal white adipose tissue. This chapter will focus primarily on changes in subcutaneous white adipose tissue that occur during and right after a session of endurance exercise.

Keywords Lipolysis · Fat oxidation · White adipose tissue · Adipocytes · Endurance exercise

7.1 Introduction

Adipose tissue is an extraordinarily complex tissue, comprised of several different cell types, including adipocytes, preadipocytes, monocytes, endothelial cells, macrophages, adipose tissue-derived stem cells, and pericytes, many of which are in constant flux. The complexity of adipose tissue is extended by the fact that there are distinct types of adipose tissue (i.e., white, brown, beige/brite) that have somewhat opposing properties, white adipose tissue being the body's most important site for energy storage and brown and beige/brite adipose tissue involved in energy

J. F. Horowitz (✉)
Substrate Metabolism Laboratory, School of Kinesiology, University of Michigan, Ann Arbor, MI, USA
e-mail: jeffhoro@umich.edu

© The Author(s), under exclusive license to Springer Nature Switzerland AG 2022

G. McConell (ed.), *Exercise Metabolism*, Physiology in Health and Disease,
https://doi.org/10.1007/978-3-030-94305-9_7

expenditure for heat generation. In this chapter, we will focus almost exclusively on the metabolic effects that occur in *white adipose tissue* during a session of *endurance exercise*, and the terms "white adipose tissue" and "adipose tissue" will be used interchangeably. Despite high interest in the expanding knowledge regarding the endocrine-like function of adipose tissue, it is important to recall that the primary role of white adipose tissue is still to store and release energy in the form of non-esterified fatty acids [NEFA] (referred to in this chapter simply as "fatty acids"). The storage and delivery of this endogenous energy source are particularly important when fasting (or even just between meals)—and it is a very important energy source during endurance/aerobic exercise. Even lean adults have enough energy stored as triacylglycerol in their adipose tissue to complete more than 25 marathon races. Although adipose tissue provides a very rich resource of potential energy, there are many regulated steps the fat stored within white adipose tissue must navigate before being able to be used during exercise in the exercising skeletal muscle. For example, to use adipose tissue-derived fat during exercise requires the coordination of many processes including lipolytic regulation, trafficking the liberated fatty acids within the circulation to the exercising muscle, as well as transport across both the sarcolemma and mitochondrial membranes of the working muscle. *The vast majority of this chapter will focus on metabolic regulation within adipose tissue during exercise and some of the key factors that can modulate this regulation (e.g., fasted* vs *fed, training status, anatomical site, sex). In addition, this chapter will also briefly address the contribution of different fat sources to energy expenditure during exercise, as well as provide a brief discussion about responses in adipose tissue after each session of exercise that may underlie some of the important health benefits of exercising regularly.*

7.2 Contribution of Fat to Energy Expenditure During Exercise

The contribution of fat to total energy expenditure during exercise depends on several factors, including exercise intensity, timing of last meal before exercise, and training status. In the overnight fasted condition, fat is the primary source of energy at rest and during low-intensity exercise (Romijn et al. 1993; Romijn et al. 2000), like walking. As the intensity of exercise increases, the relative contribution of fat decreases as the contribution of carbohydrate increases (Romijn et al. 1993). During high-intensity exercise (e.g., ~85% maximal oxygen consumption [VO_2max]), typically <20% of energy is derived from fat (Romijn et al. 1993; Romijn et al. 2000). Interestingly, artificially elevating circulating fatty acid availability via intravenous infusion of lipid and heparin during high-intensity exercise increased the rate of fat oxidation (Romijn et al. 1995), suggesting that a limitation in the availability of circulating fatty acids (derived from adipose tissue) is partly responsible for the relatively low rate of fat oxidation observed during high-intensity exercise. However, an overabundance of circulating fatty acids provided via the lipid and heparin infusion increases the rate of fat oxidation only very modestly (Romijn et al. 1995), suggesting alternative mechanisms beyond substrate availability also contribute to the relatively low rates of fat oxidation during high-intensity exercise. A candidate mechanism within the skeletal muscle that may contribute to the low rate of fat oxidation during high-intensity exercise is an inhibition of fatty acid entry into the mitochondria, perhaps through inhibition of carnitine palmitoyl transferase I (CPT-I) by malonyl-CoA (Petrick and Holloway 2019; Holloway et al. 2006; Coyle et al. 1997) and/or a reduction in free carnitine within the cytosol of the muscle cell (Stephens and Galloway 2013). However, the contributions of these mechanisms to the relatively low rates of fat oxidation during high-intensity exercise are controversial—and these issues are addressed in greater detail in the next chapter of this book (Fritzen, et al.—Chap. 8).

Eating meals or snacks in the few hours before exercise can also have a potent influence on fat use during exercise (Horowitz et al. 1997; Horowitz and Coyle 1993). Even a very modest carbohydrate content in ingested food can inhibit fat oxidation, largely through the effects of the insulin secreted in response to the increase in blood glucose after the meal. Insulin is a very potent inhibitor of adipose tissue lipolysis (Campbell et al. 1992; Nurjhan et al. 1986), and thereby elevated insulin during a meal markedly lowers the availability of circulating fatty acids (Horowitz et al. 1997; Horowitz and Coyle 1993). In addition to its anti-lipolytic effects of insulin on adipose tissue, and the resultant lower availability of circulating fatty acids available for oxidation during exercise, an insulin-induced increase in glucose flux in skeletal muscle may also trigger counter-regulatory mechanisms leading to the reduction in fat oxidation during exercise, which again is discussed in greater detail in the next chapter of this book (Fritzen, et al.—Chap. 8). It has been reported that eating meals even 4–6 hours before exercise lowered the rate of fat oxidation during exercise compared with an overnight fast, even well after plasma insulin concentration returned to pre-meal levels (Montain et al. 1991). The duration of this effect certainly depends on the quantity and composition of the meal, but because most people eat something at least every few hours throughout the day, it is likely that the rate of fat oxidation is often "under the influence" of the persistent effects stemming from the most recent meal or snack. Endurance training also impacts fuel selection during exercise. An increase in the contribution of fat to energy expenditure during exercise at the same absolute intensity (e.g., same treadmill speed and grade) is among the most classic responses to endurance exercise training (Holloszy 1967; Holloszy and Coyle 1984; Holloszy and Booth 1976). This training-induced increase in fat oxidation is proposed to be largely due to increased capacity to oxidize available fatty acids via training-induced expansion of mitochondrial density in skeletal muscle (Holloszy 1967; Holloszy and Coyle 1984).

7.3 Mobilization of Fatty Acids from Adipose Tissue During Exercise

After an overnight fast (or at least several hours removed from the last meal or snack), adipose tissue-derived fatty acids are the primary, and nearly the exclusive source of fat to cover energy (and total energy) expenditure at rest and during low-intensity exercise (Romijn et al. 1993, 2000). Even during moderate-to-vigorous exercise, blood-borne fatty acids (liberated from white adipose tissue) still contribute meaningfully to energy production (Romijn et al. 1993, 2000). The reliance on circulating fatty acids as a key energy source during exercise is intriguing considering the complex "journey" these fatty acids must make from their storage site in adipose tissue to the exercising muscle. As depicted in Fig. 7.1, these fatty acids must first be hydrolyzed from adipose tissue triacylglycerols before being

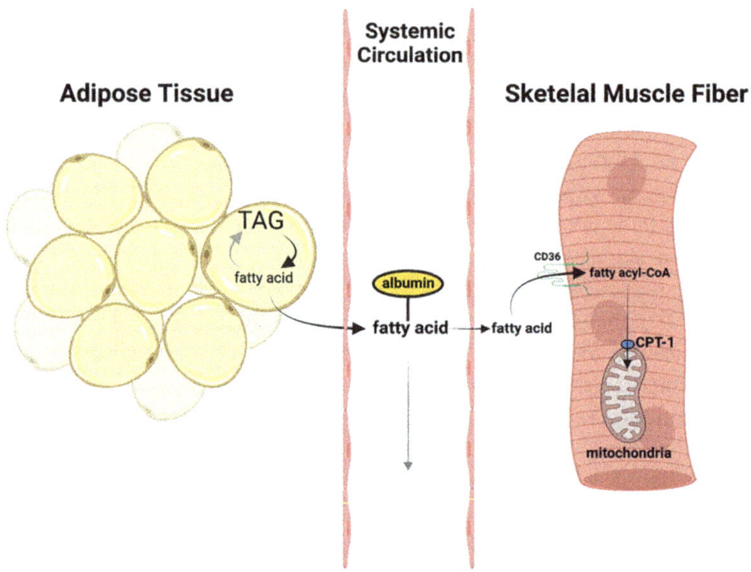

Fig. 7.1 Delivery of adipose tissue-derived fatty acids to skeletal muscle mitochondria for oxidation during exercise. Adipose tissue triacylglycerols (TAGs) must be hydrolyzed to yield glycerol and three fatty acids. Because fatty acids are hydrophobic, upon entry into the aqueous systemic circulation, they must bind to a carrier protein (albumin is the primary fatty acid carrier protein). These albumin-bound fatty acids then travel in the circulation for delivery to other tissues, such as skeletal muscle. Fatty acid entry into the myocyte (i.e., muscle fiber) is facilitated by a transport protein; cluster of differentiation 36 (CD36) is among the most prominent fatty acid transport proteins in skeletal muscle. The fatty acids are then "activated" by the enzyme fatty acyl-CoA synthetase (FACS) to form fatty acyl-CoA, which must again be escorted by carrier proteins to traverse the aqueous cytosol of the myocyte. Before entry into the mitochondria for oxidation and energy production, the fatty acyl-CoA must be converted to fatty acyl-carnitine by the enzyme carnitine palmitoyl transferase-1 (CPT-1), which is often considered the rate-limiting step of fatty acid oxidation

released into the circulation. Due to their hydrophobicity, upon entry into the bloodstream, fatty acids must bind to a carrier protein (i.e., albumin). Increased blood flow to the exercising muscle enhances delivery of circulating fatty acids to skeletal muscle—where they must be released from the carrier protein, pass through the capillary wall, transverse through the aqueous interstitial space, and transport through the muscle sarcolemma as well as through the mitochondrial membranes before being metabolized for energy (which by itself is a tightly regulated process). These wide-ranging regulatory steps required for adipose tissue-derived fatty acids to be oxidized for energy ultimately begin with the hydrolysis of adipose tissue triacylglycerol (i.e., lipolysis).

7.3.1 Lipolytic Rate During Exercise

During low-intensity exercise, such as walking, adipose tissue lipolysis is often found to increase about two- to threefold above resting levels (Romijn et al. 1993; Horowitz et al. 1997). Simultaneously, the rate of re-esterification of these liberated fatty acids to reform triacylglycerol decreases, resulting in a greater proportion of released fatty acids being delivered to skeletal muscle for oxidation (Wolfe et al. 1990). With prolonged exercise, lipolytic rate and fatty acid mobilization from adipose tissue increase progressively with exercise duration (Romijn et al. 1993; Wolfe et al. 1990) accompanying the increased contribution of fat to energy production during prolonged exercise (especially important as endogenous carbohydrate stores diminish). Interestingly, higher exercise intensities typically do not increase lipolytic rate much/any further than that found at low intensity (Romijn et al. 1993). However, despite consistent rates of lipolysis at low-, moderate-, and high-intensity exercise, fatty acid mobilization from adipose tissue is lower during vigorous exercise intensities compared with that found at lower intensities (Romijn et al. 1993, 2000) likely due in large part to a redistribution of blood flow away from adipose tissue at high intensity (Hodgetts et al. 1991). Although this reduction in this energy-rich source of fuel when energy demands are reaching their maximum may seem like a disadvantage, exercising muscles increase their reliance on readily available local intramuscular stores of carbohydrate and fat during very vigorous exercise (Romijn et al. 1993, 2000). In fact, as noted above, an artificial increase in circulating fatty acid availability to 1–2 mM via infusion of lipid and heparin during high-intensity exercise increases the rate of fat oxidation only very modestly (Romijn et al. 1995).

7.3.2 Regulation of Lipolysis

7.3.2.1 Lipolytic Proteins

Adipose tissue triacylglycerol lipolysis is regulated by enzymes (lipases) that ultimately yield the three resident fatty acids and glycerol. The main lipases involved in this process are adipose triglyceride lipase (ATGL) and hormone sensitive lipase (HSL). ATGL has very high specificity for the sn-2 position of the triacylglycerol, to yield 1 fatty acid and a diacylglycerol with fatty acids still bound in the sn-1 and sn-3 positions (Eichmann et al. 2012). ATGL appears to have limited affinity to hydrolyze diacylglycerols (Eichmann et al. 2012; Schweiger et al. 2006). Diacylglycerol is then largely hydrolyzed by HSL, resulting in the complete hydrolysis of the triacylglycerol into three fatty acids and glycerol (Eichmann et al. 2012; Schweiger et al. 2006). A third lipase, monoglycerol lipase (MGL), can also catalyze the hydrolysis of the remaining fatty acid and glycerol, but MGL is responsible for only a very small fraction (~5%) of total lipolytic rate (Schweiger et al. 2006). The activities of ATGL and HSL are highly regulated (Fig. 7.2), requiring the phosphorylation on specific amino acid residues on these proteins (e.g., ATGL, Ser404 and Thr372; HSL, Ser552, Ser649, Ser650, Ser554, and Ser589) in humans (Watt and Steinberg 2008), as well as interactions with other regulatory proteins. For example, phosphorylation of the perilipin 1 (PLIN1) on the lipid droplet appears to be essential for HSL translocation from the cytosol to the lipid droplet for its lipolytic activity (Sztalryd et al. 2003). The regulation of ATGL involves its interaction with even more regulatory proteins to modify its activity. Comparative gene identification 58 (CGI-58; also known as alpha/beta-hydrolase domain-containing protein 5 [ABHD5]) is a key co-activator for ATGL lipolytic activity. In vitro analysis indicates that adding CGI-58 to cell extracts containing ATGL increased lipolytic activity ~20-fold (Lass et al. 2006). Under basal/resting conditions, CGI-58 is found largely bound to PLIN 1, which prevents ATGL-CGI-58 binding (Subramanian et al. 2004). With the onset of exercise, PLIN 1 is phosphorylated leading to its release of CGI-58 (Ogasawara et al. 2010)—at the same time, phosphorylation of ATGL leads to its translocation to the lipid droplet (Xie et al. 2014), where it binds to CGI-58 and initiates lipolysis of triacylglycerol within the lipid droplet (Schweiger et al. 2006; Lass et al. 2006). These activation steps of the lipolytic cascade are largely initiated via adrenergic stimulation (discussed below—and Fig. 7.2). ATGL is also regulated via inhibition when bound to the protein G0/G1 switch 2 (G0S2), which both inhibits ATGL enzyme activity and attenuates its translocation to the lipid droplet (Schweiger et al. 2012; Yang et al. 2010). Other proteins are also known to regulate ATGL activity (e.g., fat-specific protein 27 (FSP27) (Grahn et al. 2014), hypoxia-inducible lipid droplet-associated protein (HILPA)) (Kulminskaya and Oberer 2020).

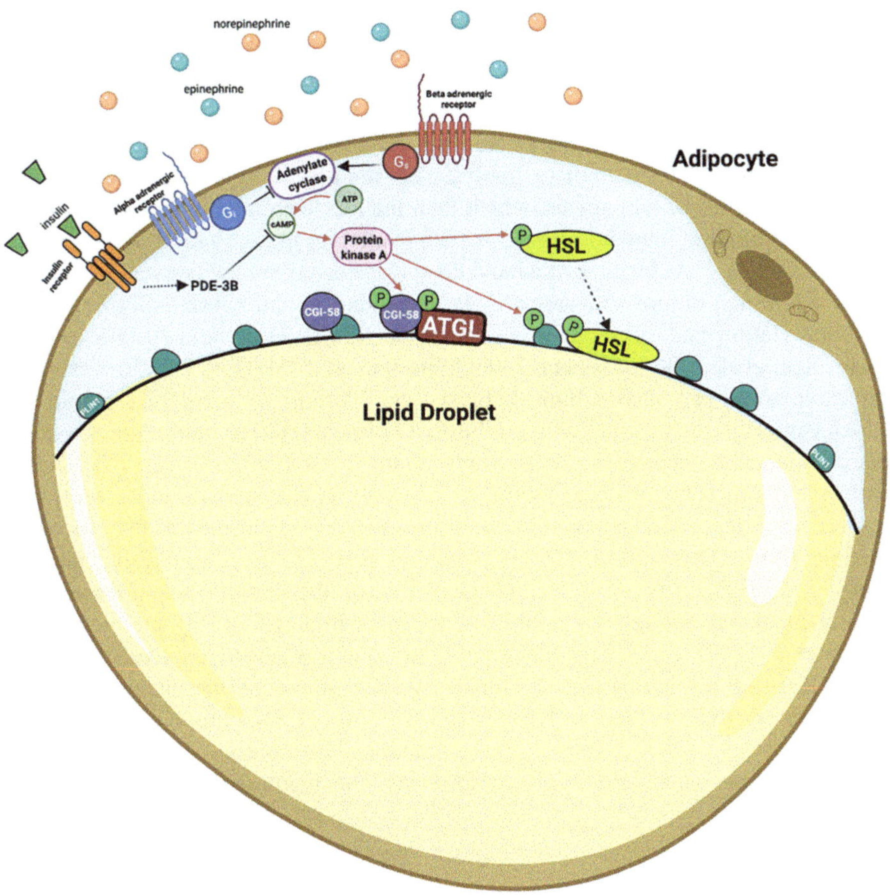

Fig. 7.2 Adrenergic regulation of lipolysis during exercise. The increase in lipolytic rate during exercise is primarily due to the increase in circulating epinephrine from the adrenal cortex and SNS-derived norepinephrine. These catecholamines bind to β-adrenoceptors on the plasma membrane of adipocytes resulting in a cascade of reactions that ultimately phosphorylate/activate the key lipase enzymes ATGL and HSL and also facilitate the translocation of these enzymes to the lipid droplet where the triacylglycerol and diacylglyerol are hydrolyzed. Insulin is a potent inhibitor of lipolysis, and the reduction in plasma insulin concentration that occurs during exercise facilitates the lipolytic activity during the exercise session. *ATGL* adipose triglyceride lipase; cAMP, *ATP* adenosine triphosphate, cyclic adenosine monophosphate; *CGI-58* comparative gene identification 58; G_i inhibitory G-protein; *Gs* stimulatory G-protein; *PDE-3B* phosphodiesterase; *PLIN1* perilipin 1

7.3.2.2 Adrenergic Regulation of Lipolysis During Exercise

Most reports suggest the adrenergic-mediated increase in lipolytic rate during exercise is primarily due to increased epinephrine (Epi) release from the adrenal cortex into the systemic circulation combined with norepinephrine (NE) spillover from the sympathetic nervous system (SNS) into the circulation (Arner 2005). However, direct SNS innervation of white adipose tissue can also play a role in lipolytic regulation (Youngstrom and Bartness 1995; Dodt et al. 1999; Bartness et al. 2014), but the contribution of direct neural stimulation on the increase in adipose tissue lipolysis during exercise in humans remains unclear. Regardless of their origin, catecholamines stimulate lipolysis by binding to β-adrenoceptors on the plasma membrane of adipocytes, which then interact with membrane-bound stimulatory GTP-binding regulatory proteins (Gs) leading to the activation of the enzyme adenylate cyclase. Activation of adenylate cyclase catalyzes the conversion of ATP to cAMP, which in turn activates cAMP-dependent protein kinase A leading to the phosphorylation and subsequent activation of ATGL, HSL, and PLIN1 proteins (Schweiger et al. 2006; Lass et al. 2006; Tansey et al. 2004) (Fig. 7.2). Interestingly, catecholamines can also inhibit lipolysis via binding to membrane-bound α_2-adrenoceptors, which couple with inhibitory G proteins (Gi) to inhibit the activation cascade described above (Fig. 7.2). The seemingly paradoxical effects of catecholamines as both activators and inhibitors of lipolysis (through β- and α-receptors, respectively) depend in large part on their concentration in plasma and their receptor-binding affinity (Arner et al. 1990; Galitzky et al. 1993). Resting plasma catecholamine concentration is low (150–500 pmol/L), and the lipolytic rate appears to be largely regulated through the inhibitory action of high-affinity α_2-adrenoreceptors (Arner et al. 1990), which keep circulating fatty acid availability relatively low when energy expenditure is low at rest. During exercise, however, the increase in circulating Epi and NE, as well as NE derived from direct SNS innervation, activates β-adrenoceptors, which override the α_2-mediated inhibition, and lipolytic rate increases (Arner et al. 1990) to accommodate the increased energy expenditure during exercise.

7.3.2.3 Insulin Regulation of Lipolysis

Adipose tissue lipolysis is also very sensitive to the anti-lipolytic effects of insulin (Campbell et al. 1992; Nurjhan et al. 1986). For example, an increase in plasma insulin concentration to ~30 μU/ml (insulin levels that a healthy adult might reach after eating a single slice of bread) can suppress lipolysis 50% below basal/overnight fasted levels (Campbell et al. 1992). Therefore, even small meals or snacks eaten in the hour or so before exercise will markedly suppress lipolytic rate during the subsequent exercise session, which in turn can be great enough to lower the rate of fat oxidation during exercise (Horowitz et al. 1997). Even though pancreatic insulin secretion is inhibited with the onset of exercise (Hirsch et al. 1991) and the elevated plasma insulin concentration from a pre-exercise meal often returns to basal levels within the first 20–30 min, there remains a persistent effect of this prior exposure to insulin that results in a suppressed lipolytic rate for at least 1 hour of moderate-to-vigorous exercise (Horowitz et al. 1997). Conversely, when exercising at least a few hours after a meal or snack, the exercise-induced reduction in plasma insulin concentration below basal levels contributes to the increase in lipolysis during exercise (Wasserman et al. 1989). Most of the inhibitory action of insulin on lipolysis has been attributed to the phosphorylation and resultant activation of phosphodiesterase (Makino et al. 1992; Lonnroth and Smith 1986), which degrades cAMP, thereby reducing the signaling cascade responsible for activating lipase activity. The anti-lipolytic activity of insulin also depends on the protein alpha-/beta-hydrolase domain-containing protein 15 (ABDH15) to help stabilize intracellular phosphodiesterase (Xia et al. 2018; Stöckli et al. 2019). Insulin phosphorylates phosphodiesterase via phosphatidylinositol 3-kinase (PI3-K) (Rahn et al. 1994), which also plays a key role in mediating insulin-stimulated glucose uptake. Therefore, much of insulin's impact on substrate metabolism (i.e., increase in carbohydrate metabolism and decrease in fat metabolism) is largely through activation of PI3-K. It is important to note that although insulin often evokes a very robust inhibitory effect on adipose tissue lipolysis and a resultant reduction in circulating fatty acid availability, the anti-lipolytic effect of insulin can be considerably blunted in adults with abdominal obesity (Nellemann et al. 2012). Additionally, it has been recently reported that not all subpopulations of adipocytes respond similarly to insulin (Bäckdahl et al. 2021).

7.3.2.4 Alternative Lipolytic Regulators

Several other hormones, peptides, and metabolites can also influence lipolytic rate (e.g., growth hormone, cortisol, interleukin 6 (IL6), natriuretic peptides, leptin, and growth differentiation factor 15 (GDF 15)). In general, the effects of these alternative lipolytic regulators are more modest and/or slower in regulating lipolysis compared with the effects of catecholamines and insulin. In addition, like catecholamines and insulin, in most cases, these factors also modify lipolytic rate through modulating intracellular cAMP levels. Here, we will briefly discuss a few of these alternative lipolytic regulators that demonstrate some intriguing tissue cross-talk during exercise.

Arterial and B-type natriuretic peptides (ANP and BNP, respectively) are primarily secreted from cardiac myocytes in response to stretch of atria and ventricle, respectively. In contrast to the effects of most lipolytic agents that act though elevation of intracellular cAMP levels, ANP and BNP stimulate lipolysis through modulation of intracellular cyclic guanosine monophosphate (cGMP), independently of cAMP (Moro et al. 2007; Sengenès et al. 2003). In a process closely resembling catecholamine-stimulated lipolysis, cGMP activates cGMP-dependent protein kinase (cGK-I, also referred to as "protein kinase G" (PKG)), which in turn phosphorylates the lipases and PLIN1 (Sengenès et al. 2003). Interestingly, the natriuretic peptide lipolytic pathway appears to be resistant to insulin-mediated inhibition (Moro et al. 2004a, b), which may contribute to the modest increase in lipolytic rate above pre-exercise levels observed after a pre-exercise meal, despite insulin's potent suppression of adrenergic-mediated lipolysis. Because these peptides are released into the circulation from cardiac muscle in response to a stretch stimulus (as occurs during exercise)—lipolytic activation stemming from increased circulating ANP and BNP may represent a communication link between the exercising heart and adipose tissue, to mobilize more energy to help fuel the body during exercise. However, circulating ANP concentrations are often found to only increase rather modestly during exercise in healthy subjects (Verboven et al. 2018; Steele et al. 1997), and BNP concentration is often found to increase even less or not at all during exercise (Steele et al. 1997; Huang et al. 2002). As such, the true contribution of these natriuretic peptides to the lipolytic response to exercise remains unclear.

Circulating IL-6 has also been found to increase adipose tissue lipolysis (van Hall et al. 2003; Petersen et al. 2005). Several tissues produce and release IL-6, but during exercise, IL-6 is released from exercising skeletal muscle generally in proportion to the decline in muscle glycogen (Chan et al. 2004; Pedersen et al. 2003). Therefore, the IL-6-mediated increase in lipolysis represents a potentially important cross-talk between skeletal muscle that is facing diminished energy reserves (in the form of glycogen) to adipose tissue, triggering the mobilization of more energy in the form of fatty acids that can be used by the exercising muscle.

GDF15 has also recently received attention for its potential to increase lipolytic rate during exercise (Laurens et al. 2020b). GDF15 is predominantly expressed in the liver, kidney, and lung (Ding et al. 2009), but also found in adipose tissue and skeletal muscle (Laurens et al. 2020b; Ding et al. 2009). Plasma GDF15 concentration increases during exercise (Laurens et al. 2020b; Kleinert et al. 2018; Galliera et al. 2014), but whether GDF15 is a "myokine" released by the exercising muscle is debated (Kleinert et al. 2018; Laurens et al. 2020a, b). Moreover, although GDF15 has been found to increase lipolytic rate in isolated adipocytes, the mechanism of lipolytic action remains unclear, and the presence of the GDF15 receptor (glial cell line-derived neurotropic factor receptor α-like; GFRAL) in adipose tissue is controversial (Laurens et al. 2020b; Tsai et al. 2014).

Identification and characterization of alternative lipolytic agents help complete our understanding about the regulation of lipolytic activity during exercise, but quantitatively these factors are likely relatively minor contributors to the overall lipolytic rate during exercise in healthy persons compared with the effects of increased catecholamines (both circulating and direct SNS innervation of adipose tissue) and the reduction in plasma insulin concentration.

7.4 Regional Differences in Adipose Tissue Metabolism and Fat Mobilization

It has been known for decades that lipolytic activity varies in different adipose tissue depots (Martin and Jensen 1991; Jensen et al. 1989). Adipocytes from intra-abdominal (i.e., visceral) adipose tissue are often reported to be more lipolytically active than adipocytes from subcutaneous adipose tissue (Mauriege et al. 1987). However, despite the high lipolytic activity of intra-abdominal adipocytes, intra-abdominal adipose tissue constitutes only a small proportion of total body fat mass—even in obese adults—and fatty acids derived from the splanchnic region contribute very little to the pool of fatty acids in the systemic circulation (Martin and Jensen 1991; Nielsen et al. 2004; Nguyen et al. 1996). In contrast, nearly 90% of circulating fatty acids are derived from subcutaneous adipose tissue (Nguyen et al. 1996, 2004). Therefore, most of the fatty acids that skeletal muscle is exposed to during exercise are derived from subcutaneous adipose tissue. Lipolytic activity is also heterogeneous in different subcutaneous adipose tissue beds. Although clearly an oversimplification, subcutaneous adipose tissue is commonly subdivided into two broad categories: upper-body/abdominal subcutaneous and lower-body/femoral+gluteal subcutaneous adipose tissue. During exercise, lipolytic rate is much greater in upper-body compared with lower-body subcutaneous adipose tissue, with lower-body adipose tissue typically contributing very little to whole-body lipolysis and systemic fatty acid delivery during exercise (Arner et al. 1990; Horowitz et al. 2000). Differences in local adipose tissue α_2- and β-adrenoceptor affinity, density, and function (Wahrenberg et al. 1991) likely underlie much of the regional heterogeneity in exercise-induced lipolytic rate between abdominal and lower-body adipose tissue beds (Frühbeck et al. 2014).

7.5 Alternative Sources of Fat Used During Exercise

In addition to the fat stored in adipose tissue—many studies indicate that fatty acids derived from lipid droplets stored within muscle cells ("intramyocellular triacylglycerol" (IMTG)) are also an important energy source during exercise (Romijn et al. 1993; Romijn et al. 2000; Loon et al. 2001). Because the fatty acids stored as IMTGs are already inside the myocyte, their ready availability and close proximity to the mitochondria (Samjoo et al. 2013; Devries et al. 2013; Shaw et al. 2008) make them a potentially effective energy source during exercise, and their contribution depends largely on the intensity of exercise (it generally increases with intensity) and training status. Endurance training has been found to increase the abundance and contribution of IMTG to energy production. In fact, much of the training-induced increase in fat oxidation commonly found after endurance training has been attributed to increased IMTG contribution (Horowitz et al. 2000; Martin et al. 1993). Lipids can also accumulate between muscle fibers, which is often associated with muscle injury, mobility impairment, aging, or disuse (Delmonico et al. 2009; Buford et al. 2012; Correa-de-Araujo et al. 2017; Song et al. 2004) and the contribution of these extramyocellular lipids to energy production during exercise is largely unknown (Sachs et al. 2019; Konopka et al. 2018; Durheim et al. 2008). Circulating triacylglycerols are another potential fat source for energy during exercise, but plasma triacylglycerols are typically not considered to be an important fuel source during exercise (Turcotte et al. 1992; Kiens and Lithell 1989; Sondergaard et al. 2011; Morio et al. 2004), although the importance of circulating triacylglycerols appears to be helpful during recovery after exercise (Morio et al. 2004; Lundsgaard et al. 2020). The next chapter in this book (Fritzen et al.—Chap. 8) provides an expanded discussion of these and other alternative sources of energy during exercise.

7.6 Endurance Training Effects on Fat Metabolism

As noted above, an increase in the relative contribution of fat to total energy expenditure during submaximal exercise at same absolute intensity is among the most robust responses to endurance exercise training (Holloszy and Coyle 1984). Interestingly, however, the primary source of this greater amount of fat used during exercise after training is not derived from adipose tissue (Horowitz et al. 2000; Martin et al. 1993). Adipose tissue lipolytic rates measured in vivo are found to be similar in endurance-trained athletes and untrained volunteers during exercise performed at the same absolute intensity (Klein et al. 1994). Additionally, in longitudinal studies, adipose tissue-derived systemic fatty acid availability during exercise does not increase (Horowitz et al. 2000) and can even decrease (Martin et al. 1993) after several weeks of endurance training, perhaps in consequence of a lower catecholamine response during exercise at the same intensity after training (Phillips et al. 1996). But even with similar catecholamine responses, lipolytic rate is often not found to be elevated after training (van Aggel-Leijssen et al. 2001; Horowitz et al. 1999). The likely source of the additional fat oxidized after training appears to be largely from IMTG (Horowitz et al. 2000; Martin et al. 1993). There is often some confusion regarding the effects of endurance training on adipose tissue lipolytic rate, perhaps in part because several classic studies demonstrated that maximal lipolytic rates in response to supraphysiological doses of catecholamines (10^{-6} to 10^{-4} mol/L) were greater in adipocytes from endurance-training vs. untrained subjects (Crampes et al. 1989; Riviere et al. 1989). But at physiological Epi concentrations (10^{-10} and 10^{-8} mol/L), lipolytic activity was the same or even slightly lower in adipocytes obtained from endurance-trained compared with untrained subjects (Crampes et al. 1989). Studies performed in situ (using microdialysis) also demonstrated no effect of training on adipose tissue Epi sensitivity (Richterova et al. 2004). So, while endurance exercise training increases maximal *lipolytic capacity* of adipose tissue, it does not appear to influence *lipolytic sensitivity* to adrenergic stimulation (Stinkens et al. 2018). Differences or changes in adiposity with exercise training can also contribute some to the confusion regarding the effects of training on lipolytic rate. The magnitude of fat mass has an important impact on lipolytic rate (Mittendorfer et al. 2009), so if exercise training is accompanied by a loss (or gain) in fat mass, lipolytic rate would indeed be affected (i.e., sustained weight loss results in lower lipolytic rate)—but this is an indirect effect of exercise training.

7.7 Sex Differences in Adipose Metabolism During Exercise

Sexual dimorphism in the proportion and distribution of adiposity in humans is well-known. The proportion of body fat mass is typically greater in women than men, and women often store more of their body fat in their lower body (femoral+gluteal), while men typically store more of their fat mass in the upper body/abdominal region.

Factors underlying these differences in body fat distribution may be related, at least in part, to differences in the abundance and response to sex steroid hormones (i.e., testosterone, estrogen) and their receptors, as well as differences in mechanisms associated with storage of dietary fat (Karastergiou et al. 2012; Karastergiou et al. 2013; Santosa et al. 2008; Santosa and Jensen 2012). Specifics regarding the putative mechanism responsible for sex differences in body fat distribution are beyond the scope of this review—this is addressed in Chap. 15 in this book (Lundsgaard, et al). Sex-related differences in adipose tissue lipolysis and fat oxidation during exercise have been widely reported, often indicating exercise lipolysis and relative contribution of fat oxidation are greater in women than men (Tarnopolsky et al. 1995; Froberg and Pedersen 1984; Friedlander et al. 1998; Chenevire et al. 2011). Because lipolytic rate is proportional to body fat mass (Mittendorfer et al. 2009), some of the higher lipolytic rate in women vs. men can be attributed to their higher proportion of body fat mass, but lipolytic rate is still elevated in women vs. men after normalizing for fat mass (Mittendorfer et al. 2002), suggesting other mechanisms underlie this difference. Interestingly, this sex difference in lipolytic rate appears to be absent in well-trained female and male endurance athletes (Roepstorff et al. 2002). Because lipolytic rate during exercise is largely driven by adrenergic regulation, sex differences in adipose tissue sensitivity to β-adrenergic stimulation or α-adrenergic inhibition of lipolysis are logical candidates. Several studies have reported that lipolytic sensitivity to physiologic concentrations of catecholamines was similar in adipocytes extracted from male and female subjects (Wahrenberg et al. 1991; Crampes et al. 1989; Mauriège et al. 1999) as well as in situ (Millet et al. 1998). Conversely, locally delivered α-adrenergic receptor blockade via microdialysis was reported to increase lipolysis during exercise in men but not women (Hellström et al. 1996), suggesting that differences in lipolytic inhibition through α-adrenergic receptor activity may contribute to sex differences in lipolytic rate during exercise.

7.8 Adipose Tissue-Derived "Adipokines"

Clearly, the primary role of adipose tissue is to store and mobilize available energy when exogenous energy availability is scarce or non-existent, but adipose tissue also acts as an endocrine organ. Adipose tissue produces several bioactive proteins and peptides, such as leptin, adiponectin, IL6, resistin, tumor necrosis factor-alpha (TNFα), retinol-binding protein 4 (RBP-4), and apelin. These adipose-derived factors, often referred to as "adipokines," can be released into the systemic circulation, delivered to other tissues (e.g., skeletal muscle, liver, heart) where they may have important biological impact. Much of the work examining the effects of exercise on adipokine release has been focused on adaptations to chronic exercise training—with specific interest on the potential impact of these adipokines on obesity-related cardiometabolic disease (Takahashi et al. 2019; Stanford et al. 2018; Lee et al. 2019). In contrast, the production, release, and impact of these adipokines *during* a session of exercise are far less clear. Leptin and adiponectin are the most well-characterized adipokines, yet the effects of acute exercise on adipose tissue leptin and adiponectin production are equivocal. Various reports indicate leptin and adiponectin mRNA expression and/or plasma concentrations remain unchanged (Ferguson et al. 2004; Varady et al. 2010; Keller et al. 2005; Bobbert et al. 2007), decrease (Højbjerre et al. 2007), or even increase modestly (Højbjerre et al. 2007; Christiansen et al. 2013; Saunders et al. 2012) during or immediately after exercise. As noted above, IL-6 release from the skeletal muscle appears to "communicate" with adipose tissue to release more fatty acids, as muscle glycogen reserves decline (Chan et al. 2004; Pedersen et al. 2003). But IL-6 is also produced in adipose tissue, and IL6 mRNA expression has been found to increase during exercise (Christiansen et al. 2013) and thereby may contribute to the circulating pool of IL-6 (but likely a relatively small contributor compared with skeletal muscle).

There now is also considerable interest in the metabolic impact of extracellular vesicles (EVs) that are released into the circulation, and the cargo within these EVs (e.g., microRNA [miRNA], nucleic acids, proteins, metabolites) can act on target tissues remote from the tissue of origin (Whitham et al. 2018; Nederveen et al. 2021). For example, EVs released during exercise have been found to have potential long-term health benefits in remote tissue, such as the liver (Whitham et al. 2018) and vascular endothelium (Wilhelm et al. 2016). There is little available evidence regarding the release of EVs from adipose tissue *during* exercise. Rigamonti et al. (Rigamonti et al. 2020) used fatty acid-binding protein (FABP) as a marker for adipose tissue-derived EVs in the circulation, and they reported these EVs did not increase significantly after 30 min of moderate-intensity exercise (60% VO_2max). There is also limited evidence regarding the effects of EVs released from other tissues on adipose tissue metabolism *during* exercise. In contrast, cargo delivered by EVs released from other tissues (e.g., skeletal muscle) during exercise, after exercise, or in response to chronic exercise training may have an important impact on adaptive changes in adipose tissue in response to regular exercise (Safdar et al. 2016; Safdar and Tarnopolsky 2018).

While the scope of this chapter is rooted on the effects of exercise on adipose metabolism *during* an exercise session, it is important to briefly touch on the responses that occur during or just after a session of exercise—that may have important impact on a wide array of longer-term fitness and/or health-related adaptations. There has been considerable interest in the possibility that exercise increases the abundance of the highly thermogenic brown/beige adipose tissue. It has been proposed that this may occur through a process initiated in the exercising muscle, where a PGC-1α-mediated increase in the membrane protein FNDC5 and the subsequent release of the cleaved protein product irisin can be taken up by white adipose tissue where it can lead to development of properties resembling highly thermogenic brown fat (i.e., "beiging") (Boström et al. 2012). The prospect of this muscle-adipose tissue cross-talk leading to adipose tissue reprograming and the subsequent health impact of this is of keen interest, but findings regarding the role of irisin are not universal, especially in human subjects (Norheim et al. 2014; Albrecht et al. 2015; Timmons et al. 2012). Other intriguing findings suggest that exercise responses in adipose tissue may lead to systemic improvements in metabolic health (Takahashi et al. 2019; Stanford et al. 2018; Stanford and Goodyear 2017). Work from Laurie Goodyear and Kristin Stanford report that exercise increased expression of over 1500 genes in white adipose tissue (Stanford et al. 2015), and studies from these researchers together, as well as in their respective laboratories, support the prospect that exercise-induced production and release of several factors from white adipose tissue, including transforming growth factor β2 (TGF-β2) and 12,13-dihydroxy-9Z-octadecenoic acid (12,13-diHOME), may have important impact on metabolic health in other tissues, such as the skeletal muscle and liver (Takahashi et al. 2019; Stanford et al. 2018; Stanford et al. 2015). Exercise training has also been reported to modify the structure and metabolic function of adipose tissue, even in the absence of weight loss. For example, adipose tissue capillarization has been found to increase after 3 months of exercise training (Walton et al. 2015). In general, adaptations to prolonged exercise training are largely due to the cumulative responses that occur with each exercise session, and work from our lab indicated the expression of vascular endothelial growth factor (VEGF; the primary transcription factor regulating angiogenesis) expression significantly increases in subcutaneous white adipose tissue in the few hours after each session of exercise (Pelt et al. 2017; Ludzki et al. 2018), which may underlie the observed increase in adipose tissue capillarization after training. Recent work from our lab also suggests that a single session of exercise rapidly modified the population of adipose tissue progenitor cells—leading to a reduction in the abundance of preadipocytes that yield adipocytes with relatively high lipolytic rates, which may in turn have important metabolic benefits in the context of obesity-related insulin resistance (Ludzki et al. 2020). Finally, a persistent elevation in fatty acid mobilization from white adipose tissue after a session of exercise not only provides an important fuel source during recovery, allowing for much of the available glucose to be resynthesized to glycogen in the skeletal muscle and liver, but the increased fatty acid availability also triggers classic adaptations to increase oxidative capacity through also signaling events initiated by fatty acids binding to transcription factors, such as peroxisome proliferator-activated receptor α (PPARα) (Lundsgaard et al. 2020).

7.9 Summary and Conclusions

Fatty acids derived from triacylglycerol stored in white adipose tissue are an essential fuel source during endurance exercise. The use of this abundant source of energy during exercise involves the complex integration of humeral, neural, and intracellular regulation to liberate the fatty acids from triacylglycerol within the adipose tissue—and is also impacted by changes in blood flow distribution during exercise to deliver the mobilized fatty acids to the exercising muscle. This process begins with the complex activation of the key lipases, ATGL and HSL, and their interaction with regulating proteins within the adipose tissue (e.g., CGI-58, PLIN,

G0S2). Lipolytic rates and regulation differ somewhat in different adipose tissue beds of the body, but the vast majority of circulating fatty acids used during exercise are derived from subcutaneous abdominal adipose tissue. Adipose tissue metabolism during exercise is affected by many factors, including exercise intensity, ingestion of meals of snacks before exercise, training status, and sex. In addition to being an abundant endogenous energy store to help fuel prolonged exercise (or other prolonged physical exertion), adipose tissue responses during and/or soon after each exercise session also underlie adaptive responses to training in the adipose tissue itself, as well as in other tissues, such as skeletal muscle and liver. There is a prevalent misconception that elevated lipolytic rate (and elevated fatty acid oxidation) can facilitate body fat loss—but in general, weight loss and fat loss cannot occur in absence of an energy deficit. However, exercise-induced modifications in adipose tissue stemming from each exercise session can lead to very meaningful improvements in metabolic health even in absence of weight loss. These beneficial health effects may be in consequence of an exercise-induced release of adipokines and/or EVs, or perhaps through exercise-mediated modifications to adipose tissue structure and function that stem from responses to *each* exercise session. Moving forward, exciting new research will not only continue to expand our understanding about the effects of exercise on adipose tissue metabolism for enhanced delivery of energy during prolonged physical activities but also provide advanced insights on exercise-induced change in adipose tissue that may improve and/or help prevent the development of many health complications.

References

Albrecht E, Norheim F, Thiede B, Holen T, Ohashi T, Schering L, Lee S, Brenmoehl J, Thomas S, Drevon CA, Erickson HP, Maak S (2015) Irisin–a myth rather than an exercise-inducible myokine. Sci Rep 5:463–410

Arner P (2005) Human fat cell lipolysis: biochemistry, regulation and clinical role. Best Pract Res Cl En 19:471–482

Arner P, Kriegholm E, Engfeldt P, Bolinder J (1990) Adrenergic regulation of lipolysis in situ at rest and during exercise. J Clin Invest 85:893–898

Bäckdahl J, Franzén L, Massier L, Li Q, Jalkanen J, Gao H, Andersson A, Bhalla N, Thorell A, Rydén M, Ståhl PL, Mejhert N (2021) Spatial mapping reveals human adipocyte subpopulations with distinct sensitivities to insulin. Cell Metab 33(9):1869–1882

Bartness TJ, Liu Y, Shrestha YB, Ryu V (2014) Neural innervation of white adipose tissue and the control of lipolysis. Front Neuroendocrinol 35:473–493

Bobbert T, Wegewitz U, Brechtel L, Freudenberg M, Mai K, Möhlig M, Diederich S, Ristow M, Rochlitz H, Pfeiffer A, Spranger J (2007) Adiponectin oligomers in human serum during acute and chronic exercise: relation to lipid metabolism and insulin sensitivity. Int J Sports Med 28:1–8

Boström P, Wu J, Jedrychowski MP, Korde A, Ye L, Lo JC, Rasbach KA, Boström EA, Choi JH, Long JZ, Kajimura S, Zingaretti MC, Vind BF, Tu H, Cinti S, Højlund K, Gygi SP, Spiegelman BM (2012) A PGC1-α-dependent myokine that drives brown-fat-like development of white fat and thermogenesis. Nature 481:463–468

Buford TW, Lott DJ, Marzetti E, Wohlgemuth SE, Vandenborne K, Pahor M, Leeuwenburgh C, Manini TM (2012) Age-related differences in lower extremity tissue compartments and associations with physical function in older adults. Exp Gerontol 47:38–44

Campbell P, Carlson M, Hill J, Nurjhan N (1992) Regulation of free fatty acid metabolism by insulin in humans: role of lipolysis and reesterification. Am J Phys 26:E1063–E1069

Chan MHS, Carey AL, Watt MJ, Febbraio MA (2004) Cytokine gene expression in human skeletal muscle during concentric contraction: evidence that IL-8, like IL-6, is influenced by glycogen availability. Am J Physiology-regulatory Integr Comp Physiology 287:R322–R327

Chenevire X, Borrani F, Sangsue D, Gojanovic B, Malatesta D (2011) Gender differences in whole-body fat oxidation kinetics during exercise. Appl Physiology Nutrition Metabolism 36:88–95

Christiansen T, Bruun JM, Paulsen SK, Ølholm J, Overgaard K, Pedersen SB, Richelsen B (2013) Acute exercise increases circulating inflammatory markers in overweight and obese compared with lean subjects. Eur J Appl Physiol 113:1635–1642

Correa-de-Araujo R, Harris-Love MO, Miljkovic I, Fragala MS, Anthony BW, Manini TM (2017) The need for standardized assessment of muscle quality in skeletal muscle function deficit and other aging-related muscle dysfunctions: a symposium report. Front Physiol 8:87

Coyle EF, Jeukendrup AE, Wagenmakers AJ, Saris WH (1997) Fatty acid oxidation is directly regulated by carbohydrate metabolism during exercise. Am J Phys 273:E268–E275

Crampes F, Riviere D, Beauville M, Marceron M, Garrigues M (1989) Lipolytic response of adipocytes to epinephrine in sedentary and exercise-trained subjects: sex-related differences. Eur J Appl Physiol O 59:249–255

Delmonico MJ, Harris TB, Visser M, Park SW, Conroy MB, Velasquez-Mieyer P, Boudreau R, Manini TM, Nevitt M, Newman AB, Goodpaster BH (2009) Longitudinal study of muscle strength, quality, and adipose tissue infiltration. Am J Clin Nutrition 90:1579–1585

Devries MC, Samjoo IA, Hamadeh MJ, McCready C, Raha S, Watt MJ, Steinberg GR, Tarnopolsky MA (2013) Endurance training modulates intramyocellular lipid compartmentalization and morphology in skeletal muscle of lean and obese women. J Clin Endocrinol Metab 98:4852–4862

Ding Q, Mracek T, Gonzalez-Muniesa P, Kos K, Wilding J, Trayhurn P, Bing C (2009) Identification of macrophage inhibitory Cytokine-1 in adipose tissue and its secretion as an Adipokine by human adipocytes. Endocrinology 150:1688–1696

Dodt C, Lönnroth P, Fehm HL, Elam M (1999) Intraneural stimulation elicits an increase in subcutaneous interstitial glycerol levels in humans. J Physiology 521:545–552

Durheim MT, Slentz CA, Bateman LA, Mabe SK, Kraus WE (2008) Relationships between exercise-induced reductions in thigh intermuscular adipose tissue, changes in lipoprotein particle size, and visceral adiposity. Am J Physiol-endoc M 295:E407–E412

Eichmann TO, Kumari M, Haas JT, Farese RV, Zimmermann R, Lass A, Zechner R (2012) Studies on the substrate and stereo/Regioselectivity of adipose triglyceride lipase, hormone-sensitive lipase, and diacylglycerol-O-acyltransferases*. J Biol Chem 287:41446–41457

Ferguson MA, White LJ, McCoy S, Kim H-W, Petty T, Wilsey J (2004) Plasma adiponectin response to acute exercise in healthy subjects. Eur J Appl Physiol 91:324–329

Friedlander A, Casazza G, Horning M, Huie M, Piacentini M, Trimmer J, Brooks G (1998) Training-induced alterations of carbohydrate metabolism in women: women respond differently from men. J Appl Physiol Respir Environ Exerc Physiol 85:1175–1186

Froberg K, Pedersen PK (1984) Sex differences in endurance capacity and metabolic response to prolonged, heavy exercise. Eur J Appl Physiol O 52:446–450

Frühbeck G, Méndez-Giménez L, Fernández-Formoso J-A, Fernández S, Rodríguez A (2014) Regulation of adipocyte lipolysis. Nutr Res Rev 27:63–93

Galitzky J, Reverte M, Portillo M, Carpene C, Lafontan M, Berlan M (1993) Coexistence of beta 1-, beta 2-, and beta 3-adrenoceptors in dog fat cells and their differential activation by catecholamines. Am J Phys 264:E402–E412

Galliera E, Lombardi G, Marazzi MG, Grasso D, Vianello E, Pozzoni R, Banfi G, Romanelli MMC (2014) Acute exercise in elite rugby players increases the circulating level of the cardiovascular biomarker GDF-15. Scand J Clin Lab Invest 74:492–499

Grahn THM, Kaur R, Yin J, Schweiger M, Sharma VM, Lee M-J, Ido Y, Smas CM, Zechner R, Lass A, Puri V (2014) Fat-specific protein 27 (FSP27) interacts with adipose triglyceride lipase (ATGL) to regulate lipolysis and insulin sensitivity in human adipocytes*. J Biol Chem 289: 12029–12039

Hellström L, Blaak E, Hagström-Toft E (1996) Gender differences in Adrenergic regulation of lipid mobilization during exercise. Int J Sports Med 17:439–447

Hirsch IB, Marker JC, Smith LJ, Spina RJ, Parvin CA, Holloszy JO, Cryer PE (1991) Insulin and glucagon in prevention of hypoglycemia during exercise in humans. Am J Physiol-endoc M 260:E695–E704

Hodgetts V, Coppack SW, Frayn KN, Hockaday TD (1991) Factors controlling fat mobilization from human subcutaneous adipose tissue during exercise. J Appl Physiol 71:445–451

Højbjerre L, Rosenzweig M, Dela F, Bruun JM, Stallknecht B (2007) Acute exercise increases adipose tissue interstitial adiponectin concentration in healthy overweight and lean subjects. Eur J Endocrinol 157:613–623

Holloszy JO (1967) Biochemical adaptations in muscle. Effects of exercise on mitochondrial oxygen uptake and respiratory enzyme activity in skeletal muscle. J Biol Chem 242:2278–2282

Holloszy JO, Booth FW (1976) Biochemical adaptations to endurance exercise in muscle. Annu Rev Physiol 38:273–291

Holloszy JO, Coyle E (1984) Adaptations of skeletal muscle to endurance exercise and their metabolic consequences. J Appl Physiol Respir Environ Exerc Physiol 56:831–838

Holloway GP, Bezaire V, Heigenhauser GJF, Tandon NN, Glatz JFC, Luiken JJFP, Bonen A, Spriet LL (2006) Mitochondrial long chain fatty acid oxidation, fatty acid translocase/CD36 content and carnitine palmitoyltransferase I activity in human skeletal muscle during aerobic exercise. J Physiology 571:201–210

Horowitz JF, Braudy RJ, Martin WH, Klein S (1999) Endurance exercise training does not alter lipolytic or adipose tissue blood flow sensitivity to epinephrine. [online]. Am J Phys 277:E325–E331. http://www.ncbi.nlm.nih.gov/entrez/query.fcgi?db=pubmed&cmd=Retrieve&dopt=AbstractPlus&list_uids=10444429

Horowitz JF, Coyle EF (1993) Metabolic responses to preexercise meals containing various carbohydrates and fat. Am J Clin Nutr 58:235–241

Horowitz JF, Leone TC, Feng W, Kelly DP, Klein S (2000) Effect of endurance training on lipid metabolism in women: a potential role for PPARalpha in the metabolic response to training. Am J Phys Endocrinol Metab 279:E348–E355

Horowitz JF, Mora-Rodriguez R, Byerley LO, Coyle EF (1997) Lipolytic suppression following carbohydrate ingestion limits fat oxidation during exercise. Am J Phys 273:E768–E775

Huang W-S, Lee M-S, Perng H-W, Yang S-P, Kuo S-W, Chang H-D (2002) Circulating brain natriuretic peptide values in healthy men before and after exercise. Metabolis 51:1423–1426

Jensen MD, Haymond M, Rizza R, Cryer PE, Miles J (1989) Influence of body fat distribution on free fatty acid metabolism in obesity. J Clin Invest 83:1168–1173

Karastergiou K, Fried SK, Xie H, Lee M-J, Divoux A, Rosencrantz MA, Chang RJ, Smith SR (2013) Distinct developmental signatures of human abdominal and gluteal subcutaneous adipose tissue depots. J Clin Endocrinol Metab 98:362–371

Karastergiou K, Smith SR, Greenberg AS, Fried SK (2012) Sex differences in human adipose tissues–the biology of pear shape. Biol Sex Differ 3:13

Keller P, Keller C, Steensberg A, Robinson LE, Pedersen BK (2005) Leptin gene expression and systemic levels in healthy men: effect of exercise, carbohydrate, interleukin-6, and epinephrine. J Appl Physiol 98:1805–1812

Kiens B, Lithell H (1989) Lipoprotein metabolism influenced by training-induced changes in human skeletal muscle. J Clin Invest 83:558–564

Klein S, Coyle E, Wolfe R (1994) Fat metabolism during low-intensity exercise in endurance trained and untrained men. Am J Phys 267:E934–E940

Kleinert M, Clemmensen C, Sjøberg KA, Carl CS, Jeppesen JF, Wojtaszewski JFP, Kiens B, Richter EA (2018) Exercise increases circulating GDF15 in humans. Molecular metabolism 9: 187–191

Konopka AR, Wolff CA, Suer MK, Harber MP (2018) Relationship between intermuscular adipose tissue infiltration and myostatin before and after aerobic exercise training. Am J Physiology-regulatory Integr Comp Physiology 315:R461–R468

Kulminskaya N, Oberer M (2020) Protein-protein interactions regulate the activity of adipose triglyceride lipase in intracellular lipolysis. Biochimie 169:62–68

Lass A, Zimmermann R, Haemmerle G, Riederer M, Schoiswohl G, Schweiger M, Kienesberger P, Strauss JG, Gorkiewicz G, Zechner R (2006) Adipose triglyceride lipase-mediated lipolysis of cellular fat stores is activated by CGI-58 and defective in Chanarin-Dorfman syndrome. Cell Metab 3:309–319

Laurens C, Bergouignan A, Moro C (2020a) Exercise-released Myokines in the control of energy metabolism. Front Physiol 11:91

Laurens C, Parmar A, Murphy E, Carper D, Lair B, Maes P, Vion J, Boulet N, Fontaine C, Marqués M-A, Larrouy D, Harant I, Thalamas C, Montastier E, Caspar-Bauguil S, Bourlier V, Tavernier G, Grolleau J-L, Bouloumié A, Langin D, Viguerie N, Bertile F, Blanc S, de Glisezinski I, O'Gorman DJ, Moro C (2020b) Growth and differentiation factor 15 is secreted by skeletal muscle during exercise and promotes lipolysis in humans. Jci Insight 5

Lee S, Norheim F, Langleite TM, Gulseth HL, Birkeland KI, Drevon CA (2019) Effects of long-term exercise on plasma adipokine levels and inflammation-related gene expression in subcutaneous adipose tissue in sedentary dysglycaemic, overweight men and sedentary normoglycaemic men of healthy weight. Diabetologia 62:1048–1064

Lonnroth P, Smith U (1986) The antilipolytic effect of insulin in human adipocytes requires activation of the phosphodiesterase. Biochem Bioph Res Co 141:1157–1161

Loon LJC, Greenhaff PL, Constantin-Teodosiu D, Saris WHM, Wagenmakers AJM (2001) The effects of increasing exercise intensity on muscle fuel utilisation in humans. J Physiology 536: 295–304

Ludzki AC, Krueger EM, Baldwin TC, Schleh MW, Porsche CE, Ryan BJ, Muir LA, Singer K, Lumeng CN, Horowitz JF (2020) Acute aerobic exercise remodels the adipose tissue progenitor cell phenotype in obese adults. Front Physiol 11:2238–2238

Ludzki AC, Pataky MW, Cartee GD, Horowitz JF (2018) Acute endurance exercise increases Vegfa mRNA expression in adipose tissue of rats during the early stages of weight gain. Appl Physiology Nutrition Metabolism 43:751–754

Lundsgaard A-M, Fritzen AM, Kiens B (2020) The importance of fatty acids as nutrients during post-exercise recovery. Nutrients 12:280

Makino H, Suzuki T, Kajinuma H, Yamazaki M, Ito H, Yoshida S (1992) The role of insulin-sensitive phosphodiesterase in insulin action. Adv Second Messenger Phosphoprotein Res 25: 185–199

Martin WH, Dalsky GP, Hurley BF, Matthews DE, Bier DM, Hagberg JM, Rogers MA, King DS, Holloszy JO (1993) Effect of endurance training on plasma free fatty acid turnover and oxidation during exercise. Am J Physiol-endoc M 265:E708–E714

Martin ML, Jensen MD (1991) Effects of body fat distribution on regional lipolysis in obesity. J Clin Invest 88:609–613

Mauriege P, Galitzky J, Berlan m, Lafontan M. (1987) Heterogeneous distribution of beta and alpha-2 adrenoceptor binding sites in human fat cells from various fat deposits: functional consequences. Eur J Clin Investig 17:156–165

Mauriège P, Imbeault P, Langin D, Lacaille M, Alméras N, Tremblay A, Després JP (1999) Regional and gender variations in adipose tissue lipolysis in response to weight loss. J Lipid Res 40:1559–1571

Millet L, Barbe P, Lafontan M, Berlan M, Galitzky J (1998) Catecholamine effects on lipolysis and blood flow in human abdominal and femoral adipose tissue. J Appl Physiol 85:181–188

Mittendorfer B, Horowitz JF, Klein S (2002) Effect of gender on lipid kinetics during endurance exercise of moderate intensity in untrained subjects. Am J Phys Endocrinol Metab 283:E58–E65

Mittendorfer B, Magkos F, Fabbrini E, Mohammed BS, Klein S (2009) Relationship between body fat mass and free fatty acid kinetics in men and women. Obesity (Silver Spring, Md) 17:1872–1877

Montain SJ, Hopper MK, Coggan AR, Coyle EF (1991) Exercise metabolism at different time intervals after a meal. J Appl Physiol 70:882–888

Morio B, Holmbäck U, Gore D, Wolfe RR (2004) Increased VLDL-TAG turnover during and after acute moderate-intensity exercise. Med Sci Sports Exerc 36:801–806

Moro C, Crampes F, Sengenes C, Glisezinski ID, Galitzky J, Thalamas C, Lafontan M, Berlan M (2004b) Atrial natriuretic peptide contributes to the physiological control of lipid mobilization in humans. FASEB J 18:908–910

Moro C, Galitzky J, Sengenes C, Crampes F, Lafontan M, Berlan M (2004a) Functional and pharmacological characterization of the natriuretic peptide-dependent Lipolytic pathway in human fat cells. J Pharmacol Exp Ther 308:984–992

Moro C, Klimcakova E, Lafontan M, Berlan M, Galitzky J (2007) Phosphodiesterase-5A and neutral endopeptidase activities in human adipocytes do not control atrial natriuretic peptide-mediated lipolysis. Brit J Pharmacol 152:1102–1110

Nederveen JP, Warnier G, Carlo AD, Nilsson MI, Tarnopolsky MA (2021) Extracellular vesicles and exosomes: insights from exercise science. Front Physiol 11:604274

Nellemann B, Gormsen LC, Sørensen LP, Christiansen JS, Nielsen S (2012) Impaired insulin-mediated Antilipolysis and lactate release in adipose tissue of upper-body obese women. Obesity 20:57–64

Nguyen TT, Mijares AH, Johnson CM, Jensen MD (1996) Postprandial leg and splanchnic fatty acid metabolism in nonobese men and women. Am J Phys 271:E965–E972

Nielsen S, Guo Z, Johnson CM, Hensrud DD, Jensen MD (2004) Splanchnic lipolysis in human obesity. J Clin Invest 113:1582–1588

Norheim F, Langleite TM, Hjorth M, Holen T, Kielland A, Stadheim HK, Gulseth HL, Birkeland KI, Jensen J, Drevon CA (2014) The effects of acute and chronic exercise on PGC-1α, irisin and browning of subcutaneous adipose tissue in humans. FEBS J 281(3):739–749

Nurjhan N, Campbell P, Kennedy F, Miles J, Gerich J (1986) Insulin dose-response characteristics for suppression of glycerol release and conversion to glucose in humans. Diabetes (Suppl 1) 35: 1326–1331

Ogasawara J, Nomura S, Rahman N, Sakurai T, Kizaki T, Izawa T, Ishida H, Haga S, Ohno H (2010) Hormone-sensitive lipase is critical mediators of acute exercise-induced regulation of lipolysis in rat adipocytes. Biochem Bioph Res Co 400:134–139

Pedersen BK, Steensberg A, Keller P, Keller C, Fischer C, Hiscock N, van Hall G, Plomgaard P, Febbraio MA (2003) Muscle-derived interleukin-6: lipolytic, anti-inflammatory and immune regulatory effects. Pflugers Arch 446:9–16

Pelt DWV, Guth LM, Wang AY, Horowitz JF (2017) Factors regulating subcutaneous adipose tissue storage, fibrosis, and inflammation may underlie low fatty acid mobilization in insulin-sensitive obese adults. Am J Phys Endocrinol Metab 313:E429–E439

Petersen EW, Carey AL, Sacchetti M, Steinberg GR, Macaulay SL, Febbraio MA, Pedersen BK (2005) Acute IL-6 treatment increases fatty acid turnover in elderly humans in vivo and in tissue culture in vitro. Am J Physiol-endoc M 288:E155–E162

Petrick HL, Holloway GP (2019) High intensity exercise inhibits carnitine palmitoyltransferase-I sensitivity to l-carnitine. Biochem J 476:547–558

Phillips S, Green H, Tarnopolsky MA, Heigenhauser G, Hill R, Grant S (1996) Effects of training duration on substrate turnover and oxidation during exercise. J Appl Physiol Respir Environ Exerc Physiol 81:2182–2191

Rahn T, Ridderstråle M, Tornqvist H, Manganiello V, Fredrikson G, Belfrage P, Degerman E (1994) Essential role of phosphatidylinositol 3-kinase in insulin-induced activation and phosphorylation of the cGMP-inhibited cAMP phosphodiesterase in rat adipocytes studies using the selective inhibitor wortmannin. FEBS Lett 350:314–318

Richterova B, Stich V, Moro C, Polak J, Klimcakova E, Majercik M, Harant I, Viguerie N, Crampes F, Langin D, Lafontan M, Berlan M (2004) Effect of endurance training on adrenergic control of lipolysis in adipose tissue of obese women. J Clin Endocrinol Metabolism 89:1325–1331

Rigamonti AE, Bollati V, Pergoli L, Iodice S, Col AD, Tamini S, Cicolini S, Tringali G, Micheli RD, Cella SG, Sartorio A (2020) Effects of an acute bout of exercise on circulating extracellular vesicles: tissue-, sex-, and BMI-related differences. Int J Obesity 44:1108–1118

Riviere D, Crampes F, Beauville M, Garrigues M (1989) Lipolytic response of fat cells to catecholamines in sedentary and exercise-trained women. J Appl Physiol 66:330–335

Roepstorff C, Steffensen CH, Madsen M, Stallknecht B, Kanstrup I-L, Richter EA, Kiens B (2002) Gender differences in substrate utilization during submaximal exercise in endurance-trained subjects. Am J Physiol-endoc M 282:E435–E447

Romijn JA, Coyle EF, Sidossis LS, Gastaldelli A, Horowitz JF, Endert E, Wolfe RR (1993) Regulation of endogenous fat and carbohydrate metabolism in relation to exercise intensity and duration. Am J Phys 265:E380–E391

Romijn JA, Coyle EF, Sidossis LS, Rosenblatt J, Wolfe RR (2000) Substrate metabolism during different exercise intensities in endurance-trained women. J Appl Physiol 88:1707–1714

Romijn JA, Coyle EF, Sidossis LS, Zhang XJ, Wolfe RR (1995) Relationship between fatty acid delivery and fatty acid oxidation during strenuous exercise. J Appl Physiol 79:1939–1945

Sachs S, Zarini S, Kahn DE, Harrison KA, Perreault L, Phang T, Newsom SA, Strauss A, Kerege A, Schoen JA, Bessesen DH, Schwarzmayr T, Graf E, Lutter D, Krumsiek J, Hofmann SM, Bergman BC (2019) Intermuscular adipose tissue directly modulates skeletal muscle insulin sensitivity in humans. Am J Phys Endocrinol Metab 316:E866–E879

Safdar A, Saleem A, Tarnopolsky MA (2016) The potential of endurance exercise-derived exosomes to treat metabolic diseases. Nat Publ Group 12:504–517

Safdar A, Tarnopolsky MA (2018) Exosomes as mediators of the systemic adaptations to endurance exercise. Cold Spring Harb Perspect Med 8:a029827–a029824

Samjoo IA, Safdar A, Hamadeh MJ, Glover AW, Mocellin NJ, Santana J, Little JP, Steinberg GR, Raha S, Tarnopolsky MA (2013) Markers of skeletal muscle mitochondrial function and lipid accumulation are moderately associated with the homeostasis model assessment index of insulin resistance in obese men. PLoS One 8:e66322

Santosa S, Hensrud DD, Votruba SB, Jensen MD (2008) The influence of sex and obesity phenotype on meal fatty acid metabolism before and after weight loss. Am J Clin Nutr 88: 1134–1141

Santosa S, Jensen MD (2012) Adipocyte fatty acid storage factors enhance subcutaneous fat storage in postmenopausal women. Diabetes 62:775–782

Saunders TJ, Palombella A, McGuire KA, Janiszewski PM, Després J-P, Ross R (2012) Acute exercise increases adiponectin levels in abdominally obese men. J Nutrition Metabolism 2012: 148729

Schweiger M, Paar M, Eder C, Brandis J, Moser E, Gorkiewicz G, Grond S, Radner FPW, Cerk I, Cornaciu I, Oberer M, Kersten S, Zechner R, Zimmermann R, Lass A (2012) G0/G1 switch gene-2 regulates human adipocyte lipolysis by affecting activity and localization of adipose triglyceride lipase. J Lipid Res 53:2307–2317

Schweiger M, Schreiber R, Haemmerle G, Lass A, Fledelius C, Jacobsen P, Tornqvist H, Zechner R, Zimmermann R (2006) Adipose triglyceride lipase and hormone-sensitive lipase are the major enzymes in adipose tissue triacylglycerol catabolism*. J Biol Chem 281:40236–40241

Sengenès C, Bouloumié A, Hauner H, Berlan M, Busse R, Lafontan M, Galitzky J (2003) Involvement of a cGMP-dependent pathway in the natriuretic peptide-mediated

hormone-sensitive lipase phosphorylation in human adipocytes*. J Biol Chem 278:48617–48626
Shaw CS, Jones DA, Wagenmakers AJM (2008) Network distribution of mitochondria and lipid droplets in human muscle fibres. Histochem Cell Biol 129:65–72
Sondergaard E, Rahbek I, Sørensen LP, Christiansen JS, Gormsen LC, Jensen MD, Nielsen S (2011) Effects of exercise on VLDL-triglyceride oxidation and turnover. AJP: Endocrinol Metab 300:E939–E944
Song M-Y, Ruts E, Kim J, Janumala I, Heymsfield S, Gallagher D (2004) Sarcopenia and increased adipose tissue infiltration of muscle in elderly African American women. Am J Clin Nutrition 79:874–880
Stanford KI, Goodyear LJ (2017) Muscle-adipose tissue cross talk. Csh Perspect Med 8:a029801
Stanford KI, Lynes MD, Takahashi H, Baer LA, Arts PJ, May FJ, Lehnig AC, Middelbeek RJW, Richard JJ, So K, Chen EY, Gao F, Narain NR, Distefano G, Shettigar VK, Hirshman MF, Ziolo MT, Kiebish MA, Tseng Y-H, Coen PM, Goodyear LJ (2018) 12,13-diHOME: an exercise-induced Lipokine that increases skeletal muscle fatty acid uptake. Cell Metab 27:1111–1120.e3
Stanford KI, Middelbeek RJW, Townsend KL, Lee M-Y, Takahashi H, So K, Hitchcox KM, Markan KR, Hellbach K, Hirshman MF, Tseng Y-H, Goodyear LJ (2015) A novel role for subcutaneous adipose tissue in exercise-induced improvements in glucose homeostasis. Diabetes 64:2002–2014
Steele IC, McDowell G, Moore A, Campbell NPS, Shaw C, Buchanan KD, Nicholls DP (1997) Responses of atrial natriuretic peptide and brain natriuretic peptide to exercise in patients with chronic heart failure and normal control subjects. Eur J Clin Investig 27:270–276
Stephens FB, Galloway SDR (2013) Carnitine and fat oxidation. Nestlé Nutrition Inst Work Ser 76: 13–23
Stinkens R, Brouwers B, Jocken JW, Blaak EE, Teunissen-Beekman KF, Hesselink MK, van Baak MA, Schrauwen P, Goossens GH (2018) Exercise training-induced effects on the abdominal subcutaneous adipose tissue phenotype in humans with obesity. J Appl Physiol Respir Environ Exerc Physiol 125:1585–1593
Stöckli J, Zadoorian A, Cooke KC, Deshpande V, Yau B, Herrmann G, Kebede MA, Humphrey SJ, James DE (2019) ABHD15 regulates adipose tissue lipolysis and hepatic lipid accumulation. Mol Metab 25:83–94
Subramanian V, Rothenberg A, Gomez C, Cohen AW, Garcia A, Bhattacharyya S, Shapiro L, Dolios G, Wang R, Lisanti MP, Brasaemle DL (2004) Perilipin a mediates the reversible binding of CGI-58 to lipid droplets in 3T3-L1 adipocytes*. J Biol Chem 279:42062–42071
Sztalryd C, Xu G, Dorward H, Tansey JT, Contreras JA, Kimmel AR, Londos C (2003) Perilipin a is essential for the translocation of hormone-sensitive lipase during lipolytic activation. J Cell Biol 161:1093–1103
Takahashi H, Alves CRR, Stanford KI, Middelbeek RJW, Nigro P, Ryan RE, Xue R, Sakaguchi M, Lynes MD, So K, Mul JD, Lee M-Y, Balan E, Pan H, Dreyfuss JM, Hirshman MF, Azhar M, Hannukainen JC, Nuutila P, Kalliokoski KK, Nielsen S, Pedersen BK, Kahn CR, Tseng Y-H, Goodyear LJ (2019) TGF-β2 is an exercise-induced adipokine that regulates glucose and fatty acid metabolism. Nat Metab 1:291–303
Tansey J, Sztalryd C, Hlavin E, Kimmel A, Londos C (2004) The central role of Perilipin a in lipid metabolism and adipocyte lipolysis. IUBMB Life 56:379–385
Tarnopolsky MA, Atkinson SA, Phillips SM, MacDougall JD (1995) Carbohydrate loading and metabolism during exercise in men and women. J Appl Physiol 78:1360–1368
Timmons JA, Baar K, Davidsen PK, Atherton PJ (2012) Is irisin a human exercise gene? Nature 488:E9-10.-discussion E10-1
Tsai VW-W, Manandhar R, Jørgensen SB, Lee-Ng KKM, Zhang HP, Marquis CP, Jiang L, Husaini Y, Lin S, Sainsbury A, Sawchenko PE, Brown DA, Breit SN (2014) The anorectic actions of the TGFβ cytokine MIC-1/GDF15 require an intact brainstem area Postrema and nucleus of the solitary tract. PLoS One 9:e100370

Turcotte L, Richter EA, Kiens B (1992) Increased plasma FFA uptake and oxidation during prolonged exercise in trained vs. untrained humans. Am J Phys 262:E791–E799

van Aggel-Leijssen D, Saris WHM, Homan M, van Baak M (2001) The effect of exercise training on beta-adrenergic stimulation of fat metabolism in obese men. Int J Obes Relat Metab Disord 25:16–23

van Hall G, Steensberg A, Sacchetti M, Fischer C, Keller C, Schjerling P, Hiscock N, Møller K, Saltin B, Febbraio MA, Pedersen BK (2003) Interleukin-6 stimulates lipolysis and fat oxidation in humans. J Clin Endocrinol Metabolism 88:3005–3010

Varady KA, Bhutani S, Church EC, Phillips SA (2010) Adipokine responses to acute resistance exercise in trained and untrained men. Medicine Sci Sports Exerc 42:456–462

Verboven K, Stinkens R, Hansen D, Wens I, Frederix I, Eijnde BO, Jocken JWE, Goossens GH, Blaak EE (2018) Adrenergically and non-adrenergically mediated human adipose tissue lipolysis during acute exercise and exercise training. Clin Sci 132:1685–1698

Wahrenberg H, Bolinder J, Arner P (1991) Adrenergic regulation of lipolysis in human fat cells during exercise. Eur J Clin Investig 21:534–541

Walton RG, Finlin BS, Mula J, Long DE, Zhu B, Fry CS, Westgate PM, Lee JD, Bennett T, Kern PA, Peterson CA (2015) Insulin-resistant subjects have normal angiogenic response to aerobic exercise training in skeletal muscle, but not in adipose tissue. Physiol Rep 3:e12415–e12415

Wasserman DH, Lacy DB, Goldstein RE, Williams PE, Cherrington AD (1989) Exercise-induced fall in insulin and increase in fat metabolism during prolonged muscular work. Diabetes 38:484–490

Watt MJ, Steinberg GR (2008) Regulation and function of triacylglycerol lipases in cellular metabolism. Biochem J 414:313–325

Whitham M, Parker BL, Friedrichsen M, Hingst JR, Hjorth M, Hughes WE, Egan CL, Cron L, Watt KI, Kuchel RP, Jayasooriah N, Estevez E, Petzold T, Suter CM, Gregorevic P, Kiens B, Richter EA, James DE, Wojtaszewski JFP, Febbraio MA (2018) Extracellular vesicles provide a means for tissue crosstalk during exercise. Cell Metab 27:237–251.e4

Wilhelm EN, González-Alonso J, Parris C, Rakobowchuk M (2016) Exercise intensity modulates the appearance of circulating microvesicles with proangiogenic potential upon endothelial cells. Am J Physiol-heart C 311:H1297–H1310

Wolfe R, Klein S, Carraro F, Weber J (1990) Role of triglyceride-fatty acid cycle in controlling fat metabolism in humans during and after exercise. Am J Phys 258:E382–E389

Xia W, Pessentheiner AR, Hofer DC, Amor M, Schreiber R, Schoiswohl G, Eichmann TO, Walenta E, Itariu B, Prager G, Hackl H, Stulnig T, Kratky D, Rülicke T, Bogner-Strauss JG (2018) Loss of ABHD15 impairs the anti-lipolytic action of insulin by altering PDE3B stability and contributes to insulin resistance. Cell Rep 23:1948–1961

Xie X, Langlais P, Zhang X, Heckmann BL, Saarinen AM, Mandarino LJ, Liu J (2014) Identification of a novel phosphorylation site in adipose triglyceride lipase as a regulator of lipid droplet localization. Am J Physiol-endoc M 306:E1449–E1459

Yang X, Lu X, Lombès M, Rha GB, Chi Y-I, Guerin TM, Smart EJ, Liu J (2010) The G(0)/G(1) switch gene 2 regulates adipose lipolysis through association with adipose triglyceride lipase. Cell Metab 11:194–205

Youngstrom TG, Bartness TJ (1995) Catecholaminergic innervation of white adipose tissue in Siberian hamsters. Am J Physiology-regulatory Integr Comp Physiology 268:R744–R751

Chapter 8
Regulation of Fatty Acid Oxidation in Skeletal Muscle During Exercise: Effect of Obesity

Andreas Mæchel Fritzen, Nicholas T. Broskey, Anne Marie Lundsgaard, G. Lynis Dohm, Joseph A. Houmard, and Bente Kiens

Abstract This chapter summarizes how fatty acid (FA) oxidation is regulated in skeletal muscle during exercise and the role of obesity in regulation of FA oxidation in skeletal muscle. The substrates fueling increased FA oxidation in skeletal muscle during exercise are mainly circulating FAs, although hydrolysis of circulating triacylglycerol (TG) in very-low-density lipoproteins (VLDL-TG) and especially lipolysis of intramuscular TG (IMTG) also appear to contribute to some extent. Several steps are involved in FA uptake and oxidation in skeletal muscle and could all be of importance in the regulation of FA oxidation during exercise. Besides trans-sarcolemmal FA uptake via fatty acid transporters, it appears that intramyocellular carnitine content plays an important regulatory step in regulation of substrate selection during exercise. Interestingly, individuals with obesity exhibit a compromised ability to oxidize FAs and to increase FA oxidation in response to lipid exposure (reduced metabolic flexibility). Skeletal muscle mitochondrial function appears to be related to this defect. It remains controversial whether this impaired FA oxidative capacity in obesity diminishes the ability to increase and properly regulate FA oxidation during an acute, single exercise bout. However,

"Andreas Mæchel Fritzen" and "Nicholas T. Broskey" are co-first authors.

A. M. Fritzen · A. M. Lundsgaard · B. Kiens (✉)
The Lipid Research Group, August Krogh Section for Molecular Physiology, Department of Nutrition, Exercise and Sports, Faculty of Science, University of Copenhagen, Copenhagen, Denmark
e-mail: bkiens@nexs.ku.dk

N. T. Broskey · J. A. Houmard
Department of Kinesiology, East Carolina University, Greenville, NC, USA

Human Performance Laboratory, East Carolina University, Greenville, NC, USA

East Carolina Diabetes and Obesity Institute, East Carolina University, Greenville, NC, USA

G. L. Dohm
East Carolina Diabetes and Obesity Institute, East Carolina University, Greenville, NC, USA

Department of Physiology, East Carolina University, Greenville, NC, USA

© The Author(s), under exclusive license to Springer Nature Switzerland AG 2022
G. McConell (ed.), *Exercise Metabolism*, Physiology in Health and Disease,
https://doi.org/10.1007/978-3-030-94305-9_8

despite these initial impairments in FA oxidation capacity in the obese situation, endurance exercise training can rescue the capacity for FA oxidation and the metabolic flexibility in the skeletal muscle of individuals with obesity at least to equivalent levels of their lean counterparts.

Keywords Fatty acid oxidation · Exercise · Skeletal muscle · Mitochondria · Obesity · Exercise training · Lipid metabolism

8.1 Introduction

The work by Krogh and Lindhard and by Christensen and Hansen in the 1920s and 1930s demonstrated from measurements of the non-protein respiratory exchange ratio (RER) that fatty acid (FA) oxidation increased five- to tenfold above resting levels during mild-to-moderate exercise and decreased with increasing exercise intensities (Krogh and Lindhard 1920; Christensen and Hansen 1939). Today, it is well recognized that FA oxidation reaches its maximum at moderate intensities between 55 and 65% of maximal oxygen uptake (VO_2peak) (Lundsgaard et al. 2018; Romijn et al. 1993). Beyond this level, a shift in fuel selection appears toward an increase in carbohydrate and a decrease in FA utilization (Lundsgaard et al. 2018; Romijn et al. 1993). Furthermore, the FA oxidation rate during mild-to-moderate exercise remains generally unchanged for about 60–90 min of exercise, but when exercise continues beyond this time point, a gradual increase in FA oxidation is induced at the expense of carbohydrate oxidation as fuel for energy (Romijn et al. 1993).

8.2 Fatty Acids as Energy Fuel in Skeletal Muscle

FAs as fuel for energy during exercise originate from three different sources: 1) Albumin-bound long-chain FAs liberated from lipolysis of triacylglycerol (TG) located in adipose tissue, 2) plasma FAs liberated from lipoprotein lipase (LPL)-dependent hydrolysis of TG situated in very-low density lipoprotein (VLDL-TG), and 3) FAs liberated by lipolysis of TG situated in lipid droplets in skeletal muscle (IMTG) (Fig. 8.1). The extent to which the different energy sources contribute during exercise is dependent on exercise intensity, duration, mode, and sex.

Findings indicate that approximately 55–65% of total whole-body FA utilization during moderate-intensity exercise, where FA oxidation is at its highest level, is derived from plasma FAs (Romijn et al. 1993; Friedlander et al. 1999; Helge et al. 2001; Roepstorff et al. 2002; van Loon et al. 2001) and that the contribution from plasma FAs to energy provision increases with time (van Loon et al. 2003).

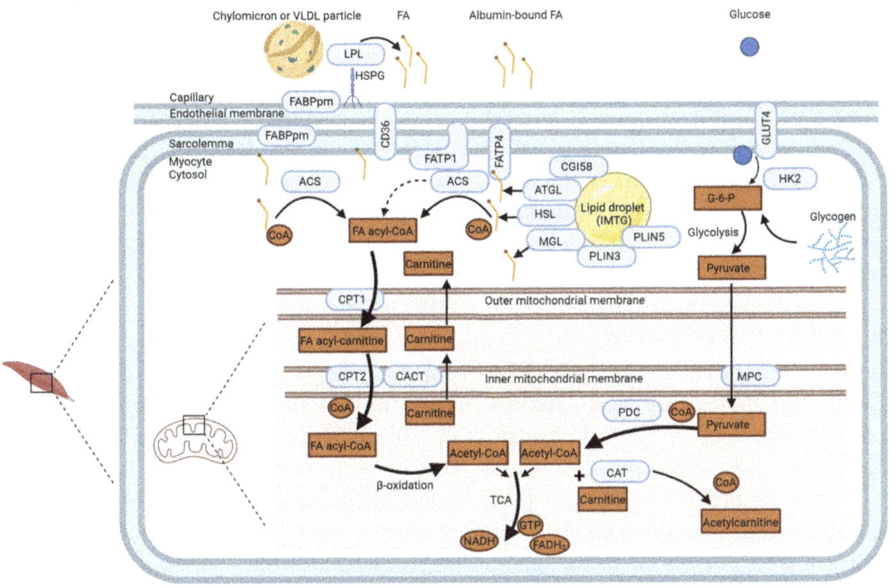

Fig. 8.1 Proposed regulation of fatty acid oxidation in skeletal muscle during exercise. This figure illustrates how we propose that fatty acid (FA) oxidation is regulated in skeletal muscle during exercise. FA oxidation during exercise is mainly covered by uptake of circulating FAs, although hydrolysis of circulating triacylglycerol (TG) in very-low-density lipoproteins (VLDL-TG) and especially lipolysis of intramuscular TG (IMTG) also appears to contribute to some extent. Several steps are involved in FA oxidation and could all be of importance in the regulation of FA oxidation during exercise. Besides trans-sarcolemmal FA uptake via fatty acid transporters, it appears that intramyocellular carnitine content plays an important regulatory step in the mitochondrial import and hence FA oxidation during exercise

8.2.1 Albumin-Bound Plasma FA

Plasma FA uptake into skeletal muscle is the product of blood flow and the arteriovenous FA concentration difference. The increase in blood flow during exercise, which increases up to 20-fold from rest to intense, dynamic exercise (Calbet and Lundby 2012; Radegran and Saltin 1998), is a main driver of the exercise-induced increased FA uptake. The other important determinant of FA uptake during exercise is the arterial FA concentration. At onset of exercise, a transient decrease in plasma FA concentration may appear followed by a slow increase (Roepstorff et al. 2002). If exercise is prolonged, the arterial concentration of FA may increase to levels approximately sixfold higher than resting levels (Romijn et al. 1993; Kiens and Richter 1998; Wolfe et al. 1990; van Hall et al. 2002; Bahr et al. 1991; Hagenfeldt and Wahren 1975). The initial drop in arterial FA concentration is caused by an imbalance between slow mobilization of FAs from the adipose tissue and a rapidly increased extraction of FAs in skeletal muscle. The following increase in plasma FA concentration during exercise is mainly caused by an increased release of FAs liberated from adipose tissue by lipolysis of TG. Accordingly, the whole-body lipolytic rate can be increased by up to fourfold compared with resting values during submaximal moderate-intensity exercise (Wolfe et al. 1990; Romijn et al. 2000). This exercise-induced increase in adipose tissue lipolysis is mediated by altered adrenergic stimulation of adipose tissue evidenced by a shift from predominant α-adrenergic suppression during rest toward predominant β-adrenergic stimulation during exercise (Arner et al. 1990). The primary adrenergic stimulus of adipose tissue during exercise originates from circulating adrenaline with only a minor contribution from noradrenaline released from sympathetic neurons (Stallknecht et al. 2001; de Glisezinski et al. 2009). Moreover, the natriuretic peptides appear to play an additional role in exercise-induced lipolysis in humans (Moro et al. 2004; de Glisezinski et al. 2009) and have been suggested to account for most of the non-adrenergic lipolytic signaling in adipose tissue during exercise (Moro et al. 2006; Lafontan et al. 2008). The regulation of adipose tissue lipolysis during exercise is described in more detail in another chapter of this eBook.

It appears that 60 to 76% in females and males, respectively, of the exercise-induced increase in whole-body FA removal from the circulation can be accounted for by uptake into working skeletal muscles (Kiens 2006). The question is how much of the FAs taken up by the working muscle is then directly oxidized? When FA uptake into skeletal muscle during exercise was directly measured, it was found that up to 100% and 84% of tracer-derived FA uptake were directly oxidized in trained females and males, respectively (Roepstorff et al. 2002; Hagenfeldt and Wahren 1968; Turcotte et al. 1992).

Another fat source generating energy during exercise may be hydrolysis of circulating TG. The liver is secreting endogenous TG enfolded into VLDL particles directly into the circulation. VLDLs are the main carriers of TG in the post-absorptive state. Hydrolysis of core TG in VLDL is mediated by the enzyme lipoprotein lipase (LPL), which in its active form is located at the luminal site of the endothelial cells in the capillary bed of various tissues such as skeletal muscle, heart, and adipose tissue (Fig. 8.1). By hydrolysis, FAs are liberated and taken up by the surrounding tissues. Oxidation of FAs from VLDL-TG may contribute to the total FA oxidation both at rest and during moderate-intensity exercise, though its relative contribution during exercise is less than albumin-bound FA oxidation. Hence, when VLDL-TG content was measured in the femoral artery and vein during knee-extensor exercise, a total net degradation of VLDL-TG was found during the 2 hours of exercise (Kiens et al. 1993), suggesting that circulating VLDL-TG contributes to the total FA oxidation during moderate exercise. Supporting these findings are recent elegant studies using infusion of labeled VLDL-TG in healthy, young humans. Data showed that FAs from labeled VLDL-TG comprised 3–6% of total energy utilization (Sondergaard et al. 2011) or 3–13% of total FA oxidation during moderate-intensity exercise in untrained and moderately trained men (Nellemann et al. 2014; Morio et al. 2004). Increased VLDL-TG hydrolysis during exercise is likely explained by the findings of increased activity of LPL in skeletal muscle in most (Kiens and Richter 1998; Perreault et al. 2004; Taskinen and Nikkila 1980), but not all, human studies (Lithell et al. 1979; Kiens et al. 1989; Søndergaard et al. 2017). The mechanisms driving translocation of newly synthesized LPL in the myocyte to the luminal side of the endothelial cells in the capillary bed during exercise remain to be elucidated. However, exercise-induced muscle LPL activity has been correlated to the IMTG and muscle glycogen concentrations (Kiens 2006), suggesting that the energy status of the myocyte might signal to LPL synthesis and translocation to the capillary to facilitate VLDL-TG hydrolysis during exercise. Importantly, LPL-derived VLDL-TG hydrolysis might be a greater contributor as substrate to cover the high FA oxidation during recovery from exercise (Kiens and Richter 1998; Morio et al. 2004; Lundsgaard et al. 2020).

8.2.2 Regulation of Fatty Acid Uptake into Skeletal Muscle

The increase in uptake of plasma FAs into skeletal muscle during exercise is regulated at several steps including transmembrane transport, cytosolic handling, mitochondrial membrane transport, and intra-mitochondrial FA oxidation (Fig. 8.1). Transport of plasma FAs across sarcolemma serves as the first step in myocellular FA uptake. Despite the fact that FAs can diffuse within biological membranes, membrane-bound lipid-binding proteins have been identified in human skeletal muscle, and evidence show that these proteins either individually or in complexes act as regulators of FA transmembrane transport (Fig. 8.1). The first suggestion for FA transporter limited FA uptake came from both human (Kiens et al. 1993) and rat studies (Turcotte et al. 1992) indicating that FA uptake is a saturable process. The FA translocase cluster of differentiation 36 (CD36) are most extensively studied for its importance in FA uptake during exercise. In the basal resting state, a large part of total muscle CD36 is stored in intramyocellular compartments (endosomes), whereas a small part is present at the sarcolemma, the outer membrane of the muscle cell, to mediate basal FA uptake (Chabowski et al. 2007). During exercise, CD36 reversibly translocates to sarcolemma (Jeppesen et al. 2011) shown in human skeletal muscle by a 75% higher sarcolemmal content of CD36 after prolonged submaximal exercise compared with rest (Bradley et al. 2012). In rat skeletal muscle, FA transport was 75% higher in contracted compared with rested rat skeletal muscle, which correlated with CD36 translocation to sarcolemma (Bonen et al. 2000). In addition, mice lacking CD36 exhibited lower FA oxidation compared with control mice during treadmill exercise at the same relative workload (McFarlan et al. 2012), whereas mice overexpressing CD36 protein exhibited greater contraction-stimulated FA oxidation than control mice (Ibrahimi et al. 1999). Together findings in both rodent and human skeletal muscle clearly show that translocation of CD36 seems vital in FA uptake in skeletal muscle during exercise.

Skeletal muscle contractions also induce translocation of other FA transporters, such as FA-binding protein at the plasma membrane (FABPpm) and FA transport 1 and 4 (FATP1/4), to the plasma membrane in human skeletal muscle (Bradley et al. 2012; Jain et al. 2009) (Fig. 8.1). Overexpression or inhibition of these proteins in rodent skeletal muscle has been shown to increase or decrease FA uptake, respectively, during resting conditions (Holloway et al. 2007b; Clarke et al. 2004; Turcotte et al. 2000). However, the role of these proteins in FA oxidation during acute exercise remains to be established.

The signaling mechanism(s) governing sarcolemmal FA transporter translocation during exercise are not clarified. Activation of various intracellular signaling pathways related to the energy status (e.g., AMP-activated protein kinase (AMPK)) (Jeppesen et al. 2011; Bonen et al. 2007), mechanical stress (e.g., extracellular signal-regulated kinase (ERK) 1/2 and p38 mitogen-activated protein (MAP) kinase signaling) (Raney and Turcotte 2006), ionic homeostasis (e.g., Ca^{2+}/calmodulin-dependent protein kinases (CaMK)) (Lally et al. 2012; McFarlan et al. 2012), and other signaling molecules such as the Rab-GTPase activating proteins TBC1 domain family member 1 and 4 (TBC1D1 and -4) (Chadt et al. 2008; Benninghoff et al. 2020; Samovski et al. 2012) have all been proposed as regulators of FA uptake via CD36 translocation to sarcolemma during resting conditions. However, exercise-stimulated CD36 translocation, FA uptake, and/or FA oxidation in skeletal muscle were not impaired, when AMPK (Jeppesen et al. 2011; Dzamko et al. 2008; Hingst et al. 2020) and TBC1D1 (Whitfield et al. 2017) were genetically deleted in rodent muscles. Besides, contraction-induced CD36 translocation occurred prior to changes in phosphorylation of ERK1/2 and other major MAP kinases in mouse muscle (Jeppesen et al. 2011). Thus, it is presently unknown what signal that stimulates CD36 and other lipid-binding proteins to translocate to sarcolemma and hereby increase FA uptake in skeletal muscle during exercise. Systemic signaling molecules from outside the muscle cells could likely also contribute, e.g., circulating levels of the lipid 12,13-dihydroxy-9Zoctadecenoic acid (12,13-diHOME) originating from brown adipose tissue (BAT) increases during exercise and are shown to be able to induce skeletal muscle FA uptake and oxidation (Stanford et al. 2018). This emphasizes that not only intracellular mechanisms but likely also systemic signals could possibly together orchestrate the increased uptake of circulating FAs into skeletal muscle during exercise.

8.3 The Intracellular FA Source

Circulating FAs taken up into skeletal muscle can either be oxidized or stored in intramyocellular lipid droplets (IMTG) and add to the plasma-delivered FAs as a potential energy source during exercise (Fig. 8.1). To what degree IMTG is used during exercise is discussed heavily in the literature for several decades, and different methodologies have contributed to evaluate the contribution of IMTG as fuel for FA oxidation during exercise. Early studies with the muscle biopsy technique found a 25–30% IMTG reduction after 99–147 min submaximal exercise in men (Costill et al. 1979; Carlson et al. 1971), and such exercise-induced decrease in IMTG in men was later supported by more sophisticated freeze-dried dissection of muscle tissue after exercise protocols of $\sim$2 h (Essén 1978; Hurley et al. 1986; Phillips et al. 1996). However, other studies did not find detectable changes in IMTG during exercise in mainly shorter 60–120 min exercise protocols (Helge et al. 2001; Roepstorff et al. 2002; Kiens and Richter 1998; Kiens et al. 1993; Bergman et al. 1999; Starling et al. 1997; Steffensen et al. 2002). Lack of breakdown of IMTG was also shown indirectly by microdialysis during knee-extensor exercise at submaximal exercise in men (Stallknecht et al. 2004). This could emphasize that exercise needs to be prolonged—likely above 90 min to acquire IMTG as a major substrate to fuel FA oxidation during exercise.

From studies, where the ^{1}H-MRS technique was applied, treadmill running at 50–70% of VO_2peak for 2 h or until exhaustion decreased IMTG signal (IMCL in MRS terminology) by 22–33% in men and women (Krssak et al. 2000; Larson-Meyer et al. 2002; Decombaz et al. 2001). However, it is hard to exclude from those studies, where it takes considerable time to be positioned in the scanner and perform the measurements, whether the lower IMCL signal is due to use of IMTG in the early recovery period rather than during exercise.

Importantly, it seems that women of widely differing training backgrounds in contrast to matched men utilize a significant greater amount of IMTG in the vastus lateralis during prolonged bicycle exercise (Steffensen et al. 2002). The role of sex in regulation of lipid metabolism during exercise is described in more detail in a separate chapter in this eBook.

Estimations from tracer-infusion studies suggest that oxidation of FAs from plasma versus IMTG and other lipid sources such as VLDL-TG and TG between fibers comprise ~60% and 30% of FA oxidation, respectively, during 2 h of moderate-intensity exercise (Romijn et al. 1993; van Loon et al. 2003). These findings support the notion that IMTG lipolysis is not likely to limit FA availability to oxidation in skeletal muscle during exercise.

Summarizing data which seems as quantitative importance of IMTG as an energy source during exercise depends on several factors as duration and intensity of exercise, exercise mode, and sex (Kiens 2006). Overall, it appears that FAs derived from hydrolysis of IMTG may contribute as energy fuel especially when exercise is prolonged—beyond 90 min, mainly when exercise is performed in the fasted state and to a larger extent in women than in men (Kiens 2006).

Interestingly, another key aspect in the necessity of IMTG as substrate for FA oxidation during exercise is a potential interdependency with the circulating FA availability. If circulating levels of FAs become limited, it could be speculated that there is an inverse interaction between plasma-derived FAs and those generated from IMTG lipolysis during exercise. Thus, when adipose tissue lipolysis was pharmacologically inhibited by nicotinic acid or acipimox prior to prolonged submaximal exercise in healthy individuals, the exercise-induced increase in plasma FA concentration was completely suppressed, while IMTG utilization was increased during exercise compared with when plasma FA availability was not limited (Watt et al. 2004b; van Loon et al. 2005). Therefore, the uptake of circulating FAs may interact with the regulation of IMTG lipolysis during exercise.

Another consideration about the importance of IMTG for FA oxidation during exercise is the fact that most investigations have focused on measuring net IMTG breakdown during exercise. From studies during resting conditions using pulse-chase methods by intravenous infusions of two distinct isotopically labelled FAs combined with mass spectrometry measurements of intramuscular lipids, it was shown that plasma FAs taken up by the muscle were not directly oxidized, but traversed the IMTG pool prior to oxidation in the resting state (Kanaley et al. 2009). At very-low-intensity exercise loads, a similar fate of circulating FA uptake undergoing esterification and then subsequent hydrolysis prior to mitochondrial entry may appear. Also, when the exercise load is low and only few muscle fibers are recruited, it could be speculated that FAs taken up into skeletal muscle are esterified and stored as IMTG in the non-recruited muscle fibers. Under such circumstances, it is difficult to evaluate IMTG breakdown during exercise with the available methods.

In terms of the molecular regulation of IMTG lipolysis during exercise and muscle contractions, this is orchestrated primarily by activation of the two lipases adipose triglyceride lipase (ATGL) (Alsted et al. 2013) and hormone-sensitive lipase (HSL) in skeletal muscle (Langfort et al. 2000; Watt et al. 2004a), which are catalyzing the conversion of TG to diacylglycerol and further to monoacylglycerol, respectively (Fig. 8.1). The signals regulating ATGL and HSL activity during exercise are complex and only scarcely understood in skeletal muscle. In this regard, ex vivo contractions in isolated skeletal muscle resulted in HSL translocation to lipid droplets (Prats et al. 2006). The translocation of HSL appears accredited to an intrinsic activation of HSL at different serine residues within the protein achieved by catecholamine-induced protein kinase A (PKA) activation (Talanian et al. 2006). The resultant translocation of HSL to the lipid droplet initiates lipolysis. In contrast, activation of ATGL does not seem to be PKA-dependent, but rather is requiring co-activation by comparative gene identification-58 (CGI-58) to achieve maximal hydrolase activity (Zechner et al. 2012) (Fig. 8.1).

The lipid droplet associated perilipins (PLINs) are also part of the lipolytic machinery in skeletal muscle, and PLIN3 and PLIN5 physically interact with HSL and ATGL (MacPherson et al. 2013; Smirnova et al. 2006) (Fig. 8.1). It remains to be further elucidated how the lipolytic regulation and/or intracellular trafficking of IMTG lipid droplets determine its quantitative importance for the FA oxidative rate in skeletal muscle during exercise.

8.4 Mitochondrial Regulation of FA Oxidation During Exercise

Plasma FAs taken into the cell or liberated from intracellular lipolysis must be activated in the cytosol to fatty acyl-CoAs by a family of acyl-CoA synthetases (ACSs) (Fig. 8.1). The active site of the ACSs has been located to the plasma membrane and mitochondria and in close proximity to lipid droplets. The isoform ACSL1 seems particularly important for partitioning FAs toward oxidation in skeletal muscle, which is emphasized by the findings in mice with muscle-specific ACSL1 deficiency exhibiting lower FA utilization during submaximal exercise compared with control mice (Li et al. 2015). To enter the mitochondria for β-oxidation, long-chain fatty acyl-CoAs (the primary FAs in humans) are converted to their fatty acyl carnitine derivatives, a reaction that requires carnitine. This reaction is catalyzed by the enzyme carnitine palmitoyl transferase 1 (CPT1), located at the outer mitochondrial membrane (Bonnefont et al. 2004) (Fig. 8.1). The importance of CPT1 for FA uptake into mitochondria during exercise is evidenced by reduced FA oxidation and increased lipid accumulation in mice with muscle-specific deletion of CPT1 (Wicks et al. 2015) and 50–90% decreased FA oxidation during ex vivo muscle contractions with concomitant pharmacological CPT1 inhibition (Dzamko et al. 2008). A role of CPT1 for regulation of long-chain FA oxidation is also displayed in humans, since oxidation of the medium-chain FA octanoate (C8), which is able to bypass CPT1, did not change when exercise intensity was shifted from 40 to 80% of VO_2peak, as was the case for oleate, a CPT1-dependent long-chain FA (Sidossis et al. 1997).

CPT1 might not be the only step in mitochondrial import of FAs. CD36 was found to be located at the outer mitochondrial membrane in some (Campbell et al. 2004), but not all studies (Jeppesen et al. 2010). Exercise has been further demonstrated to induce CD36 translocation from intracellular depots to the mitochondrial membrane in rodent and human muscle (Monaco et al. 2015; Holloway et al. 2006), and it was suggested that mitochondrial CD36 interacts with ACSs and hereby regulates fatty acyl-CoA availability to CPT1 (Smith et al. 2011). The regulatory role of CD36 in mitochondrial FA import and oxidation during exercise needs to be further investigated.

8.4.1 CPT, Carnitine, and Mitochondrial Fatty Acid Import During Exercise

Since carnitine is substrate in the CPT1 reaction, changes in the free carnitine content in skeletal muscle during exercise could contribute to the regulation of mitochondrial transmembrane FA transport and hereby FA oxidation. Acetyl-CoAs are produced both from β-oxidation of FAs and from glycolysis-derived pyruvate by pyruvate dehydrogenase complex (PDC) (Harris et al. 2002) (Fig. 8.1). Free carnitine can buffer acetyl-CoA by forming acetyl-carnitine and free CoA (Friedman and Fraenkel 1955), a reaction catalyzed by the enzyme carnitine acetyltransferase (CAT) (Fig. 8.1). This entrapment of carnitine increases, when acetyl-CoA is generated in excess of its metabolism in the Krebs cycle. Consequently, lowered amount of free carnitine to the CPT1 reaction would be expected to diminish the supply of fatty acyl-CoA for β-oxidation and hence FA oxidation.

Such mechanism could play a role in regulation of substrate selection during exercise especially at increasing exercise intensities where an increasing carbohydrate oxidation in replacement for FA oxidation takes place.

It is well established that skeletal muscle PDC activity increases rapidly during exercise in an intensity-dependent manner (Constantin-Teodosiu et al. 1991), but decreases gradually when exercise is prolonged (Watt et al. 2002). To alleviate allosteric product inhibition of PDC activity by acetyl-CoA formed in the glycolysis during increasing exercise intensities, the CAT enzyme buffers excess acetyl-CoA into acetyl-L-carnitine. This CAT-mediated acetyl-CoA buffering will reduce cellular free carnitine content and thereby limits mitochondrial FA import and hence FA oxidation allowing for a high rate of pyruvate oxidation. In agreement, in mice with a muscle-specific deletion of CAT, a higher whole-body FA oxidation rate was demonstrated during graded submaximal treadmill running (Seiler et al. 2015). As the PDC reaction rate and the concomitant CAT activity regulate mitochondrial FA import and hence substrate availability to β-oxidation, this could point to lowering of FA oxidation during increasing exercise intensities as a secondary result of increased glucose flux and accelerated glycolysis rather than due to an initial lowering of FA oxidation within lipid metabolic machinery.

This notion is supported by the findings of a one- to threefold increase in muscle acetyl-CoA and acetyl-L-carnitine content at higher exercise intensities compared with rest or low-intensity exercise in untrained individuals (Sahlin 1990; Constantin-Teodosiu et al. 1991; Harris et al. 1987). This resulted in a decreased free carnitine content from comprising ~75% of muscle total carnitine at rest to ~20% at an exercise intensity of 90–100% of VO_2peak (Sahlin 1990; Constantin-Teodosiu et al. 1991; Harris et al. 1987). Moreover, an association between lowering of free carnitine levels and increased acetyl-L-carnitine entrapment during exercise with increasing intensities is evident from several studies (van Loon et al. 2001; Odland et al. 1998; Ren et al. 2013). Importantly, when muscle carnitine content was enhanced by oral carnitine supplementation in healthy, young men, this enabled an increased use of acetyl-CoA from β-oxidation evidenced by a lower glycogen utilization and PDC activation during moderate-intensity exercise in the carnitine-supplemented state compared with the control (Wall et al. 2011). Recently, these findings were reproduced in older men, in which 25 weeks of carnitine supplementation resulted in a 20% increase in both muscle total carnitine content in total FA oxidation during a 1 h submaximal exercise bout at 50% of VO_2peak (Chee et al. 2021).

A final line of evidence highlighting carnitine availability as a regulatory mechanism for FA oxidation is derived from studies where muscle glycogen content was manipulated to be either high or low prior to a moderate-intensity exercise bout in humans. During exercise with initially high glycogen stores, RER during exercise was high indicating a high carbohydrate oxidation. In contrast, when glycogen stores were low prior to exercise, FA oxidation was high during exercise. Accordingly, when carbohydrate oxidation was high during exercise, a 49% higher muscle acetyl-CoA content, 37% higher acetyl-carnitine levels, and 55% lower free carnitine content were observed, while muscle content of acetyl-CoA, free carnitine, and acetyl-carnitine remained unchanged, when FA oxidation rate was enhanced during exercise with a low initial muscle glycogen content (Roepstorff et al. 2005). A 20% increase in muscle free carnitine content was also observed during 10 min of high-intensity exercise in a glycogen-depleted state, whereas this free-carnitine decreased 60% during exercise in the high-glycogen conditions (Constantin-Teodosiu et al. 2004).

Collectively, there is a great body of evidence suggesting carnitine as an important regulatory step in regulation of substrate selection during exercise—especially during exercise with increasing intensities. Importantly, since exercise-induced PDC activation and hereby glycolysis-derived acetyl-CoA production are lowered during prolonged exercise (Watt et al. 2002), and carnitine entrapment into acetyl-carnitine is lowered during exercise with low muscle glycogen (Roepstorff et al. 2005; Constantin-Teodosiu et al. 2004), increased amounts of free carnitine to the CPT1 reaction, in turn increasing the supply of fatty acyl-CoA for FA oxidation, could likely be a regulatory mechanism responsible for the established gradual increase in FA oxidation during prolonged exercise above ~60–90 minutes, when muscle glycogen levels are being emptied.

In addition to carnitine availability, other contributing steps regulating FA oxidation in skeletal muscle during exercise must also be considered. Carnitine-independent regulation of CPT1 could likely be an additional regulatory step for muscle FA import into the mitochondria. Activation of the cellular energy sensor, AMPK, during exercise (Wojtaszewski et al. 2000)—previously proposed to increase FA oxidation via regulation of malonyl-CoA-mediated inhibition of CPT1 (McGarry et al. 1983; Smith et al. 2012; Rasmussen and Winder 1997)—has in recent years from observations in humans (Odland et al. 1996; Odland et al. 1998; Dean et al. 2000; McConell et al. 2020) and transgenic mouse models (Dzamko et al. 2008; Hingst et al. 2020; Miura et al. 2009; Lee-Young et al. 2009; Fritzen et al. 2015; O'Neill et al. 2015) been shown not to be essential for regulation of FA oxidation in skeletal muscle during exercise (Lundsgaard et al. 2018; McConell 2020). However, fatty acyl-CoA/malonyl-CoA ratio appears to be important for CPT1 catalytic activity and FA oxidation, rather than the total malonyl-CoA content per se (Smith et al. 2012), potentially by decreasing the affinity of CPT1 for malonyl-CoA binding (Kolodziej and Zammit 1990). Moreover, malonyl-CoA inhibition kinetics of CPT1 seems modulated by interaction between the cytoskeleton and mitochondria during exercise (Miotto et al. 2017). CPT1 activity could be regulated in also malonyl-CoA-independent mechanism during exercise by a reduction in muscle pH during intense exercise (Starritt et al. 2000). Lastly, ex vivo findings suggest that the reaction rate of β-oxidative enzymes, such as the β-ketoacyl-CoA thiolase, the enzyme catalyzing the final step in the β-oxidation, is feedback regulated by mitochondrial acetyl-CoA content (Eaton 2002) and enzymes in the β-oxidation could hence also be a contributing factor in fine-tuning regulation of FA oxidation during exercise.

8.5 Summarizing Remarks on the Regulation of FA Oxidation in Skeletal Muscle During Exercise

Whole-body FA oxidation is increased several fold during prolonged moderate exercise, but a shift in substrate selection toward increased carbohydrate and decreased relative FA oxidation takes place when exercise intensity is increased beyond 55–65%. Uptake of FAs from the circulation into the skeletal muscle is a major contributing substrate to FA oxidation during submaximal exercise and mostly derived from adipose tissue lipolysis liberated FAs. However, LPL-hydrolysis of circulating VLDL-TG and lipolysis of IMTG also seem to contribute as substrate to fueling the increased FA oxidation in skeletal muscle during exercise.

The increased oxidation of plasma FAs into skeletal muscle during exercise is regulated at several steps, and regulation of FA oxidation in skeletal muscle is not allocated to one single mechanism or signaling pathway, but is apparently orchestrated by a symphony of tightly coordinated molecular events reliant on the metabolic fluxes.

The increase in FA oxidation from rest to submaximal exercise is reliant on an increased transmembrane transport of FA, in which CD36 translocation to the sarcolemma has emerged to serve a pivotal role (Fig. 8.1). The fine-tuning of FA oxidation during exercise appears allocated to the regulation within the mitochondria, where the FA import into the mitochondrial matrix, via formation of fatty acyl carnitine by CPT1, appears to be a central regulatory stage in exercise FA oxidation. This process seems regulated by the intramitochondrial acetyl-CoA homeostasis in response to exercise duration and intensity, as the acetyl-CoA content determines the free carnitine availability for CPT1 during exercise (Fig. 8.1). In this process, the rate of glycolysis appears to be a central tenet for mitochondrial acetyl-CoA availability and hence regulation of FA oxidation. Accordingly, regulation of FA oxidation in skeletal muscle includes a chain of interdependent processes, and dysfunction in any of these can lead to metabolic impairment.

8.6 Are There Impairments in Fatty Acid Oxidation in Skeletal Muscle with Obesity?

In individuals with obesity, an impairment FA oxidation is observed at least during resting conditions. In a seminal study in Pima Indians examining 24-hour whole-body RER while consuming a eucaloric diet, the incidence of weight gain (>5 kg) over a subsequent 3-year period was higher in individuals with elevated 24-hour RERs, indicative of a preference for carbohydrate over FA oxidation (Zurlo et al. 1990). Similarly, other studies have reported that a higher RER is linked with subsequent weight gain (Marra et al. 2004; Seidell et al. 1992; Rogge 2009), while tracer methodology has revealed a decreased utilization of lipid at rest in individuals with severe obesity (Thyfault et al. 2004).

A possible mechanism explaining this inability to oxidize lipid may involve skeletal muscle. We have reported an ≈50% reduction in FA oxidation in two distinct muscle groups (rectus abdominus and vastus lateralis) in individuals with severe obesity (Hulver et al. 2003; Kim et al. 2000); these findings suggest that reductions in the capacity for FA oxidation with obesity could be evident in numerous muscle groups, which could in turn affect whole-body metabolism. Others have also reported a reduction in FA oxidation in human skeletal muscle in individuals with obesity but not severe obesity (Holloway et al. 2007a). A ≈ 50% reduction in FA oxidation in primary skeletal muscle cell cultures (myotubes) derived from individuals with obesity has also been observed (Consitt et al. 2010; Hulver et al. 2005; Maples et al. 2015b; Gaster 2009; Bell et al. 2010) as well as an increased proportion of incomplete FA oxidation (Løvsletten et al. 2020). These findings in cell cultures are indicative of a persistent and perhaps inherent defect, as the muscle cells proliferate (myoblasts) and differentiate into myotubes in the absence of factors which may affect FA oxidation in vivo such as the hormonal milieu and/or differences in contractile activity. An elevated RER with obesity was also reported in vivo across a muscle bed (Kelley et al. 1999).

Deficits in the activity and/or expression of enzymes involved in FA oxidation likely contribute to this impairment in FA oxidation with obesity during resting conditions; the roles of specific proteins involved with lipid transport and oxidation are discussed in review papers (Rogge 2009; Houmard 2008; Houmard et al. 2012; Fritzen et al. 2020; Houmard et al. 2011; Holloway et al. 2009). On the organelle scale, mitochondrial mass appears to be compromised with obesity as proteins indicative of mitochondrial mass are decreased at both moderate (Simoneau et al. 1999; Kriketos et al. 1996; Kelley et al. 2002) and higher ranges (Kim et al. 2000; Holloway et al. 2007a) of obesity. Holloway et al. (Holloway et al. 2007a; Holloway et al. 2009) reported a decline in skeletal muscle FA oxidation in muscle homogenates of subjects with obesity. However, when mitochondria were isolated, FA oxidation was equivalent between lean and obese individuals, indicating that the decline at the tissue level was due to compromised mitochondrial mass. Similarly, in primary myotubes, indices of mitochondrial content and FA oxidation were reduced with obesity, but when FA oxidation was normalized to mitochondrial content (mitochondrial DNA (mtDNA), cytochrome c oxidase subunit 4 (COXIV)), there were no differences between groups suggesting that mitochondrial volume may be the driving factor in the metabolic impairments seen with obesity (Consitt et al. 2010). There is also an increased percentage of glycolytic, low mitochondrial content Type IIb fibers in the skeletal muscle of severely obese subjects (Tanner et al. 2002). The mechanism(s) responsible for the reductions in mitochondrial mass with obesity are not readily evident; it was recently reported that mitochondrial network quality and morphology were altered in primary myotubes derived from individuals with severe obesity (Gundersen et al. 2020).

There is also evidence that the mitochondrial processes involved with FA oxidation are compromised with obesity. Indeed, both FA and glucose oxidation are reduced in the skeletal muscle of obese individuals (Jones et al. 2019). The confluence of FA and glucose oxidation is the TCA cycle, and flux through the TCA cycle is impaired in skeletal muscle both with obesity (Zou et al. 2019) and type 2 diabetes (Gaster 2012). A mechanism for this reduction in TCA cycle activity could involve posttranslational modifications (i.e., phosphorylation, acetylation) which would in turn impair flux (Gaster 2012; Gaster et al. 2012; Boyle et al. 2012). In agreement, there is an increase in blood lactate concentration with obesity which would result from a reduction in the partitioning of substrate to oxidative pathways via the TCA cycle in skeletal muscle (Jones et al. 2019; Broskey et al. 2020). Together, these and other findings (Dahlmans et al. 2016; Diaz-Vegas et al. 2020) indicate that mitochondrial dysfunction and a reduction in mitochondrial mass both contribute to the dampened ability to oxidize lipid in the skeletal muscle of individuals with obesity.

Impaired FA oxidation in individuals with obesity has not only been observed during resting, fasted conditions. An inability to increase FA oxidation in response to increased lipid availability has also been reported with obesity under a variety of conditions (e.g., high-fat feedings, fasting, lipid incubations of primary myotubes, lipid infusion), and findings are summarized in more detail elsewhere (Galgani et al. 2008; Storlien et al. 2004; Goodpaster and Sparks 2017). In reference to obesity, an inability to increase FA oxidation (24 hour RER) in response to a high-fat diet was predictive of greater weight gain in the subsequent year even in lean individuals (Begaye et al. 2020). A dampened ability to increase FA oxidation in individuals with obesity is evident after as little as 2 days of a high-fat diet and subsequently contributes to positive lipid balance (Galgani et al. 2008). Thus, an inability to appropriately increase FA oxidation, or lack of metabolic flexibility, is evident with obesity and individuals prone to obesity under a variety of free-living situations.

One of the mechanisms involved with this lack of metabolic flexibility may be an impaired ability to increase the oxidative machinery of skeletal muscle upon lipid exposure. In response to lipid incubation, primary myotubes from lean individuals increased mitochondrial DNA copy number and mRNA content of genes that upregulate FA oxidation (nuclear respiratory factor (NRF)-1, NRF-2, *peroxisome proliferator-activated receptor (*PPAR)α, PPARδ, and pyruvate dehydrogenase kinase 4 (PDK4, CPT1); in contrast, cells derived from individuals with obesity did not alter or actually decreased the expression of these genes (Maples et al. 2015b; Boyle et al. 2012; Maples et al. 2015a). Similarly, a 5-day high-fat diet elicited increases in the mRNA content of key genes involved in oxidative pathways such as peroxisome proliferator-activated receptor gamma coactivator 1α (PGC1α), PDK4, mitochondrial uncoupling protein (UCP)3, and PPARα. Moreover, a single high-fat feeding increased PPARα mRNA in skeletal muscle biopsies from lean individuals, while there were no changes in subjects with obesity (Boyle et al. 2011). These findings indicate a multifaceted lack of metabolic flexibility in terms of appropriately increasing FA oxidation in the skeletal muscle of individuals with obesity.

8.7 Exercise: Do the Alterations in Fatty Acid Oxidation in Skeletal Muscle with Obesity Alter Substrate Utilization During Exercise?

A bout of mild-to-moderate endurance exercise increases the rates of both carbohydrate and FA oxidation in order to meet the energy demands of increased contractile activity in skeletal muscle. This ability to increase oxidation in response to exercise is another index of metabolic flexibility (Galgani et al. 2008; Goodpaster and Sparks 2017). As indicated in previous sections, mitochondrial content and function are compromised in the skeletal muscle of individuals with obesity, which contributes to a dampened ability to oxidize lipid as well as an impairment in metabolic flexibility. However, in spite of these mitochondrial defects, acute exercise increases the rate of FA oxidation over resting conditions both during and after the exercise bout regardless of obesity status (Thyfault et al. 2004; Fritzen et al. 2020; Arad et al. 2020; Guesbeck et al. 2001; Hansen et al. 2005). This adjustment is critical, as exercise can subsequently contribute to a state of negative lipid balance which in turn can aid in minimizing ectopic lipid accumulation. However, it is not apparent if the lack of metabolic flexibility with obesity compromises the magnitude of the increase in FA oxidation compared to lean individuals, thus resulting in a lower absolute rate of FA oxidation during exercise.

Several studies have reported an impaired ability to appropriately increase FA oxidation in response to endurance exercise with obesity. Thyfault et al. (Thyfault et al. 2004) compared FA oxidation (infused [^{14}C] palmitate and [^{14}C] acetate) during 60 minutes of cycling exercise at 50% VO_2peak in lean patients, patients with severe obesity, and patients who were previously severely obese and had lost >45 kg (bariatric surgery). FA oxidation was significantly lower in the subjects with obesity. Surprisingly, FA oxidation was also depressed during exercise in previously severely obese individuals even after profound weight loss (Thyfault et al. 2004). Similar findings were obtained in response to 10 minutes of exercise at either the same absolute (15 Watts) or relative (65% VO_2peak) exercise intensities, as FA oxidation (via indirect calorimetry) was depressed in previously severely obese women at >1 year after bariatric surgery (gastric bypass) compared to age and BMI matched controls. Eaves et al. (Eaves et al. 2012) examined pre-pubescent (≈11 years old) children of parents that were either lean or severely obese. FA oxidation/metabolic flexibility was determined during 10 minutes of mild-intensity cycle exercise. The children with a parent with severe obesity displayed a reduced rate of FA oxidation at the same absolute exercise workload of 15 watts (≈40% VO_2peak). There were, however, no differences at the higher exercise intensity (65% VO_2peak) between the two groups.

Other findings provide a more equivocal view of whether substrate utilization is altered during exercise in individuals with obesity. A comprehensive systematic review (Arad et al. 2020) concluded that majority of evidence indicated that individuals with obesity rely on the oxidation of lipid to a similar extent as lean subjects during exercise. This conclusion (Arad et al. 2020) was based upon analyses of

24 papers (out of 729 identified papers on the topic) which met rigorous inclusion criteria that included eliminating studies of older adults, individuals with chronic diseases, and/or subjects with exercise limitations. Upon further examination, ten of these 24 papers utilized incremental exercise bouts with relatively short (3–6 minutes) stages compared to those with longer stages where steady-state levels of substrate oxidation are more likely to be achieved. One study (Balcı 2012) also placed overweight and obese individuals into a single group rather than studying the effects of obesity alone. In the remaining studies, there were reduced (Mittendorfer et al. 2004), equivalent (Colberg et al. 1996; Devries et al. 2013; Ezell et al. 1999; Kanaley et al. 1993; Kanaley et al. 2001; Santiworakul et al. 2014; Slusher et al. 2015; Steffan et al. 1999), and even increased (Kanaley et al. 2001; Goodpaster et al. 2002; Horowitz and Klein 2000) rates of FA oxidation in obese compared to lean individuals, which prompted the conclusion that FA oxidation is not compromised in response to exercise with obesity.

It is difficult to determine why such disparate findings have been obtained. Factors such as duration of the exercise bout, exercise mode, exercise intensity, body fat distribution, time of day of testing, pre-exercise diet, method by which FA oxidation is determined (e.g., indirect calorimetry or tracer), comparison to appropriate control group, and numerous other factors could explain the conflicting results. There may also be racial differences as Caucasian women with obesity exhibited a reduction in FA oxidation during exercise at 65% VO_2peak compared to their lean counterparts, while there were no differences in FA oxidation between African American women with normal BMI or obesity (Hickner et al. 2001). Another confounding variable may be the degree of obesity. A decline in FA oxidation during exercise has been reported in subjects with severe (BMI $\geq$ 40 kg/m^2) obesity, subjects with severe obesity after weight loss, and the offspring of individuals with severe obesity (Thyfault et al. 2004; Guesbeck et al. 2001; Eaves et al. 2012). Studies indicating no differences (Colberg et al. 1996; Devries et al. 2013; Ezell et al. 1999; Kanaley et al. 1993, 2001; Santiworakul et al. 2014; Slusher et al. 2015; Steffan et al. 1999) or increases (Kanaley et al. 2001; Goodpaster et al. 2002; Horowitz and Klein 2000) in FA oxidation studied obese (BMI ranges of 30 to 39.9 kg/m^2) but not individuals with severe obesity. Regardless, although FA oxidation consistently increases in response to exercise, it remains to be defined if the response is compromised in individuals with obesity.

8.8 Exercise Training: An Effective Intervention for the Reduction in Fatty Acid Oxidation in the Skeletal Muscle of Individuals with Obesity?

Endurance exercise training (e.g., repeated days of aerobic exercise) classically increases the ability of skeletal muscle to oxidize lipid, primarily by increasing mitochondrial function, but also by induction of several lipid metabolic proteins related to uptake, handling, and breakdown of FAs within the myocyte (Fritzen et al. 2020). However, it is important to realize there is heterogeneity in the responses to exercise training, with some individuals displaying little to no changes in health-related variables such as VO_2peak, fasting insulin, or insulin sensitivity (Bouchard and Rankinen 2001; Bouchard et al. 2012; Stephens and Sparks 2015; Sparks 2017). In addition, individuals with obesity, particularly severe (BMI $\geq$40 kg/m^2) obesity, exhibit compromised cardiorespiratory fitness which results in low absolute exercise workloads and may not optimally stimulate adaptations to training. Supporting the concept of "exercise resistance" (Sparks 2017) with obesity, the overexpression of PGC1α, a global regulator of exercise-training mediated adaptations, did not increase FA oxidation in primary myotubes from individuals with severe obesity to the same extent as in lean subjects (Consitt et al. 2010). There are also findings suggesting that the lesions in FA oxidation and metabolic flexibility with obesity are inherent (Hulver et al. 2005; Gaster 2009; Tanner et al. 2002; Zou et al. 2019; Boyle et al. 2012) which may minimize the impact of any interventions. In support, substantive (i.e., > 50 kg) weight loss did not alter FA oxidation in skeletal muscle (Berggren et al. 2004) at rest or during exercise (Thyfault et al. 2004; Guesbeck et al. 2001) in subjects with severe obesity. A smaller magnitude of weight loss (≈14 kg) in individuals with Class I obesity (BMI of 34 kg/m^2) also did not alter resting FA oxidation determined in vivo across a muscle bed (Kelley et al. 1999). These findings suggest that the positive effects of exercise training on FA oxidation in the skeletal muscle of individuals with obesity cannot be assumed and warrant investigation.

In an attempt to minimize the confounding effects of concurrent weight loss, studies have implemented a short period (7–10 days) of exercise training (60 min/d, 50–75% VO_2peak) where the energy deficit induced by exercise is minimal and no weight loss is evident. One such study examined subjects with severe obesity and subjects that were previously severely obese who had undergone bariatric surgery (1–2 y post-surgery). The 10-day training stimulus was sufficient to increase FA oxidation in the obese groups to the same rates as in lean subjects after 10 days of exercise (Berggren et al. 2004). A similar training regimen increased FA oxidation in skeletal muscle in Caucasian and African American women with severe obesity to the same levels as lean subjects (Cortright et al. 2006). Other longer-duration studies have reported equivalent increases in the expression or activity of proteins involved with FA oxidation in the skeletal muscle of lean subjects and individuals with obesity (Devries et al. 2013; Gillen et al. 2013; de Matos et al. 2018).

In contrast, 24 hours of contractile stimuli (electrical pulse stimulation, EPS) increased FA oxidation in primary human myotubes derived from lean subjects but not in myotubes from individuals with obesity or type 2 diabetes (Løvsletten et al. 2020). The cells from the subjects with obesity initially had higher rates of incomplete oxidation and reduced expression of complexes II, III, and IV of the respiratory chain (Løvsletten et al. 2020). A comparison of these in vivo (see preceding paragraph) and in vitro (Løvsletten et al. 2020) findings suggests that factors present during whole-body exercise (i.e., hormones, myokines, neural signaling, etc.) could be important in inducing the increases in FA oxidation seen with exercise training in individuals with obesity.

In relation to metabolic flexibility, the skeletal muscle of lean individuals increased FA oxidation, but there was no change in the muscle of obese subjects in response to a 3-day eucaloric high-fat diet (Battaglia et al. 2012). Both groups then exercised for 10 consecutive days (1 hour/day, 70% of VO_2peak) and on the final 3 days of training consumed the same high-fat diet. The initial 7 days of exercise training was sufficient to increase FA oxidation in the skeletal muscle of both groups to the same extent, and both groups responded to the 3 day high-fat diet in a similar manner (Battaglia et al. 2012). Endurance exercise training also increased FA oxidation during exercise (an index of metabolic flexibility) in individuals with obesity (Fritzen et al. 2020; Devries et al. 2013; Goodpaster et al. 2003). Eaves et al. (Eaves et al. 2012) examined the effects of exercise training (4 weeks at ≈65% VO_2peak, 3 days/week for 30 minutes for week 1 progressing to 60 minutes/day for weeks 2–4), and even with this relatively mild stimulus, there were trends ($P = 0.06$) for FA oxidation to increase during exercise to an equivalent degree in children of either lean parents or parents with obesity. These findings indicate that endurance exercise training increases metabolic flexibility to a similar extent in individuals with obesity compared to their lean counterparts.

8.9 Summarizing Remarks on the Role of Obesity in Regulation of FA Oxidation During Rest and Exercise and the Counter-regulatory Effect of Exercise Training

In summary, individuals with obesity display an impaired ability to oxidize lipid and to increase FA oxidation in the face of increased lipid exposure (metabolic flexibility). At the cellular level, these defects are evident in skeletal muscle mitochondria and likely contribute, at least in part, to positive lipid balance and ectopic lipid deposition. It is equivocal if these deficits impair the ability to increase FA oxidation in response to a single exercise bout. However, despite these initial impairments, endurance exercise training can rescue FA oxidation and metabolic flexibility in the skeletal muscle of individuals with obesity at least to equivalent levels of their lean counterparts. These findings indicate the efficacy of exercise training in treating the metabolic inflexibility in human skeletal muscle metabolism evident with obesity.

References

Alsted TJ, Ploug T, Prats C, Serup AK, Hoeg L, Schjerling P, Holm C, Zimmermann R, Fledelius C, Galbo H, Kiens B (2013) Contraction-induced lipolysis is not impaired by inhibition of hormone-sensitive lipase in skeletal muscle. J Physiol 591:5141–5155

Arad AD, Basile AJ, Albu J, DiMenna FJ (2020) No influence of overweight/obesity on exercise lipid oxidation: a systematic review. Int J Mol Sci 21:1614

Arner P, Kriegholm E, Engfeldt P, Bolinder J (1990) Adrenergic regulation of lipolysis in situ at rest and during exercise. J Clin Invest 85:893–898

Bahr R, Hostmark AT, Newsholme EA, Gronnerod O, Sejersted OM (1991) Effect of exercise on recovery changes in plasma levels of FFA, glycerol, glucose and catecholamines. Acta Physiol Scand 143:105–115

Balcı SS (2012) Comparison of substrate oxidation during walking and running in normal-weight and overweight/obese men. Obes Facts 5:327–338

Battaglia GM, Zheng D, Hickner RC, Houmard JA (2012) Effect of exercise training on metabolic flexibility in response to a high-fat diet in obese individuals. Am J Physiol Endocrinol Metab 303:E1440–E1445

Begaye B, Vinales KL, Hollstein T, Ando T, Walter M, Bogardus C, Krakoff J, Piaggi P (2020) Impaired metabolic flexibility to high-fat overfeeding predicts future weight gain in healthy adults. Diabetes 69:181–192

Bell JA, Reed MA, Consitt LA, Martin OJ, Haynie KR, Hulver MW, Muoio DM, Dohm GL (2010) Lipid partitioning, incomplete fatty acid oxidation, and insulin signal transduction in primary human muscle cells: effects of severe obesity, fatty acid incubation, and fatty acid translocase/ CD36 overexpression. J Clin Endocrinol Metab 95:3400–3410

Benninghoff T, Espelage L, Eickelschulte S, Zeinert I, Sinowenka I, Müller F, Christina Schöndeling C, Batchelor H, Cames S, Zhou Z, Kotzka J, Chadt A, Al-Hasani H (2020) The Rab GAPs TBC1D1 and TBC1D4 control uptake of Long-chain fatty acids into skeletal muscle via fatty acid transporter SLC27A4/FATP4. Diabetes 69:2281–2293

Berggren JR, Hulver MW, Dohm GL, Houmard JA (2004) Weight loss and exercise: implications for muscle lipid metabolism and insulin action. Med Sci Sports Exerc 36:1191–1195

Bergman BC, Butterfield GE, Wolfel EE, Casazza GA, Lopaschuk GD, Brooks GA (1999) Evaluation of exercise and training on muscle lipid metabolism. Am J Phys 276:E106–E117

Bonen A, Han XX, Habets DD, Febbraio M, Glatz JF, Luiken JJ (2007) A null mutation in skeletal muscle FAT/CD36 reveals its essential role in insulin- and AICAR-stimulated fatty acid metabolism. Am J Physiol Endocrinol Metab 292(6):E1740–E1749

Bonen A, Luiken JJ, Arumugam Y, Glatz JF, Tandon NN (2000) Acute regulation of fatty acid uptake involves the cellular redistribution of fatty acid translocase. J Biol Chem 275:14501–14508

Bonnefont JP, Djouadi F, Prip-Buus C, Gobin S, Munnich A, Bastin J (2004) Carnitine palmitoyltransferases 1 and 2: biochemical, molecular and medical aspects. Mol Asp Med 25(5–6):495–520

Bouchard C, Blair SN, Church TS, Earnest CP, Hagberg JM, Häkkinen K, Jenkins NT, Karavirta L, Kraus WE, Leon AS, Rao DC, Sarzynski MA, Skinner JS, Slentz CA, Rankinen T (2012) Adverse metabolic response to regular exercise: is it a rare or common occurrence? PLoS One 7: e37887

Bouchard C, Rankinen T (2001) Individual differences in response to regular physical activity. Med Sci Sports Exerc 33:S446–S451

Boyle KE, Canham JP, Consitt LA, Zheng D, Koves TR, Gavin TP, Holbert D, Neufer PD, Ilkayeva O, Muoio DM, Houmard JA (2011) A high-fat diet elicits differential responses in genes coordinating oxidative metabolism in skeletal muscle of lean and obese individuals. J Clin Endocrinol Metab 96:775–781

Boyle KE, Zheng D, Anderson EJ, Neufer PD, Houmard JA (2012) Mitochondrial lipid oxidation is impaired in cultured myotubes from obese humans. Int J Obes 36:1025–1031

Bradley NS, Snook LA, Jain SS, Heigenhauser GJ, Bonen A, Spriet LL (2012) Acute endurance exercise increases plasma membrane fatty acid transport proteins in rat and human skeletal muscle. Am J Physiol Endocrinol Metab 302:E183–E189

Broskey NT, Zou K, Dohm GL, Houmard JA (2020) Plasma lactate as a marker for metabolic health. Exerc Sport Sci Rev 48:119–124

Calbet JA, Lundby C (2012) Skeletal muscle vasodilatation during maximal exercise in health and disease. J Physiol 590:6285–6296

Campbell SE, Tandon NN, Woldegiorgis G, Luiken JJ, Glatz JF, Bonen A (2004) A novel function for fatty acid translocase (FAT)/CD36: involvement in long chain fatty acid transfer into the mitochondria. J Biol Chem 279:36235–36241

Carlson LA, Ekelund LG, Fröberg SO (1971) Concentration of triglycerides, phospholipids and glycogen in skeletal muscle and of free fatty acids and beta-hydroxybutyric acid in blood in man in response to exercise. Eur J Clin Investig 1:248–254

Chabowski A, Górski J, Luiken JJ, Glatz JF, Bonen A (2007) Evidence for concerted action of FAT/CD36 and FABPpm to increase fatty acid transport across the plasma membrane. Prostaglandins Leukot Essent Fatty Acids 77:345–353

Chadt A, Leicht K, Deshmukh A, Jiang LQ, Scherneck S, Bernhardt U, Dreja T, Vogel H, Schmolz K, Kluge R, Zierath JR, Hultschig C, Hoeben RC, Schurmann A, Joost HG, Al-Hasani H (2008) Tbc1d1 mutation in lean mouse strain confers leanness and protects from diet-induced obesity. Nat Genet 40:1354–1359

Chee C, Shannon CE, Burns A, Selby AL, Wilkinson D, Smith K, Greenhaff PL, Stephens FB (2021) Increasing skeletal muscle carnitine content in older individuals increases whole-body fat oxidation during moderate-intensity exercise. Aging Cell 20:e13303

Christensen EH, Hansen O (1939) Respiratorisher quotientund O2–aufname. Skand Arch Physiol 81:160–171

Clarke DC, Miskovic D, Han XX, Calles-Escandon J, Glatz JF, Luiken JJ, Heikkila JJ, Bonen A (2004) Overexpression of membrane-associated fatty acid binding protein (FABPpm) in vivo increases fatty acid sarcolemmal transport and metabolism. Physiol Genomics 17:31–37

Colberg SR, Hagberg JM, McCole SD, Zmuda JM, Thompson PD, Kelley DE (1996) Utilization of glycogen but not plasma glucose is reduced in individuals with NIDDM during mild-intensity exercise. J Appl Physiol (1985) 81:2027–2033

Consitt LA, Bell JA, Koves TR, Muoio DM, Hulver MW, Haynie KR, Dohm GL, Houmard JA (2010) Peroxisome proliferator-activated receptor-gamma coactivator-1alpha overexpression increases lipid oxidation in myocytes from extremely obese individuals. Diabetes 59:1407–1415

Constantin-Teodosiu D, Carlin JI, Cederblad G, Harris RC, Hultman E (1991) Acetyl group accumulation and pyruvate dehydrogenase activity in human muscle during incremental exercise. Acta Physiol Scand 143:367–372

Constantin-Teodosiu D, Peirce NS, Fox J, Greenhaff PL (2004) Muscle pyruvate availability can limit the flux, but not activation, of the pyruvate dehydrogenase complex during submaximal exercise in humans. J Physiol 561:647–655

Cortright RN, Sandhoff KM, Basilio JL, Berggren JR, Hickner RC, Hulver MW, Dohm GL, Houmard JA (2006) Skeletal muscle fat oxidation is increased in African-American and white women after 10 days of endurance exercise training. Obesity (Silver Spring) 14:1201–1210

Costill DL, Fink WJ, Getchell LH, Ivy JL, Witzmann FA (1979) Lipid metabolism in skeletal muscle of endurance-trained males and females. J Appl Physiol Respir Environ Exerc Physiol 47:787–791

Dahlmans D, Houzelle A, Schrauwen P, Hoeks J (2016) Mitochondrial dynamics, quality control and mi RNA regulation in skeletal muscle: implications for obesity and related metabolic disease. Clin Sci (Lond) 130:843–852

de Glisezinski I, Larrouy D, Bajzova M, Koppo K, Polak J, Berlan M, Bulow J, Langin D, Marques MA, Crampes F, Lafontan M, Stich V (2009) Adrenaline but not noradrenaline is a determinant of exercise-induced lipid mobilization in human subcutaneous adipose tissue. J Physiol 587: 3393–3404

de Matos MA, Vieira DV, Pinhal KC, Lopes JF, Dias-Peixoto MF, Pauli JR, de Castro MF, Little JP, Rocha-Vieira E, Amorim FT (2018) High-intensity interval training improves markers of oxidative metabolism in skeletal muscle of individuals with obesity and insulin resistance. Front Physiol 9:1451

Dean D, Daugaard JR, Young ME, Saha A, Vavvas D, Asp S, Kiens B, Kim KH, Witters L, Richter EA, Ruderman N (2000) Exercise diminishes the activity of acetyl-CoA carboxylase in human muscle. Diabetes 49(8):1295–1300

Decombaz J, Schmitt B, Ith M, Decarli B, Diem P, Kreis R, Hoppeler H, Boesch C (2001) Postexercise fat intake repletes intramyocellular lipids but no faster in trained than in sedentary subjects. Am J Physiol Regul Integr Comp Physiol 281:R760–R769

Devries MC, Samjoo IA, Hamadeh MJ, McCready C, Raha S, Watt MJ, Steinberg GR, Tarnopolsky MA (2013) Endurance training modulates intramyocellular lipid compartmentalization and morphology in skeletal muscle of lean and obese women. J Clin Endocrinol Metab 98:4852–4862

Diaz-Vegas A, Sanchez-Aguilera P, Krycer JR, Morales PE, Monsalves-Alvarez M, Cifuentes M, Rothermel BA, Lavandero S (2020) Is mitochondrial dysfunction a common root of noncommunicable chronic diseases? Endocr Rev 41:491–517

Dzamko N, Schertzer JD, Ryall JG, Steel R, Macaulay SL, Wee S, Chen ZP, Michell BJ, Oakhill JS, Watt MJ, Jorgensen SB, Lynch GS, Kemp BE, Steinberg GR (2008) AMPK-independent pathways regulate skeletal muscle fatty acid oxidation. J Physiol 586(Pt 23):5819–5831

Eaton S (2002) Control of mitochondrial beta-oxidation flux. Prog Lipid Res 41:197–239

Eaves AD, Colon A, Dubose KD, Collier D, Houmard JA (2012) Substrate utilization during submaximal exercise in children with a severely obese parent. Nutr Metab (Lond) 9:38–39

Essén B (1978) Studies on the regulation of metabolism in human skeletal muscle using intermittent exercise as an experimental model. Acta Physiol Scand Suppl 454:1–32

Ezell DM, Geiselman PJ, Anderson AM, Dowdy ML, Womble LG, Greenway FL, Zachwieja JJ (1999) Substrate oxidation and availability during acute exercise in non-obese, obese, and post-obese sedentary females. Int J Obes Relat Metab Disord 23:1047–1056

Friedlander AL, Casazza GA, Horning MA, Usaj A, Brooks GA (1999) Endurance training increases fatty acid turnover, but not fat oxidation, in young men. J Appl Physiol (1985) 86: 2097–2105

Friedman S, Fraenkel G (1955) Reversible enzymatic acetylation of carnitine. Arch Biochem Biophys 59:491–501

Fritzen AM, Lundsgaard AM, Jeppesen J, Christiansen ML, Bienso R, Dyck JR, Pilegaard H, Kiens B (2015) 5'-AMP activated protein kinase alpha2 controls substrate metabolism during post-exercise recovery via regulation of pyruvate dehydrogenase kinase 4. J Physiol 593:4765–4780

Fritzen AM, Lundsgaard AM, Kiens B (2020) Tuning fatty acid oxidation in skeletal muscle with dietary fat and exercise. Nat Rev Endocrinol 16:683–696

Galgani JE, Moro C, Ravussin E (2008) Metabolic flexibility and insulin resistance. Am J Physiol Endocrinol Metab 295:E1009–E1017

Gaster M (2009) Reduced lipid oxidation in myotubes established from obese and type 2 diabetic subjects. Biochem Biophys Res Commun 382:766–770

Gaster M (2012) Reduced TCA flux in diabetic Myotubes: determined by single defects? Biochem Res Int 2012:716056

Gaster M, Nehlin JO, Minet AD (2012) Impaired TCA cycle flux in mitochondria in skeletal muscle from type 2 diabetic subjects: marker or maker of the diabetic phenotype? Arch Physiol Biochem 118:156–189

Gillen JB, Percival ME, Ludzki A, Tarnopolsky MA, Gibala MJ (2013) Interval training in the fed or fasted state improves body composition and muscle oxidative capacity in overweight women. Obesity (Silver Spring) 21:2249–2255

Goodpaster BH, Katsiaras A, Kelley DE (2003) Enhanced fat oxidation through physical activity is associated with improvements in insulin sensitivity in obesity. Diabetes 52:2191–2197

Goodpaster BH, Sparks LM (2017) Metabolic flexibility in health and disease. Cell Metab 25: 1027–1036

Goodpaster BH, Wolfe RR, Kelley DE (2002) Effects of obesity on substrate utilization during exercise. Obes Res 10:575–584

Guesbeck NR, Hickey MS, Mac Donald KG, Pories WJ, Harper I, Ravussin E, Dohm GL, Houmard JA (2001) Substrate utilization during exercise in formerly morbidly obese women. J Appl Physiol (1985) 90:1007–1012
Gundersen AE, Kugler BA, McDonald PM, Veraksa A, Houmard JA, Zou K (2020) Altered mitochondrial network morphology and regulatory proteins in mitochondrial quality control in myotubes from severely obese humans with or without type 2 diabetes. Appl Physiol Nutr Metab 45:283–293
Hagenfeldt L, Wahren J (1968) Human forearm muscle metabolism during exercise. II. Uptake, release and oxidation of individual FFA and glycerol. Scand J Clin Lab Invest 21:263–276
Hagenfeldt L, Wahren J (1975) Turnover of free fatty acids during recovery from exercise. J Appl Physiol 39:247–250
Hansen K, Shriver T, Schoeller D (2005) The effects of exercise on the storage and oxidation of dietary fat. Sports Med 35:363–373
Harris RA, Bowker-Kinley MM, Huang B, Wu P (2002) Regulation of the activity of the pyruvate dehydrogenase complex. Adv Enzym Regul 42:249–259
Harris RC, Foster CV, Hultman E (1987) Acetylcarnitine formation during intense muscular contraction in humans. J Appl Physiol (1985) 63:440–442
Helge JW, Watt PW, Richter EA, Rennie MJ, Kiens B (2001) Fat utilization during exercise: adaptation to a fat-rich diet increases utilization of plasma fatty acids and very low density lipoprotein-triacylglycerol in humans. J Physiol 537:1009–1020
Hickner RC, Privette J, McIver K, Barakat H (2001) Fatty acid oxidation in African-American and Caucasian women during physical activity. J Appl Physiol (1985) 90:2319–2324
Hingst JR, Kjøbsted R, Birk JB, Jørgensen NO, Larsen MR, Kido K, Larsen JP, Kjeldsen SAS, Fentz J, Frøsig C, Holm S, Fritzen AM, Dohlman TL, Larsen S, Foretz M, Viollet B, Schjerling P, Overby P, Halling JF, Pilegaard H, Hellsten Y, Wojtaszewski JF (2020) Inducible deletion of skeletal muscle AMPKα reveals that AMPK is required fornucleotide balance but dispensable for muscle glucose uptake and fat oxidation during exercise. Mol Metab 40:101028
Holloway GP, Bezaire V, Heigenhauser GJ, Tandon NN, Glatz JF, Luiken JJ, Bonen A, Spriet LL (2006) Mitochondrial long chain fatty acid oxidation, fatty acid translocase/CD36 content and carnitine palmitoyltransferase I activity in human skeletal muscle during aerobic exercise. J Physiol 571:201–210
Holloway GP, Bonen A, Spriet LL (2009) Regulation of skeletal muscle mitochondrial fatty acid metabolism in lean and obese individuals. Am J Clin Nutr 89:455S–462S
Holloway GP, Lally J, Nickerson JG, Alkhateeb H, Snook LA, Heigenhauser GJ, Calles-Escandon-J, Glatz JF, Luiken JJ, Spriet LL, Bonen A (2007b) Fatty acid binding protein facilitates sarcolemmal fatty acid transport but not mitochondrial oxidation in rat and human skeletal muscle. J Physiol 582(Pt 1):393–405
Holloway GP, Thrush AB, Heigenhauser GJ, Tandon NN, Dyck DJ, Bonen A, Spriet LL (2007a) Skeletal muscle mitochondrial FAT/CD36 content and palmitate oxidation are not decreased in obese women. Am J Physiol Endocrinol Metab 292:E1782–E1789
Horowitz JF, Klein S (2000) Oxidation of nonplasma fatty acids during exercise is increased in women with abdominal obesity. J Appl Physiol (1985) 89:2276–2282
Houmard JA (2008) Intramuscular lipid oxidation and obesity. Am J Physiol Regul Integr Comp Physiol 294:R1111–R1116
Houmard JA, Pories WJ, Dohm GL (2011) Is there a metabolic program in the skeletal muscle of obese individuals? J Obes 2011:250496
Houmard JA, Pories WJ, Dohm GL (2012) Severe obesity: evidence for a deranged metabolic program in skeletal muscle? Exerc Sport Sci Rev 40:204–210
Hulver MW, Berggren JR, Carper MJ, Miyazaki M, Ntambi JM, Hoffman EP, Thyfault JP, Stevens R, Dohm GL, Houmard JA, Muoio DM (2005) Elevated stearoyl-CoA desaturase-1 expression in skeletal muscle contributes to abnormal fatty acid partitioning in obese humans. Cell Metab 2:251–261

Hulver MW, Berggren JR, Cortright RN, Dudek RW, Thompson RP, Pories WJ, Mac Donald KG, Cline GW, Shulman GI, Dohm GL, Houmard JA (2003) Skeletal muscle lipid metabolism with obesity. Am J Physiol Endocrinol Metab 284:E741–E747

Hurley BF, Nemeth PM, Martin WH III, Hagberg JM, Dalsky GP, Holloszy JO (1986) Muscle triglyceride utilization during exercise: effect of training. J Appl Physiol (1985) 60:562–567

Ibrahimi A, Bonen A, Blinn WD, Hajri T, Li X, Zhong K, Cameron R, Abumrad NA (1999) Muscle-specific overexpression of FAT/CD36 enhances fatty acid oxidation by contracting muscle, reduces plasma triglycerides and fatty acids, and increases plasma glucose and insulin. J Biol Chem 274(38):26761–26766

Jain SS, Chabowski A, Snook LA, Schwenk RW, Glatz JF, Luiken JJ, Bonen A (2009) Additive effects of insulin and muscle contraction on fatty acid transport and fatty acid transporters, FAT/CD36, FABPpm, FATP1, 4 and 6. FEBS Lett 583:2294–2300

Jeppesen J, Albers PH, Rose AJ, Birk JB, Schjerling P, Dzamko N, Steinberg GR, Kiens B (2011) Contraction-induced skeletal muscle FAT/CD36 trafficking and FA uptake is AMPK independent. J Lipid Res 52(4):699–711

Jeppesen J, Mogensen M, Prats C, Sahlin K, Madsen K, Kiens B (2010) FAT/CD36 is localized in sarcolemma and in vesicle-like structures in subsarcolemma regions but not in mitochondria. J Lipid Res 51:1504–1512

Jones TE, Pories WJ, Houmard JA, Tanner CJ, Zheng D, Zou K, Coen PM, Goodpaster BH, Kraus WE, Dohm GL (2019) Plasma lactate as a marker of metabolic health: implications of elevated lactate for impairment of aerobic metabolism in the metabolic syndrome. Surgery 166:861–866

Kanaley JA, Cryer PE, Jensen MD (1993) Fatty acid kinetic responses to exercise. Effects of obesity, body fat distribution, and energy-restricted diet. J Clin Invest 92:255–261

Kanaley JA, Shadid S, Sheehan MT, Guo Z, Jensen MD (2009) Relationship between plasma free fatty acid, intramyocellular triglycerides and long-chain acylcarnitines in resting humans. J Physiol 587:5939–5950

Kanaley JA, Weatherup-Dentes MM, Alvarado CR, Whitehead G (2001) Substrate oxidation during acute exercise and with exercise training in lean and obese women. Eur J Appl Physiol 85:68–73

Kelley DE, Goodpaster B, Wing RR, Simoneau JA (1999) Skeletal muscle fatty acid metabolism in association with insulin resistance, obesity, and weight loss. Am J Phys 277:E1130–E1141

Kelley DE, He J, Menshikova EV, Ritov VB (2002) Dysfunction of mitochondria in human skeletal muscle in type 2 diabetes. Diabetes 51:2944–2950

Kiens B (2006) Skeletal muscle lipid metabolism in exercise and insulin resistance. Physiol Rev 86(1):205–243

Kiens B, Essen-Gustavsson B, Christensen NJ, Saltin B (1993) Skeletal muscle substrate utilization during submaximal exercise in man: effect of endurance training. J Physiol 469:459–478

Kiens B, Lithell H, Mikines KJ, Richter EA (1989) Effects of insulin and exercise on muscle lipoprotein lipase activity in man and its relation to insulin action. J Clin Invest 84(4): 1124–1129

Kiens B, Richter EA (1998) Utilization of skeletal muscle triacylglycerol during postexercise recovery in humans. Am J Phys 275:E332–E337

Kim JY, Hickner RC, Cortright RL, Dohm GL, Houmard JA (2000) Lipid oxidation is reduced in obese human skeletal muscle. Am J Physiol Endocrinol Metab 279:E1039–E1044

Kolodziej MP, Zammit VA (1990) Re-evaluation of the interaction of malonyl-CoA with the rat liver mitochondrial carnitine palmitoyltransferase system by using purified outer membranes. Biochem J 267:85–90

Kriketos AD, Pan DA, Lillioja S, Cooney GJ, Baur LA, Milner MR, Sutton JR, Jenkins AB, Bogardus C, Storlien LH (1996) Interrelationships between muscle morphology, insulin action, and adiposity. Am J Phys 270:R1332–R1339

Krogh A, Lindhard J (1920) The relative value of fat and carbohydrate as sources of muscular energy: with appendices on the correlation between standard metabolism and the respiratory quotient during rest and work. Biochem J 14(3–4):290–363

Krssak M, Petersen KF, Bergeron R, Price T, Laurent D, Rothman DL, Roden M, Shulman GI (2000) Intramuscular glycogen and intramyocellular lipid utilization during prolonged exercise and recovery in man: a 13C and 1H nuclear magnetic resonance spectroscopy study. J Clin Endocrinol Metab 85:748–754
Lafontan M, Moro C, Berlan M, Crampes F, Sengenes C, Galitzky J (2008) Control of lipolysis by natriuretic peptides and cyclic GMP. Trends Endocrinol Metab 19:130–137
Lally JS, Jain SS, Han XX, Snook LA, Glatz JF, Luiken JJ, McFarlan J, Holloway GP, Bonen A (2012) Caffeine-stimulated fatty acid oxidation is blunted in CD36 null mice. Acta Physiol (Oxf) 205:71–81
Langfort J, Ploug T, Ihlemann J, Holm C, Galbo H (2000) Stimulation of hormone-sensitive lipase activity by contractions in rat skeletal muscle. Biochem J 351:207–214
Larson-Meyer DE, Newcomer BR, Hunter GR (2002) Influence of endurance running and recovery diet on intramyocellular lipid content in women: a 1H NMR study. Am J Physiol Endocrinol Metab 282:E95–E106
Lee-Young RS, Griffee SR, Lynes SE, Bracy DP, Ayala JE, McGuinness OP, Wasserman DH (2009) Skeletal muscle AMP-activated protein kinase is essential for the metabolic response to exercise in vivo. J Biol Chem 284(36):23925–23934
Li LO, Grevengoed TJ, Paul DS, Ilkayeva O, Koves TR, Pascual F, Newgard CB, Muoio DM, Coleman RA (2015) Compartmentalized acyl-CoA metabolism in skeletal muscle regulates systemic glucose homeostasis. Diabetes 64:23–35
Lithell H, Hellsing K, Lundqvist G, Malmberg P (1979) Lipoprotein-lipase activity of human skeletal-muscle and adipose tissue after intensive physical exercise. Acta Physiol Scand 105(3): 312–315
Løvsletten NG, Rustan AC, Laurens C, Thoresen GH, Moro C, Nikolić N (2020) Primary defects in lipid handling and resistance to exercise in myotubes from obese donors with and without type 2 diabetes. Appl Physiol Nutr Metab 45:169–179
Lundsgaard AM, Fritzen AM, Kiens B (2018) Molecular regulation of fatty acid oxidation in skeletal muscle during aerobic exercise. Trends Endocrinol Metab 29:18–30
Lundsgaard AM, Fritzen AM, Kiens B (2020) The importance of fatty acids as nutrients during post-exercise recovery. Nutrients 12:nu12020280
MacPherson RE, Vandenboom R, Roy BD, Peters SJ (2013) Skeletal muscle PLIN3 and PLIN5 are serine phosphorylated at rest and following lipolysis during adrenergic or contractile stimulation. Physiol Rep 1:e00084
Maples JM, Brault JJ, Shewchuk BM, Witczak CA, Zou K, Rowland N, Hubal MJ, Weber TM, Houmard JA (2015a) Lipid exposure elicits differential responses in gene expression and DNA methylation in primary human skeletal muscle cells from severely obese women. Physiol Genomics 47:139–146
Maples JM, Brault JJ, Witczak CA, Park S, Hubal MJ, Weber TM, Houmard JA, Shewchuk BM (2015b) Differential epigenetic and transcriptional response of the skeletal muscle carnitine palmitoyltransferase 1B (CPT1B) gene to lipid exposure with obesity. Am J Physiol Endocrinol Metab 309:E345–E356
Marra M, Scalfi L, Contaldo F, Pasanisi F (2004) Fasting respiratory quotient as a predictor of long-term weight changes in non-obese women. Ann Nutr Metab 48:189–192
McConell GK (2020) It's well and truly time to stop stating that AMPK regulates glucose uptake and fat oxidation during exercise. Am J Physiol Endocrinol Metab 318:E564–E567
McConell GK, Wadley GD, Le PK, Linden KC (2020) Skeletal muscle AMPK is not activated during 2h of moderate intensity exercise at 65% VO_2 peak in endurance trained men. J Physiol 598:3859–3870
McFarlan JT, Yoshida Y, Jain SS, Han XX, Snook LA, Lally J, Smith BK, Glatz JF, Luiken JJ, Sayer RA, Tupling AR, Chabowski A, Holloway GP, Bonen A (2012) In vivo, fatty acid translocase (CD36) critically regulates skeletal muscle fuel selection, exercise performance, and training-induced adaptation of fatty acid oxidation. J Biol Chem 287(28):23502–23516

McGarry JD, Mills SE, Long CS, Foster DW (1983) Observations on the affinity for carnitine, and malonyl-CoA sensitivity, of carnitine palmitoyltransferase I in animal and human tissues. Demonstration of the presence of malonyl-CoA in non-hepatic tissues of the rat. Biochem J 214:21–28

Miotto PM, Steinberg GR, Holloway GP (2017) Controlling skeletal muscle CPT-I malonyl-CoA sensitivity: the importance of AMPK-independent regulation of intermediate filaments during exercise. Biochem J 474:557–569

Mittendorfer B, Fields DA, Klein S (2004) Excess body fat in men decreases plasma fatty acid availability and oxidation during endurance exercise. Am J Physiol Endocrinol Metab 286: E354–E362

Miura S, Kai Y, Kamei Y, Bruce CR, Kubota N, Febbraio MA, Kadowaki T, Ezaki O (2009) Alpha 2-AMPK activity is not essential for an increase in fatty acid oxidation during low-intensity exercise. Am J Physiol Endocrinol Metab 296(1):E47–E55

Monaco C, Whitfield J, Jain SS, Spriet LL, Bonen A, Holloway GP (2015) Activation of AMPKalpha2 is not required for mitochondrial FAT/CD36 accumulation during exercise. PLoS One 10:e0126122

Morio B, Holmback U, Gore D, Wolfe RR (2004) Increased VLDL-TAG turnover during and after acute moderate-intensity exercise. Med Sci Sports Exerc 36:801–806

Moro C, Crampes F, Sengenes C, de Glisezinski I, Galitzky J, Thalamas C, Lafontan M, Berlan M (2004) Atrial natriuretic peptide contributes to physiological control of lipid mobilization in humans. FASEB J 18:908–910

Moro C, Polak J, Hejnova J, Klimcakova E, Crampes F, Stich V, Lafontan M, Berlan M (2006) Atrial natriuretic peptide stimulates lipid mobilization during repeated bouts of endurance exercise. Am J Physiol Endocrinol Metab 290:E864–E869

Nellemann B, Sondergaard E, Jensen J, Pedersen SB, Jessen N, Jorgensen JO, Nielsen S (2014) Kinetics and utilization of lipid sources during acute exercise and acipimox. Am J Physiol Endocrinol Metab 307:E199–E208

Odland LM, Heigenhauser GJ, Lopaschuk GD, Spriet LL (1996) Human skeletal muscle malonyl-CoA at rest and during prolonged submaximal exercise. Am J Phys 270:E541–E544

Odland LM, Howlett RA, Heigenhauser GJ, Hultman E, Spriet LL (1998) Skeletal muscle malonyl-CoA content at the onset of exercise at varying power outputs in humans. Am J Phys 274: E1080–E1085

O'Neill HM, Lally JS, Galic S, Pulinilkunnil T, Ford RJ, Dyck JR, van Denderen BJ, Kemp BE, Steinberg GR (2015) Skeletal muscle ACC2 S212 phosphorylation is not required for the control of fatty acid oxidation during exercise. Physiol Rep 3(7):e12444

Perreault L, Lavely JM, Kittelson JM, Horton TJ (2004) Gender differences in lipoprotein lipase activity after acute exercise. Obes Res 12:241–249

Phillips SM, Green HJ, Tarnopolsky MA, Heigenhauser GJ, Grant SM (1996) Progressive effect of endurance training on metabolic adaptations in working skeletal muscle. Am J Phys 270:E265–E272

Prats C, Donsmark M, Qvortrup K, Londos C, Sztalryd C, Holm C, Galbo H, Ploug T (2006) Decrease in intramuscular lipid droplets and translocation of HSL in response to muscle contraction and epinephrine. J Lipid Res 47:2392–2399

Radegran G, Saltin B (1998) Muscle blood flow at onset of dynamic exercise in humans. Am J Phys 274:H314–H322

Raney MA, Turcotte LP (2006) Regulation of contraction-induced FA uptake and oxidation by AMPK and ERK1/2 is intensity dependent in rodent muscle. Am J Physiol Endocrinol Metab 291:E1220–E1227

Rasmussen BB, Winder WW (1997) Effect of exercise intensity on skeletal muscle malonyl-CoA and acetyl-CoA carboxylase. J Appl Physiol (1985) 83:1104–1109

Ren J, Lakoski S, Haller RG, Sherry AD, Malloy CR (2013) Dynamic monitoring of carnitine and acetylcarnitine in the trimethylamine signal after exercise in human skeletal muscle by 7T 1H-MRS. Magn Reson Med 69:7–17

Roepstorff C, Halberg N, Hillig T, Saha AK, Ruderman NB, Wojtaszewski JF, Richter EA, Kiens B (2005) Malonyl-CoA and carnitine in regulation of fat oxidation in human skeletal muscle during exercise. Am J Physiol Endocrinol Metab 288:E133–E142
Roepstorff C, Steffensen CH, Madsen M, Stallknecht B, Kanstrup IL, Richter EA, Kiens B (2002) Gender differences in substrate utilization during submaximal exercise in endurance-trained subjects. Am J Physiol Endocrinol Metab 282:E435–E447
Rogge MM (2009) The role of impaired mitochondrial lipid oxidation in obesity. Biol Res Nurs 10: 356–373
Romijn JA, Coyle EF, Sidossis LS, Gastaldelli A, Horowitz JF, Endert E, Wolfe RR (1993) Regulation of endogenous fat and carbohydrate metabolism in relation to exercise intensity and duration. Am J Phys 265(3 Pt 1):E380–E391
Romijn JA, Coyle EF, Sidossis LS, Rosenblatt J, Wolfe RR (2000) Substrate metabolism during different exercise intensities in endurance-trained women. J Appl Physiol (1985) 88:1707–1714
Sahlin K (1990) Muscle carnitine metabolism during incremental dynamic exercise in humans. Acta Physiol Scand 138:259–262
Samovski D, Su X, Xu Y, Abumrad NA, Stahl PD (2012) Insulin and AMPK regulate FA translocase/CD36 plasma membrane recruitment in cardiomyocytes via Rab GAP AS160 and Rab8a Rab GTPase. J Lipid Res 53:709–717
Santiworakul A, Chuaychoo B, Kriengsinyos W, Saengsirisuwan V, Jalayondeja W (2014) Substrate utilization during and after high intensity exercise in healthy lean and obese men. J Med Assoc Thail 97(Suppl 7):S50–S54
Seidell JC, Muller DC, Sorkin JD, Andres R (1992) Fasting respiratory exchange ratio and resting metabolic rate as predictors of weight gain: the Baltimore longitudinal study on aging. Int J Obes Relat Metab Disord 16:667–674
Seiler SE, Koves TR, Gooding JR, Wong KE, Stevens RD, Ilkayeva OR, Wittmann AH, DeBalsi KL, Davies MN, Lindeboom L, Schrauwen P, Schrauwen-Hinderling VB, Muoio DM (2015) Carnitine acetyltransferase mitigates metabolic inertia and muscle fatigue during exercise. Cell Metab 22:65–76
Sidossis LS, Gastaldelli A, Klein S, Wolfe RR (1997) Regulation of plasma fatty acid oxidation during low- and high-intensity exercise. Am J Phys 272:E1065–E1070
Simoneau JA, Veerkamp JH, Turcotte LP, Kelley DE (1999) Markers of capacity to utilize fatty acids in human skeletal muscle: relation to insulin resistance and obesity and effects of weight loss. FASEB J 13:2051–2060
Slusher AL, Whitehurst M, Zoeller RF, Mock JT, Maharaj A, Huang CJ (2015) Brain-derived neurotrophic factor and substrate utilization following acute aerobic exercise in obese individuals. J Neuroendocrinol 27:370–376
Smirnova E, Goldberg EB, Makarova KS, Lin L, Brown WJ, Jackson CL (2006) ATGL has a key role in lipid droplet/adiposome degradation in mammalian cells. EMBO Rep 7:106–113
Smith BK, Jain SS, Rimbaud S, Dam A, Quadrilatero J, Ventura-Clapier R, Bonen A, Holloway GP (2011) FAT/CD36 is located on the outer mitochondrial membrane, upstream of long-chain acyl-CoA synthetase, and regulates palmitate oxidation. Biochem J 437:125–134
Smith BK, Perry CG, Koves TR, Wright DC, Smith JC, Neufer PD, Muoio DM, Holloway GP (2012) Identification of a novel malonyl-CoA IC (50) for CPT-I: implications for predicting in vivo fatty acid oxidation rates. Biochem J 448:13–20
Søndergaard E, Andersen IR, Sørensen LP, Gormsen LC, Nielsen S (2017) Lipoprotein lipase activity does not predict very low-density lipoprotein-triglyceride fatty acid oxidation during exercise. Scand J Med Sci Sports 27:474–481
Sondergaard E, Rahbek I, Sorensen LP, Christiansen JS, Gormsen LC, Jensen MD, Nielsen S (2011) Effects of exercise on VLDL-triglyceride oxidation and turnover. Am J Physiol Endocrinol Metab 300:E939–E944
Sparks LM (2017) Exercise training response heterogeneity: physiological and molecular insights. Diabetologia 60:2329–2336

Stallknecht B, Kiens B, Helge JW, Richter EA, Galbo H (2004) Interstitial glycerol concentrations in human skeletal muscle and adipose tissue during graded exercise. Acta Physiol Scand 180: 367–377

Stallknecht B, Lorentsen J, Enevoldsen LH, Bulow J, Biering-Sorensen F, Galbo H, Kjaer M (2001) Role of the sympathoadrenergic system in adipose tissue metabolism during exercise in humans. J Physiol 536:283–294

Stanford KI, Lynes MD, Takahashi H, Baer LA, Arts PJ, May FJ, Lehnig AC, Middelbeek RJW, Richard JJ, So K, Chen EY, Gao F, Narain NR, Distefano G, Shettigar VK, Hirshman MF, Ziolo MT, Kiebish MA, Tseng YH, Coen PM, Goodyear LJ (2018) 12,13-diHOME: an exercise-induced Lipokine that increases skeletal muscle fatty acid uptake. Cell Metab 27:1111–1120

Starling RD, Trappe TA, Parcell AC, Kerr CG, Fink WJ, Costill DL (1997) Effects of diet on muscle triglyceride and endurance performance. J Appl Physiol (1985) 82:1185–1189

Starritt EC, Howlett RA, Heigenhauser GJ, Spriet LL (2000) Sensitivity of CPT I to malonyl-CoA in trained and untrained human skeletal muscle. Am J Physiol Endocrinol Metab 278:E462–E468

Steffan HG, Elliott W, Miller WC, Fernhall B (1999) Substrate utilization during submaximal exercise in obese and normal-weight women. Eur J Appl Physiol Occup Physiol 80:233–239

Steffensen CH, Roepstorff C, Madsen M, Kiens B (2002) Myocellular triacylglycerol breakdown in females but not in males during exercise. Am J Physiol Endocrinol Metab 282:E634–E642

Stephens NA, Sparks LM (2015) Resistance to the beneficial effects of exercise in type 2 diabetes: are some individuals programmed to fail? J Clin Endocrinol Metab 100:43–52

Storlien L, Oakes ND, Kelley DE (2004) Metabolic flexibility. Proc Nutr Soc 63:363–368

Talanian JL, Tunstall RJ, Watt MJ, Duong M, Perry CG, Steinberg GR, Kemp BE, Heigenhauser GJ, Spriet LL (2006) Adrenergic regulation of HSL serine phosphorylation and activity in human skeletal muscle during the onset of exercise. Am J Physiol Regul Integr Comp Physiol 291:R1094–R1099

Tanner CJ, Barakat HA, Dohm GL, Pories WJ, Mac Donald KG, Cunningham PR, Swanson MS, Houmard JA (2002) Muscle fiber type is associated with obesity and weight loss. Am J Physiol Endocrinol Metab 282:E1191–E1196

Taskinen MR, Nikkila EA (1980) Effect of acute vigorous exercise on lipoprotein lipase activity of adipose tissue and skeletal muscle in physically active men. Artery 6:471–483

Thyfault JP, Kraus RM, Hickner RC, Howell AW, Wolfe RR, Dohm GL (2004) Impaired plasma fatty acid oxidation in extremely obese women. Am J Physiol Endocrinol Metab 287:E1076–E1081

Turcotte LP, Richter EA, Kiens B (1992) Increased plasma FFA uptake and oxidation during prolonged exercise in trained vs. untrained humans. Am J Phys 262:E791–E799

Turcotte LP, Swenberger JR, Tucker MZ, Yee AJ, Trump G, Luiken JJ, Bonen A (2000) Muscle palmitate uptake and binding are saturable and inhibited by antibodies to FABP (PM). Mol Cell Biochem 210:53–63

van Hall G, Sacchetti M, Radegran G, Saltin B (2002) Human skeletal muscle fatty acid and glycerol metabolism during rest, exercise and recovery. J Physiol 543:1047–1058

van Loon LJ, Greenhaff PL, Constantin-Teodosiu D, Saris WH, Wagenmakers AJ (2001) The effects of increasing exercise intensity on muscle fuel utilisation in humans. J Physiol 536:295–304

van Loon LJ, Koopman R, Stegen JH, Wagenmakers AJ, Keizer HA, Saris WH (2003) Intramyocellular lipids form an important substrate source during moderate intensity exercise in endurance-trained males in a fasted state. J Physiol 553:611–625

van Loon LJ, Thomason-Hughes M, Constantin-Teodosiu D, Koopman R, Greenhaff PL, Hardie DG, Keizer HA, Saris WH, Wagenmakers AJ (2005) Inhibition of adipose tissue lipolysis increases intramuscular lipid and glycogen use in vivo in humans. Am J Physiol Endocrinol Metab 289:E482–E493

Wall BT, Stephens FB, Constantin-Teodosiu D, Marimuthu K, Macdonald IA, Greenhaff PL (2011) Chronic oral ingestion of L-carnitine and carbohydrate increases muscle carnitine content and alters muscle fuel metabolism during exercise in humans. J Physiol 589:963–973

Watt MJ, Heigenhauser GJ, Dyck DJ, Spriet LL (2002) Intramuscular triacylglycerol, glycogen and acetyl group metabolism during 4 h of moderate exercise in man. J Physiol 541(Pt 3):969–978

Watt MJ, Holmes AG, Steinberg GR, Mesa JL, Kemp BE, Febbraio MA (2004b) Reduced plasma FFA availability increases net triacylglycerol degradation, but not GPAT or HSL activity, in human skeletal muscle. Am J Physiol Endocrinol Metab 287(1):E120–E127

Watt MJ, Steinberg GR, Chan S, Garnham A, Kemp BE, Febbraio MA (2004a) Beta-adrenergic stimulation of skeletal muscle HSL can be overridden by AMPK signaling. FASEB J 18(12): 1445–1446

Whitfield J, Paglialunga S, Smith BK, Miotto PM, Simnett G, Robson HL, Jain SS, Herbst EAF, Desjardins EM, Dyck DJ, Spriet LL, Steinberg GR, Holloway GP (2017) Ablating TBC1D1 impairs contraction-induced sarcolemmal glucose transporter type 4 redistribution, but not insulin-mediated responses in rats. J Biol Chem 292(40):16653–16664

Wicks SE, Vandanmagsar B, Haynie KR, Fuller SE, Warfel JD, Stephens JM, Wang M, Han X, Zhang J, Noland RC, Mynatt RL (2015) Impaired mitochondrial fat oxidation induces adaptive remodeling of muscle metabolism. Proc Natl Acad Sci U S A 112:E3300–E3309

Wojtaszewski JF, Nielsen P, Hansen BF, Richter EA, Kiens B (2000) Isoform-specific and exercise intensity-dependent activation of 5'-AMP-activated protein kinase in human skeletal muscle. J Physiol 528(Pt 1):221–226

Wolfe RR, Klein S, Carraro F, Weber JM (1990) Role of triglyceride-fatty acid cycle in controlling fat metabolism in humans during and after exercise. Am J Phys 258(2 Pt 1):E382–E389

Zechner R, Zimmermann R, Eichmann TO, Kohlwein SD, Haemmerle G, Lass A, Madeo F (2012) FAT SIGNALS–lipases and lipolysis in lipid metabolism and signaling. Cell Metab 15:279–291

Zou K, Hinkley JM, Park S, Zheng D, Jones TE, Pories WJ, Hornby PJ, Lenhard J, Dohm GL, Houmard JA (2019) Altered tricarboxylic acid cycle flux in primary myotubes from severely obese humans. Int J Obes 43:895–905

Zurlo F, Lillioja S, Esposito-Del PA, Nyomba BL, Raz I, Saad MF, Swinburn BA, Knowler WC, Bogardus C, Ravussin E (1990) Low ratio of fat to carbohydrate oxidation as predictor of weight gain: study of 24-h RQ. Am J Phys 259:E650–E657

Chapter 9
Skeletal Muscle Protein Metabolism During Exercise

Sophie Joanisse, James McKendry, Everson A. Nunes, Aaron C. Q. Thomas, and Stuart M. Phillips

Abstract Skeletal muscle is essential in locomotion and plays a role in whole-body metabolism, particularly during exercise. Skeletal muscle is the largest 'reservoir' of amino acids, which can be released for fuel and as a precursor for gluconeogenesis. During exercise, whole-body, and more specifically skeletal muscle, protein catabolism is increased, but protein synthesis is suppressed. Metabolism of skeletal muscle proteins can support energy demands during exercise, and persistent exercise (i.e. training) results in skeletal muscle protein remodelling. Exercise is generally classified as being either 'strength' or 'aerobic/endurance' in nature, and the type of exercise will reflect the phenotypic and metabolic adaptations of the muscle. In this chapter, we describe the impact of various exercise modes on protein metabolism during and following exercise.

Keywords Amino acid · Protein turnover · Branched-chain amino acid · Proteolysis · Protein synthesis

9.1 General Introduction

Skeletal muscle is a highly 'plastic' tissue that adapts its phenotype with increased use or disuse. Constituting up to ~40% of total body mass and containing 60–75% of all body proteins, skeletal muscle is one the largest contributors to basal metabolic rate; it enables locomotion and is critical for sporting performance endeavours. By converting chemical to mechanical energy, skeletal muscle is a substantial metabolic sink for storage and oxidation (when active) of substrates such as glucose and lipids (Frontera and Ochala 2015). Given that there is no place of dedicated storage for

S. Joanisse
Department of Sport and Exercise Sciences, Manchester Metropolitan University, Manchester, UK

J. McKendry · E. A. Nunes · A. C. Q. Thomas · S. M. Phillips (✉)
Department of Kinesiology, McMaster University, Hamilton, ON, Canada
e-mail: phillis@mcmaster.ca

© The Author(s), under exclusive license to Springer Nature Switzerland AG 2022
G. McConell (ed.), *Exercise Metabolism*, Physiology in Health and Disease,
https://doi.org/10.1007/978-3-030-94305-9_9

amino acids in the body, excluding fully functional bodily proteins (e.g. within the skeletal muscle as part of the contractile machinery), the role of skeletal muscle also has an important protective role by providing a crucial reserve of amino acids. Thus, through proteolysis, muscle-bound amino acids can be liberated to provide substrates for energy provision or provide precursors for gluconeogenesis in situations of catabolism (Biolo et al. 2002; Felig et al. 1969; Wolfe 2006). The aim of this chapter is to discuss muscle protein (amino acid) metabolism during exercise performance and highlight the influence that different exercise modalities—aerobic and strength—have on amino acid metabolism.

9.2 Muscle Protein Synthesis and Breakdown

Numerous exogenous stimuli can modify the muscle proteome and affect muscle protein metabolism. Two widely investigated stimuli that influence muscle mass and protein metabolism are physical activity (exercise) and food intake (nutrition). The total amount of skeletal muscle mass is dictated by the balance between two processes: muscle protein synthesis (MPS) and muscle protein breakdown (MPB) (Phillips et al. 1997). Both MPS and MPB occur continuously and simultaneously, exchanging amino acids between bound protein and the free amino acid pool; however, it is the net balance between the two (MPS minus MPB), which changes in response to exercise and nutrition, ultimately determining whether muscle protein is accrued or lost over time. For example, strength-based exercise (SE), which typically describes high(er) contractile forces working against an increased external load, is most commonly associated with muscle size and strength increases. In contrast, aerobic exercise (AE), typically characterized by low(er) contractile forces and repetitive/cyclical contractions, elicits non-hypertrophic remodelling of skeletal muscle and instead promotes an enhanced oxidative capacity. Exercise, whether aerobic or strength-based, has been shown to augment mixed-MPS and MPB in the post-exercise period (Harber et al. 2010; Phillips et al. 1997), an effect that can persist for up to 24–48 h (Burd et al. 2010; Phillips et al. 1997). However, in the absence of protein ingestion and the ensuing hyperaminoacidemia, the balance between MPS and MPB remains negative (MPB > MPS) (Biolo et al. 1995). Hyperaminoacidemia exerts a direct stimulatory effect on MPS (Moore et al. 2009), and net muscle protein balance becomes, transiently, positive (MPS > MPB) (Atherton et al. 2010a). When prior exercise and protein nutrition are combined, they exert a synergistic effect on MPS and result in a small suppression of MPB, and muscle protein balance is even more positive (MPS > > MPB) (Biolo et al. 1997), which underpins a longer (versus feeding alone) period of positive net protein balance (Biolo et al. 1997; Greenhaff et al. 2008). The combination of exercise and protein ingestion practiced over time would drive the net accrual of proteins specific to the exercise modality.

9.3 The Influence of Exercise

Exercise results in the remodelling of skeletal muscle, and this occurs in an exercise mode-specific manner. Expectedly, the amino acid utilization for both AE and SE differs substantially. Unlike AE, SE leads to minimal amino acid oxidation during exercise; however, amino acid provision following exercise supports an elevated synthetic rate of the myofibrillar protein fraction and hypertrophic muscle remodelling. Conversely, AE results in a greater rate of amino acid oxidation during and after the bout of exercise compared to SE. Importantly, amino acid provision following AE primarily facilitates the mitochondrial protein fraction remodelling to enhance muscle oxidative capacity (Churchward-Venne et al. 2020). Repeated engagement in AE or SE for an extended period (e.g. exercise training) can also alter the protein synthetic response of the skeletal muscle to an acute bout of exercise. The skeletal muscle protein synthetic response to an exercise stimulus becomes 'refined' over time, however, and this specificity dictates phenotypic adaptation (Damas et al. 2019; Wilkinson et al. 2008). These specific signalling events will be discussed in greater detail in subsequent sections of the chapter.

9.4 The Influence of Nutrition

Protein is an essential component of any healthy diet providing amino acid precursors for protein synthesis and other metabolic fates. However, MPB can provide ~80–90% of amino acids to maintain the intramuscular amino acid pool for muscle protein resynthesis/remodelling and the formation of other important metabolic intermediates; thus, protein turnover in a day is 3–4 times the total loss of amino acids. Therefore, the discrepancy of amino acid supply hinges on sufficient dietary protein intake to replenish the intramuscular amino acid pool. Currently, the recommended dietary allowance (RDA) for the daily intake of protein for adults is 0.8 g/kg of body mass (Institute of Medicine (US) and Institute of Medicine (US) 2005). However, this recommendation is a minimum protein intake and is based on achieving nitrogen (protein) balance in a relatively small cohort of individuals and is not a target for optimal intake (Rand et al. 2003). While the daily relative protein intake in Irish adults aged 18–35 years has recently been reported to be greater than the RDA (~1.3 g/kg/d), this declines with advancing age (Hone et al. 2020). Given that older adults often display reduced responsiveness of MPS to protein nutrition, termed 'anabolic resistance', and highly active individuals who require a greater abundance of amino acids to support training adaptations (e.g. increased muscle mass), skeletal muscle remodelling, and, in many cases, fuel provision, the protein requirements for these cohorts have been repeatedly shown to be significantly greater than the RDA (Bauer et al. 2013; Deutz et al. 2014; Mazzulla et al. 2019; Morton et al. 2018).

9.4.1 Amino Acids

Following the consumption of protein-containing food sources, proteins are hydrolysed into smaller peptides and constituent amino acids. In addition to the dietary provision of amino acids, tissue (e.g. skeletal muscle) degradation of existing protein structures occurs via the ubiquitin-proteasomal system, lysosomal autophagy, calpain Ca2 +-dependent or caspase-mediated degradation. Amino acid synthesis, of non-essential amino acids (NEAA), in the liver can also provide amino acids as the necessary substrates to support the synthesis of body proteins and other important amino-containing biological compounds (Blanco and Blanco 2017).

Under normal physiological conditions, amino acids are either retained in the desired tissue or enter the systemic circulation and transported to the cells and tissues that require them. Amino acids have one of four metabolic fates within the body (see Fig. 9.1) (Papachristodoulou et al. 2018). First, and most commonly, amino acids remain unmodified and are utilized to synthesize new proteins (e.g. contractile, structural, plasma proteins, enzymes, or haem proteins). Second, amino acids can produce non-protein nitrogen-containing compounds such as hormones, neurotransmitters, and

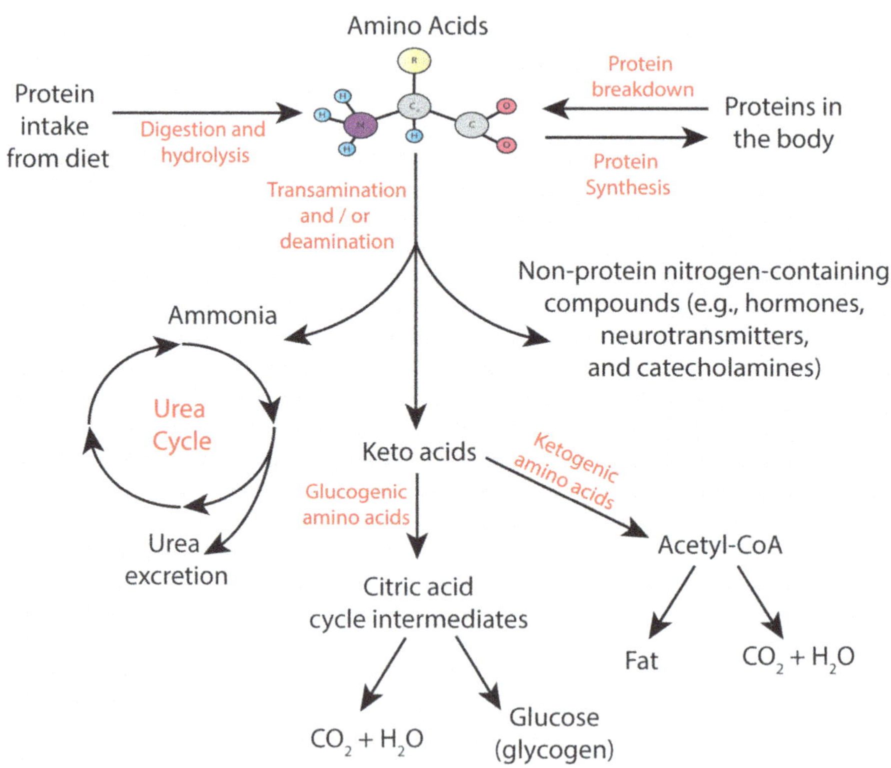

Fig. 9.1 Simplified schematic representing amino acid metabolism in the human body

catecholamines. Third, amino acids undergo transamination/deamination to separate the amino group from the carbon skeleton and form non-nitrogen-containing organic acids. The organic acid carbon skeleton that remains following transamination can then be used in gluconeogenesis to form glucose or form a metabolic intermediate for the citric acid cycle. Certain amino acids (isoleucine) can be lipogenic or ketogenic (e.g. acetyl-CoA) metabolic intermediate for oxidation in the mitochondria and subsequent energy provision. Finally, it needs to be emphasized that amino acids provided over the immediate requirements for protein synthesis and other amino acid-utilizing pathways invariably undergo deamination to produce ammonia and urea for removal by excretion (Blanco and Blanco 2017).

When considering skeletal muscle protein (amino acid) metabolism, not all amino acids are created equal. Broadly, amino acids can be subdivided into two categories: essential (indispensable) amino acids (EAA) and non-essential (dispensable) amino acids (NEAA). The main distinguishing feature between the two groups of amino acids is that EAA cannot be synthesized endogenously and therefore must be obtained from protein nutrition or, in some rare cases, endogenous protein breakdown. In contrast, NEAA can be synthesized endogenously. Skeletal muscle contains the full complement of EAA, and the myofibrillar proteins actin and myosin are the primary components of muscle protein, making up ~65% of skeletal muscle protein. Myofibrillar proteins contain a significantly higher than expected proportion of the branched-chain amino acids (BCAA), a subgroup of EAA that includes leucine, isoleucine, and valine. BCAA, in particular leucine, plays a pivotal role in the regulation of muscle protein metabolism.

At rest, skeletal muscle can metabolize six of the twenty amino acids, the BCAA: leucine, isoleucine, valine, asparagine, aspartate, and glutamate. These six amino acids provide the amino groups and carbon skeletons to synthesize TCA-cycle intermediates, glutamine, and alanine (Wagenmakers 1998a). However, only leucine and part of the isoleucine molecule can undergo complete degradation by transamination and decarboxylation, acetyl-CoA, and subsequently be oxidized, contributing a small amount to energy production. In addition, leucine is the primary amino acid responsible for stimulating an increase in MPS (Atherton et al. 2010b).

9.5 The Measurement of Protein Turnover

The assessment of protein metabolism at rest, during, and after exercise has been a topic of intense scientific inquiry. From a methodological standpoint, the assessment tools used to assess protein metabolism have significantly improved (Wilkinson 2018; Wilkinson et al. 2017). Early studies of protein requirements employed the nitrogen balance technique, which involves quantifying all of the ingested protein (nitrogen) and all of the nitrogen that is removed from the body (via urine, faeces, sweat, and various minor routes: hair, skin, nails, menstrual loss) and the use of

3-methylhistidine (3-MH) as an assessment of myofibrillar proteolysis. However, both the nitrogen balance and 3-MH techniques have several limitations, and therefore, stable isotope tracers have been the preferred technique for studying amino acid metabolism for several decades (Wilkinson 2018; Wilkinson et al. 2017). Since the inception of stable isotopes in metabolic research, and as a result of the significant advances in the field (Brook et al. 2017), the knowledge of protein metabolism has significantly improved, including, but not limited to, protein requirements (Bauer et al. 2013; Deutz et al. 2014; Moore 2019; Moore et al. 2015), optimal protein dose/timing/distribution to maximize MPS (Areta et al. 2013; Moore et al. 2015; Moore 2019), and amino acid oxidation (Mazzulla et al. 2019). Importantly, skeletal muscle protein metabolism is a modifiable process and one on which exercise exerts a substantial influence.

9.6 The Concept of Exercise Specificity

It is well-known that AE can stimulate mitochondrial protein synthesis within skeletal muscle, a contributor to mitochondrial biogenesis (Joyner and Coyle 2008; Wilkinson et al. 2008). In comparison, the myofibrillar fraction of muscle protein is stimulated preferentially and synthesized following SE, promoting muscular hypertrophy and strength gains (Mitchell et al. 2013; Murphy et al. 2018; Wilkinson et al. 2008). Phenotypic adaptations to training are directly related to the action, frequency, and intensity of the exercise performed (Hawley 2002), commonly referred to as the 'exercise specificity principle'. It is relatively well established that the divergent phenotypes characterizing strength and aerobic training are due to stimulation of the distinct muscle fractions—either myofibrillar or mitochondrial protein synthesis (Wilkinson et al. 2008).

Regular SE, or skeletal muscle overload, induces a signalling cascade that converges on the mammalian or main target of rapamycin (mTORC), resulting in its phosphorylation and ultimately activating the muscle protein synthetic response (Hawley 2002). Contrarily, AE elicits a molecular response upregulating adenosine monophosphate-activated protein kinase (AMPK) activity converging on peroxisome proliferator-activated receptor gamma coactivator 1-alpha (PGC-1α) and signalling for muscle oxidative adaptation (Baar 2014). However, despite the dogma of preferential exercise adaption in response to contrasting training modalities, it has been observed in training-naïve individuals that the post-translational signalling response to an acute bout of exercise is rather non-specific when comparisons of SE and AE have been made (Di Donato et al. 2014; Wilkinson et al. 2008). These findings (Di Donato et al. 2014; Wilkinson et al. 2008) have led to a hypothesis that proposes an initially non-specific exercise stimulus that is, with repeated exposures, 'fine-tuned' with refinements of the muscle transcriptome, and subsequently the proteome, in response to either chronic AE or SE stimuli (Coffey and Hawley 2017; Hoffman et al. 2015; Wilkinson et al. 2008). The hypothesis generally suggests that the muscle protein synthetic response following exercise is specific to the modality of exercise performed, allowing for recovery and adaptation to structural and metabolic stress; additionally, the specificity of de novo proteins following regular exercise is calibrated to respond and adapt when subjected to a familiar exercise stimulus (Hawley 2002; Wilkinson et al. 2008). An illustration of this hypothesis is clear when observing well-trained or master athletes who have trained either an aerobic or strength-type discipline for a substantial period (years), demonstrating enhanced myofibrillar protein synthesis (MyoPS) and mitochondrial content (Coffey et al. 2006; McKendry et al. 2019; Tang et al. 2008). Therefore, the phenotypic outcomes arising from the practice of aerobic and strength training are divergent from one another. In the following sections, we discuss the effects of AE, SE, and concurrent exercise on muscle protein metabolism.

9.6.1 Aerobic Exercise

It is widely accepted that energy provision during a bout of AE comes predominantly from carbohydrates and lipids, whereas the oxidation of amino acids provides a minimal contribution. Although the overall provision of energy stemming from amino acid oxidation is minimal during AE, the oxidation of amino acids increases significantly during exercise versus resting levels (Gibala 2007). The BCAA, such as leucine and isoleucine, can be metabolized to acetyl-CoA and ultimately provide ATP to the working muscle (Newsholme et al. 2019). Amino acids ‘indirectly’ provide energy to the skeletal muscle as precursors for the synthesis of intermediates of the TCA-cycle, a major energy-providing pathway during AE, in addition to providing the carbon skeleton necessary for the synthesis of glutamine and alanine (Wagenmakers 1998a, b). For instance, the transamination of alanine to α-ketoglutarate via alanine aminotransferase results in pyruvate and glutamate formation. This reaction is an equilibrium reaction that, during AE, contributes to maintaining the concentration of TCA intermediates (e.g. α-ketoglutarate) (Newsholme et al. 2019).

Glutamine is the most abundant amino acid in the free amino acid pool within skeletal muscle and blood. Glutamine is important to the metabolic function of tissues throughout the body and, along with alanine, transfers amino-derived nitrogen from the muscle to the liver (Wagenmakers 1998b). Glutamine can be converted to glutamate in a reaction catalysed by the enzyme glutamine synthase in the skeletal muscle (Wagenmakers 1998b). There is a significant reduction in glutamate concentration in the skeletal muscle’s free amino acid pool within the first 10 mins of the onset of AE (cycling 75% VO_2 max or 90 min of knee extension at an intensity of 60–65% max power). Such a reduction in skeletal muscle glutamate persists for up to 90 mins and is mirrored by an increase in glutamate uptake from systemic circulation (van Hall et al. 1995; Sahlin et al. 1990). In contrast, alanine concentration in skeletal muscle is increased in the first 10 min of AE and gradually decreases, returning to basal concentrations by 90 min (van Hall et al. 1995; Sahlin et al. 1990), with a concomitant increase in alanine release into the circulation. The rapid increase in

TCA-cycle intermediates at the onset of exercise is believed to be due to the alanine-aminotransferase reaction consuming alanine and providing α-ketoglutarate (Wagenmakers 1998b). Due to increased glycolysis, there is a marked rise in muscle pyruvate content at the immediate onset of exercise, which increases α-ketoglutarate and alanine (Wagenmakers 1998b). However, in an energy-depleted state, the deamination of BCAA provides the skeletal muscle with TCA-cycle intermediates to allow continued flux through the cycle, but this is reduced compared to a non-depleted state (Wagenmakers 1998a).

Amino acid utilization during AE was first described using nitrogen balance studies; however, results from such studies provided equivocal evidence due to high variability (Millward 2001). More recently, researchers have described the effect of exercise on protein metabolism by determining amino acid oxidation during various AE interventions via the use of isotopically labelled amino acids. Common practice involves the infusion of ^{13}C-labelled leucine to determine leucine flux, oxidation, and indirect synthesis estimates (Wolfe et al. 1982). Leucine oxidation is determined using the labelled enrichment in expired CO_2 and the plasma enrichment of α-ketoisocaproic acid (α-KIC), the ketoacid of leucine formed following transamination, as a more accurate representation of the precursor pool leucine enrichment (Wolfe et al. 1982). Also, estimates of protein breakdown, measured as the rate of appearance (Ra) of leucine into the bloodstream from intact proteins (Wolfe et al. 1982) and synthesis (by difference), via the nonoxidative portion of leucine disposal (NOLD) (Wolfe et al. 1984), can be calculated. The use of the so-called leucine reciprocal pool method has consistently demonstrated an increase in leucine oxidation during endurance exercise of various duration and intensity (Lamont et al. 1999, 1995; McKenzie et al. 2000; Phillips et al. 1993; Rennie et al. 1981; Wolfe et al. 1982), indicating an increase in the use of this amino acid as a fuel source during endurance exercise (Gibala 2007). Notably, we have no direct estimates of amino acid oxidation of other BCAA (isoleucine and valine); however, lysine oxidation, in contrast to leucine, is not oxidized to an appreciable extent during exercise (Wolfe et al. 1984). The BCAA are likely the only amino acids oxidized, to an appreciable extent, during exercise, which is likely because their oxidation readily yields acetyl units. The highly regulated enzyme controlling BCAA oxidation, branched-chain ketoacid dehydrogenase (BCKAD), has a Km for each ketoacid by which would dictate that leucine and its ketoacid, α-KIC, would be oxidized over the other BCAA (Wagenmakers et al. 1990). AE has also been shown to increase the indirect measure of protein breakdown (leucine Ra) (Phillips et al. 1993; Rennie et al. 1981; Wolfe et al. 1982) and either suppresses (Mazzulla et al. 2017; McKenzie et al. 2000; Wolfe et al. 1984) or does not affect (Phillips et al. 1993) the indirect measure of protein synthesis (NOLD). These results suggest that AE results in a negative protein balance predominantly via an increase in amino acid flux toward deamination, amino acid oxidation, and a suppression of protein synthesis. Importantly, the reciprocal pool model of leucine oxidation (Lamont et al. 1999, 1995; McKenzie et al. 2000; Phillips et al. 1993; Rennie et al. 1981; Wolfe et al. 1982) can only assess whole-body protein turnover and cannot delineate between different tissue sites of metabolism (e.g. muscle and liver amino acid oxidation) without biopsies.

A central regulator of BCAA oxidation is the BCKAD, considered the rate-limiting step in BCAA oxidation. Analogous in structure to pyruvate dehydrogenase, BCKAD activity is increased following an acute bout of moderate-intensity (60–75%) continuous exercise lasting from 50 to 90 min (Howarth et al. 2007; Jackman et al. 1997; McKenzie et al. 2000). It has been postulated that BCKAD enzyme activity may be related to glycogen content (van Hall et al. 1996; Wagenmakers et al. 1990). The extent of BCKAD activity is associated with 'lower' muscle glycogen (Jackman et al. 1997), such that lower glycogen availability for use as fuel increases BCAA oxidation via elevations in BCKAD activity. Accordingly, BCKAD activation is correlated with leucine oxidation during endurance exercise (McKenzie et al. 2000). These findings could have important implications for leucine oxidation and the training-induced reduction in amino acid oxidation as is discussed below.

9.6.1.1 The Effect of Aerobic Training Status on Skeletal Muscle Protein Metabolism During Exercise

Skeletal muscle has the remarkable ability to adapt and remodel following different types of stimuli, such as exercise training. It should, therefore, come as no surprise that exercise training modifies the acute responses to exercise. Lamont et al. (1999) originally reported that AE at 50% VO_2 max for 60 min resulted in greater leucine oxidation and a reduced Ra of leucine (an indirect measure of protein breakdown) in trained compared to untrained individuals (Lamont et al. 1999). However, when values were expressed relative to fat-free mass, no differences were observed between trained and untrained individuals (Lamont et al. 1999). Conversely, McKenzie et al. (2000) showed that leucine oxidation was increased following a bout of exercise at 60% VO_2 max for 90 min in untrained participants; however, following 38 days of endurance exercise training, this increase was no longer apparent (McKenzie et al. 2000). Leucine flux and NOLD were reduced during exercise, which was not affected by training status (McKenzie et al. 2000), which is somewhat congruent with observations in highly trained individuals (Mazzulla et al. 2017; Phillips et al. 1993). Nonetheless, NOLD has been reported to remain unchanged (Phillips et al. 1993) or be suppressed (Mazzulla et al. 2017) with endurance training. Although not conclusive, there may be fundamental differences in muscle metabolism in highly trained versus untrained individuals even when trained, particularly with respect to amino acid oxidation. The absence of an increase in leucine oxidation following AE training may explain the extent of activation of BCKAD activity (Howarth et al. 2007). Training with AE results in an attenuation of the degree of activation of BCKAD with exercise (Howarth et al. 2007; McKenzie et al. 2000), which was associated with a reduction in leucine oxidation (McKenzie et al. 2000). AE training also increases BCKAD kinase content (Howarth et al.

2007), which can phosphorylate BCKAD, rendering it inactive and, in theory, also reducing leucine oxidation.

9.6.1.2 Aerobic Exercise and Muscle Protein Turnover

Although understanding the oxidation of amino acids *during* exercise is essential to understand muscle metabolism, researchers have also focused on AE training on basal muscle protein synthesis. Four weeks of AE training (3–5d/wk., 30–45 min/session, 65–85% HRmax) resulted in an increase in muscle fractional synthesis rate (FSR) in a fasted state (Pikosky et al. 2006). Short et al. (2004) also report an increase in basal mixed MPS following 16 weeks of AE training (Short et al. 2004). Although AE training is not routinely associated with increases in muscle size, which is associated with increases in MyoPS, it does elicit remodelling of the skeletal muscle proteome. Such AE training proteome remodelling is characterized by increases in mitochondrial volume and increased capillarization, postulated as the main reasons behind post-exercise elevations in MPS (Burd et al. 2019). In recent years, researchers have also sought to describe the acute protein synthetic response following AE in hopes of maximizing the effects of exercise and ensuring sufficient protein is consumed in proximity to exercise to support increased metabolic demands (Churchward-Venne et al. 2020; Lin et al. 2020). The next section of this chapter will summarize these data.

Using stable isotopically labelled tracers, initial studies demonstrated acute increases in mixed muscle MPS following low-intensity exercise of 40% of VO_2 max for short (45 min) (Sheffield-Moore et al. 2004) and long (4 h) (Carraro et al. 1990) durations in untrained young and older adults. However, in highly trained female swimmers, deltoid muscle MPS was not increased following a high-intensity swimming session (Tipton et al. 1996). The discrepancy in results is likely due to several reasons, including the muscle studied, exercise type, and the individuals' training status. The swimmers studied by Tipton et al. (Tipton et al. 1996) were well trained and thus likely already adapted to the intensity of the AE training session, which may not have been true of other subjects. Some differences also exist in the duration of the increased muscle protein synthetic response following exercise. Lower-intensity exercise, 40% VO_2 max for 45 min, resulted in an increase in MPS in the first hour after exercise followed by a decline in the following 2 h (Sheffield-Moore et al. 2004); however, more intense exercise >65% VO_2 max for 60 min resulted in an increase of FSR from 90 to 180 min (Mascher et al. 2011) and up to 6 h (Harber et al. 2010) following exercise. In untrained individuals, AE stimulates mixed MPS (Carraro et al. 1990; Harber et al. 2010; Mascher et al. 2011; Sheffield-Moore et al. 2004), and more intense exercise results in a sustained MPS response (Harber et al. 2010; Mascher et al. 2011).

Researchers have investigated the synthetic rate of different pools of proteins such as myofibrillar, sarcoplasmic, and mitochondrial fractions. Of particular interest is mitochondrial protein synthesis following AE as this is the main site of fuel metabolism within muscle cells, and AE training commonly elicits increases in

mitochondrial pool size. Following AE (45 min at 75% of VO_2 max), MyoPS was not increased compared to rest, but mitochondrial protein synthesis increased in both the trained and untrained states from 0 to 4 h post-exercise (Wilkinson et al. 2008). This concept was extended to untrained middle-aged men who completed 40 min of AE at 55% of peak aerobic power output, where both MyoPS and mitochondrial protein synthesis were increased 1 to 5 h after exercise (Donges et al. 2012). As such, age may modify the influence of AE on MPS in a protein subfraction-specific manner. The intensity of AE can also influence the skeletal muscle protein synthetic response. A bout of low (60 min at 30% watt max) or high (30 min at 60% watt max) intensity AE results in increased MyoPS immediately (0.5–4.5 h) after exercise and remains elevated only following high-intensity exercise 24–28 h post-exercise (Di Donato et al. 2014). Mitochondrial protein synthesis was only increased following high-intensity exercise 24–28 h after exercise (Di Donato et al. 2014). These studies have provided valuable insight into the effects of different types of endurance exercise on skeletal muscle protein synthesis and, more recently, the specific subfractions of skeletal muscle.

The molecular regulation of protein metabolism following AE has again not been as extensively studied as in SE. However, similar to SE, following acute AE, there is an increase in mTOR phosphorylation (Benziane et al. 2008; Camera et al. 2010; Di Donato et al. 2014; Mascher et al. 2011, 2007; Wilkinson et al. 2008). Further to this, the familiar downstream target of mTORC1, the ribosomal protein of 70 kDa S6 kinase 1 ($p70^{S6K1}$) is also reported to be phosphorylated following AE (Benziane et al. 2008; Mascher et al. 2011, 2007; Wilkinson et al. 2008), although this finding is not consistent (Camera et al. 2010; Coffey et al. 2006; Di Donato et al. 2014). Additionally, AE stimulates AMPK phosphorylation, which can inhibit mTORC1 activity. How the activation of divergent signalling pathways may impact MPS is discussed in the following sections (see Fig. 9.2). Although not commonly associated with increases in muscle size, endurance exercise in untrained individuals does stimulate MyoPS, and more intense exercise results in a stimulation of mitochondrial protein synthesis. These increases likely occur to support remodelling and organelle biogenesis rather than increased muscle fibre size associated with AE. Although not the focus of this chapter, it is also important to consider nutritional and feeding strategies when exploring skeletal muscle protein turnover following AE (Moore 2015).

9.6.2 *Strength Exercise*

SE is a known strategy to increase muscle mass and improve function (McGlory et al. 2017). The current knowledge on this field points to the direction that skeletal muscle hypertrophy induced by RE results from repeated intermittent increases in MPS, especially of the myofibrillar protein pool (Brook et al. 2015). Increasing the myofibrillar content in skeletal muscle cells is a relatively slow process demanding approximately 6 weeks of training to reflect any detectable changes in muscle fibre

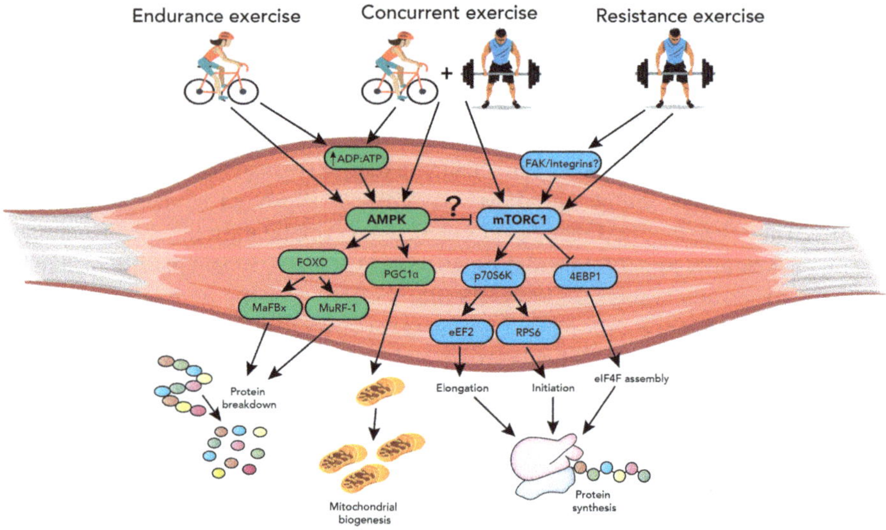

Fig. 9.2 A simplified schematic of the molecular pathways in response to exercise training. Strength exercise leads to mechanical and metabolic perturbations that stimulate the mTORC1 signalling cascade and increased myofibrillar protein synthesis. Endurance exercise modifies the cellular energy status and stimulates AMPK and PGC1α and predominately leads to mitochondrial biogenesis. Concurrent training stimulates both pathways and may produce an interference effect, whereby the AMPK pathway inhibits the activation of the mTORC1 pathway. Here, simplified linear pathways are shown, but these pathways undoubtedly display a large degree of dependence, crosstalk, interference, and redundancy. *ADP* adenosine diphosphate; *ATP* adenosine triphosphate; *Akt* protein kinase B; *AMPK* 5′ adenosine monophosphate-activated protein kinase; *mTORC1* mammalian target of rapamycin complex 1; *FOXO* forkhead box-containing subfamily; *PGC1 α* peroxisome proliferator-activated receptor gamma coactivator 1-alpha; *p70S6K* ribosomal protein 70-kDa S6 kinase 1; *4EBP1* 4E-binding protein-1; *MaFBx* muscle atrophy F-box; *MuRF1* muscle ring finger-1; *eEF2* eukaryotic elongation factor 2; *RPS6* ribosomal protein S6

cross-sectional area (fCSA) (Goreham et al. 1999; McGlory et al. 2017). Short-term changes in muscle mass resulting from muscle swelling are not due to the increase of contractile elements in skeletal muscle cells and are poorly correlated with the increment in MPS (Damas et al. 2016a, b). However, factors such as protein ingestion, an individual's genetic predisposition to muscle hypertrophy, and training status can change the magnitude and the time course of adaptations to SE (Churchward-Venne et al. 2015, 2012; Damas et al. 2019; Pescatello et al. 2006; Riechman et al. 2004).

Phillips et al. (1997) showed that SE increased mixed MPS for up to 48 h after the exercise bout (Phillips et al. 1997). Nevertheless, MPB was also significantly increased post-SE, resulting in an increased (less negative) muscle but no net protein incorporation when subjects were kept in fasting conditions (Phillips et al. 1997). The increment in the MPB response seems to be a physiological response to SE that may be underpinned by a need to degrade proteins exposed to mechanical, oxidative, or misfolding due to the SE bout. Furthermore, most amino acids released from the breakdown of damaged proteins can be recycled and used to synthesize new proteins (Moore et al. 2007). Because MPB can counterbalance MPS after RE, nutrition (e.g. protein ingestion) seems to be a pivotal factor in the long-term response to SE, as explored by other sections and chapters in this book. When a protein-containing meal is ingested in close temporal proximity to a SE bout, there is an expressive increment in MPS, leading to a positive balance and the increment of the protein pool (Phillips et al. 1997). Because the increments in MPS are 4–6-fold higher than the changes in MPB (Biolo et al. 1995; Wilkinson et al. 2007), MPS is the most relevant aspect to account for when focusing on skeletal muscle hypertrophy in healthy subjects.

9.6.2.1 The Effect of Strength Exercise Training Status on Skeletal Muscle Protein Metabolism

Interestingly, the acute increase in MPS caused by SE seems to change over time according to the subject's training state (Damas et al. 2015; Morton et al. 2015). In SE-untrained subjects, mixed MPS response is at least 50% above baseline levels for more than 40 h after the SE bout. Conversely, in SE-trained subjects, mixed MPS response is shorter, with the values returning to ~40–50% above baseline values 10 h after the exercise bout (Damas et al. 2015). MyoPS seems to be higher in untrained subjects, reaching 60% above baseline values at 4 h, during the acute response over the first 16 h after the SE bout (Damas et al. 2015; Kim et al. 2005). However, the current data indicate that after SE bouts performed during the first few weeks of training, the MPS response leads to skeletal muscle remodelling without significant changes in fCSA (Damas et al. 2015). On this topic, Damas et al. (2016a) investigated the time course of MyoPS and muscle damage response during 10 weeks of SE (Damas et al. 2016a). The authors showed that MyoPS was 'refined' and correlated to muscle hypertrophy after the first 3 weeks of a continuous SE training programme. The mechanism involved in SE-induced increment in MPS involves activating several intracellular pathways related to the regulation of protein synthesis at the transcriptional and translational levels. Furthermore, several cell growth-related pathways seem to contribute to the regulation of MPS during the response to SE.

9.6.2.2 Molecular Regulation of Skeletal Muscle Protein Synthesis in Response to Strength Exercise

Several different signalling pathways regulate protein synthesis within the muscle cell. Overall, MPS rates depend on the activation of the translational machinery and transcriptional templates that are translated. An increase in ribosomal content and other necessary protein translation molecules are also involved in the regulation of MPS (Brook et al. 2019; Chaillou 2019; Figueiredo and McCarthy 2019; McGlory et al. 2017). SE can drive MPS mainly by two distinct mechanisms (Goodman 2019;

McGlory et al. 2017). The first mechanism involves the signalling response to the mechanotransductive signal in the myofibre caused by muscle contraction (Drummond et al. 2009; You et al. 2012). A second relevant mechanism seems to be linked to the increased concentration of growth factors in the muscle due to skeletal muscle contraction. Insulin growth factor 1 (IGF-1), or its various splice isoforms, is an important growth factor that would further contribute to the activation of membrane receptors in skeletal muscle cells and, as a result, increase the MPS response (Drummond et al. 2009; Figueiredo and McCarthy 2019).

Both growth factors and mechanical stimulation can activate mTOR, one of the main regulatory proteins activated by SE, to stimulate MPS (Goodman 2019; You et al. 2019). The mTORC1 protein complex, consisting of mTOR and four other proteins, is the primary regulator of the muscle protein synthetic response to an acute SE bout. The five constituent proteins of the mTORC1 are mTOR, the regulatory protein associated with mTOR (Raptor), mammalian lethal with Sec13 protein 8 (mLST8, also known as GβL), DEP domain-containing mTOR interacting protein (DEPTOR), and proline-rich Akt substrate of 40 kDa (PRAS40) (Goodman 2019). mTORC1 activation increases MPS by enhancing the activity of downstream kinases such as the $p70^{S6K1}$ and 4E-binding protein-1 (4EBP1), leading to ribosomal binding to mRNA enhancing protein synthesis (Goodman 2019). It has been shown that mTORC1 activation by SE upregulates transcription of other components involved in translational machinery itself (e.g. mRNA, ribosomes) (Figueiredo and McCarthy 2019), thus increasing the translational capacity of the skeletal muscle cells over time (Brook et al. 2019; Figueiredo and McCarthy 2019).

The signalling mediated by the mTORC1 complex has been an intense field of study in the past 20 years, and it has been identified as one of the major pathways regulating cell growth and metabolism. The mTORC1 complex is commonly found in the skeletal muscle cell cytoplasm. It has been shown that mTORC1 complex activation is mandatory to maintain skeletal muscle mass (Goodman 2019). However, there is a necessary interaction between mTORC1 and the lysosome membrane in order for mTORC1 to be activated. Therefore, the translocation of the mTORC1 complex to the lysosome and the interaction with other proteins are necessary. A protein called small guanosine triphosphatase (GTPase) Ras homolog enriched in brain (Rheb) is a crucial component in the mTORC1 activation due to the interaction with the lysosome. Rheb-mediated activation of mTORC1 is orchestrated by an upstream GTPase activating protein named tuberous sclerosis complex 2 (TSC2). TSC2 is a central protein in the regulation of mTORC1 activation since it can be phosphorylated by distinct proteins such as protein kinase b (PKB or Akt), the mitogen-activated protein kinase (MAPK), also called extracellular signal-regulated kinase (ERK). When Akt- or ERK-related pathways phosphorylate TSC2, there is a reduction of the inhibitory effect that TSC2 exerts over mTORC1 (Saxton and Sabatini 2017).

Akt is a classic member of the growth factor-related signalling pathways. Akt exists in three distinct isoforms (Akt1, Akt2, and Akt3), while Akt2 seems to be the main isoform expressed in the human skeletal muscle (Matheny et al. 2018). SE induces local production of IGF-1 and other growth factors in the muscle. Local

IGF-1 production, acting in an autocrine fashion in the muscle tissue, but not systemically, is linked to the SE-induced MPS (Ahtiainen et al. 2016; Psilander et al. 2003). This hypothesis has been reinforced since systemic levels of several hormones are uncorrelated with skeletal muscle hypertrophy and strength gains (Morton et al. 2016; West et al. 2009, p. 200). Locally produced IGF-1 has been shown to act as a potent activator of protein synthesis by activating the Akt/mTORC1 pathway. When IGF-1 binds to its receptor on the muscle cell membrane, it activates phosphoinositide 3-kinase (PI3-K) to generate phosphatidylinositol (4,5)-bisphosphate. The result is an increased concentration of phosphatidylinositol 3,4,5-trisphosphate (PIP3). PIP3 binds to phosphoinositide-dependent kinase-1 (PDK1), which activates Akt by the phosphorylation on threonine 308 residue. When active, Akt phosphorylates other targets, including TSC2, leading to mTORC1 activation (Saxton and Sabatini 2017).

Mechanical stimuli (e.g. skeletal muscle contraction during SE) have repeatedly been shown to increase mTORC1 activity resulting in activation of MPS (Drummond et al. 2009; Goodman 2019). Cell signalling mediated by mechanotransduction in the muscle has been an intense field of investigation. However, the complete series of events involved in this process is not known. Focal adhesion kinase (FAK) is an integrin. Integrins are involved in several processes regarding the communication and signalling between the extracellular matrix, cell membrane, and cytoskeleton. It has been shown that FAK is mechanically sensitive and activates mTORC1 complex through a TSC2-dependent mechanism (Camera et al. 2016). Another molecule involved in the mechano-activation of mTORC1 is phosphatidic acid (PA), a lipid signalling molecule that binds to the FKBP-rapamycin-binding-domain of mTORC1, leading to its activation (Yoon et al. 2011). An enzyme named diacylglycerol kinase-ζ (DGK ζ) transforms diacylglycerol (DAG) into phosphatidic acid (PA) in response to muscle contraction. Therefore, the accumulation of PA in the cytosol and mTORC1 activation increases MPS (You et al. 2012); however, the exact chain of events involved in this process are unknown and requires further investigation.

Several proteins have been identified as downstream targets of mTORC1 and are involved in different steps regulating MPS. Some relevant proteins in the pathway regulating protein synthesis are $p70^{S6K1}$ and 4E-BP1, both mTORC1 complex direct targets. 4E-BP1 and $p70^{S6K1}$ are essential regulators of the translation initiation and elongation processes and are commonly applied as proxies for mTORC1 activity (Drummond et al. 2009; Wang and Proud 2006). $p70^{S6K1}$ is phosphorylated on Thr389 once mTORC1 is activated. Then, $p70^{S6K1}$ phosphorylates further downstream effectors, including eukaryotic elongation factor 2 kinase (eEF2k) and ribosomal protein S6 (RPS6), which facilitate the elongation process and ribosomal biogenesis (Wang and Proud 2006). 4E-BP1 is also phosphorylated by mTORC1 on Thr37 and Thr46, causing its dissociation from the eukaryotic initiation factor 4E (eIF4E) complex facilitating translation initiation.

Therefore, the regulation of MPS by SE seems to involve several different components having mTORC1 complex activation as a common convergency point. mTORC1 activation facilitates the initiation and elongation phases of the translation. Ribosome biogenesis is increased by mTORC1 activation, increasing translational capacity over time. Still, it is essential to mention that research applying SE in rodent models while inhibiting mTORC1 activity shows significant muscle hypertrophic responses over time, revealing that other mTORC1 independent signalling pathways also contribute to the stimulation of MPS resulting in muscle hypertrophy (Goodman 2019) (see Fig. 9.2).

9.6.3 Concurrent Exercise and Muscle Protein Metabolism

Concurrent exercise can be defined as utilizing both aspects of SE and AE in the same training regime (Baar 2014). To our knowledge, only two studies have directly compared the muscle protein synthetic response of concurrent to SE or AE (Carrithers et al. 2007; Donges et al. 2012). Examining the acute response following SE and concurrent exercise demonstrated no significant differences in mixed MPS or MyoPS, suggesting an equivocal response between the two exercise stimuli (Carrithers et al. 2007; Donges et al. 2012). These findings demonstrate that an acute bout of concurrent exercise is sufficient and capable of inducing a post-exercise rise in MPS comparable to SE alone. Consistent in highly trained female swimmers, an exercise session consisting of intense swimming followed by SE increased MPS 5 h afterwards (Tipton et al. 1996). As previously stated, a concurrent exercise design incorporates an aerobic and strength stimulus into the same training regime (Murach and Bagley 2016); beyond this common feature, there is a high degree of heterogeneity in study design, chiefly due to the additional number of exercise-related variables that could be manipulated (Eddens 2019). A study examining the anabolic response to concurrent exercise in mice manipulated whether AE was performed before or after SE (Ogasawara et al. 2014). The research team demonstrated a post-exercise stimulation of MPS following both concurrent exercise modalities; however, MPS was significantly greater in the group that performed AE before SE (Ogasawara et al. 2014). The researchers concluded that 'anabolic signalling stimulated via resistance exercise is susceptible to interference from a subsequent bout of endurance exercise and metabolic stress' (Ogasawara et al. 2014).

9.6.3.1 Concurrent Exercise Training and Potential Interference

The demands of many sports require an athlete's physiology be optimized for both endurance, strength, and power to be successful (Moore et al. 2014). Accordingly, many athletes incorporate AE and SE concomitantly in their training regime (concurrent training), aiming to maximize their muscular endurance, strength, and power (Baar 2014). Early studies of AE and SE performed concomitantly suggested a dampening of muscular hypertrophy with high volume training, which was dubbed the interference effect (Dudley and Djamil 1985; Hickson 1980). However, since those early studies, multiple randomized control trials have examined the effects of concurrent exercise training and shown that frequency, intensity, volume, timing, modality, and nutritional variables used in study design profoundly influence outcomes (Baar 2014; Coffey and Hawley 2017). Utilization of both AE and SE training in the same training regime was first examined in 1980 by Hickson et al., who, through personal anecdotal evidence, first described and further demonstrated the attenuation of strength development with concurrent exercise training (Hickson 1980). Subsequent studies appeared to confirm and expand upon these initial findings (Hickson 1980), in which combined SE and AE training elicited similar attenuation of strength development and the hypertrophic response (Dudley and Djamil 1985; Kraemer et al. 1995). However, more recently, a review by Murach and Bagley (2016) concluded that, in ideal circumstances involving adequate recovery, concurrent exercise does not interfere with adaptation to SE and may even facilitate the hypertrophic response at least in short-term training periods.

Concurrent exercise interference is characterized by attenuation in strength and hypertrophic gains when SE is completed alongside AE in the same training regime. Contrarily, concurrent exercise is not thought to substantially interfere with AE-induced adaptations, e.g. mitochondrial biogenesis, angiogenesis, and myofibre oxidative capacity (Baar 2014). It is well established that SE is typically followed by a pronounced stimulation of MyoPS (MacDougall et al. 1995), culminating in the accumulation of contractile protein and inducing muscular hypertrophy. Thus, the efficacy of concurrent exercise to elicit an equivalent hypertrophic response may be evaluated by comparing the acute rise in MyoPS during or immediately following exercise. There is, however, scant evidence from studies contrasting the muscle protein synthetic response and much less the myofibrillar fraction following concurrent exercise to other modalities such as AE or SE. Furthermore, a major barrier noted by researchers while studying the outcomes of concurrent exercise training is the overwhelming number of novel variables introduced. The addition of a second training modality into a study design means having to consider the sequence of the exercises, the timing between unique exercise stimuli, the modality of exercise chosen, and the feeding or fasting pattern of the study subjects (Wilkinson et al. 2008). Despite these complicating design considerations, it appears that a significant increase in mixed muscle MPS and MyoPS does occur following an acute bout of concurrent exercise training (Churchward-Venne et al. 2019; Donges et al. 2012; Parr et al. 2014), suggesting that in the acute hours following concurrent exercise, hypertrophic signalling is sufficient to induce a robust protein synthetic response.

9.6.3.2 Molecular Regulation of Skeletal Muscle Protein Synthesis in Response to Concurrent Exercise

Multiple stimuli are associated with both AE and SE, eliciting complex downstream signalling mechanisms targeted towards optimal cellular adaptation (Egan et al. 2016). Anabolic signalling is primarily mediated through the IGF-1Akt-mTORC1 axis as described in detail above (Yoon 2017). Contrarily, AE depletes cellular energy stores, stimulating AMPK and downstream effectors known to be largely involved in AE-induced adaptation (Baar 2014; Ogasawara et al. 2014). Activation of AMPK is a known inhibitor of mTORC1 activity via RAPTOR suppression, negatively affecting MyoPS (Lantier et al. 2010; Mounier et al. 2011). This mechanism is hypothesized to be why AE performed after SE may suppress mTORC1 activity from maximally stimulating MyoPS (see Fig. 9.2). The proximity of the AE before the performance of SE may be best determined by the duration of activation of AMPK following an acute session of AE, which could be for up to 3 h post-exercise (Wojtaszewski et al. 2000). Nonetheless, the initial observations of multimodal close-proximity exercise in human muscle demonstrated no meaningful differences in myogenic signalling pathway activation regardless of exercise order (Coffey et al. 2009a, b). More recently, researchers have confirmed that when a 'control' (SE only) group was included, no differences existed between the acute signalling response of either alternate concurrent exercise order and SE (Jones et al. 2016). Furthermore, some have hypothesized that the molecular response following combined AE and SE training is augmented compared to unimodal exercise (Kazior et al. 2016; Lundberg et al. 2014; Lundberg et al. 2012).

Although the variables regulating concurrent exercise protein synthesis are still mostly unknown, practical training recommendations that consider current molecular data for individuals hoping to maximize their muscular adaptation are directed to read reviews on concurrent exercise training (Baar 2014). The molecular basis of training adaptation is unique to each exercise modality and dynamic in response to chronic stimuli. Characterization of how the muscle proteome alters in response to prolonged AE, SE, and concurrent exercise training is necessary to elucidate the optimal stimuli for muscle adaptation to exercise.

9.7 Conclusion

Exercise is one of the main contributors to acute and long-term changes in protein metabolism. The research established in the past 30 years has focused on understanding several intrinsic complex layers within the field. The MPS response seems to be the main relevant component of protein metabolism affected by exercise in healthy subjects. Besides some expected generic changes in individuals initiating exercise training, most long-term adaptations in protein metabolism seem to involve changes specific to the training programme. AE induces an initial remodelling phase that includes contractile elements. However, AE induces increased expression of proteins handling energy production processes and mitochondrial proteins. In contrast, SE induces skeletal muscle remodelling, emphasizing the synthesis of new contractile elements. While the response to an exercise bout is attenuated in trained subjects, it prioritized MyoPS. Concurrent training induces both AE and SE phenotypic MPS characteristics. Still, the overlapping intracellular signalling responses due to concurrent exercise training show two-way inhibitory signalling, therefore limiting, at some point, the achievement of the full potential of specific phenotypical MPS responses.

References

Ahtiainen JP, Hulmi JJ, Lehti M, Kraemer WJ, Nyman K, Selänne H, Alen M, Komulainen J, Kovanen V, Mero AA, Philippou A, Laakkonen EK, Häkkinen K (2016) Effects of resistance training on expression of IGF-I splice variants in younger and older men. Eur J Sport Sci 16: 1055–1063

Areta JL, Burke LM, Ross ML, Camera DM, West DWD, Broad EM, Jeacocke NA, Moore DR, Stellingwerff T, Phillips SM, Hawley JA, Coffey VG (2013) Timing and distribution of protein ingestion during prolonged recovery from resistance exercise alters myofibrillar protein synthesis. J Physiol 591:2319–2331

Atherton PJ, Etheridge T, Watt PW, Wilkinson D, Selby A, Rankin D, Smith K, Rennie MJ (2010a) Muscle full effect after oral protein: time-dependent concordance and discordance between human muscle protein synthesis and mTORC1 signaling. Am J Clin Nutr 92:1080–1088

Atherton PJ, Smith K, Etheridge T, Rankin D, Rennie MJ (2010b) Distinct anabolic signalling responses to amino acids in C2C12 skeletal muscle cells. Amino Acids 38:1533–1539

Baar K (2014) Using molecular biology to maximize concurrent training. Sports Med 44:117–125

Bauer J, Biolo G, Cederholm T, Cesari M, Cruz-Jentoft AJ, Morley JE, Phillips S, Sieber C, Stehle P, Teta D, Visvanathan R, Volpi E, Boirie Y (2013) Evidence-based recommendations for optimal dietary protein intake in older people: a position paper from the PROT-AGE study group. J Am Med Dir Assoc 14:542–559

Benziane B, Burton TJ, Scanlan B, Galuska D, Canny BJ, Chibalin AV, Zierath JR, Stepto NK (2008) Divergent cell signaling after short-term intensified endurance training in human skeletal muscle. Am J Physiol Endocrinol Metab 295:E1427–E1438

Biolo G, Maggi SP, Williams BD, Tipton KD, Wolfe RR (1995) Increased rates of muscle protein turnover and amino acid transport after resistance exercise in humans. Am J Phys 268:E514–E520

Biolo G, Tipton KD, Klein S, Wolfe RR (1997) An abundant supply of amino acids enhances the metabolic effect of exercise on muscle protein. Am J Phys 273:E122–E129

Biolo G, Fleming RYD, Maggi SP, Nguyen TT, Herndon DN, Wolfe RR (2002) Inverse regulation of protein turnover and amino acid transport in skeletal muscle of hypercatabolic patients. J Clin Endocrinol Metab 87:3378–3384

Blanco G, Blanco A (2017) Medical biochemistry. Academic Press

Brook MS, Wilkinson DJ, Mitchell WK, Lund JN, Szewczyk NJ, Greenhaff PL, Smith K, Atherton PJ (2015) Skeletal muscle hypertrophy adaptations predominate in the early stages of resistance exercise training, matching deuterium oxide-derived measures of muscle protein synthesis and mechanistic target of rapamycin complex 1 signaling. FASEB. J. Off. Publ. Fed. Am. Soc. Exp. Biol. 29:4485–4496

Brook MS, Wilkinson DJ, Atherton PJ, Smith K (2017) Recent developments in deuterium oxide tracer approaches to measure rates of substrate turnover: implications for protein, lipid, and nucleic acid research. Curr Opin Clin Nutr Metab Care 20:375–381

Brook MS, Wilkinson DJ, Smith K, Atherton PJ (2019) It's not just about protein turnover: the role of ribosomal biogenesis and satellite cells in the regulation of skeletal muscle hypertrophy. Eur J Sport Sci 19:952–963

Burd NA, West DWD, Staples AW, Atherton PJ, Baker JM, Moore DR, Holwerda AM, Parise G, Rennie MJ, Baker SK, Phillips SM (2010) Low-load high volume resistance exercise stimulates muscle protein synthesis more than high-load low volume resistance exercise in young men. PLoS One 5:e12033

Burd NA, McKenna CF, Salvador AF, Paulussen KJM, Moore DR (2019) Dietary protein quantity, quality, and exercise are key to healthy living: a muscle-centric perspective across the lifespan. Front Nutr 6:83

Camera DM, Edge J, Short MJ, Hawley JA, Coffey VG (2010) Early time course of Akt phosphorylation after endurance and resistance exercise. Med Sci Sports Exerc 42:1843–1852

Camera DM, Smiles WJ, Hawley JA (2016) Exercise-induced skeletal muscle signaling pathways and human athletic performance. Free Radic Biol Med 98:131–143
Carraro F, Stuart CA, Hartl WH, Rosenblatt J, Wolfe RR (1990) Effect of exercise and recovery on muscle protein synthesis in human subjects. Am J Phys 259:E470–E476
Carrithers JA, Carroll CC, Coker RH, Sullivan DH, Trappe TA (2007) Concurrent exercise and muscle protein synthesis: implications for exercise countermeasures in space. Aviat Space Environ Med 78:457–462
Chaillou T (2019) Ribosome specialization and its potential role in the control of protein translation and skeletal muscle size. J Appl Physiol 1985 127:599–607
Churchward-Venne TA, Burd NA, Phillips SM (2012) Nutritional regulation of muscle protein synthesis with resistance exercise: strategies to enhance anabolism. Nutr Metab 9:40
Churchward-Venne TA, Tieland M, Verdijk LB, Leenders M, Dirks ML, de Groot LCPGM, van Loon LJC (2015) There are no nonresponders to resistance-type exercise training in older men and women. J Am Med Dir Assoc 16:400–411
Churchward-Venne TA, Pinckaers PJM, Smeets JSJ, Peeters WM, Zorenc AH, Schierbeek H, Rollo I, Verdijk LB, van Loon LJC (2019) Myofibrillar and mitochondrial protein synthesis rates do not differ in Young men following the ingestion of carbohydrate with whey, soy, or leucine-enriched soy protein after concurrent resistance- and endurance-type exercise. J Nutr 149:210–220
Churchward-Venne TA, Pinckaers PJM, Smeets JSJ, Betz MW, Senden JM, Goessens JPB, Gijsen AP, Rollo I, Verdijk LB, van Loon LJC (2020) Dose-response effects of dietary protein on muscle protein synthesis during recovery from endurance exercise in young men: a double-blind randomized trial. Am J Clin Nutr 112:303–317
Coffey VG, Hawley JA (2017) Concurrent exercise training: do opposites distract? J Physiol 595: 2883–2896
Coffey VG, Zhong Z, Shield A, Canny BJ, Chibalin AV, Zierath JR, Hawley JA (2006) Early signaling responses to divergent exercise stimuli in skeletal muscle from well-trained humans. FASEB J 20:190–192
Coffey VG, Jemiolo B, Edge J, Garnham AP, Trappe SW, Hawley JA (2009a) Effect of consecutive repeated sprint and resistance exercise bouts on acute adaptive responses in human skeletal muscle. Am J Physiol Regul Integr Comp Physiol 297:R1441–R1451
Coffey VG, Pilegaard H, Garnham AP, O'Brien BJ, Hawley JA (2009b) Consecutive bouts of diverse contractile activity alter acute responses in human skeletal muscle. J. Appl. Physiol 1985 106:1187–1197
Damas F, Phillips S, Vechin FC, Ugrinowitsch C (2015) A review of resistance training-induced changes in skeletal muscle protein synthesis and their contribution to hypertrophy. Sports Med 45:801–807
Damas F, Phillips SM, Lixandrão ME, Vechin FC, Libardi CA, Roschel H, Tricoli V, Ugrinowitsch C (2016a) Early resistance training-induced increases in muscle cross-sectional area are concomitant with edema-induced muscle swelling. Eur J Appl Physiol 116:49–56
Damas F, Phillips SM, Lixandrão ME, Vechin FC, Libardi CA, Roschel H, Tricoli V, Ugrinowitsch C (2016b) An inability to distinguish edematous swelling from true hypertrophy still prevents a completely accurate interpretation of the time course of muscle hypertrophy. Eur J Appl Physiol 116:445–446
Damas F, Angleri V, Phillips SM, Witard OC, Ugrinowitsch C, Santanielo N, Soligon SD, Costa LAR, Lixandrão ME, Conceição MS, Libardi CA (2019) Myofibrillar protein synthesis and muscle hypertrophy individualized responses to systematically changing resistance training variables in trained young men. J. Appl. Physiol 1985 127:806–815
Deutz NEP, Bauer JM, Barazzoni R, Biolo G, Boirie Y, Bosy-Westphal A, Cederholm T, Cruz-Jentoft A, Krznariç Z, Nair KS, Singer P, Teta D, Tipton K, Calder PC (2014) Protein intake and exercise for optimal muscle function with aging: recommendations from the ESPEN expert group. Clin Nutr Edinb Scotl 33:929–936

Di Donato DM, West DWD, Churchward-Venne TA, Breen L, Baker SK, Phillips SM (2014) Influence of aerobic exercise intensity on myofibrillar and mitochondrial protein synthesis in young men during early and late postexercise recovery. Am J Physiol-Endocrinol Metab 306: E1025–E1032

Donges CE, Burd NA, Duffield R, Smith GC, West DWD, Short MJ, Mackenzie R, Plank LD, Shepherd PR, Phillips SM, Edge JA (2012) Concurrent resistance and aerobic exercise stimulates both myofibrillar and mitochondrial protein synthesis in sedentary middle-aged men. J. Appl. Physiol 1985 112:1992–2001

Drummond MJ, Dreyer HC, Fry CS, Glynn EL, Rasmussen BB (2009) Nutritional and contractile regulation of human skeletal muscle protein synthesis and mTORC1 signaling. J Appl Physiol 1985 106:1374–1384

Dudley GA, Djamil R (1985) Incompatibility of endurance- and strength-training modes of exercise. J Appl Physiol 1985 59:1446–1451

Eddens L (2019) Physiological and molecular responses to concurrent training in endurance-trained athletes [WWW Document]. undefined. URL/paper/Physiological-and-Molecular-Responses-to-Concurrent-Eddens/edffcf22ba4454ad0d4b417fa3e9c9e20ee31f33. Accessed 11.27.20

Egan B, Hawley JA, Zierath JR (2016) SnapShot: exercise metabolism. Cell Metab 24:342–342.e1

Felig P, Owen OE, Wahren J, Cahill GF (1969) Amino acid metabolism during prolonged starvation. J Clin Invest 48:584–594

Figueiredo VC, McCarthy JJ (2019) Regulation of ribosome biogenesis in skeletal muscle hypertrophy. Physiol 34:30–42

Frontera WR, Ochala J (2015) Skeletal muscle: a brief review of structure and function. Calcif Tissue Int 96:183–195

Gibala MJ (2007) Protein metabolism and endurance exercise. Sports Med 37:337–340

Goodman CA (2019) Role of mTORC1 in mechanically induced increases in translation and skeletal muscle mass. J Appl Physiol 1985 127:581–590

Goreham C, Green HJ, Ball-Burnett M, Ranney D (1999) High-resistance training and muscle metabolism during prolonged exercise. Am J Phys 276:E489–E496

Greenhaff PL, Karagounis LG, Peirce N, Simpson EJ, Hazell M, Layfield R, Wackerhage H, Smith K, Atherton P, Selby A, Rennie MJ (2008) Disassociation between the effects of amino acids and insulin on signaling, ubiquitin ligases, and protein turnover in human muscle. Am J Physiol Endocrinol Metab 295:E595–E604

Harber MP, Konopka AR, Jemiolo B, Trappe SW, Trappe TA, Reidy PT (2010) Muscle protein synthesis and gene expression during recovery from aerobic exercise in the fasted and fed states. Am J Physiol Regul Integr Comp Physiol 299:R1254–R1262

Hawley JA (2002) Adaptations of skeletal muscle to prolonged, intense endurance training. Clin Exp Pharmacol Physiol 29:218–222

Hickson RC (1980) Interference of strength development by simultaneously training for strength and endurance. Eur J Appl Physiol 45:255–263

Hoffman NJ, Parker BL, Chaudhuri R, Fisher-Wellman KH, Kleinert M, Humphrey SJ, Yang P, Holliday M, Trefely S, Fazakerley DJ, Stöckli J, Burchfield JG, Jensen TE, Jothi R, Kiens B, Wojtaszewski JFP, Richter EA, James DE (2015) Global Phosphoproteomic analysis of human skeletal muscle reveals a network of exercise-regulated kinases and AMPK substrates. Cell Metab 22:922–935

Hone M, Nugent AP, Walton J, McNulty BA, Egan B (2020) Habitual protein intake, protein distribution patterns and dietary sources in Irish adults with stratification by sex and age. J Hum Nutr Diet 33(4):465–476

Howarth KR, Burgomaster KA, Phillips SM, Gibala MJ (2007) Exercise training increases branched-chain oxoacid dehydrogenase kinase content in human skeletal muscle. Am J Physiol Regul Integr Comp Physiol 293:R1335–R1341

Institute of Medicine (U.S.), Institute of Medicine (U.S.) (Eds.), (2005) Dietary reference intakes for energy, carbohydrate, fiber, fat, fatty acids, cholesterol, protein, and amino acids. National Academies Press, Washington, D.C.

Jackman ML, Gibala MJ, Hultman E, Graham TE (1997) Nutritional status affects branched-chain oxoacid dehydrogenase activity during exercise in humans. Am J Phys 272:E233–E238

Jones TW, Walshe IH, Hamilton DL, Howatson G, Russell M, Price OJ, Gibson ASC, French DN (2016) Signaling responses after varying sequencing of strength and endurance training in a fed state. Int J Sports Physiol Perform 11:868–875

Joyner MJ, Coyle EF (2008) Endurance exercise performance: the physiology of champions. J Physiol 586:35–44

Kazior Z, Willis SJ, Moberg M, Apró W, Calbet JAL, Holmberg H-C, Blomstrand E (2016) Endurance exercise enhances the effect of strength training on muscle fiber size and protein expression of Akt and mTOR. PLoS One 11:e0149082

Kim PL, Staron RS, Phillips SM (2005) Fasted-state skeletal muscle protein synthesis after resistance exercise is altered with training. J Physiol 568:283–290

Kraemer WJ, Patton JF, Gordon SE, Harman EA, Deschenes MR, Reynolds K, Newton RU, Triplett NT, Dziados JE (1995) Compatibility of high-intensity strength and endurance training on hormonal and skeletal muscle adaptations. J Appl Physiol 1985 78:976–989

Lamont LS, McCullough AJ, Kalhan SC (1995) Beta-adrenergic blockade heightens the exercise-induced increase in leucine oxidation. Am J Phys 268:E910–E916

Lamont LS, McCullough AJ, Kalhan SC (1999) Comparison of leucine kinetics in endurance-trained and sedentary humans. J Appl Physiol 1985 86:320–325

Lantier L, Mounier R, Leclerc J, Pende M, Foretz M, Viollet B (2010) Coordinated maintenance of muscle cell size control by AMP-activated protein kinase. FASEB. J Off Publ Fed Am Soc Exp Biol 24:3555–3561

Lin Y-N, Tseng T-T, Knuiman P, Chan WP, Wu S-H, Tsai C-L, Hsu C-Y (2020) Protein supplementation increases adaptations to endurance training: a systematic review and meta-analysis. Clin Nutr 40(5):3123–3132

Lundberg TR, Fernandez-Gonzalo R, Gustafsson T, Tesch PA (2012) Aerobic exercise alters skeletal muscle molecular responses to resistance exercise. Med Sci Sports Exerc 44:1680–1688

Lundberg TR, Fernandez-Gonzalo R, Tesch PA (2014) Exercise-induced AMPK activation does not interfere with muscle hypertrophy in response to resistance training in men. J Appl Physiol 1985 116:611–620

MacDougall JD, Gibala MJ, Tarnopolsky MA, MacDonald JR, Interisano SA, Yarasheski KE (1995) The time course for elevated muscle protein synthesis following heavy resistance exercise. Can J Appl Physiol Rev Can Physiol Appl 20:480–486

Mascher H, Andersson H, Nilsson P-A, Ekblom B, Blomstrand E (2007) Changes in signalling pathways regulating protein synthesis in human muscle in the recovery period after endurance exercise. Acta Physiol Oxf Engl 191:67–75

Mascher H, Ekblom B, Rooyackers O, Blomstrand E (2011) Enhanced rates of muscle protein synthesis and elevated mTOR signalling following endurance exercise in human subjects. Acta Physiol. Oxf. Engl. 202:175–184

Matheny RW, Geddis AV, Abdalla MN, Leandry LA, Ford M, McClung HL, Pasiakos SM (2018) AKT2 is the predominant AKT isoform expressed in human skeletal muscle. Physiol Rep 6: e13652

Mazzulla M, Parel JT, Beals JW, Van Vliet S, Abou Sawan S, West DWD, Paluska SA, Ulanov AV, Moore DR, Burd NA (2017) Endurance exercise attenuates postprandial whole-body leucine balance in trained men. Med Sci Sports Exerc 49:2585–2592

Mazzulla M, Sawan SA, Williamson E, Hannaian SJ, Volterman KA, West DWD, Moore DR (2019) Protein intake to maximize whole-body anabolism during Postexercise recovery in resistance-trained men with high habitual intakes is Severalfold greater than the current recommended dietary allowance. J Nutr 150(3):505–511

McGlory C, Devries MC, Phillips SM (2017) Skeletal muscle and resistance exercise training; the role of protein synthesis in recovery and remodeling. J Appl Physiol 1985 122:541–548

McKendry J, Joanisse S, Baig S, Liu B, Parise G, Greig CA, Breen L (2019) Superior aerobic capacity and indices of skeletal muscle morphology in chronically trained master endurance athletes compared with untrained older adults. J Gerontol A Biol Sci Med Sci 75(6):1079–1088
McKenzie S, Phillips SM, Carter SL, Lowther S, Gibala MJ, Tarnopolsky MA (2000) Endurance exercise training attenuates leucine oxidation and BCOAD activation during exercise in humans. Am J Physiol Endocrinol Metab 278:E580–E587
Millward DJ (2001) Methodological considerations. Proc Nutr Soc 60:3–5
Mitchell CJ, Churchward-Venne TA, Bellamy L, Parise G, Baker SK, Phillips SM (2013) Muscular and systemic correlates of resistance training-induced muscle hypertrophy. PLoS One 8:e78636
Moore DR (2015) Nutrition to support recovery from endurance exercise: optimal carbohydrate and protein replacement. Curr Sports Med Rep 14(4):294–300
Moore DR (2019) Maximizing post-exercise anabolism: the case for relative protein intakes. Front Nutr 6:147
Moore DR, Del Bel NC, Nizi KI, Hartman JW, Tang JE, Armstrong D, Phillips SM (2007) Resistance training reduces fasted- and fed-state leucine turnover and increases dietary nitrogen retention in previously untrained young men. J Nutr 137:985–991
Moore DR, Robinson MJ, Fry JL, Tang JE, Glover EI, Wilkinson SB, Prior T, Tarnopolsky MA, Phillips SM (2009) Ingested protein dose response of muscle and albumin protein synthesis after resistance exercise in young men. Am J Clin Nutr 89:161–168
Moore DR, Camera DM, Areta JL, Hawley JA (2014) Beyond muscle hypertrophy: why dietary protein is important for endurance athletes. Appl Physiol Nutr Metab Physiol Appl Nutr Metab 39:987–997
Moore DR, Churchward-Venne TA, Witard O, Breen L, Burd NA, Tipton KD, Phillips SM (2015) Protein ingestion to stimulate myofibrillar protein synthesis requires greater relative protein intakes in healthy older versus younger men. J Gerontol A Biol Sci Med Sci 70:57–62
Morton RW, McGlory C, Phillips SM (2015) Nutritional interventions to augment resistance training-induced skeletal muscle hypertrophy. Front Physiol 6:245
Morton RW, Oikawa SY, Wavell CG, Mazara N, McGlory C, Quadrilatero J, Baechler BL, Baker SK, Phillips SM (2016) Neither load nor systemic hormones determine resistance training-mediated hypertrophy or strength gains in resistance-trained young men. J Appl Physiol 121: 129–138
Morton RW, Murphy KT, McKellar SR, Schoenfeld BJ, Henselmans M, Helms E, Aragon AA, Devries MC, Banfield L, Krieger JW, Phillips SM (2018) A systematic review, meta-analysis and meta-regression of the effect of protein supplementation on resistance training-induced gains in muscle mass and strength in healthy adults. Br J Sports Med 52:376–384
Mounier R, Lantier L, Leclerc J, Sotiropoulos A, Foretz M, Viollet B (2011) Antagonistic control of muscle cell size by AMPK and mTORC1. Cell Cycle Georget Tex 10:2640–2646
Murach KA, Bagley JR (2016) Skeletal muscle hypertrophy with concurrent exercise training: contrary evidence for an interference effect. Sports Med Auckl NZ 46:1029–1039
Murphy CH, Shankaran M, Churchward-Venne TA, Mitchell CJ, Kolar NM, Burke LM, Hawley JA, Kassis A, Karagounis LG, Li K, King C, Hellerstein M, Phillips SM (2018) Effect of resistance training and protein intake pattern on myofibrillar protein synthesis and proteome kinetics in older men in energy restriction. J Physiol 596:2091–2120
Newsholme P, Stenson L, Sulvucci M, Sumayao R, Krause M (2019) Amino acid metabolism, in: comprehensive biotechnology, scientific fundamentals of biotechnology. Elsevier, pp 3–14
Ogasawara R, Sato K, Matsutani K, Nakazato K, Fujita S (2014) The order of concurrent endurance and resistance exercise modifies mTOR signaling and protein synthesis in rat skeletal muscle. Am J Physiol Endocrinol Metab 306:E1155–E1162
Parr EB, Camera DM, Areta JL, Burke LM, Phillips SM, Hawley JA, Coffey VG (2014) Alcohol ingestion impairs maximal post-exercise rates of myofibrillar protein synthesis following a single bout of concurrent training. PLoS One 9:e88384
Papachristodoulou D, Snape A, Elliot WH, Elliot DC (2018) Biochemistry and molecular biology, 6th edn. Oxford University Press

Pescatello LS, Kostek MA, Gordish-Dressman H, Thompson PD, Seip RL, Price TB, Angelopoulos TJ, Clarkson PM, Gordon PM, Moyna NM, Visich PS, Zoeller RF, Devaney JM, Hoffman EP (2006) ACE ID genotype and the muscle strength and size response to unilateral resistance training. Med Sci Sports Exerc 38:1074–1081
Phillips SM, Atkinson SA, Tarnopolsky MA, MacDougall JD (1993) Gender differences in leucine kinetics and nitrogen balance in endurance athletes. J Appl Physiol 1985 75:2134–2141
Phillips SM, Tipton KD, Aarsland A, Wolf SE, Wolfe RR (1997) Mixed muscle protein synthesis and breakdown after resistance exercise in humans. Am J Phys 273:E99–E107
Pikosky MA, Gaine PC, Martin WF, Grabarz KC, Ferrando AA, Wolfe RR, Rodriguez NR (2006) Aerobic exercise training increases skeletal muscle protein turnover in healthy adults at rest. J Nutr 136:379–383
Psilander N, Damsgaard R, Pilegaard H (2003) Resistance exercise alters MRF and IGF-I mRNA content in human skeletal muscle. J Appl Physiol 1985 95:1038–1044
Rand WM, Pellett PL, Young VR (2003) Meta-analysis of nitrogen balance studies for estimating protein requirements in healthy adults. Am J Clin Nutr 77:109–127
Rennie MJ, Edwards RH, Krywawych S, Davies CT, Halliday D, Waterlow JC, Millward DJ (1981) Effect of exercise on protein turnover in man. Clin Sci Lond Engl 1979(61):627–639
Riechman SE, Balasekaran G, Roth SM, Ferrell RE (2004) Association of interleukin-15 protein and interleukin-15 receptor genetic variation with resistance exercise training responses. J Appl Physiol 1985 97:2214–2219
Sahlin K, Katz A, Broberg S (1990) Tricarboxylic acid cycle intermediates in human muscle during prolonged exercise. Am J Phys 259:C834–C841
Saxton RA, Sabatini DM (2017) mTOR signaling in growth, metabolism, and disease. Cell 168: 960–976
Sheffield-Moore M, Yeckel CW, Volpi E, Wolf SE, Morio B, Chinkes DL, Paddon-Jones D, Wolfe RR (2004) Postexercise protein metabolism in older and younger men following moderate-intensity aerobic exercise. Am J Physiol Endocrinol Metab 287:E513–E522
Short KR, Vittone JL, Bigelow ML, Proctor DN, Nair KS (2004) Age and aerobic exercise training effects on whole body and muscle protein metabolism. Am J Physiol Endocrinol Metab 286: E92–E101
Tang JE, Perco JG, Moore DR, Wilkinson SB, Phillips SM (2008) Resistance training alters the response of fed state mixed muscle protein synthesis in young men. Am J Physiol Regul Integr Comp Physiol 294:R172–R178
Tipton KD, Ferrando AA, Williams BD, Wolfe RR (1996) Muscle protein metabolism in female swimmers after a combination of resistance and endurance exercise. J Appl Physiol 1985 81: 2034–2038
van Hall G, van der Vusse GJ, Söderlund K, Wagenmakers AJ (1995) Deamination of amino acids as a source for ammonia production in human skeletal muscle during prolonged exercise. J Physiol 489:251–261
van Hall G, MacLean DA, Saltin B, Wagenmakers AJ (1996) Mechanisms of activation of muscle branched-chain alpha-keto acid dehydrogenase during exercise in man. J Physiol 494:899–905
Wagenmakers AJ (1998a) Protein and amino acid metabolism in human muscle. Adv Exp Med Biol 441:307–319
Wagenmakers AJ (1998b) Muscle amino acid metabolism at rest and during exercise: role in human physiology and metabolism. Exerc Sport Sci Rev 26:287–314
Wagenmakers AJ, Coakley JH, Edwards RH (1990) Metabolism of branched-chain amino acids and ammonia during exercise: clues from McArdle's disease. Int J Sports Med 11(Suppl 2): S101–S113
Wang X, Proud CG (2006) The mTOR pathway in the control of protein synthesis. Physiol 21:362–369
West DWD, Kujbida GW, Moore DR, Atherton P, Burd NA, Padzik JP, De Lisio M, Tang JE, Parise G, Rennie MJ, Baker SK, Phillips SM (2009) Resistance exercise-induced increases in

putative anabolic hormones do not enhance muscle protein synthesis or intracellular signalling in young men. J Physiol 587:5239–5247
Wilkinson DJ (2018) Historical and contemporary stable isotope tracer approaches to studying mammalian protein metabolism. Mass Spectrom Rev 37:57–80
Wilkinson SB, Tarnopolsky MA, Macdonald MJ, Macdonald JR, Armstrong D, Phillips SM (2007) Consumption of fluid skim milk promotes greater muscle protein accretion after resistance exercise than does consumption of an isonitrogenous and isoenergetic soy-protein beverage. Am J Clin Nutr 85:1031–1040
Wilkinson SB, Phillips SM, Atherton PJ, Patel R, Yarasheski KE, Tarnopolsky MA, Rennie MJ (2008) Differential effects of resistance and endurance exercise in the fed state on signalling molecule phosphorylation and protein synthesis in human muscle. J Physiol 586:3701–3717
Wilkinson DJ, Brook MS, Smith K, Atherton PJ (2017) Stable isotope tracers and exercise physiology: past, present and future. J Physiol 595:2873–2882
Wojtaszewski JFP, Nielsen P, Hansen BF, Richter EA, Kiens B (2000) Isoform-specific and exercise intensity-dependent activation of 5′-AMP-activated protein kinase in human skeletal muscle. J Physiol 528:221–226
Wolfe RR (2006) The underappreciated role of muscle in health and disease. Am J Clin Nutr 84: 475–482
Wolfe RR, Goodenough RD, Wolfe MH, Royle GT, Nadel ER (1982) Isotopic analysis of leucine and urea metabolism in exercising humans. J Appl Physiol 52:458–466
Wolfe RR, Wolfe MH, Nadel ER, Shaw JH (1984) Isotopic determination of amino acid-urea interactions in exercise in humans. J Appl Physiol 56:221–229
Yoon M-S (2017) mTOR as a key regulator in maintaining skeletal muscle mass. Front Physiol 8:788
Yoon M-S, Sun Y, Arauz E, Jiang Y, Chen J (2011) Phosphatidic acid activates mammalian target of rapamycin complex 1 (mTORC1) kinase by displacing FK506 binding protein 38 (FKBP38) and exerting an allosteric effect. J Biol Chem 286:29568–29574
You JS, Frey JW, Hornberger TA (2012) Mechanical stimulation induces mTOR signaling via an ERK-independent mechanism: implications for a direct activation of mTOR by phosphatidic acid. PLoS One 7:e47258
You JS, McNally RM, Jacobs BL, Privett RE, Gundermann DM, Lin KH, Steinert ND, Goodman CA, Hornberger TA (2019) The role of raptor in the mechanical load-induced regulation of mTOR signaling, protein synthesis, and skeletal muscle hypertrophy. FASEB J 33:4021–4034

Chapter 10
The Effect of Training on Skeletal Muscle and Exercise Metabolism

Martin J. MacInnis, Brendan Egan, and Martin J. Gibala

Abstract This chapter reviews the molecular and metabolic responses in human skeletal muscle to exercise training. Acute changes in various stimuli that trigger adaptations largely depend on the type of exercise performed and particularly the intensity and duration of discrete sessions. These stimuli are linked to the activation and/or repression of an array of intracellular signal transduction pathways, pre- and posttranscriptional processes, and the regulation of protein translation. Given the considerable overlap in these underlying molecular processes, the mechanistic basis for how repeated, acute changes are translated into specific training responses remains a topic of much investigation. Endurance training is primarily associated with an enhanced capacity for oxidative energy provision and a shift in substrate utilization, from carbohydrate to lipid, at a given absolute exercise intensity. Strength training mainly results in increased muscle size, force-generating capacity, and enhanced capacity for non-oxidative energy provision. Sprint training also increases the capacity for non-oxidative energy provision, but can elicit a range of responses, including some that resemble endurance or strength training. Training generally enhances fatigue resistance and performance in a manner that is specific, but not exclusive, to the type of exercise performed. These improvements are owed, in part to training-induced changes in both the maximal capacity for, and the specific utilization of, various substrates during exercise.

Keywords Aerobic · Endurance · Gene expression · Mitochondria · Resistance · Sprint · Signal transduction · Strength · Substrate utilization · Hypertrophy

---

M. J. MacInnis
Faculty of Kinesiology, University of Calgary, Calgary, AB, Canada

B. Egan
School of Health and Human Performance, Dublin City University, Dublin, Ireland

M. J. Gibala (✉)
Department of Kinesiology, McMaster University, Hamilton, ON, Canada
e-mail: gibalam@mcmaster.ca

© The Author(s), under exclusive license to Springer Nature Switzerland AG 2022
G. McConell (ed.), *Exercise Metabolism*, Physiology in Health and Disease,
https://doi.org/10.1007/978-3-030-94305-9_10

10.1 Introduction

Training refers to repeated sessions of exercise that are maintained from as little as a few days to many years. This process elicits physiological remodeling that is shaped by many factors, including the type of exercise performed, the intensity and duration of discrete sessions, and the total number of sessions performed. Multiple physiological systems interact to determine the holistic response to training, including those that regulate whole-body oxygen transport and substrate metabolism, neurohumoral control, acid-base balance, and thermoregulation (Hawley et al. 2014). This chapter focuses on the metabolic and molecular responses in human skeletal muscle to exercise training. Three distinct types of exercise are considered: endurance, strength, and sprint training. After characterizing each type, the chapter provides an overview of metabolic regulation during exercise, and the molecular basis of training-induced changes. It then considers the major metabolic responses to each type of exercise, characteristics of the trained state, molecular basis of training adaptations, and changes in exercise metabolism after training. The present review is informed by many previous works that have comprehensively examined the effect of training on skeletal muscle (Booth and Thomason 1991; Saltin and Gollnick 1983) or characterized responses to specific types of training (Abernethy et al. 1994; Holloszy and Coyle 1984; Ross and Leveritt 2001).

10.2 The Nature of the Exercise Stimulus: Endurance, Strength, and Sprint Training

In their seminal review on skeletal muscle adaptability, Saltin and Gollnick (1983) categorized exercise as being one of three basic types: (1) "endurance training"; (2) "strength training"; and (3) "varying intermediate combinations (e.g., sprint training)." Endurance and strength training are often conceptualized as falling on the two ends of a spectrum in terms of relative contractile force versus contractile duration (Fig. 10.1), with the molecular processes that are triggered in response to these two stimuli also being conceptualized as distinct and specific (Egan and Zierath 2013; Hawley et al. 2014). However, there is also a large degree of overlap in the molecular and adaptive responses, as evidenced by the that fact that sprint training—which is generally recognized as brief but intense exercise, with the goal to develop speed and power—can elicit a broad range of muscle adaptations, including some that resemble traditional endurance or strength training (Ross and Leveritt 2001), as discussed further below.

Endurance training is primarily associated with an enhanced capacity for oxidative energy provision in skeletal muscle and a shift toward increased contribution from lipid oxidation—and decreased carbohydrate oxidation—at the same absolute exercise intensity as before training (Holloszy and Coyle 1984). Training intensity is often expressed relative to maximal oxygen uptake (VO_{2max}) or the power output

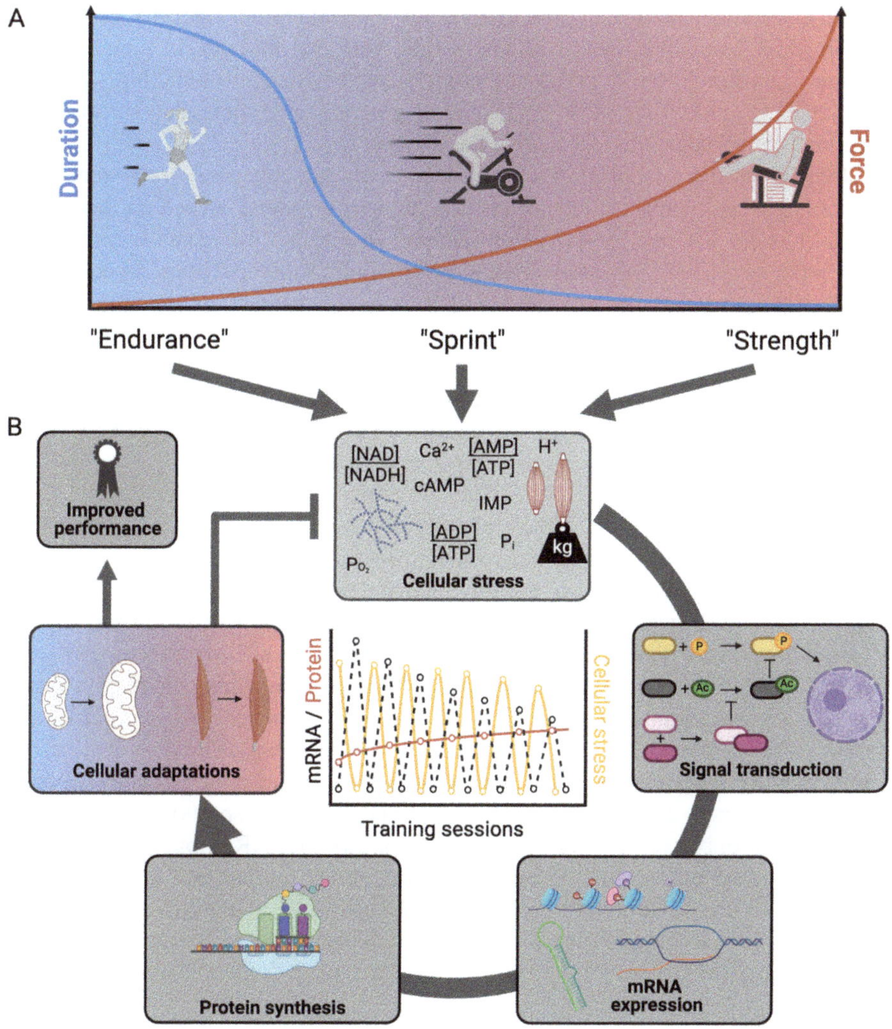

Fig. 10.1 The general process through which exercise training induces skeletal muscle adaptation. Panel A illustrates the dramatic differences in the duration and force of endurance and strength exercise and the positioning of sprint exercise somewhere between these two types of training. Panel B illustrates the molecular process through which exercise-induced cellular stress activates signal transduction pathways to alter mRNA expression and rates of protein synthesis. With repeated bouts of exercise (i.e., training), changes in protein content or enzymatic activity yield cellular adaptations that blunt cellular stress during future exercise sessions and improve exercise performance. Created with BioRender.com

(W_{max}) or velocity (vVO_{2max}) that corresponds to this parameter. Training prescription is also commonly anchored to measurements of lactate threshold (LT) and critical power (CP) and based on the three recognized intensity domains: moderate (below LT), heavy (between LT and CP), and severe (above CP) (Poole and Jones

2012). Endurance training can be continuous (i.e., moderate intensity continuous training, MICT) or intermittent in nature (i.e., "interval training"). A high overall volume of training, mainly at low- to-moderate intensities but including ~15–20% high-intensity intervals (≥85% $\dot{V}O_{2max}$), is generally recommended to maximize endurance performance in athletes (Seiler 2010).

Strength training is primarily associated with an enhanced force-generating capacity, a greater capacity for non-oxidative energy provision, and an increase in skeletal muscle size resulting from an increase in muscle fiber cross-sectional area (Abernethy et al. 1994). Neural adaptations, including the activation, synchronization, and maintenance of muscle fiber recruitment, are also particularly important for the enhanced force-generating capacity of skeletal muscle after strength training (Ross et al. 2001; Sale 1988). Training intensity and prescription is typically expressed relative to one repetition maximum (1-RM), the highest load than can be lifted or moved once. Repeated sets performed to volitional failure at ~80% 1-RM are generally recommended to increase muscle mass, as typically practiced by bodybuilders (Hackett et al. 2013). Training using relatively low loads (~30% 1-RM) is also effective to stimulate growth provided that volitional failure is achieved, at least in less-trained individuals (McGlory et al. 2017).

Sprint training is mainly associated with an enhanced capacity for non-oxidative energy provision, but skeletal muscle responses are highly dependent on the duration of sprint bouts, recovery between bouts, and total volume within sessions (Ross and Leveritt 2001). Sprint training involves a wide range of exercise modes, including those used for endurance (e.g., running) and strength training (e.g., plyometrics), as well as novel combinations (e.g., resisted sled sprint training) (Petrakos et al. 2016). Sprint training for performance enhancement often simulates the demands of competition (e.g., track and field events or "stop-and-go" team sports), in which most efforts last for only a few seconds or involve short distances (Rumpf et al. 2016). The capacity to maintain maximal velocity or a high power output for a longer duration (speed endurance) or perform multiple short sprints with brief recovery periods (repeated sprint ability) is also important for many athletes (Rumpf et al. 2016; Bishop et al. 2011).

10.3 Overview of Skeletal Muscle Metabolic Regulation During Exercise

Skeletal muscle stores relatively small amounts of ATP, yet [ATP] is remarkably steady during most forms of exercise, decreasing by only ~20–40% during intense exercise (Parolin et al. 1999; Black et al. 2017; Sahlin et al. 1989). These findings are striking, as maximal in vivo rates of ATP hydrolysis could theoretically deplete total muscle ATP stores within ~2 s (Parolin et al. 1999). Underpinning the ability of skeletal muscle to withstand large declines in [ATP] is a complex, interacting set of metabolic pathways that resynthesize ATP quickly, robustly, and proportionally in response to the metabolic demand placed on the muscle (Brooks 2012; Hargreaves and Spriet 2020). Although exercise training does not affect resting skeletal muscle [ATP] (Green et al. 1991; Leblanc et al. 2004), the activation of these pathways and their contributions to the overall ATP resynthesis rate during exercise are heavily dependent on the biochemical composition of the contracting skeletal muscle fibers, which is largely a product of an individual's fitness and training status (Holloszy and Coyle 1984; Booth and Thomason 1991; Saltin and Gollnick 1983). Importantly, there are also differences in the human skeletal muscle fiber-type specific response to acute exercise, which may be masked when responses are assessed based on mixed skeletal muscle biopsy samples (Ball-Burnett et al. 1991; Greenhaff et al. 1994; Hultman et al. 1991); however, that topic is beyond the scope of this chapter.

Collectively, metabolic, neural, and hormonal signals coordinate the resynthesis of ATP via various pathways in an attempt to match ATP demand. Among the many factors that regulate ATP resynthesis are calcium (Ca^{2+}); ADP, AMP, creatine, and Pi concentrations; epinephrine; ratios of oxidized to reduced coenzymes (e.g., $[NAD^+]/[NADH]$ and $[FAD]/[FADH_2]$); and substrate/product concentrations (Hargreaves and Spriet 2020; Brooks 2012). In general, Ca^{2+}, a feedforward signal of muscle contraction, provides gross control of metabolic regulation, and feedback related to ATP demand provides fine-tuning (Hargreaves and Spriet 2020).

Through ATP hydrolysis and the adenylate kinase (ADK) reaction, [ADP], [AMP], and [Pi] increase in proportion to the relative exercise intensity (Sahlin et al. 1987; Black et al. 2017; Howlett et al. 1998). Catalyzed by the near-equilibrium creatine kinase (CK) reaction, phosphocreatine (PCr) provides an immediate supply of ATP by transferring a phosphate to ADP as [ADP] increases. Accordingly, [PCr] declines in proportion to relative exercise intensity (Howlett et al. 1998; Sahlin et al. 1987). For moderate and heavy intensity exercise, [PCr] stabilizes (Black et al. 2017; Howlett et al. 1998), as the aerobic ATP resynthesis rate eventually increases to reach an equilibrium with the rate of PCr hydrolysis. Yet, [PCr] continually decreases if the exercise intensity is severe (Sahlin et al. 1989; Howlett et al. 1998).

Glycogenolysis and glycolysis increase during the initial seconds of exercise (Parolin et al. 1999; Hultman et al. 1991), yielding ATP through substrate phosphorylation and resulting in lactate production, again in proportion to the relative exercise intensity (Howlett et al. 1998; Sahlin et al. 1987). Briefly, the activation of glycogen phosphorylase (PHOS) and phosphofructokinase (PFK) collectively increases glycolytic flux, the (reversible) lactate dehydrogenase (LDH) enzyme converts pyruvate to lactate, and pyruvate dehydrogenase (PDH), which regulates the entry of pyruvate into the mitochondria for oxidation, converts pyruvate to acetyl CoA (the regulation of these enzymes is reviewed in detail by Hargreaves and Spriet 2020). Accordingly, muscle lactate concentrations increase when the rate of glycolysis exceeds the rate of PDH flux. Lactate, if not converted back to pyruvate for oxidation, can be shuttled out of the muscle cell via monocarboxylate transporters in the plasma membrane to undergo oxidation in a variety of other cells, including skeletal muscle, or to undergo gluconeogenesis in the liver (Brooks 2018).

Inside the mitochondria, some of the acetyl CoA generated by PDH enters the tricarboxylic acid (TCA) cycle, with the remainder shunted to acetylcarnitine

(Constantin-Teodosiu et al. 1991). Direct ATP production from the TCA cycle is minimal, but the reduction of NAD^+ and FAD molecules—forming NADH and $FADH_2$—and subsequent passage of electrons to oxygen via the electron transport chain are the main driver of ATP resynthesis through oxidative phosphorylation. Key enzymes of the TCA cycle are activated by Ca^{2+}, linking the TCA cycle to muscle contraction, and inhibited by NADH, preventing flux through the TCA cycle when demand for electrons is low (reviewed in Hargreaves and Spriet 2020). The control of oxidative phosphorylation is complex but largely dependent on [ADP] at rest. During exercise, the phosphorylation potential, ([ATP] / [ADP][Pi]), and the mitochondrial redox potential ([NADH] / [NAD^+]), which decline in proportion to relative exercise intensity (Sahlin et al. 1987), facilitate the matching of ATP resynthesis and ATP hydrolysis rates.

Free fatty acids (FFAs) derived from triglycerides stored in adipose tissue and skeletal muscle are the other main sources of fuel for oxidative metabolism. After undergoing lipolysis in adipose tissue, fatty acids transported in plasma bound to albumin are taken up into skeletal muscle via fatty acid transporters, e.g., CD36 and fatty acid binding protein (FABPpm), or through diffusion. Alternatively, intramuscular triglyceride (IMTG) lipolysis, controlled by the integrated actions of adipose triglyceride lipase (ATGL) and hormone-sensitive lipase (HSL), and several other proteins, including the perilipin (PLIN) family, releases FFAs in the sarcoplasm (Watt and Cheng 2017). Regardless of source, FFAs, chaperoned by FABPc, are converted to fatty acyl CoA molecules via acyl-CoA synthetase (ACS) and transported into the mitochondria via carnitine palmitoyltransferase 1 and 2 (CPT1/CPT2), subject to the availability of free carnitine (as described later). Here, the beta-oxidation pathway yields acetyl CoA, NADH, and $FADH_2$, which contribute to ATP resynthesis via oxidative phosphorylation, as described above for carbohydrate.

Contributions from the three energy systems encompassed by the pathways described above (i.e., ATP-PCr, glycolytic, and aerobic) depend on multiple factors, including the intensity and duration of exercise (Howlett et al. 1998; Romijn et al. 1993; van Loon et al. 2001). In general, the anaerobic processes have small capacities for ATP resynthesis but can yield ATP at high rates, whereas the aerobic system resynthesizes ATP at a lower rate but with a much larger capacity. Anaerobic pathways are an important source of ATP for transitions to higher endurance exercise intensities (including from rest) and for severe intensities of exercise (i.e., endurance exercise above CP as well as strength training and sprinting); however, net ATP resynthesis is almost entirely supported by aerobic metabolism during steady-state exercise. Although this simplified framework of metabolic regulation is described independent of skeletal muscle fitness, exercise training leads to whole-body and intramuscular adaptations that alter the dynamics of these pathways, leading to changes in substrate utilization during exercise (Fig. 10.2).

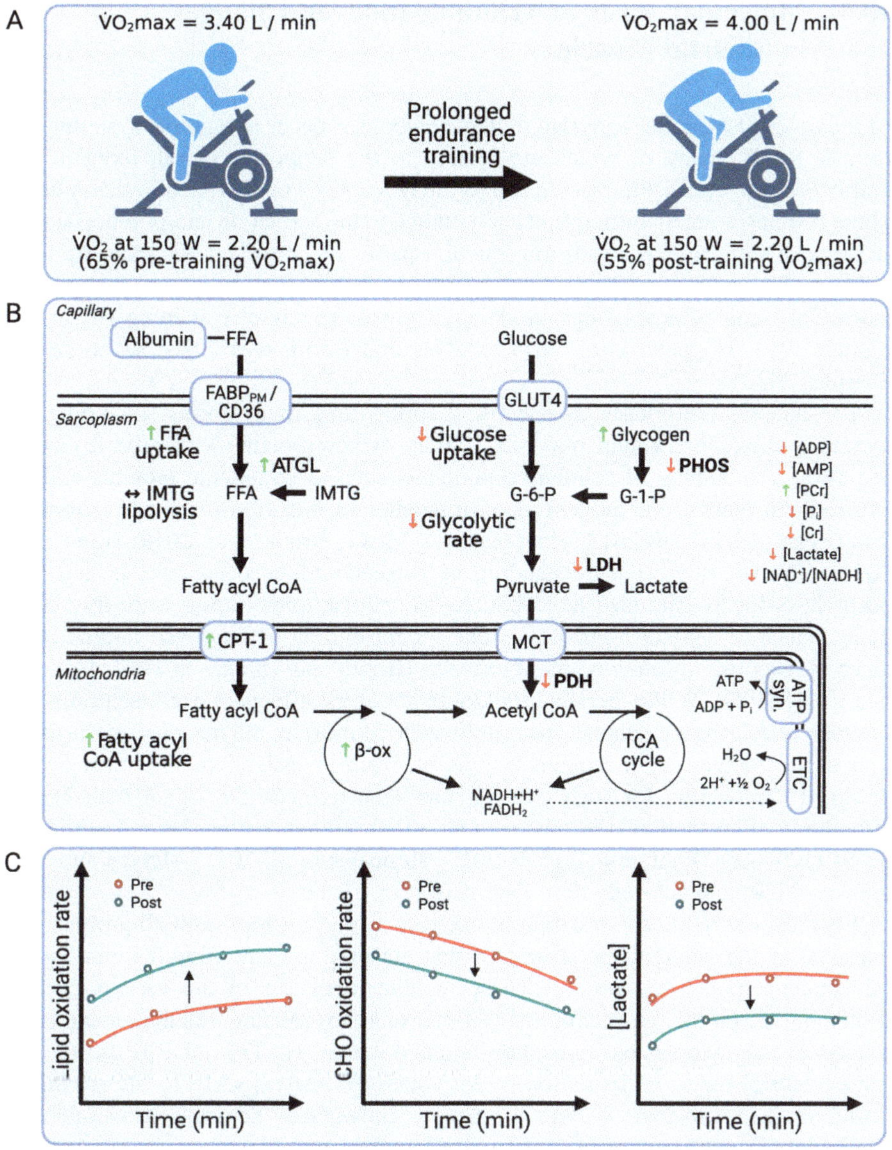

Fig. 10.2 The effects of endurance training on skeletal muscle metabolism. Panel A reports VO_{2max} and the VO_2 elicited by exercise at the same power output (i.e., 150 W) before and after prolonged endurance training. In this example, VO_{2max} increased with training, but the absolute VO_2 at 150 W was essentially unchanged, meaning that this power output elicited 65% of the pre-training VO_{2max} but only 55% of the post-training VO_{2max}. Panel B illustrates the key metabolic changes that occur in exercising skeletal muscle at this fixed power output. Arrows indicate changes relative to the pre-training state. Panel C illustrates the changes in lipid oxidation rates, carbohydrate (CHO) oxidation rates, and lactate concentration during exercise at this fixed power output. Created with BioRender.com

10.4 Molecular Basis of Training-Induced Changes in Skeletal Muscle

At the onset of exercise, a myriad of acute responses occur at multiple systemic and cellular levels, many of which are related to the supply of blood, oxygen, and nutrients to the exercising muscle and to the resynthesis of ATP within that muscle. These responses are, in turn, intrinsically linked to the activation and/or repression of an array of intracellular signal transduction pathways, pre- and posttranscriptional processes, and the regulation of protein translation, which together underpin the molecular basis of adaptations in skeletal muscle to exercise training (Egan and Zierath 2013; Hoppeler et al. 2011; McGee and Hargreaves 2020; Seaborne and Sharples 2020).

The adaptive responses to exercise training take many forms, including the increased abundance and/or maximal activity of key proteins with roles in energy provision, remodeling of cellular components such as contractile proteins and the extracellular matrix, and biogenesis of organelles such as ribosomes and mitochondria (Egan and Zierath 2013; Hoppeler et al. 2011; Brook et al. 2019; Hood et al. 2016). The teleological understanding of these coordinated changes is that they occur in order to minimize perturbations to cellular homeostasis, with this better maintenance of cellular homeostasis likely contributing to improved fatigue resistance in response to future bouts of exercise (Booth and Thomason 1991; Holloszy and Coyle 1984). While there is little doubt about the effects of exercise training to produce wide-ranging adaptations within skeletal muscle, the mechanistic basis for *how* these changes occur remains a topic of much investigation. The most widely accepted explanation for such changes has been termed "the signal transduction hypothesis of adaptation" (Burniston et al. 2014). This model has been described in detail elsewhere (Egan and Zierath 2013; Hoppeler et al. 2011; McGee and Hargreaves 2020) and is briefly described here (Fig. 10.1).

First, the onset and continuation of exercise result in responses both intrinsic and extrinsic to the active skeletal muscle that act as important stimuli initiating the molecular response to exercise. These include regulators of the various energy-producing pathways as mentioned above, such as calcium release, metabolites related to the cytoplasmic phosphorylation potential ([ATP]/[ADP][Pi]), and the mitochondrial reduction/oxidation (redox) state ([NADH]/[NAD^+]), the prevailing hormonal and substrate milieu, electrolyte imbalances across cell membranes, declining pH, reduced partial pressure of oxygen, and elevations in oxygen free radical production, muscle temperature, mechanical load, and sarcolemmal disruption (Fig. 10.1). These exercise-induced responses represent some, but unlikely all, of the responses that constitute perturbations to cellular homeostasis and the initiation of signal transduction pathways.

Next, these stimuli and consequent perturbations activate and/or repress a variety of cellular sensors capable of amplifying this initial signal through the activation and/or repression of signal transduction pathways (Fig. 10.1). For example, the protein deacetylase, sirtuin 1 (SIRT1), and the heterotrimeric serine/threonine kinase, AMP-activated protein kinase (AMPK), are sensors of acute exercise-induced changes in redox state ([NADH]/[NAD$^+$]) and ATP turnover ([ATP]/[ADP][Pi] and [AMP]), respectively; calmodulin is a sensor of Ca^{2+} release in the sarcoplasmic reticulum during muscle contraction and leads to the activation of Ca^{2+}/calmodulin-dependent protein kinases (CaMKs); and mechanosensors, such as focal adhesion kinase (FAK) proteins, are sensors that detect alterations mechanical load and force transduction, which converge on the activation of the mechanistic target of rapamycin complexes (mTORC) (Ogasawara et al. 2019; Egan and Zierath 2013; McGee and Hargreaves 2020). Of the many sensors, AMPK is the best-studied target linking cellular metabolism to adaptive changes in skeletal muscle. Through its action as a kinase, AMPK modulates cellular metabolism—through phosphorylation of metabolic enzymes—and transcription and translation, through phosphorylation of transcription factors and other signaling proteins (Kjøbsted et al. 2018). Increased AMPK phosphorylation and enzymatic activity after acute exercise often occurs in an intensity-dependent manner, probably reflecting intensity-dependent effects of exercise on ATP turnover, [ADP] and [AMP], and depletion of muscle glycogen (Howlett et al. 1998; Sahlin et al. 1987). Acute AMPK activation acts to mitigate a decline in [ATP] by stimulating catabolic pathways to restore cellular energy stores while simultaneously inhibiting biosynthetic pathways and anabolic pathways (Kjøbsted et al. 2018). Conversely, in experimental models that utilize the repeated and chronic activation of AMPK (acting as a surrogate for exercise training), adaptive changes are observed in skeletal muscle that are similar to those seen with endurance training, such as an increase number and size of mitochondria and improved exercise capacity (Bergeron et al. 2001; Garcia-Roves et al. 2008).

Much of the focus in skeletal muscle has been on "classical" signal transduction (i.e., protein-protein interactions, posttranslational modifications such as phosphorylation and acetylation, and protein translocation), particularly as it relates to the activities of transcriptional and translational regulators (Egan and Zierath 2013; Hoppeler et al. 2011; Burniston et al. 2014). Thus, downstream of the activation and/or repression of various kinases, phosphatases, deactylases, and other enzymes that catalyze posttranslational modifications is a gamut of regulators, or effector proteins, and their associated processes (Fig. 10.1). Historically, the activity of transcription factors, co-activators, and repressors were the subject of much interest. This interest was driven by the observation that changes in tissue form and function as an adaptive process are driven by transcript-level adjustments prior to changes in proteins that provoke gradual structural remodeling and long-term functional adjustments (Hoppeler et al. 2011). While transcription of exercise-responsive genes remains central to the model of exercise-induced adaptations in skeletal muscle (Pillon et al. 2020), in the last decade, the discovery of novel roles for pre- and posttranscriptional processes particularly as they relate to epigenetics, namely, histone deacetylation (McGee and Hargreaves 2020), DNA methylation (Seaborne and Sharples 2020), and miRNA (Massart et al. 2016), has added further and intriguing complexity to the model (Fig. 10.1). Specifically, transient changes in histone modifications (McGee et al. 2009), and gene-specific changes in DNA methylation status (Barres et al. 2012), are induced by acute exercise and precede changes in gene expression in the post-exercise period (McGee et al. 2009; Barres et al. 2012). For example, exercise can produce an increase in histone acetylation by inhibition of the activity of histone deacetylases (HDACs) (McGee et al. 2009), which typically results in increased accessibility of chromatin and activation of transcription. Similarly, increased accessibility of chromatin and activation of transcription occurs with hypomethylation of GC-rich consensus binding sequences of DNA in specific genes and has been observed for various exercise-responsive genes (Barres et al. 2012).

The molecular networks that link exercise to epigenetic regulation are presently a matter of intense research interest, with a large number of enzymes regulating acetylation/deacetylation and methylation/demethylation. Many of these enzymes are linked to metabolic pathways and signal transduction pathways that are established as being exercise-responsive (e.g., AMPK, CaMK). This suggests the model for exercise-induced signal transduction in skeletal muscle should be updated to include regulators of epigenetic modifications (McGee and Hargreaves 2020; Seaborne and Sharples 2020). Lastly, the regulation of protein translation, the biogenesis and activity of organelles including ribosomes and lysosomes, and the importance of skeletal muscle satellite cells for adaptation are some of the new vistas in the field (Brook et al. 2019).

The processes described above occur in a temporal manner, such that homeostatic perturbations, signal transduction, and pre-transcriptional regulation occur during exercise and the early phase of recovery (minutes to hours), whereas alterations in mRNA and protein abundance occur in the hours and day that follow. As shown in Fig. 10.1, changes that ultimately result in functional improvements in exercise capacity and performance occur in the following days, weeks, and months consequent to cumulative effect of frequent, repeated bouts of exercise (Egan and Zierath 2013; Hoppeler et al. 2011). Notably, the effect of training status or a period of exercise training on the molecular responses to acute exercise suggests that the magnitude of activation of signal transduction pathways or changes in mRNA abundance are attenuated in moderate to well-trained individuals (Coffey et al. 2006; McConell et al. 2020), or during and after exercise training interventions (Benziane et al. 2008; Mallinson et al. 2020). In studies investigating the timecourse of training adaptations, the molecular response assessed by acute changes in mRNA abundance is also dampened by interventions with both high-intensity interval training (Perry et al. 2010) and strength training (Mallinson et al. 2020).

Using endurance and strength training as conceptual extremes on the exercise continuum, there are stimuli, signaling molecules, and downstream pathways, targets, and processes that exhibit differential responses to these divergent types of exercise (Coffey and Hawley 2017). Central to the signal transduction hypothesis of adaptation is that the type of exercise stimulus is reflected in the specificity of the molecular networks that are activated, which in turn explain the divergent adaptations to exercise training manifested, for example, by the vastly different phenotypical appearance, muscle morphology, and performance parameters of well-trained marathon runners compared to elite power athletes. The specifics of the molecular response to the different types of exercise are considered in later sections. However, the agreement between the specifics of an exercise bout and the elicited molecular responses is not absolute, and the molecular responses, as presently understood, fail to fully explain divergent outcomes in response to training (Coffey and Hawley 2017). For example, as described above SIT, despite consisting of a small number of repeated, intermittent sets of high (force) power output, all-out sprint activity demonstrates a remarkable capacity to produce an endurance phenotype in skeletal muscle and improve endurance performance (MacInnis and Gibala 2017), which is consequent to SIT eliciting molecular responses similar to those associated with endurance exercise (Gibala et al. 2009; Granata et al. 2017; Little et al. 2011).

These points highlight that a simplified view of exercise training adaptations being drawn along extremes of the exercise continuum as a result of discrete and specific signaling pathways is erroneous. In fact, these pathways demonstrate some degree of dependence, cross talk, interference, and redundancy in their regulation, making the exact contribution of each signaling pathway to measured changes in gene expression difficult to isolate. A multiple signal transduction-to-transcription-coupled control system with inherent redundancy allows for fine-tuning of the adaptive responses to exercise training dependent on the exercise mode, nutrition status, and trained state, among many other factors.

10.5 Skeletal Muscle Responses to Training

The effect of training on exercise metabolism is most often inferred by comparing responses between groups of trained and untrained individuals, or by studying the same group of individuals before and after a defined intervention. Responses are commonly based on comparisons made at the same absolute power output, such that exercise is performed at a lower relative intensity in the trained state, e.g., as a percentage of VO_{2max} or another measure of maximal exercise capacity (Fig. 10.2). This approach provides insight into how training alters the regulation of metabolic pathways that attempt to match ATP supply with ATP demand, and assumes there is little to no change in energetic efficiency, i.e., the rate of ATP hydrolysis required for a given amount of contractile work. Comparisons are sometimes made at the same relative intensity, which informs whether the proportional contribution of various metabolic fuels is changed at a given percentage of maximal capacity in the trained versus untrained state. This chapter will mainly consider the results from longitudinal training studies, in which comparisons were made in the same group of individuals at the same absolute power output, before and after an intervention that typically lasted from a few weeks to up to several months (Fig. 10.2). The following sections summarize the major metabolic and molecular responses to endurance, strength, and sprint training.

10.6 Endurance Training

10.6.1 Acute Metabolic Response

The metabolic demands of endurance exercise are sustained by aerobic ATP resynthesis, which primarily involves the oxidation of carbohydrates and lipids. The seminal studies of Romijn et al. (1993) and van Loon et al. (2001) indicate that increases in exercise intensity led to increased carbohydrate utilization and dependence on intramuscular fuel sources, notably glycogen. In contrast, increases in exercise duration at a fixed intensity led to increased lipid oxidation and dependence on extramuscular fuel sources, mainly free fatty acids derived from adipose tissue (Romijn et al. 1993; van Loon et al. 2001).

The "choice" of substrate is complex, and multiple hypotheses have been proposed to explain the regulation of substrate utilization. As reviewed by Lundsgaard et al. (2018), glycolysis may regulate free fatty acid oxidation by dictating the availability of free carnitine to transport fatty acyl CoA into the mitochondria. Briefly, at high exercise intensities, increased glycolysis elevates acetyl-CoA concentrations in the mitochondria, leading initially to inhibition of PDH flux; however, the conversion of excess acetyl-CoA to acetylcarnitine by carnitine-acetyl transferase facilitates increased pyruvate oxidation while simultaneously reducing concentrations of free carnitine, limiting fatty acyl CoA uptake into the mitochondria. Alternative explanations for reduced lipid oxidation at higher intensities include decreased blood flow to adipose tissue; decreased release of free fatty acids into plasma; decreased delivery to skeletal muscle; decreased transport into skeletal muscle; decreased IMTG breakdown; and decreased delivery to mitochondria (Spriet 2014).

10.6.2 Molecular Basis of Adaptations

An increasingly well-defined network of signaling pathways, transcription factors, and co-regulator proteins has emerged as important regulators of the skeletal muscle phenotype in response to endurance training. Unsurprisingly, there are complex interactions between various kinases/phosphatases/acetylases/deacetylases and transcription factors and transcriptional coactivators that mediate exercise effects on gene expression. The activity of these factors and coactivators can be modified by their expression, their phosphorylation, and/or acetylation status and by their intracellular localization. For example, the peroxisome proliferator-activated receptor γ coactivator 1α (PGC-1α) acts as a transcriptional coactivator through recruitment and co-regulation of multiple transcription factors including nuclear respiratory factor 1 (NRF-1), NRF-2, estrogen-related receptor α (ERRα), and mitochondrial transcription factor A (Tfam), which regulate skeletal muscle gene expression, particularly the expression of metabolic and mitochondrial genes. PGC-1α activity is highly regulated by numerous posttranslational modifications, including phosphorylation and deacetylation (Martínez-Redondo et al. 2015). Key upstream kinases and deacetylases regulating these modifications—including AMPK, p38 MAPK, and SIRT1—are activated by acute exercise, coincident with alterations in protein stability, functional activity, and subcellular localization of PGC-1α. In this manner, the convergence of signal transduction pathways upon the regulation of PGC-1α and the consequent regulation of metabolic and mitochondrial gene expression programs is an illustration of the coupling of the exercise stimulus to changes in skeletal muscle phenotype.

PGC-1α has garnered much attention over the past two decades because of marked effects on skeletal muscle in murine models where PGC-1α knockout disrupts basal mitochondrial content and function, and where PGC-1α overexpression produces endurance-like outcomes such as angiogenesis, mitochondrial biogenesis, and improved metabolic flexibility (Martínez-Redondo et al. 2015). In human skeletal muscle, a single bout of exercise reliably and robustly increases PGC-1α mRNA abundance across various exercise protocols and increases in PGC-1α protein occur with endurance training (see detailed review by Granata et al. 2018). These points suggest that PGC-1α gene expression is responsive to exercise, but additional evidence is suggestive of changes in PGC-1α activity in response to exercise. For example, exercise results in translocation of PGC-1α to the nucleus and mitochondria in skeletal muscle (Little et al. 2010; Little et al. 2011), thereby promoting nuclear-mitochondrial communication and increasing nuclear MEF2-PGC-1α association (McGee and Hargreaves 2004), which are key regulatory aspects of the activity of PGC-1α (Martínez-Redondo et al. 2015).

PGC-1α has often been considered a "master regulator" of mitochondrial biogenesis and central to exercise-induced adaptations in skeletal muscle (Martínez-Redondo et al. 2015). However, both of these contentions have been challenged as outlined in detail elsewhere (Islam et al. 2018; Hood et al. 2016). To briefly summarize, firstly the observational nature of human exercise studies means that no cause-and-effect relationship can be established for exercise-induced changes in PGC-1α mRNA or protein with exercise-mediated changes in skeletal muscle phenotype. Secondly, the observations in mice that either whole-body (Leick et al. 2008) or muscle-specific (Rowe et al. 2012) PGC-1α knockout does not impair exercise training-induced changes in skeletal muscle indicate that PGC 1 α-independent networks contribute to adaptive responses in skeletal muscle. Several other molecular processes have also emerged as regulators of mitochondrial adaptation in skeletal muscle, including mitochondrial fission-fusion dynamics, the mitochondrial unfolded protein response, and mitochondrial quality control through mitophagy (Hood et al. 2016). Given that there are immediate and transient changes in these processes in the post-exercise period, the current model suggests that this response initiates turnover of the mitochondrial pool within skeletal muscle in a coordinated process of removal of dysfunctional mitochondria, in collaboration with the activation of biogenesis.

10.6.3 Metabolic Characteristics and Responses in the Trained State

The endurance-trained state is principally characterized by an increased maximal capacity of skeletal muscle to oxidize carbohydrates and lipids and a shift in the relative utilization of these two substrates that promotes the latter at a given submaximal exercise intensity (Saltin and Gollnick 1983; Brooks 2012; Holloszy and Coyle 1984). Endurance training can also increase the maximal capacity for amino acid oxidation in skeletal muscle (McKenzie et al. 2000), but protein is a minor fuel during most forms of exercise (Brooks 2012) and will not be further considered here. The endurance training-induced increase in skeletal muscle oxidative capacity is largely owed to an increase in mitochondria form and function, as reflected by an increased volume density and respiratory capacity (Lundby and Jacobs 2016; Hoppeler et al. 2011; Holloszy and Coyle 1984) as a consequence of mitochondrial biogenesis (Hood et al. 2016). The increased mitochondrial volume density involves both an increase in cross-sectional area and longitudinal growth (Lundby and Jacobs 2016). Training-induced increases in total mitochondrial protein content include a higher maximal activity of the two enzyme complexes regarded as rate-determining for skeletal muscle carbohydrate and lipid oxidation, respectively: PDH (Leblanc et al. 2004) and CPT (Starritt et al. 2000). Endurance training also increases the skeletal muscle content of proteins that facilitate the sarcolemmal transport of glucose (Houmard et al. 1993) and fatty acids (Kiens et al. 1997). These responses help facilitate a higher maximal capacity to take up glucose (Kristiansen et al. 2000) and FFA (Kiens et al. 1993) from the circulation during exercise. The changes also contribute to an enhanced capacity to store higher amounts of intramuscular glycogen (Greiwe et al. 1999) and triglyceride (van Loon and Goodpaster 2006), the extent of which is also influenced by nutrient intake (Hawley et al. 2011). Endurance training also increases the skeletal muscle content of membrane proteins involved in lactate transport and pH regulation (Juel 2006). Training-induced increases in the maximal capacity for carbohydrate oxidation primarily facilitate a higher power output during intense endurance exercise, whereas the increased capacity for lipid oxidation largely facilitates the ability to sustain power output during prolonged submaximal exercise.

In general, when exercise performed at the same absolute intensity is compared before and after training, there is a decreased reliance on anaerobic pathways at the onset of exercise and an increased reliance on lipid oxidation over carbohydrate oxidation during steady-state conditions (Fig. 10.2). Indeed, greater rates of lipid oxidation (and lower rates of carbohydrate oxidation) at the same absolute exercise intensity, considered a hallmark adaptation to exercise training (Holloszy and Coyle 1984), have been reported by many studies, using a variety of techniques including pulmonary gas exchange, metabolic tracers, skeletal muscle biopsies, and arterial and venous catheterization (Henriksson 1977; Green et al. 1991; Kiens et al. 1993; Martin 3rd et al. 1993; Bergman et al. 1999b). These changes in fuel selection are largely underpinned by cellular adaptations within skeletal muscle, as discussed below.

After endurance exercise training, the net contribution of PCr hydrolysis to ATP resynthesis at the onset of exercise is lower (Chesley et al. 1996; Green et al. 1991; Leblanc et al. 2004) because oxidative phosphorylation increases more rapidly and is more responsive to pulsatile increases in free AMP and ADP when mitochondrial content is augmented (Phillips et al. 1995). Thus, because the rate of ATP resynthesis via oxidative phosphorylation more quickly reaches equilibrium with ATP hydrolysis to prevent further declines in [PCr], a higher [PCr] is maintained during steady-state exercise post-training at the same absolute intensity (Chesley et al. 1996; Green et al. 1991, 1995; Leblanc et al. 2004; Karlsson et al. 1972). Similarly, the rate at which PCr is resynthesized following a bout of exercise or between bouts of exercise (i.e., interval exercise) is largely dependent on the oxidative capacity of the muscle (Paganini et al. 1997). Accordingly, PCr recovery after exercise (or between bouts of exercise) is improved following exercise training (Forbes et al. 2008).

Training-induced attenuations in metabolite perturbations have consequences for the regulation of carbohydrate oxidation (Fig. 10.2). With a greater density of mitochondria in trained skeletal muscle, the rate of aerobic ATP resynthesis in the mitochondrial reticulum is greater for a given [ADP], resulting in smaller increases in [ADP], [AMP], [Pi], and [IMP] for the same absolute intensity of exercise (Green et al. 1991, 1995; Leblanc et al. 2004; Phillips et al. 1996). Although systemic epinephrine concentrations are reduced following training (Phillips et al. 1996; Chesley et al. 1996), exercise training does not seem to effect the transformation of PHOS to its active form during exercise, indicating that it is the reduction in metabolic perturbations (i.e., allosteric regulation) that reduces the activity of PHOS and the rate of glycogenolysis post-training (Chesley et al. 1996). The sparing of muscle glycogen is commonly reported post-training (Chesley et al. 1996; Leblanc et al. 2004; Green et al. 1995; Karlsson et al. 1972) and is a hallmark adaptation of endurance training (Holloszy and Coyle 1984). The reduction in glycolytic flux during submaximal exercise at a given absolute intensity (i.e., moderate and heavy intensity) after training results in reduced pyruvate production (Green et al. 1995; Leblanc et al. 2004), which could be explained by an increased abundance or activity of PDH kinase, which suppresses PDH, or reduced allosteric activation of PDH (Leblanc et al. 2004). Regardless of mechanism, there is a better matching of glycolytic flux and PDH flux at a given absolute intensity of exercise in trained skeletal muscle, leading to reduced lactate production relative to untrained skeletal muscle (Chesley et al. 1996; Leblanc et al. 2004; Karlsson et al. 1972; Saltin et al. 1976; Phillips et al. 1996). Consequently, exercise training blunts the rate of appearance for lactate in the blood, but it also increases the metabolic clearance rate for lactate when exercise is performed at the same absolute exercise intensity (Bergman et al. 1999b). Cross-sectional data have also demonstrated greater capacities for lactate production, disposal, and clearance in trained compared to untrained cyclists exercising at their individual lactate thresholds (Messonnier et al. 2013). Subsequently, as shown in Fig. 10.2, blood lactate concentrations are lower at the same absolute exercise intensity post-training (Karlsson et al. 1972; Saltin et al. 1976). Finally, glucose uptake by active skeletal muscle at a given intensity is decreased with exercise training (Saltin et al. 1976; Bergman et al. 1999b; Coggan et al. 1990), demonstrating that glucose uptake does not compensate for skeletal muscle glycogen sparing. This reduction in glucose uptake may result from reduced GLUT4 translocation to the plasma membrane at the same absolute exercise intensity post-training (Richter et al. 1998). Maximal rates of glucose uptake, however, are higher in trained versus untrained muscle when compared at 100% of thigh peak relative power output and likely due to higher GLUT4 content in trained muscle (Kristiansen et al. 2000). Collectively, this series of metabolic responses leads to a reduction in the overall rate of carbohydrate oxidation at the same absolute exercise intensity following a period of training (Fig. 10.2).

A reduction in the contribution of ATP resynthesis from carbohydrate oxidation necessitates an increase in lipid oxidation at the same absolute submaximal intensity of exercise (Fig. 10.2). As compared to carbohydrate and glycogen metabolism, less is known about training-induced changes in the regulation of lipid and IMTG catabolism, and results are mixed. Kiens et al. (1993) reported that, following 8 weeks of unilateral knee extensor training, respiratory quotient (RQ) over the exercising thigh muscle was lower compared to the untrained leg when the limbs were tested 1 week apart. The reduced carbohydrate oxidation rate was attributed to lower muscle glycogenolysis since muscle glucose flux was not different during exercise after training (Kiens et al. 1993). The increased lipid oxidation was attributed to an enhanced muscle FFA uptake, which was increased during exercise after training. IMTG content was not different following exercise (Kiens et al. 1993), but other studies have reported greater IMTG utilization during exercise after endurance training (Hurley et al. 1986; Shepherd et al. 2013). In contrast, Bergman et al. (1999a) reported similar RQ values measured over the thigh and unchanged leg FFA uptake in males before and after 9 weeks of cycling; however, they also reported a reduced respiratory exchange ratio (i.e., increased whole-body lipid oxidation). Yet, the same group reported greater FFA oxidation post-training in females (Friedlander et al. 1998). Although discrepancies in results are difficult to reconcile, reduced glycolytic flux would be expected to better maintain free carnitine concentrations, facilitating sustained mitochondrial lipid transport. CPT1 activity is higher in trained compared to untrained individuals (Starritt et al. 2000), and ATGL is increased in human skeletal muscle after endurance training (Watt and Cheng 2017). That trained individuals have a greater abundance of proteins involved in the storage, mobilization, and oxidation of lipids can be explained largely by trained individuals having a greater percentage of type I fibers, which have significantly greater abundances of key lipid regulatory proteins, such as ATGL, HSL, PLIN2, and PLIN5 (Shaw et al. 2020).

10.7 Strength Training

10.7.1 Acute Metabolic Response

In contrast to the well-described metabolic response to endurance exercise, the metabolic response to strength exercise is not as well-studied. Several studies have examined changes in muscle metabolites (ATP, PCr) and substrates (glycogen, intramuscular triglyceride) in response to an acute bout of strength exercise (Hokken et al. 2020; MacDougall et al. 1999; Essen-Gustavsson and Tesch 1990; Koopman et al. 2006). Declines in PCr concentrations are large (i.e., >60%) but dependent on the intensity and volume of the exercises performed (MacDougall et al. 1999). However in broad terms, there is a large reliance on glycolytic metabolism during strength exercise, with as high as 80% of energy provision from these pathways when single sets are performed to failure (MacDougall et al. 1999). Declines in muscle glycogen concentration are typically in the range of 25 to 40% of resting concentrations (MacDougall et al. 1999; Hokken et al. 2020; Essen-Gustavsson and Tesch 1990; Koopman et al. 2006). On a fiber-type specific level, larger declines in muscle glycogen are observed in type II muscle fibers (Koopman et al. 2006; Hokken et al. 2020). Lastly, although data are limited, intramuscular triglyceride declines by ~25 to 30% on average during strength exercise regardless of training status (Essen-Gustavsson and Tesch 1990; Koopman et al. 2006). The source of intramuscular triglyceride was primarily from type I muscle fibers (Koopman et al. 2006), but in both studies, large interindividual variation was noted in both resting concentrations of intramuscular triglyceride and the declines observed during the training session (Essen-Gustavsson and Tesch 1990; Koopman et al. 2006).

10.7.2 Molecular Basis of Adaptations

Strength training outcomes have tended to focus on the molecular regulation of muscle hypertrophy as opposed to changes in substrate metabolism. The control of muscle mass is proposed to be determined by the balance between processes of muscle protein synthesis (MPS) and muscle protein breakdown/degradation (MPB), with hypertrophy occurring when the former exceeds the latter for an extended period of time (Figueiredo 2019). The regulation of MPB is primarily dependent on the activity of the ubiquitin-proteasome pathway (Sandri 2008). MPB occurs via two muscle-specific E3 ubiquitin ligases, muscle atrophy F-box (atrogin-1/MAFbx) and muscle RING finger 1 (MuRF1), which are key regulators of skeletal muscle proteolysis under catabolic conditions. MPB is challenging to measure and likely to be less consequential to regulation of strength training-induced muscle hypertrophy, which has resulted in the focus of molecular networks in this paradigm predominantly being on the activation of critical regulators of MPS, namely, mTORC, ribosomal protein S6K ($p70^{S6K}$), and several downstream targets (Figueiredo

2019; Ogasawara et al. 2019). Despite the dominant focus on MPS and muscle hypertrophy in response to strength training, it is worth noting that strength exercise also has marked effects on the skeletal muscle transcriptome in both acute and chronic contexts (Mallinson et al. 2020; Pillon et al. 2020).

High force contractions during strength exercise serve as the stimulus for the activation of a network of signaling events that can be broadly considered as mechanotransduction via a variety of mechanosensory regulators. These regulators include increases in the concentration of the membrane phospholipid, phosphatidic acid (PA) through activation of phospholipase D (PLD), and activation of FAK proteins, a class of transmembrane receptors that act as protein tyrosine kinases. Both PA and FAK can activate MPS through mTOR-dependent and mTOR-independent mechanisms (Ogasawara et al. 2019).

An important point in the overall discussion of exercise adaptation is the delineation of the term MPS. In the broadest sense, this term refers to *mixed* muscle protein synthesis. While it is true that muscle hypertrophy will require a greater volume of muscle protein per se, the process of adaptation to any kind of exercise requires an increase in protein synthesis, i.e., the synthesis of new proteins. Thus, focus on MPS as a process as being only relevant to strength training would be an obvious misconception; for example, sprint interval training is a potent stimulus to promote skeletal muscle anabolism in a general sense (Callahan et al. 2021). Rather, the specificity of exercise adaptation to different modes of exercise resides at the level of differential responses of different protein fractions (myofibrillar, sarcoplasmic, and mitochondrial) as well as individual proteins. Additionally, rates of protein synthesis are primarily dependent on translational efficiency and capacity. Both increased translational efficiency (protein synthesis per unit RNA) and elevated translational capacity (total RNA content per unit tissue) as a consequence of ribosome biogenesis have emerged as important regulators of the adaptive response to exercise (Brook et al. 2019). While there has been little focus on these pathways in the context of endurance exercise, a single bout of strength training increases markers of ribosomal DNA transcription (Stec et al. 2016), whereas an increase in RNA content is a rapid (Bickel et al. 2005) and sustained (Brook et al. 2015) response to strength training. Moreover, "high responders" to strength training have been demonstrated to exhibit greater increases in translational capacity compared to low- and non-responders (Stec et al. 2016). A similar phenomenon has been observed for skeletal muscle satellite cells such that large ranges of interindividual variability in the magnitude of hypertrophic response to strength training are explained by the relative ability to mobilize satellite cells and add myonuclei to existing muscle fibers (Petrella et al. 2006; Petrella et al. 2008). However, whether ribosomal biogenesis and myonuclear addition are *obligatory* for strength training-induced muscle hypertrophy in humans remains an open question (Brook et al. 2019).

10.7.3 Metabolic Characteristics and Exercise Responses in the Trained State

The strength-trained state is principally characterized by skeletal muscle hypertrophy owing to an increased muscle fiber cross-sectional area (Abernethy et al. 1994; Tesch 1988; McGlory et al. 2017). Longitudinal studies employing heavy strength training protocols have generally reported either no change or a decrease in skeletal muscle mitochondrial volume density and maximal activities of marker enzymes such as citrate synthase (MacDougall et al. 1979; Tesch et al. 1987). For example, Green et al. (1999) reported no effect of 12 weeks of strength training on succinate dehydrogenase content in any fiber type. There are equivocal data regarding the effect of resistance training on the maximal activities of enzymes involved in non-oxidative energy provision such as PFK, CK, and myokinase (Abernethy et al. 1994). It has been suggested that changes in enzyme density may depend on the extent of skeletal muscle hypertrophy (Tesch 1988). Heavy strength training is also associated with an increased intramuscular content of glycogen (MacDougall et al. 1977; Tesch et al. 1987). There are limited data regarding the effect of strength training on skeletal muscle metabolite transport proteins (Juel 2006).

Whether the acute metabolic response in skeletal muscle is altered after a period of strength training is largely unexplored, with potential changes based on indirect evidence. Despite equivocal data on adaptations in muscle fiber capillarization, oxygen extraction, fiber conversion, enzyme activity, and substrate levels to strength training, like endurance exercise training, an improvement in fatigue resistance is often observed (Abernethy et al. 1994). For example, the maximum number of repetitions of a bench press exercise performed at 70% 1RM before and after training was increased by 31% after 3 to 4 months of training (Guezennec et al. 1986). Likewise, there is some evidence of changes in substrate utilization after a period of strength training, albeit measured during endurance exercise. Heavy lower limb strength training altered the metabolic response to submaximal cycle exercise after 4 weeks (Goreham et al. 1999). The changes included reduced PCr and glycogen degradation and an attenuated increase in muscle lactate. The latter changes are consistent with a reduced accumulation of metabolites that would otherwise activate PHOS and PFK, which thereby reduces reliance on glycolytic pathways (Holloszy and Coyle 1984).

10.8 Sprint Training

10.8.1 Acute Metabolic Response

PCr hydrolysis and non-oxidative glycolysis provided the majority of ATP resynthesis during brief, maximal sprint efforts lasting $\leq$10–15 s (Parolin et al. 1999; Hultman et al. 1991). The energy contribution from oxidative phosphorylation increases during longer sprints with aerobic metabolism estimated to provide half of the ATP resynthesized during the latter half of a single 30 s maximal effort (Bogdanis et al. 1996; Parolin et al. 1999). The increased contribution from aerobic metabolism during longer sprints comes at the expense of a lower power output, as the contribution from non-oxidative ATP resynthesis declines due to reduced PCr availability and impaired glycolysis (Bogdanis et al. 1996; Parolin et al. 1999). Similar to heavy strength exercise, acute sprinting is characterized by marked reductions in muscle PCr and glycogen, with decreases of >75% and > 20% compared to rest, respectively, reported following a single 30 s all-out effort (Bogdanis et al. 1996; Parolin et al. 1999). Energy provision during repeated sprint efforts depends heavily on the duration individuals sprint and also the recovery period in between. During multiple 30 s all-out sprints with ~4 min of recovery, anaerobic energy provision is reduced owing to impaired glycogenolysis and incomplete recovery of PCr between bouts (Bogdanis et al. 1996; Parolin et al. 1999). Although the relative contribution of aerobic pathways to the ATP turnover rate was increased across multiple sprints, the absolute contribution was similar, leading to lower total ATP turnover and reduced power output (Parolin et al. 1999).

10.8.2 Molecular Basis of Adaptations

In comparison to endurance and strength exercise, less is known regarding the acute signaling events that mediate skeletal muscle remodeling in response to sprint exercise. Acute sprint interval exercise can activate molecular signaling pathways linked to mitochondrial biogenesis similar to endurance training (Gibala et al. 2009; Granata et al. 2017; Little et al. 2011). Gibala et al. (2009) showed that a single session involving four 30 s Wingate tests interspersed by 4 min of recovery increased the phosphorylation of AMPK and CaMK following exercise and the subsequent expression of PGC-1α. Like endurance exercise, expression of several genes related to substrate metabolism and structural remodeling is changed after a single session of sprint interval exercise (3x20 s sprints with 2 min recovery) (Skelly et al. 2017), but notably almost 900 differentially expressed genes were identified after a session involving three 30 s Wingate tests interspersed by 20 min of recovery (Rundqvist et al. 2019). Additionally, increases in Akt/mTOR signaling have been reported after a single session of the same effort (Esbjörnsson et al. 2012), which is consistent with observations that sprint interval exercise produces an anabolic response in skeletal muscle (Callahan et al. 2021).

10.8.3 Metabolic Characteristics and Exercise Responses in the Trained State

The sprint-trained state is associated with an enhanced maximal capacity for non-oxidative energy provision in skeletal muscle. This is most commonly reflected by an increased maximal activity of PFK, the rate-determining enzyme for glycolysis (Jacobs et al. 1987; MacDougall et al. 1998). Other reported enzymatic changes include increases in the maximal activities of CK, PHOS, and LDH, although data are equivocal in this regard (Ross and Leveritt 2001). It has also long been recognized that sprint interval training (e.g., repeated "all-out" 30 s Wingate tests interspersed with a few minutes of recovery) is a potent stimulus to elicit increases in skeletal muscle oxidative capacity, as indicated by the maximal activities and protein content of various mitochondrial enzymes including citrate synthase, cytochrome oxidase, and succinate dehydrogenase (Jacobs et al. 1987; MacDougall et al. 1998; Saltin et al. 1976). Such changes can occur relatively quickly and with a surprisingly small total volume of exercise (MacInnis and Gibala 2017). For example, as little as six sessions of sprint interval training over 2 weeks, involving a total of ~2–3 min of intense exercise per session, have been shown to increase the content of mitochondrial enzymes by ~30% (Burgomaster et al. 2005; Gibala et al. 2006). Sprint training is also a potent stimulus to increase the skeletal muscle content of membrane proteins involved in lactate and ion regulation (Juel 2006; McKenna et al. 1993; Hostrup and Bangsbo 2017). While increased muscle fiber size is sometimes observed after sprint training (Ross and Leveritt 2001), it is not generally associated with marked skeletal muscle hypertrophy (McGlory et al. 2017). This may be owing to the fact that the absolute load during sprinting is modest compared to heavy strength training. For example, while direct comparisons are difficult, it has been estimated that the maximal load during a Wingate test is ~10% of the 1RM during leg press exercise (Baar 2009).

Sprint training increases non-oxidative energy provision calculated from changes in muscle metabolites (e.g., glycogen, lactate) during brief "all-out" efforts lasting ≤30 seconds (Nevill et al. 1989). This response in part is believed to facilitate increased peak and mean power output during sprinting, in addition to changes in ion handling and fatigue development (Hostrup and Bangsbo 2017). Sprint interval training has also been shown to elicit metabolic adjustments during exercise that resemble those associated with traditional endurance training (Burgomaster et al. 2006; Burgomaster et al. 2008). For example, glycogen and PCr utilization during a 60 minute bout of moderate-intensity exercise were reduced after 6 weeks of both sprint interval and endurance training (Burgomaster et al. 2008). Burgomaster et al. (2006) also reported a higher active fraction of PDH during submaximal exercise and reduced muscle lactate accumulation, after 2 weeks of sprint interval training. This finding suggests a closer matching between rates of pyruvate production and oxidation, similar to the metabolic adjustments induced by endurance training and described above. Six weeks of sprint interval training was also reported to increase net IMTG breakdown during moderate-intensity cycling similar to moderate-intensity continuous training (Shepherd et al. 2013). A greater concentration of the lipid droplet-associated proteins, PLIN 2 and 5 was observed following both training modes and suggested to contribute to the increases in net IMTG breakdown following training.

10.9 Conclusion

The effect of training on skeletal muscle and exercise metabolism is an enormous topic, and the complex interplay of factors that determine responses cannot be captured in a single chapter. The present review attempted to summarize major, well-established responses to endurance, strength, and sprint training and highlight areas of contemporary research focus. Our current understanding of the molecular regulation of skeletal muscle remodeling is based mainly on studies involving either endurance or strength training, with comparatively little examination of the response to sprint training. With respect to the regulation of substrate metabolism during exercise, studies have mainly focused on changes mediated by endurance training, with a greater emphasis on the regulation of carbohydrate as compared to lipid utilization. One important area of future investigation is further consideration of the physiological diversity of exercise responses, including but not limited to the role of biological sex in mediating training adaptations (Ansdell et al. 2020).

Acknowledgments The authors would like to acknowledge the constructive comments on early versions of this chapter from Dr. Chris McGlory (Queen's University), Dr. Donal O'Gorman and Mr. Ian Darragh (both Dublin City University), and Prof. Gianni Parise (McMaster University) as well as the insightful feedback from Prof. Paul Greenhaff (University of Nottingham) on the penultimate version of this chapter. Many of the works from Dr. MacInnis and Prof. Gibala cited herein were supported by funding from the Natural Sciences and Engineering Research Council (Canada).

References

Abernethy PJ, Jürimäe J, Logan PA, Taylor AW, Thayer RE (1994) Acute and chronic response of skeletal muscle to resistance exercise. Sports Med 17(1):22–38

Ansdell P, Thomas K, Hicks KM, Hunter SK, Howatson G, Goodall S (2020) Physiological sex differences affect the integrative response to exercise: acute and chronic implications. Exp Physiol 105(12):2007–2021

Baar K (2009) The signaling underlying FITness. Appl Physiol Nutr Metab 34(3):411–419

Ball-Burnett M, Green HJ, Houston ME (1991) Energy metabolism in human slow and fast twitch fibres during prolonged cycle exercise. J Physiol 437:257–267

Barres R, Yan J, Egan B, Treebak JT, Rasmussen M, Fritz T, Caidahl K, Krook A, O'Gorman DJ, Zierath JR (2012) Acute exercise remodels promoter methylation in human skeletal muscle. Cell Metab 15(3):405–411

Benziane B, Burton TJ, Scanlan B, Galuska D, Canny BJ, Chibalin AV, Zierath JR, Stepto NK (2008) Divergent cell signaling after short-term intensified endurance training in human skeletal muscle. Am J Physiol Endocrinol Metab 295(6):E1427–E1438

Bergeron R, Ren JM, Cadman KS, Moore IK, Perret P, Pypaert M, Young LH, Semenkovich CF, Shulman GI (2001) Chronic activation of AMP kinase results in NRF-1 activation and mitochondrial biogenesis. Am J Physiol Endocrinol Metab 281(6):E1340–E1346

Bergman BC, Butterfield GE, Wolfel EE, Casazza GA, Lopaschuk GD, Brooks GA (1999a) Evaluation of exercise and training on muscle lipid metabolism. Am J Phys 276(1):E106–E117

Bergman BC, Wolfel EE, Butterfield GE, Lopaschuk GD, Casazza GA, Horning MA, Brooks GA (1999b) Active muscle and whole body lactate kinetics after endurance training in men. J Appl Physiol (1985) 87(5):1684–1696
Bickel CS, Slade J, Mahoney E, Haddad F, Dudley GA, Adams GR (2005) Time course of molecular responses of human skeletal muscle to acute bouts of resistance exercise. J Appl Physiol (1985) 98(2):482–488
Bishop D, Girard O, Mendez-Villanueva A (2011) Repeated-sprint ability–part II: recommendations for training. Sports Med 41(9):741–756
Black MI, Jones AM, Blackwell JR, Bailey SJ, Wylie LJ, McDonagh ST, Thompson C, Kelly J, Sumners P, Mileva KN, Bowtell JL, Vanhatalo A (2017) Muscle metabolic and neuromuscular determinants of fatigue during cycling in different exercise intensity domains. J Appl Physiol (1985) 122(3):446–459
Bogdanis GC, Nevill ME, Boobis LH, Lakomy HK (1996) Contribution of phosphocreatine and aerobic metabolism to energy supply during repeated sprint exercise. J Appl Physiol (1985) 80(3):876–884
Booth FW, Thomason DB (1991) Molecular and cellular adaptation of muscle in response to exercise: perspectives of various models. Physiol Rev 71(2):541–585
Brook MS, Wilkinson DJ, Mitchell WK, Lund JN, Szewczyk NJ, Greenhaff PL, Smith K, Atherton PJ (2015) Skeletal muscle hypertrophy adaptations predominate in the early stages of resistance exercise training, matching deuterium oxide-derived measures of muscle protein synthesis and mechanistic target of rapamycin complex 1 signaling. FASEB J 29(11):4485–4496
Brook MS, Wilkinson DJ, Smith K, Atherton PJ (2019) It's not just about protein turnover: the role of ribosomal biogenesis and satellite cells in the regulation of skeletal muscle hypertrophy. Eur J Sport Sci 19(7):952–963
Brooks GA (2012) Bioenergetics of exercising humans. Compr Physiol 2(1):537–562
Brooks GA (2018) The science and translation of lactate shuttle theory. Cell Metab 27(4):757–785
Burgomaster KA, Heigenhauser GJ, Gibala MJ (2006) Effect of short-term sprint interval training on human skeletal muscle carbohydrate metabolism during exercise and time-trial performance. J Appl Physiol (1985) 100(6):2041–2047
Burgomaster KA, Howarth KR, Phillips SM, Rakobowchuk M, Macdonald MJ, McGee SL, Gibala MJ (2008) Similar metabolic adaptations during exercise after low volume sprint interval and traditional endurance training in humans. J Physiol 586(1):151–160
Burgomaster KA, Hughes SC, Heigenhauser GJ, Bradwell SN, Gibala MJ (2005) Six sessions of sprint interval training increases muscle oxidative potential and cycle endurance capacity in humans. J Appl Physiol (1985) 98(6):1985–1990
Burniston JG, Towler M, Wackerhage H (2014) Signal transduction and adaptation to exercise: background and methods. In: Wackerhage H (ed) Molecular exercise physiology: an introduction, 1st edn. Routledge, Oxon, UK, pp 52–78
Callahan MJ, Parr EB, Hawley JA, Camera DM (2021) Can high-intensity interval training promote skeletal muscle anabolism? Sports Med 51(3):405–421
Chesley A, Heigenhauser GJ, Spriet LL (1996) Regulation of muscle glycogen phosphorylase activity following short-term endurance training. Am J Phys 270(2 Pt 1):E328–E335
Coffey VG, Hawley JA (2017) Concurrent exercise training: do opposites distract? J Physiol 595(9):2883–2896
Coffey VG, Zhong Z, Shield A, Canny BJ, Chibalin AV, Zierath JR, Hawley JA (2006) Early signaling responses to divergent exercise stimuli in skeletal muscle from well-trained humans. FASEB J 20(1):190–192
Coggan AR, Kohrt WM, Spina RJ, Bier DM, Holloszy JO (1990) Endurance training decreases plasma glucose turnover and oxidation during moderate-intensity exercise in men. J Appl Physiol (1985) 68(3):990–996
Constantin-Teodosiu D, Carlin JI, Cederblad G, Harris RC, Hultman E (1991) Acetyl group accumulation and pyruvate dehydrogenase activity in human muscle during incremental exercise. Acta Physiol Scand 143(4):367–372

Egan B, Zierath JR (2013) Exercise metabolism and the molecular regulation of skeletal muscle adaptation. Cell Metab 17(2):162–184
Esbjörnsson M, Rundqvist HC, Mascher H, Österlund T, Rooyackers O, Blomstrand E, Jansson E (2012) Sprint exercise enhances skeletal muscle p70S6k phosphorylation and more so in women than in men. Acta Physiol (Oxf) 205(3):411–422
Essen-Gustavsson B, Tesch PA (1990) Glycogen and triglyceride utilization in relation to muscle metabolic characteristics in men performing heavy-resistance exercise. Eur J Appl Physiol Occup Physiol 61(1–2):5–10
Figueiredo VC (2019) Revisiting the roles of protein synthesis during skeletal muscle hypertrophy induced by exercise. Am J Physiol Regul Integr Comp Physiol 317(5):R709–r718
Forbes SC, Slade JM, Meyer RA (2008) Short-term high-intensity interval training improves phosphocreatine recovery kinetics following moderate-intensity exercise in humans. Appl Physiol Nutr Metab 33(6):1124–1131
Friedlander AL, Casazza GA, Horning MA, Buddinger TF, Brooks GA (1998) Effects of exercise intensity and training on lipid metabolism in young women. Am J Phys 275(5):E853–E863
Garcia-Roves PM, Osler ME, Holmstrom MH, Zierath JR (2008) Gain-of-function R225Q mutation in AMP-activated protein kinase {gamma}3 subunit increases mitochondrial biogenesis in glycolytic skeletal muscle. J Biol Chem 283(51):35724–35734
Gibala MJ, Little JP, van Essen M, Wilkin GP, Burgomaster KA, Safdar A, Raha S, Tarnopolsky MA (2006) Short-term sprint interval versus traditional endurance training: similar initial adaptations in human skeletal muscle and exercise performance. J Physiol 575(Pt 3):901–911
Gibala MJ, McGee SL, Garnham AP, Howlett KF, Snow RJ, Hargreaves M (2009) Brief intense interval exercise activates AMPK and p38 MAPK signaling and increases the expression of PGC-1alpha in human skeletal muscle. J Appl Physiol (1985) 106(3):929–934
Goreham C, Green HJ, Ball-Burnett M, Ranney D (1999) High-resistance training and muscle metabolism during prolonged exercise. Am J Phys 276(3):E489–E496
Granata C, Jamnick NA, Bishop DJ (2018) Principles of exercise prescription, and how they influence exercise-induced changes of transcription factors and other regulators of mitochondrial biogenesis. Sports Med 48(7):1541–1559
Granata C, Oliveira RS, Little JP, Renner K, Bishop DJ (2017) Sprint-interval but not continuous exercise increases PGC-1α protein content and p53 phosphorylation in nuclear fractions of human skeletal muscle. Sci Rep 7:44227
Green H, Goreham C, Ouyang J, Ball-Burnett M, Ranney D (1999) Regulation of fiber size, oxidative potential, and capillarization in human muscle by resistance exercise. Am J Phys 276(2):R591–R596
Green HJ, Jones S, Ball-Burnett M, Farrance B, Ranney D (1995) Adaptations in muscle metabolism to prolonged voluntary exercise and training. J Appl Physiol (1985) 78(1):138–145
Green HJ, Jones S, Ball-Burnett ME, Smith D, Livesey J, Farrance BW (1991) Early muscular and metabolic adaptations to prolonged exercise training in humans. J Appl Physiol (1985) 70(5): 2032–2038
Greenhaff PL, Nevill ME, Soderlund K, Bodin K, Boobis LH, Williams C, Hultman E (1994) The metabolic responses of human type I and II muscle fibres during maximal treadmill sprinting. J Physiol 478(Pt 1):149–155
Greiwe JS, Hickner RC, Hansen PA, Racette SB, Chen MM, Holloszy JO (1999) Effects of endurance exercise training on muscle glycogen accumulation in humans. J Appl Physiol (1985) 87(1):222–226
Guezennec Y, Leger L, Lhoste F, Aymonod M, Pesquies PC (1986) Hormone and metabolite response to weight-lifting training sessions. Int J Sports Med 7(2):100–105
Hackett DA, Johnson NA, Chow CM (2013) Training practices and ergogenic aids used by male bodybuilders. J Strength Cond Res 27(6):1609–1617
Hargreaves M, Spriet LL (2020) Skeletal muscle energy metabolism during exercise. Nat Metab 2(9):817–828

Hawley JA, Burke LM, Phillips SM, Spriet LL (2011) Nutritional modulation of training-induced skeletal muscle adaptations. J Appl Physiol (1985) 110(3):834–845
Hawley JA, Hargreaves M, Joyner MJ, Zierath JR (2014) Integrative biology of exercise. Cell 159(4):738–749
Henriksson J (1977) Training induced adaptation of skeletal muscle and metabolism during submaximal exercise. J Physiol 270(3):661–675
Hokken R, Laugesen S, Aagaard P, Suetta C, Frandsen U, Ørtenblad N, Nielsen J (2020) Subcellular localization- and fibre type-dependent utilization of muscle glycogen during heavy resistance exercise in elite power and Olympic weightlifters. Acta Physiol (Oxf) e13561
Holloszy JO, Coyle EF (1984) Adaptations of skeletal muscle to endurance exercise and their metabolic consequences. J Appl Physiol Respir Environ Exerc Physiol 56(4):831–838
Hood DA, Tryon LD, Carter HN, Kim Y, Chen CC (2016) Unravelling the mechanisms regulating muscle mitochondrial biogenesis. Biochem J 473(15):2295–2314
Hoppeler H, Baum O, Lurman G, Mueller M (2011) Molecular mechanisms of muscle plasticity with exercise. Compr Physiol 1(3):1383–1412
Hostrup M, Bangsbo J (2017) Limitations in intense exercise performance of athletes–effect of speed endurance training on ion handling and fatigue development. J Physiol 595(9):2897–2913
Houmard JA, Shinebarger MH, Dolan PL, Leggett-Frazier N, Bruner RK, McCammon MR, Israel RG, Dohm GL (1993) Exercise training increases GLUT-4 protein concentration in previously sedentary middle-aged men. Am J Phys 264(6 Pt 1):E896–E901
Howlett RA, Parolin ML, Dyck DJ, Hultman E, Jones NL, Heigenhauser GJ, Spriet LL (1998) Regulation of skeletal muscle glycogen phosphorylase and PDH at varying exercise power outputs. Am J Phys 275(2):R418–R425
Hultman E, Greenhaff PL, Ren JM, Söderlund K (1991) Energy metabolism and fatigue during intense muscle contraction. Biochem Soc Trans 19(2):347–353
Hurley BF, Nemeth PM, Martin WH 3rd, Hagberg JM, Dalsky GP, Holloszy JO (1986) Muscle triglyceride utilization during exercise: effect of training. J Appl Physiol (1985) 60(2):562–567
Islam H, Edgett BA, Gurd BJ (2018) Coordination of mitochondrial biogenesis by PGC-1α in human skeletal muscle: a re-evaluation. Metabolism 79:42–51
Jacobs I, Esbjörnsson M, Sylvén C, Holm I, Jansson E (1987) Sprint training effects on muscle myoglobin, enzymes, fiber types, and blood lactate. Med Sci Sports Exerc 19(4):368–374
Juel C (2006) Training-induced changes in membrane transport proteins of human skeletal muscle. Eur J Appl Physiol (1985) 96(6):627–635
Karlsson J, Nordesjö LO, Jorfeldt L, Saltin B (1972) Muscle lactate, ATP, and CP levels during exercise after physical training in man. J Appl Physiol (1985) 33(2):199–203
Kiens B, Essen-Gustavsson B, Christensen NJ, Saltin B (1993) Skeletal muscle substrate utilization during submaximal exercise in man: effect of endurance training. J Physiol 469:459–478
Kiens B, Kristiansen S, Jensen P, Richter EA, Turcotte LP (1997) Membrane associated fatty acid binding protein (FABPpm) in human skeletal muscle is increased by endurance training. Biochem Biophys Res Commun 231(2):463–465
Kjøbsted R, Hingst JR, Fentz J, Foretz M, Sanz MN, Pehmøller C, Shum M, Marette A, Mounier R, Treebak JT, Wojtaszewski JFP, Viollet B, Lantier L (2018) AMPK in skeletal muscle function and metabolism. FASEB J 32(4):1741–1777
Koopman R, Manders RJ, Jonkers RA, Hul GB, Kuipers H, van Loon LJ (2006) Intramyocellular lipid and glycogen content are reduced following resistance exercise in untrained healthy males. Eur J Appl Physiol 96(5):525–534
Kristiansen S, Gade J, Wojtaszewski JF, Kiens B, Richter EA (2000) Glucose uptake is increased in trained vs. untrained muscle during heavy exercise. J Appl Physiol (1985) 89(3):1151–1158
Leblanc PJ, Howarth KR, Gibala MJ, Heigenhauser GJ (2004) Effects of 7 wk of endurance training on human skeletal muscle metabolism during submaximal exercise. J Appl Physiol (1985) 97(6):2148–2153

Leick L, Wojtaszewski JF, Johansen ST, Kiilerich K, Comes G, Hellsten Y, Hidalgo J, Pilegaard H (2008) PGC-1alpha is not mandatory for exercise- and training-induced adaptive gene responses in mouse skeletal muscle. Am J Physiol Endocrinol Metab 294(2):E463–E474

Little JP, Safdar A, Bishop D, Tarnopolsky MA, Gibala MJ (2011) An acute bout of high-intensity interval training increases the nuclear abundance of PGC-1α and activates mitochondrial biogenesis in human skeletal muscle. Am J Physiol Regul Integr Comp Physiol 300(6): R1303–R1310

Little JP, Safdar A, Cermak N, Tarnopolsky MA, Gibala MJ (2010) Acute endurance exercise increases the nuclear abundance of PGC-1alpha in trained human skeletal muscle. Am J Physiol Regul Integr Comp Physiol 298(4):R912–R917

Lundby C, Jacobs RA (2016) Adaptations of skeletal muscle mitochondria to exercise training. Exp Physiol 101(1):17–22

Lundsgaard AM, Fritzen AM, Kiens B (2018) Molecular regulation of fatty acid oxidation in skeletal muscle during aerobic exercise. Trends Endocrinol Metab 29(1):18–30

MacDougall JD, Hicks AL, MacDonald JR, McKelvie RS, Green HJ, Smith KM (1998) Muscle performance and enzymatic adaptations to sprint interval training. J Appl Physiol (1985) 84(6): 2138–2142

MacDougall JD, Ray S, Sale DG, McCartney N, Lee P, Garner S (1999) Muscle substrate utilization and lactate production. Can J Appl Physiol 24(3):209–215

MacDougall JD, Sale DG, Moroz JR, Elder GC, Sutton JR, Howald H (1979) Mitochondrial volume density in human skeletal muscle following heavy resistance training. Med Sci Sports 11(2):164–166

MacDougall JD, Ward GR, Sale DG, Sutton JR (1977) Biochemical adaptation of human skeletal muscle to heavy resistance training and immobilization. J Appl Physiol Respir Environ Exerc Physiol 43(4):700–703

MacInnis MJ, Gibala MJ (2017) Physiological adaptations to interval training and the role of exercise intensity. J Physiol 595(9):2915–2930

Mallinson JE, Taylor T, Constantin-Teodosiu D, Billeter-Clark R, Constantin D, Franchi MV, Narici MV, Auer D, Greenhaff PL (2020) Longitudinal hypertrophic and transcriptional responses to high-load eccentric-concentric vs concentric training in males. Scand J Med Sci Sports 30(11):2101–2115

Martin WH 3rd, Dalsky GP, Hurley BF, Matthews DE, Bier DM, Hagberg JM, Rogers MA, King DS, Holloszy JO (1993) Effect of endurance training on plasma free fatty acid turnover and oxidation during exercise. Am J Phys 265(5 Pt 1):E708–E714

Martínez-Redondo V, Pettersson AT, Ruas JL (2015) The hitchhiker's guide to PGC-1α isoform structure and biological functions. Diabetologia 58(9):1969–1977

Massart J, Katayama M, Krook A (2016) microManaging glucose and lipid metabolism in skeletal muscle: role of microRNAs. Biochim Biophys Acta 1861(12 Pt B):2130–2138

McConell GK, Wadley GD, Le Plastrier K, Linden KC (2020) Skeletal muscle AMPK is not activated during 2 h of moderate intensity exercise at ~65% VO2peak in endurance trained men. J Physiol 598(18):3859–3870

McGee SL, Fairlie E, Garnham AP, Hargreaves M (2009) Exercise-induced histone modifications in human skeletal muscle. J Physiol 587(Pt 24):5951–5958

McGee SL, Hargreaves M (2004) Exercise and myocyte enhancer factor 2 regulation in human skeletal muscle. Diabetes 53(5):1208–1214

McGee SL, Hargreaves M (2020) Exercise adaptations: molecular mechanisms and potential targets for therapeutic benefit. Nat Rev Endocrinol 16(9):495–505

McGlory C, Devries MC, Phillips SM (2017) Skeletal muscle and resistance exercise training; the role of protein synthesis in recovery and remodeling. J Appl Physiol (1985) 122(3):541–548

McKenna MJ, Schmidt TA, Hargreaves M, Cameron L, Skinner SL, Kjeldsen K (1993) Sprint training increases human skeletal muscle Na(+)-K(+)-ATPase concentration and improves K+ regulation. J Appl Physiol (1985) 75(1):173–180

McKenzie S, Phillips SM, Carter SL, Lowther S, Gibala MJ, Tarnopolsky MA (2000) Endurance exercise training attenuates leucine oxidation and BCOAD activation during exercise in humans. Am J Physiol Endocrinol Metab 278(4):E580–E587
Messonnier LA, Emhoff CA, Fattor JA, Horning MA, Carlson TJ, Brooks GA (2013) Lactate kinetics at the lactate threshold in trained and untrained men. J Appl Physiol (1985) 114(11): 1593–1602
Nevill ME, Boobis LH, Brooks S, Williams C (1989) Effect of training on muscle metabolism during treadmill sprinting. J Appl Physiol (1985) 67(6):2376–2382
Ogasawara R, Jensen TE, Goodman CA, Hornberger TA (2019) Resistance exercise-induced hypertrophy: a potential role for rapamycin-insensitive mTOR. Exerc Sport Sci Rev 47(3): 188–194
Paganini AT, Foley JM, Meyer RA (1997) Linear dependence of muscle phosphocreatine kinetics on oxidative capacity. Am J Phys 272(2 Pt 1):C501–C510
Parolin ML, Chesley A, Matsos MP, Spriet LL, Jones NL, Heigenhauser GJ (1999) Regulation of skeletal muscle glycogen phosphorylase and PDH during maximal intermittent exercise. Am J Phys 277(5):E890–E900
Perry CG, Lally J, Holloway GP, Heigenhauser GJ, Bonen A, Spriet LL (2010) Repeated transient mRNA bursts precede increases in transcriptional and mitochondrial proteins during training in human skeletal muscle. J Physiol 588(Pt 23):4795–4810
Petrakos G, Morin JB, Egan B (2016) Resisted sled Sprint training to improve Sprint performance: a systematic review. Sports Med 46(3):381–400
Petrella JK, Kim JS, Cross JM, Kosek DJ, Bamman MM (2006) Efficacy of myonuclear addition may explain differential myofiber growth among resistance-trained young and older men and women. Am J Physiol Endocrinol Metab 291(5):E937–E946
Petrella JK, Kim JS, Mayhew DL, Cross JM, Bamman MM (2008) Potent myofiber hypertrophy during resistance training in humans is associated with satellite cell-mediated myonuclear addition: a cluster analysis. J Appl Physiol (1985) 104(6):1736–1742
Phillips SM, Green HJ, MacDonald MJ, Hughson RL (1995) Progressive effect of endurance training on VO2 kinetics at the onset of submaximal exercise. J Appl Physiol (1985) 79(6): 1914–1920
Phillips SM, Green HJ, Tarnopolsky MA, Heigenhauser GF, Hill RE, Grant SM (1996) Effects of training duration on substrate turnover and oxidation during exercise. J Appl Physiol (1985) 81(5):2182–2191
Pillon NJ, Gabriel BM, Dollet L, Smith JAB, Sardón Puig L, Botella J, Bishop DJ, Krook A, Zierath JR (2020) Transcriptomic profiling of skeletal muscle adaptations to exercise and inactivity. Nat Commun 11(1):470
Poole DC, Jones AM (2012) Oxygen uptake kinetics. Compr Physiol 2(2):933–996
Richter EA, Jensen P, Kiens B, Kristiansen S (1998) Sarcolemmal glucose transport and GLUT-4 translocation during exercise are diminished by endurance training. Am J Phys 274(1):E89–E95
Romijn JA, Coyle EF, Sidossis LS, Gastaldelli A, Horowitz JF, Endert E, Wolfe RR (1993) Regulation of endogenous fat and carbohydrate metabolism in relation to exercise intensity and duration. Am J Phys 265(3 Pt 1):E380–E391
Ross A, Leveritt M (2001) Long-term metabolic and skeletal muscle adaptations to short-sprint training: implications for sprint training and tapering. Sports Med 31(15):1063–1082
Ross A, Leveritt M, Riek S (2001) Neural influences on sprint running: training adaptations and acute responses. Sports Med 31(6):409–425
Rowe GC, El-Khoury R, Patten IS, Rustin P, Arany Z (2012) PGC-1alpha is dispensable for exercise-induced mitochondrial biogenesis in skeletal muscle. PLoS One 7(7):e41817
Rumpf MC, Lockie RG, Cronin JB, Jalilvand F (2016) Effect of different Sprint training methods on Sprint performance over various distances: a brief review. J Strength Cond Res 30(6): 1767–1785
Rundqvist HC, Montelius A, Osterlund T, Norman B, Esbjornsson M, Jansson E (2019) Acute sprint exercise transcriptome in human skeletal muscle. PLoS One 14(10):e0223024

Sahlin K, Broberg S, Ren JM (1989) Formation of inosine monophosphate (IMP) in human skeletal muscle during incremental dynamic exercise. Acta Physiol Scand 136(2):193–198
Sahlin K, Katz A, Henriksson J (1987) Redox state and lactate accumulation in human skeletal muscle during dynamic exercise. Biochem J 245(2):551–556
Sale DG (1988) Neural adaptation to resistance training. Med Sci Sports Exerc 20(5 Suppl):S135–S145
Saltin B, Gollnick PD (1983) Skeletal muscle adaptability: significance for metabolism and performance. In: Peachy LD, Adrian RH, Geiger SR (eds) Handbook of physiology, section 10: skeletal muscle. American Physiological Society, Bethesda, MD, pp 555–631
Saltin B, Nazar K, Costill DL, Stein E, Jansson E, Essén B, Gollnick D (1976) The nature of the training response; peripheral and central adaptations of one-legged exercise. Acta Physiol Scand 96(3):289–305
Sandri M (2008) Signaling in muscle atrophy and hypertrophy. Physiology (Bethesda) 23:160–170
Seaborne RA, Sharples AP (2020) The interplay between exercise metabolism, epigenetics, and skeletal muscle remodeling. Exerc Sport Sci Rev 48(4):188–200
Seiler S (2010) What is best practice for training intensity and duration distribution in endurance athletes? Int J Sports Physiol Perform 5(3):276–291
Shaw CS, Swinton C, Morales-Scholz MG, McRae N, Erftemeyer T, Aldous A, Murphy RM, Howlett KF (2020) Impact of exercise training status on the fiber type-specific abundance of proteins regulating intramuscular lipid metabolism. J Appl Physiol (1985) 128(2):379–389
Shepherd SO, Cocks M, Tipton KD, Ranasinghe AM, Barker TA, Burniston JG, Wagenmakers AJ, Shaw CS (2013) Sprint interval and traditional endurance training increase net intramuscular triglyceride breakdown and expression of perilipin 2 and 5. J Physiol 591(3):657–675
Skelly LE, Gillen JB, MacInnis MJ, Martin BJ, Safdar A, Akhtar M, MacDonald MJ, Tarnopolsky MA, Gibala MJ (2017) Effect of sex on the acute skeletal muscle response to sprint interval exercise. Exp Physiol 102(3):354–365
Spriet LL (2014) New insights into the interaction of carbohydrate and fat metabolism during exercise. Sports Med 44(Suppl 1):S87–S96
Starritt EC, Howlett RA, Heigenhauser GJ, Spriet LL (2000) Sensitivity of CPT I to malonyl-CoA in trained and untrained human skeletal muscle. Am J Physiol Endocrinol Metab 278(3):E462–E468
Stec MJ, Kelly NA, Many GM, Windham ST, Tuggle SC, Bamman MM (2016) Ribosome biogenesis may augment resistance training-induced myofiber hypertrophy and is required for myotube growth in vitro. Am J Physiol Endocrinol Metab 310(8):E652–e661
Tesch PA (1988) Skeletal muscle adaptations consequent to long-term heavy resistance exercise. Med Sci Sports Exerc 20(5 Suppl):S132–S134
Tesch PA, Komi PV, Häkkinen K (1987) Enzymatic adaptations consequent to long-term strength training. Int J Sports Med 8(Suppl 1):66–69
van Loon LJ, Goodpaster BH (2006) Increased intramuscular lipid storage in the insulin-resistant and endurance-trained state. Pflugers Arch 451(5):606–616
van Loon LJ, Greenhaff PL, Constantin-Teodosiu D, Saris WH, Wagenmakers AJ (2001) The effects of increasing exercise intensity on muscle fuel utilisation in humans. J Physiol 536(Pt 1): 295–304
Watt MJ, Cheng Y (2017) Triglyceride metabolism in exercising muscle. Biochim Biophys Acta Mol Cell Biol Lipids 1862(10 Pt B):1250–1259

Chapter 11
Role and Regulation of Hepatic Metabolism During Exercise

Elijah Trefts and David H. Wasserman

Abstract The liver is a rechargeable battery. It releases stored energy at times of high metabolic demand such as exercise and replenishes energy stores in response to a meal. The liver is a recycler. It converts metabolites and amino acids into glucose. The liver is a detoxifier. It removes nitrogenous molecules, hemoglobin, hormones, foreign substances, immunoglobulin, and other compounds from the circulation. The muscle contracts, the adipose tissue stores fat, and the heart pumps blood. The functions of the liver are far too diverse to define by a single dominant process. The underlying role of the functions of the liver is that they make broad contributions to arterial homeostasis and thereby homeostasis of numerous cell types. Physical exercise poses a unique challenge to the liver as metabolic demands of working muscles require the liver to mobilize energy stores, recycle metabolites, and convert compounds that are toxic in excess to innocuous forms. The focus of this review will be on how the liver adapts to the metabolic demands of physical exercise.

Keywords Liver · Gluconeogenesis · Ketogenesis · Glucose · Amino acids

The liver supplies glucose to tissues, recycles carbons, and counters disequilibrium by extracting metabolites and toxins from the blood. Blood flow to the liver and splanchnic bed is reduced to allow for a greater fraction of the cardiac output to be diverted to muscle. These functions of the liver are essential for sustained physical activity. Moreover, regular physical activity causes adaptations at the liver that improve metabolic health. There is a considerable amount known about the role and regulation of nutrient fluxes during exercise. Despite the essential role of the liver during exercise and the wealth of data demonstrating control of macronutrient

E. Trefts
Molecular and Cellular Biological Laboratory, Salk Institute for Biological Studies, La Jolla, CA, USA

D. H. Wasserman (✉)
Department of Molecular Physiology and Biophysics, Vanderbilt University School of Medicine, Nashville, TN, USA
e-mail: david.wasserman@vanderbilt.edu

© The Author(s), under exclusive license to Springer Nature Switzerland AG 2022

G. McConell (ed.), *Exercise Metabolism*, Physiology in Health and Disease,
https://doi.org/10.1007/978-3-030-94305-9_11

flux during exercise, the liver does not receive commensurate attention. This is largely due to the difficulty in performing well-controlled direct studies of the liver in exercising humans. Studies in rodent models have been extremely valuable. However, studies of the liver in rodent models have been largely limited to endpoint measurements of excised liver and measurements of peripheral blood. Much of the foundation of regulation of liver nutrient metabolism during exercise and in the post-exercise state is based on experiments in large animals. Large animal models permit implantation of catheters in the vessels that both perfuse (portal vein) and drain (hepatic vein) the liver during exercise. The measurement of arteriovenous differences combined with isotopic techniques provides a powerful model system to gain a comprehensive picture of the mechanisms that control liver nutrient fluxes during exercise. A compilation of studies in humans, large animals, and rodents is necessary for a more comprehensive understanding of regulation of liver metabolism during exercise.

Here we focus on the integrated control of liver nutrient metabolism that is essential to accommodate the accelerated metabolic demands of exercise. Important advances have been made leading to a broader definition of an endocrine gland. This has led to the recognition of new proteins released into the circulation that may have actions at the liver. While much still remains to be learned, the major regulators of liver metabolism in response to moderate sustained exercise will not likely change. Changes in glucagon and insulin can quantitatively account for the increased hepatic nutrient metabolism. These hormones are secreted from pancreatic islets into the hepatic portal circulation which directly perfuses the liver. This is a highly efficient anatomical arrangement. This chapter focuses on the diverse metabolic demands on the liver during exercise, how liver metabolism is regulated during and after exercise, and how the liver adapts to regular physical activity.

11.1 Metabolic Demands of Exercise on the Liver

Exercise results in elevated demand for metabolic substrates by working skeletal muscle. This demand is primarily met by accelerated carbohydrate and lipid metabolism, with only a small contribution from branched chain amino acids. Specifically, the utilization of intramuscular glycogen and triglycerides stores and the uptake of glucose and lipids from the circulation are increased to meet muscle energy demands during exercise (Hargreaves and Spriet 2020). The relative contribution of these substrate pools to energy provision during exercise is dependent on the length, modality, and intensity of exercise (Felig et al. 1982; Romijn et al. 1993). The potential energy from circulating substrates and muscle glycogen stores is limited. As such, mobilization of glucose and lipids from extra-myocellular sources is required for sustained exercise. In contrast to mobilization of non-esterified fatty acids (NEFAs) and triglycerides which can undergo large acute swings in circulating concentrations, glucose homeostasis must be tightly maintained to preserve function of the central nervous system. To preserve arterial glucose homeostasis the rate of glucose release from the liver into the blood must increase to match the increased glucose uptake by working muscle (Wasserman 2009). With increasing exercise intensity, muscle consumes more circulating glucose. This is matched by an increase in the rate of glucose production by the liver.

Exercise-induced increases in hepatic glucose production require mobilization of hepatic glycogen (glycogenolysis) and the de novo synthesis of glucose from metabolic intermediates (gluconeogenesis). The fraction of glucose production derived from liver glycogen increases with work intensity, while the rate that is gluconeogenic increases with exercise duration. We developed a novel technique for measuring hepatic metabolic fluxes over time in mice using venous infusions of $[^{13}C_3]$propionate, $[^2H_2]$water, and $[6,6\text{-}^2H_2]$glucose isotopes and arterial sampling (Hasenour et al. 2015) before and during exercise (Hughey et al. 2017). GC-MS and mass isotopomer distribution (MID) analysis of glucose was performed on arterial glucose samples. Model-based regression of hepatic glucose production, gluconeogenesis, tricarboxylic acid (TCA) cycle, and anaplerosis-related fluxes was performed using a comprehensive isotopomer model to measure reaction rates within a defined metabolic network. This approach showed the coordination of reactions that comprise hepatic metabolic fluxes during exercise (Hughey et al. 2017). Anaplerosis, TCA cycling, and pyruvate cycling in the liver were all increased in synchrony with flux through the key gluconeogenic enzyme enolase (Fig. 11.1). These data characterize how gluconeogenic precursors are scavenged from the circulation and recycled to glucose via gluconeogenesis. The working muscle and adipose tissue represent major sources for these gluconeogenic precursors (Ahlborg et al. 1974). Fascinatingly, lactate from non-working muscle (Ahlborg and Felig 1982) and amino acids from the gastrointestinal tract (Wasserman et al. 1991a; Williams et al. 1996) are also significant sources of gluconeogenic substrates during exercise (Fig. 11.2).

The recycling of metabolites to glucose is evidence of the efficient mechanisms that have evolved to maintain glucose homeostasis. Gluconeogenesis requires energy. Oxidation of fatty acids supplies energy by providing reducing equivalents and ATP for gluconeogenesis (Wahren et al. 1984; Wasserman et al. 1989a, b). Ketone bodies (e.g., β-hydroxybutyrate and acetoacetate) are a by-product of accelerated fatty acid oxidation (Evans et al. 2017) and to a small extent from specific amino acid oxidation (Thomas et al. 1982). Ketogenesis is liver-specific and an increase in circulating ketones is a marker of elevated hepatic fat oxidation (Wasserman et al. 1989b; Evans et al. 2017), such as during prolonged exercise or poorly controlled diabetes. Ketones can be utilized for energy in oxidative tissues such as the heart. Ketones produced by the liver during exercise are not typically a primary energy source. However, the increasing consumption of diets with low carbohydrate and high fat content, termed ketogenic diets, has led to increased interest of the impact of ketone metabolism (Puchalska and Crawford 2017).

Increased amino acid metabolism and AMP deamination occur in working muscle leading to increased formation of NH_3 (Eriksson et al. 1985). Moreover, increased skeletal muscle (Felig and Wahren 1971; Van Loon 2014) or gastrointestinal tract (Wasserman et al. 1991a; Williams et al. 1996) protein breakdown results

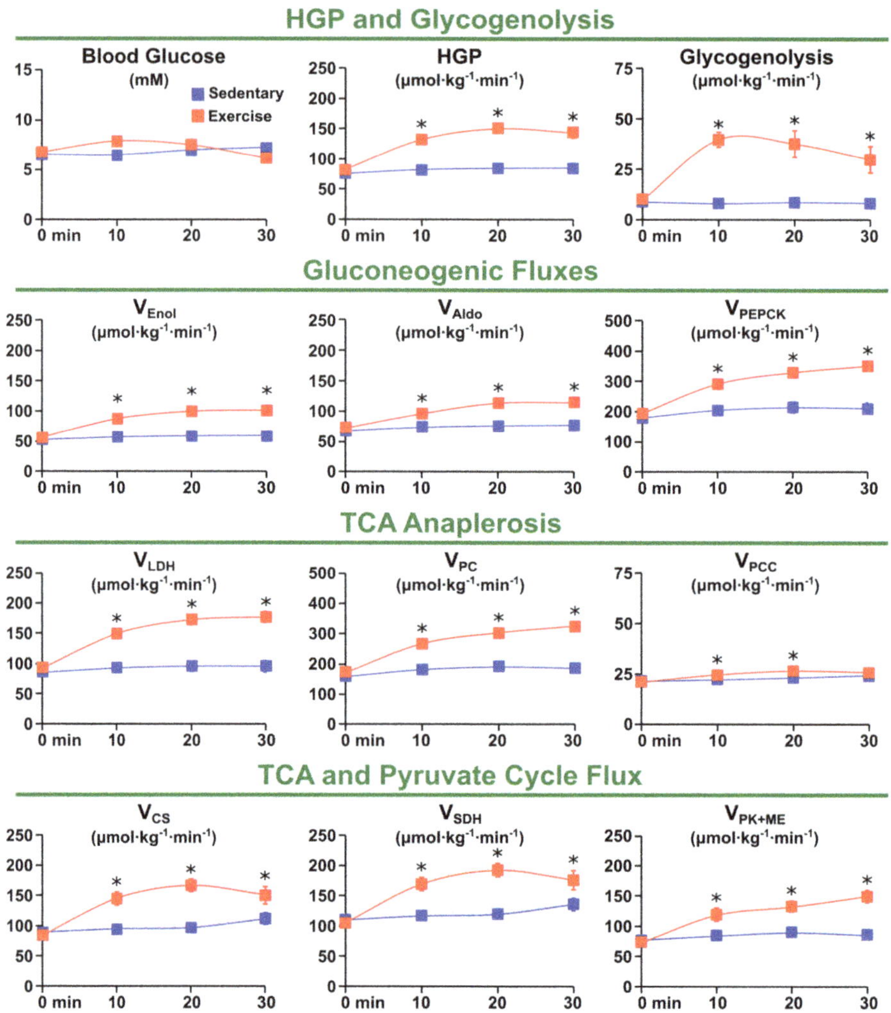

Fig. 11.1 Metabolic flux analysis during exercise in the mouse. Mice were infused with ^{13}C-propionate, $^{2}H_2O$, and 6.6-[^{2}H]glucose and mass isotopomer labeling in plasma glucose was determined by gas chromatography mass spectrometry. Fluxes were calculated by the best fit of regression analysis of mass balance equations using INCA software (Hasenour et al. 2015). HGP is hepatic glucose production; V_{Enol} is flux through enolase; V_{Aldo} is the flux through aldolase; V_{PEPCK} is flux though PEPCK; V_{LDH} is the flux from lactate to pyruvate; V_{PC} is flux through pyruvate carboxylase; V_{PCC} is flux through propionyl CoA carboxylase; V_{CS} is flux through citrate synthase; V_{SDH} is flux through succinate dehydrogenase; and V_{PK+ME} is combined flux through pyruvate kinase and malic enzyme and is indicative of pyruvate cycling. Data are mean +/– SE. * $p < 0.05$ compared to sedentary controls

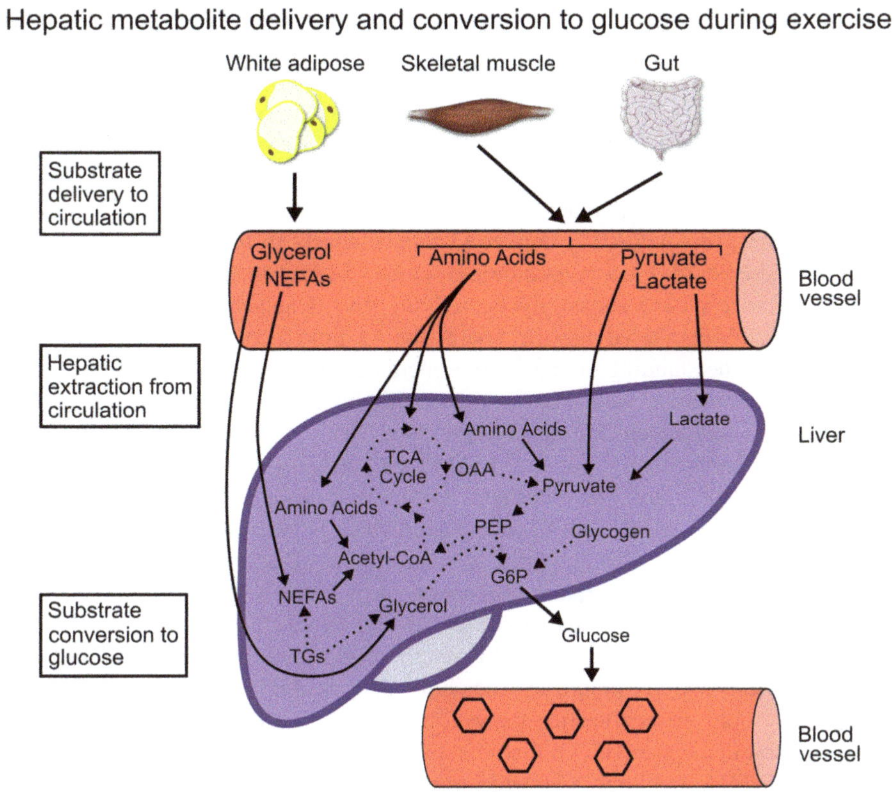

Fig. 11.2 Liver nutrient metabolism and gluconeogenesis are regulated by (1) substrate delivery to the liver by the circulation; (2) extraction of nutrients by the liver; and (3) activation of metabolic pathways that lead to formation of glucose in the liver. Modified from Trefts et al. (2015)

in the release of amino acids into the circulation. Alanine, glutamine, glutamate, serine, threonine, and valine are delivered to the liver where the carbons can be recycled into glucose. The scavenging of amino acid carbons in the liver requires the disposal of nitrogen. This is accomplished in large part by accelerated urea formation and release into the hepatic vein during exercise. Urea is filtered in the kidney and excreted in the urine.

Glycogenolysis represents a relatively low energy barrier to the release of glucose from the liver requiring 1 ATP per liberated glucose moiety. In contrast, gluconeogenesis requires six high-energy phosphate bonds (4 ATP, 2 GTP) per molecule of glucose produced (Blackman 1982). In addition to the direct energy cost of gluconeogenesis, the metabolic processes that support gluconeogenesis (e.g., ureagenesis, fatty acid activation) add to the energy demand. In this way increased energy demands of exercise extend to the energy requirement of liver. This is reflected by a 50% increase in liver O_2 uptake of dogs after 2 h of exercise (Wasserman et al. 1992). Liver uptake and output of a circulating factor cannot be measured directly in humans. However, a greater than twofold increase in splanchnic O_2 uptake has been shown to occur after 90 min of exercise in humans (Ahlborg et al. 1974). The energy demands of the liver are also evident from decreased adenosine triphosphate (ATP) and increased adenosine monophosphate (AMP), which compound to decrease the overall energy charge of the liver (Camacho et al. 2006; Berglund et al. 2009). Further studies have utilized 5'-aminoimidazole-4-carboxamide-1-beta-D-ribofuranoside (AICAR) to elucidate nucleotide-mediated mechanisms of metabolic control in the liver. AICAR is converted intracellularly to the AMP analog ZMP. Elevating hepatic ZMP concentrations to match exercise-mediated levels of AMP using AICAR result in a robust glycogenolytic drive (Camacho et al. 2005; Pencek et al. 2005). With altered hepatic energy being linked to a number of clinical pathologies, understanding the nucleotide-mediated mechanisms regulating hepatic metabolism in exercise can also offer therapeutic insights going forward.

AMP-activated protein kinase (AMPK) is a highly conserved regulator of cell metabolism (Garcia and Shaw 2017). AMPK transduces cellular energy status to downstream effects on macronutrient metabolism and mitochondrial function (Herzig and Shaw 2018). During exercise AMPK is activated in a manner that corresponds to decreased energy charge (Camacho et al. 2006). Given the energetically intensive nature of hepatic gluconeogenesis, AMPK has been posited as a critical regulator of this pathway. However, several recent studies have demonstrated that mice lacking expression of AMPK catalytic subunits in hepatocytes maintain gluconeogenesis and the supporting metabolic pathways (Hasenour et al. 2017). By applying isotopic metabolic flux analysis to mice with a deletion of both AMPK catalytic subunits, it was shown that this enzyme is not required for hepatic gluconeogenesis, TCA cycling, anaplerosis, and pyruvate cycling during exercise (Hughey et al. 2017). In contrast to hepatic gluconeogenesis, loss of liver AMPK impacts the capacity for glycogenolysis, the maintenance of hepatic energy charge, and the maintenance of oxidative phosphorylation during exercise (Hasenour et al. 2014; Hughey et al. 2017). It is possible that AMPK is required for the support of overall mitochondrial "tone" within the liver through turnover of dysfunctional mitochondria by autophagy/mitophagy and stimulation of mitochondrial biogenesis (Egan et al. 2011). Whether these roles of AMPK in mitochondrial regulation play a direct role in exercise-mediated metabolic adaptations is an active area of research.

11.2 Endocrine Regulation of the Liver During Exercise

Exercise presents a challenge to the homeostasis of multiple systems including those that regulate arterial pH, blood gases, core body temperature, and blood pressure, in addition to blood glucose. A broad neuro-endocrine response counters challenges to these homeostatic processes. The study of the processes that control liver glucose production have been challenging as the liver is primarily perfused by the portal vein, which is inaccessible in conscious humans (Fig. 11.3). On the other hand, rodent models are not suited to abdominal catheterizations or the blood sample

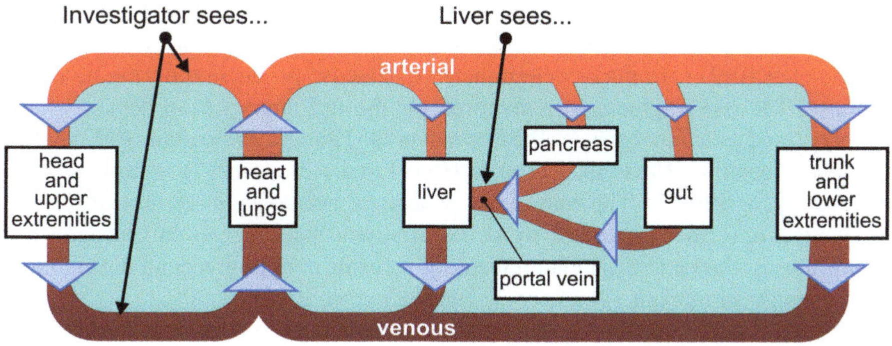

Fig. 11.3 Experiments in human subjects are unable to measure the glucagon and insulin levels at the liver. The pancreas is directly upstream of the liver and the liver extracts a high percentage of these hormones. The glucagon and insulin entering the vena cava from the hepatic vein have been dampened and delayed in relation to the concentrations that enter the liver (Wasserman et al. 1993; Coker et al. 1999b). Studies in exercising human subjects that have used arterial or peripheral blood samples to describe the content of blood perfusing the liver grossly mischaracterized the concentration of glucagon at the liver

volume needed for sampling from multiple sites. This leads to a reliance on metabolite and hormone levels measured in the peripheral circulation to draw conclusions on liver physiology in human subjects. As peripheral glucagon concentrations are not particularly responsive to exercise a dogma of glucagon being "unimportant" prevailed among a few clinical researchers. Since peripheral glucagon concentrations have little to do with glucagon concentrations that enter and perfuse the liver, large animal models have proven to be useful for gaining insights into liver function. In these models, the portal vein, which drains the pancreas and perfuses the liver, and the hepatic vein, which drains the liver, can be accessed using catheters implanted weeks in advance of a study. Glucagon and insulin released from the pancreas first perfuse the liver where extraction of these hormones dampens and delays their entry into the peripheral circulation. This effect is readily apparent during exercise as a robust decrease in portal vein insulin and an increase in glucagon occur, while changes in systemic blood concentrations are dampened and delayed due to hepatic extraction of these hormones (Wasserman et al. 1993, Coker et al. 1999b). This contributes to an approximately tenfold increase in the gradient of glucagon concentrations between the portal vein and arterial circulation during exercise (Wasserman et al. 1993; Coker et al. 1999b). The placement of the liver between the pancreas and general circulation is efficient as it allows for increased glucagon in the blood perfusing the liver without high glucagon secretion rates needed to rapidly fill the general circulation.

There is overwhelming evidence showing the exquisite sensitivity of the liver to glucagon in humans particularly in the presence of a small decline in insulin (Lins et al. 1983). Studies conducted in exercising dogs and humans defined the specific roles of insulin and glucagon in control of hepatic glucose output (Tuttle et al. 1988;

Lavoie et al. 1997). Studies have shown that the rise in glucagon (Wasserman et al. 1989c; Lavoie et al. 1997) and the fall in insulin (Wasserman et al. 1989a, 1991b; Lavoie et al. 1997) are major determinants of glucose production during moderate exercise. The rise in glucagon is required for the full increment in hepatic glycogenolysis and gluconeogenesis (Wasserman et al. 1989c; Lavoie et al. 1997), while the fall in insulin (Wasserman et al. 1989a; Lavoie et al. 1997) is necessary for hepatic glycogenolysis. The role of the pancreatic hormones in control of glucose production is further supported by demonstrations that changes in glucagon and insulin, using the pancreatic clamp technique, result in hypoglycemia during moderate exercise in humans (Wolfe et al. 1986; Kjaer et al. 1993).

The unanswered question is what causes the robust exercise-induced decrease in insulin secretion and increase in glucagon secretion (Wasserman 2009). Afferent nerves originating at the working limb, a deficit in fuel availability, and a neural feed-forward mechanism have all been postulated to mediate the pancreatic hormone response (Wasserman 2009). Afferent sensors in the carotid sinus area are required for the full increases in glucagon and norepinephrine during exercise (Koyama et al. 2001). Surprisingly, denervation of the pancreas does not impair the glucagon and insulin responses to exercise (Coker et al. 1999a). The myokine IL-6 is released in response to physical exercise (Pedersen and Febbraio 2012) and has been shown to stimulate glucagon release from the pancreatic alpha cell under stressful conditions (Barnes et al. 2014). The pancreas is highly sensitive to small changes in blood glucose during exercise (Wasserman et al. 1984, 1991b; Jenkins et al. 1985; Berger et al. 1994). It is possible that small changes in glucose that are undetectable provide feedback to the pancreas leading to changes in hormone secretion. This does not appear to be the case as preventing the increase in working muscle glucose uptake by deleting the glucose transporter, GLUT4, in mice does not prevent the glucagon and insulin responses and increase in hepatic glucose production (Fig. 11.4) (Fueger et al. 2007). The impediment in exercise-stimulated glucose uptake and the uninhibited increase in glucose production resulted in a marked hyperglycemia. This suggests that exercise-induced changes in glucagon and insulin not only do not require a small deficit in glycemia but occur in the presence of the inhibitory effects of hyperglycemia. This study suggests that the endocrine and hepatic responses to exercise are not due to a feedback signal from accelerated blood glucose removal by muscle as once proposed.

An increase in glucagon during exercise is considerably more potent than the equivalent increase under sedentary experimental conditions (Fig. 11.5). A twofold increase in glucagon causes a fivefold greater increase in glucose production during exercise than during rest (Wasserman 2009). Glucagon action is fully manifested during exercise because the increased glucose utilization of working muscle prevents hyperglycemia that would be expected to antagonize glucose production at the liver and accompanies an experimental increase in glucagon. In addition, as mentioned earlier prolonged exercise creates a physiological environment that supports gluconeogenesis by mobilizing gluconeogenic substrates from muscle, adipose, and intestine (Wasserman et al. 1992). Finally, exercise causes a fall in insulin that potently sensitizes the liver to glucagon (Wasserman 2009). Unger originally

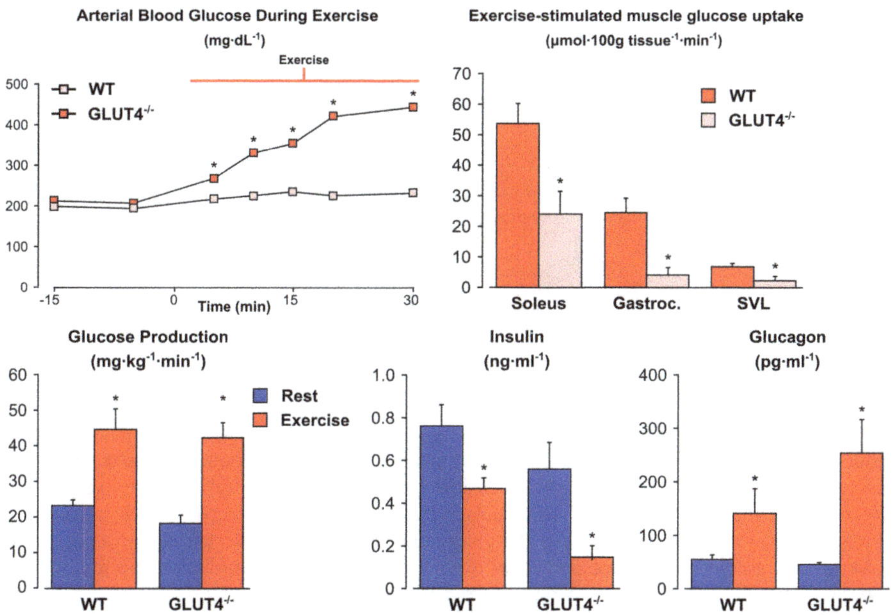

Fig. 11.4 Pancreatic and hepatic responses to exercise are not reliant on feedback resulting from the increase in muscle glucose uptake. Studies show the response to exercise in mice with genetic deletion of the GLUT4 transporter and wild-type littermates (Fueger et al. 2007). Mice with GLUT4 deletion exhibit a gradual increase in arterial glucose resulting from a dampened muscle glucose uptake in the presence of a normal exercise-induced increase in glucose production. Consistent with the increase in glucose production is the presence of the normal exercise-induced increase in glucagon and fall in insulin

proposed and demonstrated experimentally that it is the interaction of glucagon and insulin rather than the effect of either hormone individually that is most important to control glucose production (Unger and Orci 1976).

As described previously, increased gluconeogenesis requires an integrated response to metabolism. Figure 11.6 illustrates how a twofold increase in glucagon stimulates hepatic gluconeogenesis from alanine (Wasserman et al. 1989c), fat oxidation as reflected by ketone body production (Wasserman et al. 1989b), and glutamine amide nitrogen incorporation into urea (Krishna et al. 2000). In addition, glucagon plays a vital role in stimulating the hepatic extraction of gluconeogenic precursors during exercise (Wasserman et al. 1988, 1989b; Krishna et al. 2000). Glucagon stimulates the N (Nissim et al. 1999) and A (Cariappa and Kilberg 1992) amino acid transport systems activating transport of amino acids into the liver.

A robust increase in circulating catecholamine concentrations occurs during exercise. This led to the premise that norepinephrine and epinephrine participate in control of hepatic glucose production (Christensen and Galbo 1983). It is now clear that the increase in circulating catecholamines does not quantitatively translate to changes in concentrations at the liver. Hepatic norepinephrine spillover (reflecting sympathetic drive) is not increased during moderate exercise, and the portal vein

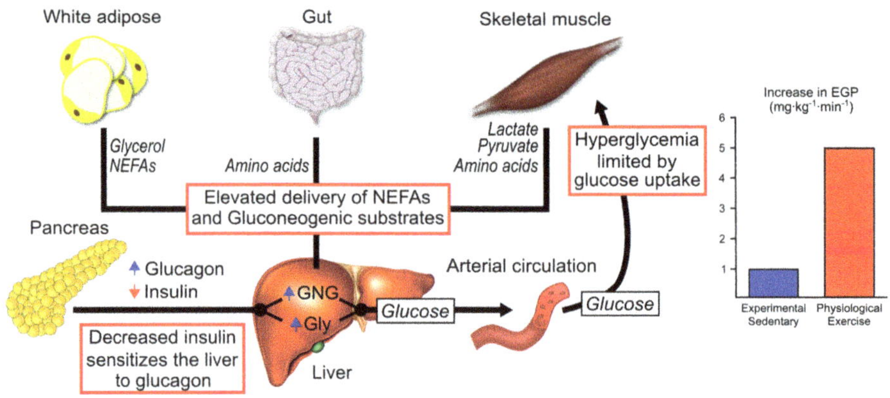

Fig. 11.5 A comparison of an experimental glucagon increase in the sedentary state to an exercise-induced glucagon increase of similar magnitude. Exercise creates conditions that are conducive to glucagon action at the liver. Glucagon is sensitized by a decrease in insulin during exercise. An experimental glucagon increase is associated with an increase in glucose which antagonizes glucagon action in the sedentary state. The glucose response is more effective due to an exercise-induced increase in delivery of substrates for gluconeogenesis. Data derived primarily from experiments in the conscious dog. Modified from Wasserman (2009)

epinephrine concentration at the liver is markedly attenuated as gastrointestinal tract extraction is 50% (Coker et al. 1997b). In line with this, a broad range of experimental approaches have failed to show an appreciable direct effect of either hepatic sympathetic nerves or circulating epinephrine in direct stimulation of hepatic glucose production during moderate exercise. There are instances such as high-intensity exercise where the adrenergic response is unusually high (Sigal et al. 2004). Circulating blood norepinephrine and epinephrine can increase by 10–20-fold (Marliss et al. 1992), whereas the increase in the glucagon to insulin ratio in *peripheral* blood is considerably less and in some cases undetectable (Kjaer et al. 1993). Moreover, when high-intensity exercise is performed during a pancreatic clamp (peripheral insulin and glucagon are fixed at basal), hepatic glucose output may still increase (Sigal et al. 1996). Despite these observations, no direct role of the catecholamines in control of hepatic glucose production has been demonstrated even during high-intensity exercise. Studies that have assessed the role for catecholamines during high-intensity exercise using adrenergic receptor blockade have uniformly been without an effect on hepatic glucose production (Marliss et al. 1992; Kjaer et al. 1993; Sigal et al. 1994). Such studies are difficult to interpret due to the lack of tissue specificity of these receptor blockers. Intraportal propranolol and phentolamine infusion has been used in a dog model to create selective hepatic adrenergic blockade (Coker et al. 1997a, 2000). This successfully causes hepatic adrenergic blockade without extrahepatic effects. Hepatic adrenergic blockade did not impair the increase in hepatic glucose production or affect glucose homeostasis during high-intensity exercise (Coker et al. 1997a). Similar results are seen in dogs treated with the β-cell toxin, alloxan (Coker et al. 2000). Alloxan-diabetic dogs in poor metabolic control

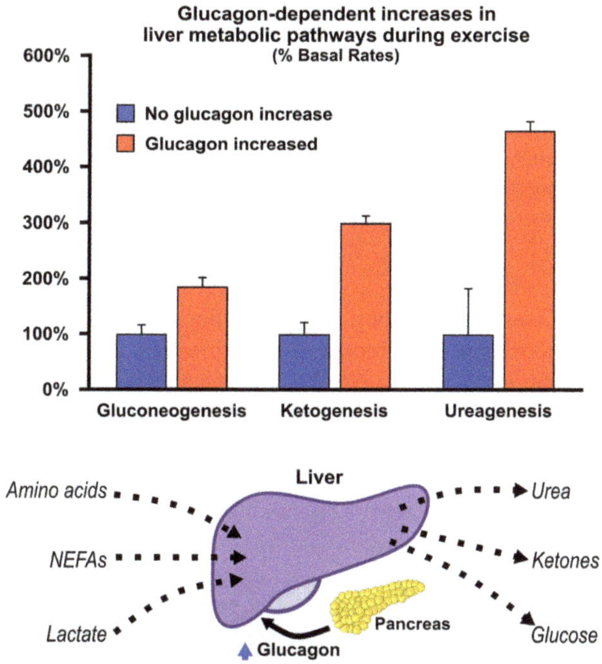

Fig. 11.6 The increase in glucagon orchestrates liver nutrient metabolism and production. The percent change in tracer-determined gluconeogenesis from alanine (Wasserman et al. 1989b), the net hepatic balance of beta-hydroxybutyrate and acetoacetate (Wasserman et al. 1989a), and the transfer of isotopic glutamine amide nitrogen to urea after 150 min of exercise (Krishna et al. 2000). Somatostatin was infused to suppress endogenous pancreatic hormone release and portal vein insulin replaced to simulate the insulin response to exercise and glucagon replaced to (a) fix at basal levels during exercise and (b) increased to simulate the response to exercise at resting levels. Experiments were performed in dogs with portal vein catheters implanted for hormone infusion and to obtain inflowing blood to the liver and hepatic vein catheters were implanted to obtain samples of outflowing blood. Data are mean +/– SE

have sevenfold higher rates of hepatic norepinephrine spillover than non-diabetic dogs during moderate exercise. Even then selective hepatic adrenergic receptor blockade did not attenuate hepatic glucose production during exercise (Coker et al. 2000). Thus, hepatic glucose production in humans and the dog model is not reliant on adrenergic receptor stimulation even during heavy exercise.

One can speculate that some myokines such as IL-6 (Febbraio et al. 2004), retinol binding protein 4 (Besse-Patin et al. 2014), apelin (Besse-Patin et al. 2014), or myonectin (Seldin et al. 2013) are required for the full actions of changes in glucagon and insulin on the exercise-induced increment in hepatic glucose production or during high-intensity exercise. At this point, there is little basis for this speculation. To date a convincing role for other hormones, hepatic nerves, or cytokines in direct control of glucose production during exercise have been demonstrated.

11.3 Liver Glycogen Repletion After Exercise

There is limited ability to measure liver glucose uptake in humans. Large animals have been used to study the liver following exercise as it is possible to access blood perfusing and draining the liver (Galassetti et al. 1999a). Prior glycogen-depleting exercise causes a twofold increase in hepatic glucose uptake in response to a twofold increase in hepatic glucose load and hyperinsulinemia load in the dog (Galassetti et al. 1999a) (Fig. 11.7). These data show that the improved glucose tolerance in the post-exercise state is due to adaptations at the liver as well as the muscle. Indirect assessments in the anesthetized rabbit support findings in the dog, showing that liver deposition of a glucose analog is increased after muscles are stimulated to contract electrically (Matsuhisa et al. 1998). Studies using magnetic resonance spectroscopy showed that ingestion of ^{13}C-glucose immediately after completion of prolonged moderate exercise in humans increased liver glycogen resynthesis by ~0.7 mg kg^{-1} min^{-1} over a period of 4 h of post-exercise recovery (Casey et al. 2000).

An increase in liver insulin sensitivity contributes to the increased ability of the liver to consume and store glucose after exercise. Insulin suppresses net hepatic glucose output to a greater extent following prolonged exercise in humans (Koyama et al. 2002). Moreover, the increase in net hepatic glucose uptake and fractional glucose extraction in the presence of a simulated meal (portal vein glucose infusion)

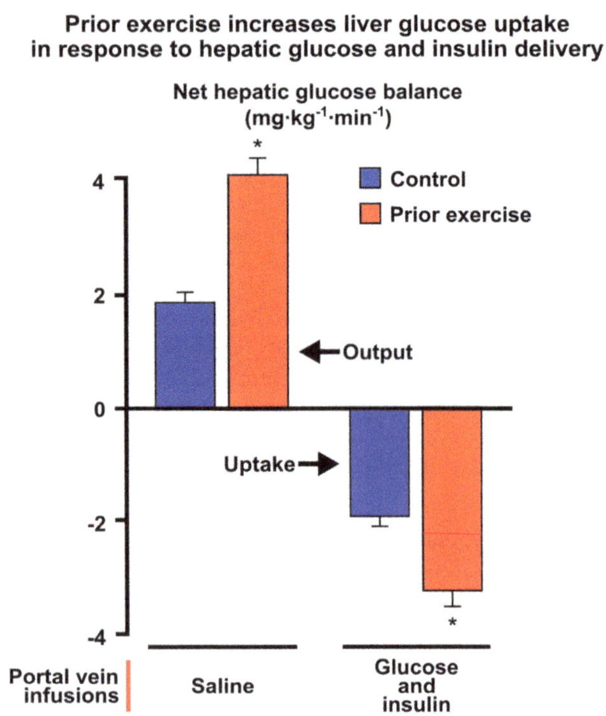

Fig. 11.7 Prior exercise increases the ability of the liver to consume glucose in response to a physiological increase in portal venous glucose and insulin concentrations of twofold and threefold, respectively, in the conscious dog (Galassetti et al. 1999a). Mean +/– SE. $*p < 0.05$

was ~50% greater in exercised compared to sedentary dogs in response to a physiological increment in insulin (Pencek et al. 2003a). Findings show that a greater fraction of the glucose taken up by the liver after exercise is metabolized non-oxidatively (Hamilton et al. 1996). The response to exercise is similar to the response to a prolonged, glycogen-depleting fast (Galassetti et al. 1999b). These findings were supported by subsequent studies in mice showing that exercise induced a rapid transcriptional effect in the liver, and increased expression of IRS proteins leading to improved cellular insulin signaling (Hoene et al. 2009). These studies provide yet another reason to recommend exercise for patients with pre-diabetes.

As described previously changes in glucagon and decrease in insulin are the major stimuli for the accelerated mobilization of glucose from the liver during exercise. The role of exercise-induced changes in glucagon and insulin during exercise to the adaptations of the liver after exercise was tested in a large animal model. Somatostatin was used to suppress endogenous glucagon and insulin and pancreatic hormones were either replaced (a) at basal rates or (b) exercise-simulated rates (Pencek et al. 2004). Preventing the glucagon and insulin responses to exercise prevented hepatic glucose output and glycogen breakdown during exercise as expected, while simulation of the pancreatic hormone response to exercise restored the increase in hepatic glucose production and glycogen breakdown. Despite the distinct responses to exercise by the liver, it is remarkable that hepatic glucose uptake is increased equally in response to a glucose load and hyperinsulinemia. The fate of the glucose consumed by the liver is where a difference exists. When pancreatic hormone responses were simulated a greater fraction of the glucose consumed by the liver was directed to glycogen (Pencek et al. 2004). These experiments showed that the replenishment of glycogen but not the increase in hepatic glucose uptake is reliant on the pancreatic hormone response during exercise.

Exercise also leads to a number of other endocrine changes. It is possible, for example, that the exercise-stimulated glucocorticoid response may prime the liver to take up more glucose since this hormone can stimulate hepatic glycogen deposition (Long et al. 1940). Myokines, hepatokines, and adipokines may also have implications for the post-exercise increase in liver insulin sensitivity and glycogen repletion. Adiponectin and irisin released from adipocytes (Bouassida et al. 2010) and myocytes (Arias-Loste et al. 2014), respectively, have insulin-sensitizing effects at the liver. It is notable that the effects of prior exercise cause persistent effects on processes and enzymes involved in liver glucose metabolism that are sustained well after the cessation of exercise (Dohm et al. 1985). Prior exercise also has effects on the intestinal tract that may facilitate glycogen repletion. Prior exercise increases intestinal absorption of intragastric or ingested glucose (Maehlum et al. 1978; Hamilton et al. 1996; Pencek et al. 2003b) by passive absorption (Pencek et al. 2003b).

11.4 Hepatic Adaptation to Exercise Training

Habitual exercise places chronic demand on the integrative metabolic functions of the liver. As such, the liver responds through a host of adaptations to enhance exercise capacity. Importantly, these hepatic adaptations can improve pathology-associated metabolic indices such as hepatic lipid content, thereby reducing susceptibility to nonalcoholic fatty liver disease (NAFLD) (Trefts et al. 2015; Thyfault and Rector 2020). Clinical studies of hepatic responses to exercise have established epidemiological associations between health outcomes, biochemical markers of liver function, and hepatic lipid content with physical activity (Perseghin et al. 2007; Caldwell and Lazo 2009). These studies demonstrate correlations of regular physical activity with decreased hepatic lipid content and decreased incidence of NAFLD (Church et al. 2006; Zelber-Sagi et al. 2008). However, a more complete understanding of the mechanistic underpinnings of these relationships is still needed.

As previously noted, studying hepatic metabolism and responses in humans is generally limited by access to the organ and its circulatory components. Risks associated with obtaining liver biopsy samples limit the opportunity for mechanistic insights to effects of exercise in humans. As such, a variety of animal models have proven valuable for defining hepatic adaptations to exercise and the implications of these adaptations to clinically relevant pathologies such as NAFLD. Among these models, the Otsuka Long-Evans Tokushima Fatty (OLETF) rat presents with characteristics that mirror clinical progression of the metabolic syndrome from peripheral insulin resistance to the loss of beta cell function (Kawano et al. 1992). Exercise training has been applied to the OLETF rat as a preventative and interventional strategy for metabolic dysfunction and progression of NAFLD relevant phenotypes. Regardless of modality, exercise training of OLETF rats improves overall metabolic health indicated by improved glucose tolerance and calculated insulin sensitivity. Modified indicators of hepatic lipid and glucose metabolism represent potential contributors to these shifts in overall metabolic health. Specifically, exercise-induced reductions in steatosis are likely tied to a decrease in lipid synthesis coupled with increased lipid oxidation (Rector et al. 2008, 2011; Borengasser et al. 2012; Linden et al. 2014). Decreased levels of lipogenic enzymes such as fatty acid synthase (FAS) and acetyl-CoA carboxylase (Acc) coupled with inhibitory phosphorylation of Acc likely combine to limit contributions of lipid synthesis to steatosis. Exercise-induced reduction of hepatic FAS protein was also observed in swim-trained C57Bl6/J mice fed either a standard chow or a high-fat diet (Schultz et al. 2012). The conservation of training-induced reductions in hepatic lipid synthesis enzymes across multiple species and dietary regimens reinforces this effect as a primary metabolic adaptation to exercise. Endurance exercise training also promotes hepatic lipid oxidation as indicated by increased oxidation of palmitate to CO_2 in the OLETF model and a decrease in incomplete palmitate oxidation in livers of Sprague-Dawley rats on a low-fat chow diet (Linden et al. 2014; Fletcher et al. 2014). Elevated mitochondrial content and/or function may underlie these exercise-mediated improvements in hepatic lipid oxidation. This is indicated by elevated hepatic mitochondrial respiratory function, cytochrome *c* oxidase levels, CPT-1 activity, β-HAD activity, citrate synthase activity, and TR4 protein in response to training (Rector et al. 2008; Haase et al. 2011; Borengasser et al. 2012; Lezi et al. 2013; Fletcher et al. 2014; Linden et al. 2015). Interestingly, elevated indicators of hepatic lipid oxidation have also been observed in rats bred for an intrinsically higher aerobic capacity (Morris et al. 2014). Models of exercise training and increased intrinsic aerobic capacity are linked by enhanced hepatic mitochondrial integrity during NAFLD progression (Gonçalves et al. 2014a, b; Morris et al. 2017). Therefore, the importance of mitochondrial quality control processes such as biogenesis, networking between mitochondria, and turnover may represent targets to begin harnessing the therapeutic efficacy of exercise in the liver.

Many training-mediated shifts in hepatic metabolism described occur in the presence of weight loss or prevention of weight gain during obesogenic stimuli. This makes it difficult to distinguish training-mediated from weight-mediated effects. Several clinical studies have addressed this issue and demonstrated a weight loss independent component of regular exercise that reduces hepatic lipid content (Shojaee-Moradie et al. 2007; Johnson et al. 2009; van der Heijden et al. 2010; Hallsworth et al. 2011; Sullivan et al. 2012). While the ability of exercise to reduce circulating triglycerides and free fatty acids appears to be dependent on weight loss (Tamura et al. 2005; Rector et al. 2008, 2011; Johnson et al. 2009; Marques et al. 2010; Hallsworth et al. 2011; Borengasser et al. 2012; Jenkins et al. 2012; Schultz et al. 2012; Linden et al. 2014), weight loss independent shifts from visceral to peripheral adipose depot fat storage have been reported (Johnson et al. 2009; van der Heijden et al. 2010). This may indicate a dynamic shift in hepatic lipid fluxes with decreased provision of lipids from visceral adipose depots directly to the portal vein that perfuses the liver. Training-induced decreases in hepatic lipid content are not accompanied by altered VLDL-TG or VLDL-ApoB100 secretion rates (Sullivan et al. 2012). This would predictably cause a redistribution of lipid to peripheral depots. Additionally, exercise training also seems to stimulate the use of lipid substrates in humans (Hallsworth et al. 2011; Fealy et al. 2012). Whether these increases in lipid metabolism are due, in part, to hepatic shifts in lipid oxidation, like those observed in rodent models, is yet to be determined. Exercise also induces weight loss independent improvements in hepatic insulin sensitivity, which has implications for other obesity-related pathologies including the metabolic syndrome and type 2 diabetes (Coker et al. 2009).

Acknowledgments The authors acknowledge grants DK050277, DK054902, and DK059637 from the National Institutes of Health.

References

Ahlborg G, Felig P (1982) Lactate and glucose exchange across the forearm, legs, and splanchnic bed during and after prolonged leg exercise. J Clin Invest 69:45–54

Ahlborg G, Felig P, Hagenfeldt L et al (1974) Substrate turnover during prolonged exercise in man. Splanchnic and leg metabolism of glucose, free fatty acids, and amino acids. J Clin Invest 53: 1080–1090
Barnes TM, Otero YF, Elliott AD et al (2014) Interleukin-6 amplifies glucagon secretion: coordinated control via the brain and pancreas. Am J Physiol Endocrinol Metab 307:E896–E905
Berger CM, Sharis PJ, Bracy DP et al (1994) Sensitivity of exercise-induced increase in hepatic glucose production to glucose supply and demand. Am J Physiol Endocrinol Metab 267: E411–E421
Berglund ED, Lee-young RS, Lustig DG et al (2009) Hepatic energy state is regulated by glucagon receptor signaling in mice. J Cin Investig 119:2412–2422
Besse-Patin A, Montastier E, Vinel C et al (2014) Effect of endurance training on skeletal muscle myokine expression in obese men: identification of apelin as a novel myokine. Int J Obes 38: 707–713
Blackman D (1982) The economics of gluconeogenesis. Biochem Educ 10:141
Borengasser SJ, Rector RS, Uptergrove GM et al (2012) Exercise and omega-3 polyunsaturated fatty acid supplementation for the treatment of hepatic steatosis in hyperphagic OLETF rats. J Nutr Metab 2012:268680
Bouassida A, Chamari K, Zaouali M et al (2010) Review on leptin and adiponectin responses and adaptations to acute and chronic exercise. Br J Sports Med 44:620–630
Caldwell S, Lazo M (2009) Is exercise an effective treatment for NASH? Knowns and unknowns. Ann Hepatol Off J Mex Assoc Hepatol 8 Suppl 1:S60–S66
Camacho RC, Pencek RR, Lacy DB et al (2005) Portal venous 5-aminoimidazole-4-carboxamide-1-beta-D-ribofuranoside infusion overcomes hyperinsulinemic suppression of endogenous glucose output. Diabetes 54:373–382
Camacho RC, Donahue EP, James FD et al (2006) Energy state of the liver during short-term and exhaustive exercise in C57BL/6J mice. Am J Physiol Endocrinol Metab 290:E405–E408
Cariappa R, Kilberg MS (1992) Plasma membrane domain localization and transcytosis of the glucagon-induced hepatic system A carrier. Am J Physiol Metab 263:E1021–E1028
Casey A, Mann R, Banister K et al (2000) Effect of carbohydrate ingestion on glycogen resynthesis in human liver and skeletal muscle, measured by 13C MRS. Am J Physiol Endocrinol Metab 278:65–75
Christensen NJ, Galbo H (1983) Sympathetic nervous activity during exercise. Annu Rev Physiol 45:139–153
Church TS, Kuk JL, Ross R et al (2006) Association of cardiorespiratory fitness, body mass index, and waist circumference to nonalcoholic fatty liver disease. Gastroenterology 130:2023–2030
Coker RH, Krishna MG, Brooks Lacy D et al (1997a) Role of hepatic α- and β-adrenergic receptor stimulation on hepatic glucose production during heavy exercise. Am J Physiol Endocrinol Metab 273:831–838
Coker RH, Krishna MG, Zinker BA et al (1997b) Sympathetic drive to liver and nonhepatic splanchnic tissue during prolonged exercise is increased in diabetes. Metabolism 46:1327–1332
Coker RH, Koyama Y, Lacy DB et al (1999a) Pancreatic innervation is not essential for exercise-induced changes in glucagon and insulin or glucose kinetics. Am J Physiol Endocrinol Metab 277:E1122–E1129
Coker RH, Lacy DB, Krishna MG, Wasserman DH (1999b) Splanchnic glucagon kinetics in exercising alloxan-diabetic dogs. J Appl Physiol 86:1626–1631
Coker RH, Lacy DB, Williams PE, Wasserman DH (2000) Hepatic α- and β-adrenergic receptors are not essential for the increase in R(a) during exercise in diabetes. Am J Physiol Endocrinol Metab 278:444–451
Coker RH, Williams RH, Yeo SE et al (2009) The impact of exercise training compared to caloric restriction on hepatic and peripheral insulin resistance in obesity. J Clin Endocrinol Metab 94: 4258–4266
Dohm GL, Kasperek GJ, Barakat HA (1985) Time course of changes in gluconeogenic enzyme activities during exercise and recovery. Am J Physiol 249:E6–E11

Egan DF, Shackelford DB, Mihaylova MM et al (2011) Phosphorylation of ULK1 (hATG1) by AMP-activated protein kinase connects energy sensing to mitophagy. Science 331:456–461

Eriksson LS, Broberg S, Björkman O, Wahren J (1985) Ammonia metabolism during exercise in man. Clin Physiol 5:325–336

Evans M, Cogan KE, Egan B (2017) Metabolism of ketone bodies during exercise and training: physiological basis for exogenous supplementation. J Physiol 595:2857–2871

Fealy CE, Haus JM, Solomon TPJ et al (2012) Short-term exercise reduces markers of hepatocyte apoptosis in nonalcoholic fatty liver disease. J Appl Physiol 113:1–6

Febbraio MA, Hiscock N, Sacchetti M et al (2004) Interleukin-6 is a novel factor mediating glucose homeostasis during skeletal muscle contraction. Diabetes 53:1643–1648

Felig P, Wahren J (1971) Amino acid metabolism in exercising man. J Clin Invest 50:2703–2714

Felig P, Cherif A, Minagawa A, Wahren J (1982) Hypoglycemia during prolonged exercise in normal men. N Engl J Med 306:895–900

Fletcher JA, Meers GM, Linden MA et al (2014) Impact of various exercise modalities on hepatic mitochondrial function. Med Sci Sports Exerc 46:1089–1097

Fueger PT, Li CY, Ayala JE et al (2007) Glucose kinetics and exercise tolerance in mice lacking the GLUT4 glucose transporter. J Physiol 582:801–812

Galassetti P, Coker RH, Lacy DB et al (1999a) Prior exercise increases net hepatic glucose uptake during a glucose load. Am J Physiol Endocrinol Metab 276:E1022–E1029

Galassetti P, Hamilton KS, Gibbons FK et al (1999b) Effect of fast duration on disposition of an intraduodenal glucose load in the conscious dog. Am J Physiol Endocrinol Metab 276:543–552

Garcia D, Shaw RJ (2017) AMPK: mechanisms of cellular energy sensing and restoration of metabolic balance. Mol Cell 66:789–800

Gonçalves IO, Maciel E, Passos E et al (2014a) Exercise alters liver mitochondria phospholipidomic profile and mitochondrial activity in non-alcoholic steatohepatitis. Int J Biochem Cell Biol 54:163–173

Gonçalves IO, Passos E, Rocha-Rodrigues S et al (2014b) Physical exercise prevents and mitigates non-alcoholic steatohepatitis-induced liver mitochondrial structural and bioenergetics impairments. Mitochondrion 15:40–51

Haase TN, Ringholm S, Leick L et al (2011) Role of PGC-1 in exercise and fasting-induced adaptations in mouse liver. AJP Regul Integr Comp Physiol 301:R1501–R1509

Hallsworth K, Fattakhova G, Hollingsworth KG et al (2011) Resistance exercise reduces liver fat and its mediators in non-alcoholic fatty liver disease independent of weight loss. Gut 60:1278–1283

Hamilton KS, Gibbons FK, Bracy DP et al (1996) Effect of prior exercise on the partitioning of an intestinal glucose load between splanchnic bed and skeletal muscle. J Clin Invest 98:125–135

Hargreaves M, Spriet LL (2020) Skeletal muscle energy metabolism during exercise. Nat Metab 2: 817–828

Hasenour CM, Ridley DE, Hughey CC et al (2014) 5-Aminoimidazole-4-carboxamide-1-β-D-ribofuranoside (AICAR) effect on glucose production, but not energy metabolism, is independent of hepatic AMPK in vivo. J Biol Chem 309(2):E191–E203

Hasenour CM, Wall ML, Ridley DE et al (2015) Mass spectrometry-based microassay of (2)H and (13)C plasma glucose labeling to quantify liver metabolic fluxes in vivo. Am J Physiol Endocrinol Metab 309:E191–E203

Hasenour CM, Ridley DE, James FD et al (2017) Liver AMP-activated protein kinase is unnecessary for gluconeogenesis but protects energy state during nutrient deprivation. PLoS One 12:1–18

Herzig S, Shaw RJ (2018) AMPK: guardian of metabolism and mitochondrial homeostasis. Nat Rev Mol Cell Biol 19:121–135

Hoene M, Lehmann R, Hennige AM et al (2009) Acute regulation of metabolic genes and insulin receptor substrates in the liver of mice by one single bout of treadmill exercise. J Physiol 587: 241–252

Hughey CC, James FD, Bracy DP et al (2017) Loss of hepatic AMP-activated protein kinase impedes the rate of glycogenolysis but not gluconeogenic fluxes in exercising mice. J Biol Chem 292:20125–20140
Jenkins AB, Chisholm DJ, James DE et al (1985) Exercise-induced hepatic glucose output is precisely sensitive to the rate of systemic glucose supply. Metabolism 34:431–436
Jenkins NT, Padilla J, Arce-Esquivel AA et al (2012) Effects of endurance exercise training, metformin, and their combination on adipose tissue leptin and IL-10 secretion in OLETF rats. J Appl Physiol 113(12):1873–1883
Johnson NA, Sachinwalla T, Walton DW et al (2009) Aerobic exercise training reduces hepatic and visceral lipids in obese individuals without weight loss. Hepatology 50:1105–1112
Kawano K, Hirashima T, Mori S et al (1992) Spontaneous long-term hyperglycemic rat with diabetic complications. Otsuka Long-Evans Tokushima Fatty (OLETF) strain. Diabetes 41: 1422–1428
Kjaer M, Engfred K, Fernandes A et al (1993) Regulation of hepatic glucose production during exercise in humans: role of sympathoadrenergic activity. Am J Physiol Endocrinol Metab 265: E275–E283
Koyama Y, Coker RH, Denny JC et al (2001) Role of carotid bodies in control of the neuroendocrine response to exercise. Am J Physiol Endocrinol Metab 281:E742–E748
Koyama Y, Galassetti P, Coker RH et al (2002) Prior exercise and the response to insulin-induced hypoglycemia in the dog. Am J Physiol Endocrinol Metab 282:1128–1138
Krishna MG, Coker RH, Brooks Lacy D et al (2000) Glucagon response to exercise is critical for accelerated hepatic glutamine metabolism and nitrogen disposal. Am J Physiol Endocrinol Metab 279:E638–E645
Lavoie C, Ducros F, Bourque J et al (1997) Glucose metabolism during exercise in man: the role of insulin and glucagon in the regulation of hepatic glucose production and gluconeogenesis. Can J Physiol Pharmacol 75:26–35
Lezi E, Lu J, Burns JM, Swerdlow RH (2013) Effect of exercise on mouse liver and brain bioenergetic infrastructures. Exp Physiol 98:207–219
Linden MA, Fletcher JA, Morris EM et al (2014) Combining metformin and aerobic exercise training in the treatment of type 2 diabetes and NAFLD in OLETF rats. Am J Physiol Endocrinol Metab 306:E300–E310
Linden MA, Fletcher JA, Morris EM et al (2015) Treating NAFLD in OLETF rats with vigorous-intensity interval exercise training. Med Sci Sports Exerc 47:556–567
Lins PE, Wajngot A, Adamson U et al (1983) Minimal increases in glucagon levels enhance glucose production in man with partial hypoinsulinemia. Diabetes 32:633–636
Long CNH, Katzin B, Fry EG (1940) The adrenal cortex and carbohydrate metabolism. Endocrinology 26:309–344
Maehlum S, Felig P, Wahren J (1978) Splanchnic glucose and muscle glycogen metabolism after glucose feeding during postexercise recovery. Am J Physiol Endocrinol Metab Gastrointest Physiol 4:E255
Marliss EB, Purdon C, Halter JB et al (1992) Glucoregulation during and after intense exercise in control and diabetic subjects. In: Diabetes mellitus and exercise. Smith-Gordon, London, pp 173–188
Marques CMM, Motta VF, Torres TS, Aguila MB (2010) Beneficial effects of exercise training (treadmill) on insulin resistance and nonalcoholic fatty liver disease in high-fat fed C57BL/6 mice. Med Biol 43:467–475
Matsuhisa M, Nishizawa H, Ikeda M et al (1998) Prior muscular contraction enhances disposal of glucose analog in the liver and muscle. Metabolism 47:44–49
Morris EM, Jackman MR, Johnson GC et al (2014) Intrinsic aerobic capacity impacts susceptibility to acute high-fat diet-induced hepatic steatosis. Am J Physiol Endocrinol Metab 307:E355–E364
Morris EM, McCoin CS, Allen JA et al (2017) Aerobic capacity mediates susceptibility for the transition from steatosis to steatohepatitis. J Physiol 595:4909–4926

Nissim I, Brosnan ME, Yudkoff M et al (1999) Studies of hepatic glutamine metabolism in the perfused rat liver with 15 N-labeled glutamine. J Biol Chem 274:28958–28965
Pedersen BK, Febbraio MA (2012) Muscles, exercise and obesity: skeletal muscle as a secretory organ. Nat Rev Endocrinol 8:457–465
Pencek RR, James F, Lacy DB et al (2003a) Interaction of insulin and prior exercise in control of hepatic metabolism of a glucose load. Diabetes 52:1897–1903
Pencek RR, Koyama Y, Lacy DB et al (2003b) Prior exercise enhances passive absorption of intraduodenal glucose. J Appl Physiol 95:1132–1138
Pencek RR, James FD, Lacy DB et al (2004) Exercise-induced changes in insulin and glucagon are not required for enhanced hepatic glucose uptake after exercise but influence the fate of glucose within the liver. Diabetes 53:3041–3047
Pencek RR, Shearer J, Camacho RC et al (2005) 5-Aminoimidazole-4-carboxamide-1-beta-D-ribofuranoside causes acute hepatic insulin resistance in vivo. Diabetes 54:355–360
Perseghin G, Lattuada G, De Cobelli F et al (2007) Habitual physical activity is associated with intrahepatic fat content in humans. Diabetes Care 30:683–688
Puchalska P, Crawford PA (2017) Multi-dimensional roles of ketone bodies in fuel metabolism, signaling, and therapeutics. Cell Metab 25:262–284
Rector RS, Thyfault JP, Morris RT et al (2008) Daily exercise increases hepatic fatty acid oxidation and prevents steatosis in Otsuka Long-Evans Tokushima Fatty rats. Am J Physiol Gastrointest Liver Physiol 294:G619–G626
Rector RS, Uptergrove GM, Morris EM et al (2011) Daily exercise vs. caloric restriction for prevention of nonalcoholic fatty liver disease in the OLETF rat model. Am J Physiol Gastrointest Liver Physiol 300:G874–G883
Romijn JA, Coyle EF, Sidossis LS et al (1993) Regulation of endogenous fat and carbohydrate metabolism in relation to exercise intensity and duration. Am J Physiol 265:E380–E391
Schultz A, Mendonca LS, Aguila MB, Mandarim-de-Lacerda CA (2012) Swimming training beneficial effects in a mice model of nonalcoholic fatty liver disease. Exp Toxicol Pathol 64: 273–282
Seldin MM, Lei X, Tan SY et al (2013) Skeletal muscle-derived myonectin activates the mammalian target of rapamycin (mTOR) pathway to suppress autophagy in liver. J Biol Chem 288: 36073–36082
Shojaee-Moradie F, Baynes KCR, Pentecost C et al (2007) Exercise training reduces fatty acid availability and improves the insulin sensitivity of glucose metabolism. Diabetologia 50:404–413
Sigal RJ, Fisher S, Marliss EB (1994) Catecholamines are the primary regulators of glucose production (Ra) during intense exercise: definition using the islet-cell clamp (IC). Diabetes 43:159A
Sigal RJ, Fisher S, Halter JB et al (1996) The roles of catecholamines in glucoregulation in intense exercise as defined by the islet cell clamp technique. Diabetes 45:148–156
Sigal RJ, Kenny GP, Wasserman DH, Castaneda-Sceppa C (2004) Physical activity/exercise and type 2 diabetes. Diabetes Care 27:2518–2539
Sullivan S, Kirk EP, Mittendorfer B et al (2012) Randomized trial of exercise effect on intrahepatic triglyceride content and lipid kinetics in nonalcoholic fatty liver disease. Hepatology 55:1738–1745
Tamura Y, Tanaka Y, Sato F et al (2005) Effects of diet and exercise on muscle and liver intracellular lipid contents and insulin sensitivity in type 2 diabetic patients. J Clin Endocrinol Metab 90:3191–3196
Teresa Arias-Loste M, Ranchal I, Romero-Gómez M, Crespo J (2014) Irisin, a link among fatty liver disease, physical inactivity and insulin resistance. Int J Mol Sci 15:23163–23178
Thomas LK, Ittmann M, Cooper C (1982) The role of leucine in ketogenesis in starved rats. Biochem J 204:399–403
Thyfault JP, Scott Rector R (2020) Exercise combats hepatic steatosis: potential mechanisms and clinical implications. Diabetes 69:517–524

Trefts E, Williams AS, Wasserman DH (2015) Exercise and the regulation of hepatic metabolism. Prog Mol Biol Transl Sci 135:203–225
Tuttle KR, Marker JC, Dalsky GP et al (1988) Glucagon, not insulin, may play a secondary role in defense against hypoglycemia during exercise. Am J Physiol Endocrinol Metab 254:713–719
Unger RH, Orci L (1976) Physiology and pathophysiology of glucagon. Physiol Rev 56:778–826
van der Heijden G-J, Wang ZJ, Chu ZD et al (2010) A 12-week aerobic exercise program reduces hepatic fat accumulation and insulin resistance in obese, Hispanic adolescents. Obesity (Silver Spring) 18:384–390
Van Loon LJC (2014) Is there a need for protein ingestion during exercise? Sport Med 44:105–111
Wahren J, Sato Y, Ostman J et al (1984) Turnover and splanchnic metabolism of free fatty acids and ketones in insulin-dependent diabetics at rest and in response to exercise. J Clin Invest 73:1367–1376
Wasserman DH (2009) Four grams of glucose. Am J Physiol Endocrinol Metab 296:11–21
Wasserman DH, Lickley HLA, Vranic M (1984) Interactions between glucagon and other counterregulatory hormones during normoglycemic and hypoglycemic exercise in dogs. J Clin Invest 74:1404–1413
Wasserman DH, Williams PE, Lacy DB et al (1988) Importance of intrahepatic mechanisms to gluconeogenesis from alanine during exercise and recovery. Am J Physiol Metab 254:E518–E525
Wasserman DH, Lacy DB, Goldstein RE et al (1989a) Exercise-induced fall in insulin and increase in fat metabolism during prolonged muscular work. Diabetes 38:484–490
Wasserman DH, Spalding JA, Bracy D et al (1989b) Exercise-induced rise in glucagon and ketogenesis during prolonged muscular work. Diabetes 38:799–807
Wasserman DH, Spalding JA, Lacy DB et al (1989c) Glucagon is a primary controller of hepatic glycogenolysis and gluconeogenesis during muscular work. Am J Physiol Endocrinol Metab 257:E108–E117
Wasserman DH, Geer RJ, Williams PE et al (1991a) Interaction of gut and liver in nitrogen metabolism during exercise. Metabolism 40:307–314
Wasserman DH, Lacy DB, Colburn CA et al (1991b) Efficiency of compensation for absence of fall in insulin during exercise. Am J Physiol Endocrinol Metab 261:E811–E824
Wasserman DH, Lacy DB, Bracy D, Williams PE (1992) Metabolic regulation in peripheral tissues and transition to increased gluconeogenic mode during prolonged exercise. Am J Physiol Endocrinol Metab 263:345–354
Wasserman DH, Lacy DB, Bracy DP (1993) Relationship between arterial and portal vein immunoreactive glucagon during exercise. J Appl Physiol 75:724–729
Williams BD, Wolfe RR, Bracy DP, Wasserman DH (1996) Gut proteolysis contributes essential amino acids during exercise. Am J Physiol 270:E85–E90
Wolfe RR, Nadel ER, Shaw JHF, Stephenson LA (1986) Role of changes in insulin and glucagon in glucose homeostasis in exercise. J Clin Invest 77:900–907
Zelber-Sagi S, Nitzan-Kaluski D, Goldsmith R et al (2008) Role of leisure-time physical activity in nonalcoholic fatty liver disease: a population-based study. Hepatology 48:1791–1798

Chapter 12
Influence of Exercise on Cardiac Metabolism and Resilience

Andrew A. Gibb, Kyle Fulghum, Bradford G. Hill, John Quindry, and Gary D. Lopaschuk

Abstract The numerous health benefits that derive from exercise are associated with remarkable changes in metabolism. With respect to the heart, exercise acutely increases cardiac energy demand, which, in concert with alterations in circulating hormones and energy substrates, augments cardiac ATP production by increasing the utilization of several substrates. However, reliance of the heart on different energy substrates varies as a product of nutritional state as well as the type, intensity, and duration of exercise. Chronic aerobic and resistance exercise training is associated with eccentric or concentric cardiac growth, metabolic adaptations such as mitochondrial biogenesis, and enhanced hormonal signaling, which collectively support a physiological state of adaptation. The exercise-adapted state is also associated with significant protection of the heart from a range of biological stressors, including myocardial ischemia-reperfusion injury. The cardioprotected state of exercise is due in part to upregulation of antioxidant enzymes, preservation of mitochondrial integrity, and metabolic resilience. Further understanding of how exercise-mediated changes in metabolism orchestrate cardiac adaptation and protection could facilitate therapeutic strategies to maximize the benefits of exercise and improve cardiac health.

Keywords Exercise · Heart · Mitochondria · Cell signaling · Cardioprotection

---

A. A. Gibb (✉)
Center for Translational Medicine, Lewis Katz School of Medicine, Temple University, Philadelphia, PA, USA
e-mail: andrew.gibb@temple.edu

K. Fulghum · B. G. Hill
Division of Environmental Medicine, Department of Medicine, Christina Lee Brown Envirome Institute, Diabetes and Obesity Center, University of Louisville, Louisville, KY, USA

J. Quindry
School of Integrative Physiology and Athletic Training, University of Montana, Missoula, MT, USA

G. D. Lopaschuk
Cardiovascular Research Centre, Department of Pediatrics, University of Alberta, Edmonton, AB, Canada

© The Author(s), under exclusive license to Springer Nature Switzerland AG 2022
G. McConell (ed.), *Exercise Metabolism*, Physiology in Health and Disease,
https://doi.org/10.1007/978-3-030-94305-9_12

12.1 Introduction

Although physical inactivity is a health risk, it is a modifiable one. The first documented study of the beneficial effects of an active lifestyle was in 1953. This study showed that those in more physically active jobs had a lower incidence of cardiovascular events and a lower early mortality rate than their sedentary counterparts (Morris et al. 1953). Along with this observation, the benefits of physical activity have only become more noticeable over the years. Regular exercise has been shown to promote cardiovascular health (Blair et al. 1996; Joyner and Green 2009; Mora et al. 2007), augment skeletal muscle function (Egan and Zierath 2013), and increase both health span (Egan and Zierath 2013; de Cabo et al. 2014; Mercken et al. 2012) and life span (Blair et al. 1989, 1996; Myers et al. 2002; Paffenbarger Jr et al. 1986). Physical fitness, commonly measured by exercise capacity, is a more powerful predictor of mortality than other established cardiovascular risk factors such as hypertension, blood lipids, and smoking (Myers et al. 2002). Exercise-induced changes in the heart and vasculature are in part responsible for improvements in health and in resilience to a range of stressors (Nystoriak and Bhatnagar 2018). Similar to its effects on skeletal muscle (Egan and Zierath 2013), exercise alters metabolism in the heart and promotes structural and functional adaptations. The adapted state following training is also associated with significant protection of the heart from myocardial ischemia (Andrews Portes et al. 2009; Barboza et al. 2013), pressure overload (Xu et al. 2015), and diabetes (Stolen et al. 2009). In this chapter, we discuss how exercise influences the heart, with particular focus on how exercise influences cardiac metabolism and resilience.

12.2 Integrated Cardiac Responses to Exercise

12.2.1 Acute Changes in Cardiac Physiology

Acute changes in cardiac physiology during exercise coordinate with alterations in several peripheral tissues. During exercise, increases in cardiac output, along with physiological adjustments in both ventilation and vascular resistance, help to sustain skeletal muscle work (Heinonen et al. 2014). Changes in cardiac function occur immediately and commence with an increase in heart rate and stroke volume (Weiner and Baggish 2012; No Authors 1967); however, after prolonged moderate- to high-intensity aerobic exercise (e.g., >20 min), heart rate tends to increase further and stroke volume begins to drop. This phenomenon, commonly termed cardiovascular drift, helps to maintain cardiac output (Rowell et al. 1969; Rowell 1986; Coyle and Gonzalez-Alonso 2001; Rowland 2008). These coordinated changes, along with changes in vascular function, integrate to increase blood flow to skeletal muscle, with cardiac output distribution to working muscle tracking with exercise intensity (Plowman and Smith 2017; Fulghum and Hill 2018).

Resistance exercise typically increases blood pressure, due in part to mechanical restriction of blood flow during contraction. Compared with aerobic exercise, these features of resistance exercise cause markedly different cardiac responses. For example, the modest increase in cardiac output during resistance exercise is predominantly due to higher heart rate, with little change in stroke volume (Clausen 1976; Bezucha et al. 1982). Heart rate tends to increase with the number of repetitions, leading to larger increases in cardiac output (Hill and Butler 1991). During heavy weightlifting, the heart must also work against a much higher afterload because spikes in blood pressure can transiently reach levels of 320/250 mmHg (MacDougall et al. 1985) or higher. Overall, the degree to which blood pressure increases during resistance exercise seems to be a function of physical effort, muscle mass, and the Valsalva maneuver breathing pattern performed during strength training (Hill and Butler 1991; MacDougall et al. 1992).

12.2.2 Cardiac Energy Metabolism During Exercise

Exercise increases cardiac contractile power and oxygen consumption up to tenfold above resting rates (Lopaschuk et al. 2010; Olver et al. 2015; Lombardo et al. 1953). This requires a rapid increase in energy substrate utilization and ATP production. Even under resting conditions, the heart must produce high amounts of ATP to sustain contractile function. Mitochondrial oxidative phosphorylation provides approximately 95% of the heart's ATP requirements, with the remainder of the ATP production originating from glycolysis. Even under resting conditions, mitochondria use a variety of energy substrates to produce ATP, which include fatty acids, pyruvate originating from glucose and lactate, ketone bodies, and amino acids (Fig. 12.1). The increase in contractile power observed with exercise is accompanied by a dramatic increase in ATP production from both mitochondrial oxidative metabolism and glycolysis (Oram et al. 1973). Mitochondrial oxidative metabolism of all energy substrates may also increase, although the relative contribution of the various energy substrates to ATP production can vary quite dramatically (Fig. 12.1). It is also possible and important to consider that although some substrates may increase in their uptake during exercise, they are not destined for oxidation and their percent contribution to total ATP production may change. As detailed below, this regulation of cardiac metabolism is a product of changes in circulating energy substrates, circulating hormones, and the molecular changes that control energy pathways.

Energy Substrate Availability With respect to metabolic substrate availability, acute exercise increases the circulating levels of multiple substrates (Fig. 12.2) (Gertz et al. 1988; Kaijser and Berglund 1992; Kemppainen et al. 2002; Lassers et al. 1971; Goodwin and Taegtmeyer 2000; Goodwin et al. 1998a). Increases in circulating free fatty acids (FFA) during exercise are caused primarily by activation of lipolysis in adipose tissue, which can increase blood FFA concentrations up to 2.4 mM (Rodahl et al. 1964), thereby increasing their uptake and utilization (Lassers

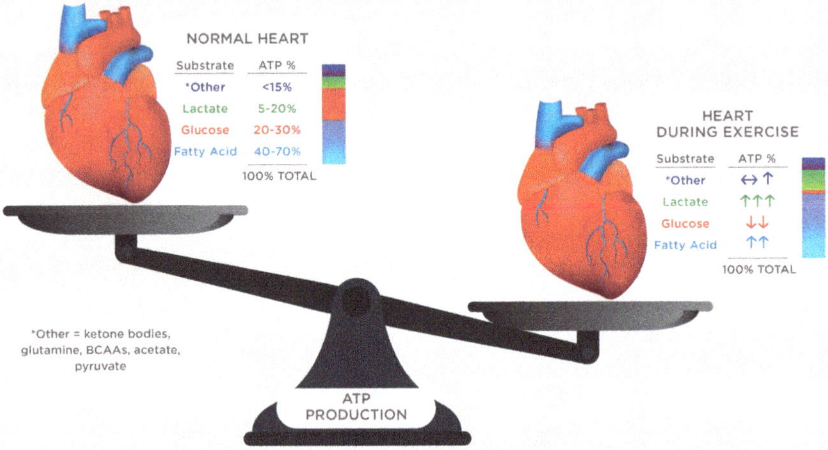

Fig. 12.1 Substrate contributions to cardiac ATP production during rest and exercise. Mitochondrial oxidative phosphorylation is responsible for meeting ~95% of the heart's ATP demand. Under normal resting conditions, the heart primarily oxidizes fatty acids and glucose to meet this energetic demand, with the remaining contributions derived from lactate, pyruvate, ketone bodies, branched chain amino acids (BCAAs), and other anaplerotic amino acids. During exercise, the heart's demand for ATP can increase tenfold, which is dependent upon the mode, duration, and intensity of the exercise bout. This results in dramatic remodeling of substrate use for energy provisions, specifically with an increase in lactate and fatty acid oxidation and an apparent decrease in glucose oxidation

et al. 1972a, b; Wisneski et al. 1987). The increase in fatty acid oxidation rates seen in the heart during exercise seem to primarily occur due to the increase in circulating fatty acid levels and to mass-action effects of fatty acid oxidation accelerating due to decreases in downstream fatty acid intermediates (Oram et al. 1973). Malonyl CoA, a potent inhibitor of carnitine palmitoyltransferase 1, is an important regulator of fatty acid oxidation in the heart (Lopaschuk et al. 2010). In skeletal muscle, a decrease in malonyl CoA, due to an exercise-induced activation of AMP-activated protein kinase (AMPK, which decreases malonyl CoA due to phosphorylation and inhibition of acetyl CoA decarboxylase), can increase fatty acid oxidation rates (Kuhl et al. 2006; Saha et al. 1995). However, while AMPK is activated in the heart during exercise (Musi et al. 2005), this is not accompanied by a decrease in malonyl CoA levels (Goodwin and Taegtmeyer 2000). As a result, it is unlikely that the increase in cardiac fatty acid oxidation rates seen during exercise is due to alterations in malonyl CoA control of mitochondrial fatty acid uptake. However, higher circulating FFAs appear to only be partly responsible for increasing fatty acid oxidation because higher cardiac workloads are sufficient to increase fatty acid oxidation in the heart (Oram et al. 1973; Bergman et al. 2009a). Cardiac triacylglycerol (TAG) utilization also increases with exercise (Lassers et al. 1971) and is further stimulated by lactate availability (de Groot et al. 1993), highlighting the importance of endogenous fatty acid stores of the heart. Interestingly, it seems

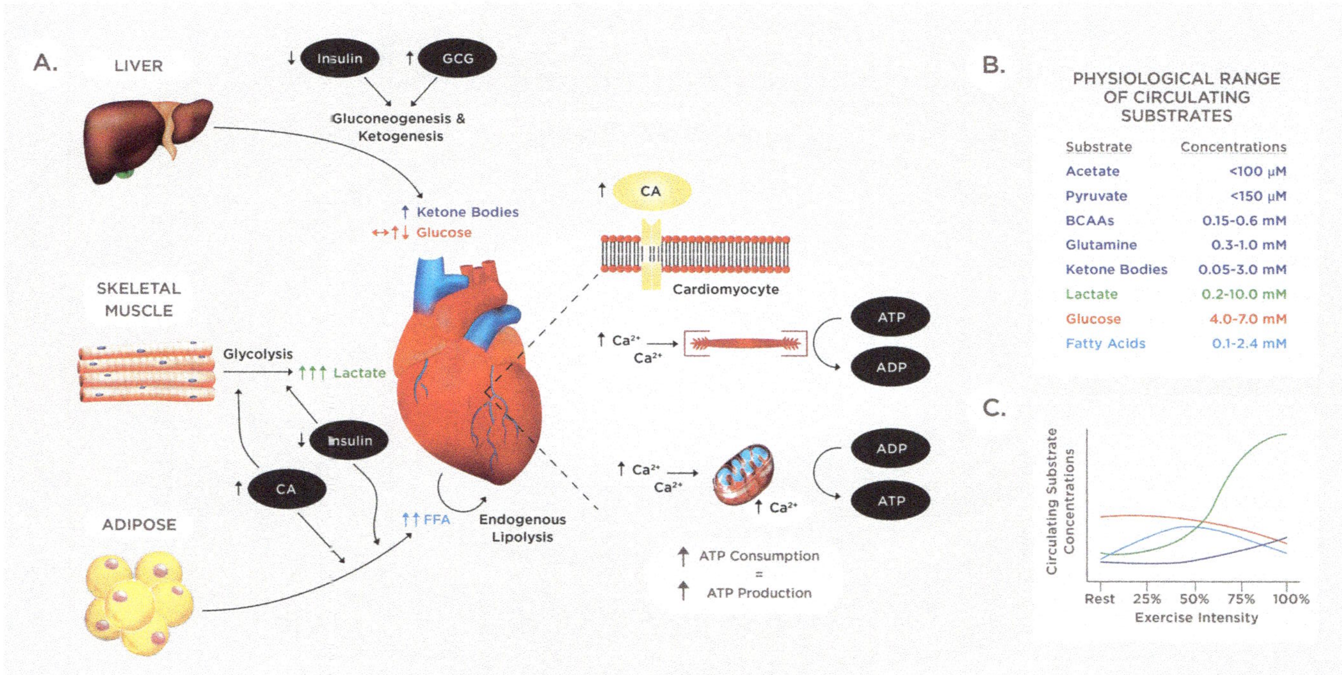

Fig. 12.2 Hormonal control of energy substrate availability and utilization during exercise. Acute exercise liberates substrates from peripheral and endogenous stores for cardiac energy provisions. (**a**) Adipose tissue and endogenous cardiac TAG store lipolysis increases the bioavailability of FFAs for uptake and oxidation. Increases in exercising skeletal muscle produce substantial amounts of lactate that is extruded and made available for uptake and utilization by the heart. Circulating blood glucose levels and gluconeogenesis seem to diminish during exercise, while the activation of liver ketogenesis may increase circulating ketone bodies under certain conditions (e.g., reactive hypoglycemia, extreme strenuous exercise). Changes in substrate availability are mediated in part by circulating hormones which not only mediate peripheral tissue metabolism and subsequent change in circulating substrates, but also increase cardiac contractility, increasing ATP demand and thus mitochondrial ATP production. These coordinated systemic metabolic changes control cardiac substrate availability and contribute to the regulation of substrate utilization in response to exercise. These substrates can be in μM–mM concentrations under these

that although cardiac TAG lipolysis increases with exercise, the synthesis of TAGs also increases, which leads to a relatively rapid recovery of TAGs after an exercise bout (Podbielski and Palmer 1989).

Although most studies suggest that circulating glucose levels are relatively stable during exercise, some studies show that glucose levels decrease (Rodahl et al. 1964; Coyle 2000) or increase during exercise (Fig. 12.2) (Kemppainen et al. 2002). Such changes may depend on several factors including the intensity, duration, and mode of exercise and the current nutritional state of the individual. Although increases in cardiac workload and moderate exercise are associated with elevations in myocardial glucose uptake and oxidation (Gertz et al. 1988; Kemppainen et al. 2002; Goodwin and Taegtmeyer 2000; Goodwin et al. 1998a), elevations in competing substrates (e.g., lactate, FFAs, ketone bodies) brought upon by exercise may in fact decrease glucose catabolism (Kemppainen et al. 2002; Lassers et al. 1971; Goodwin and Taegtmeyer 2000; Takala et al. 1983). Indeed, studies in humans and animal models suggest that exercise lowers oxygen extraction ratios for glucose and decreases myocardial glucose utilization (Lassers et al. 1971; Takala et al. 1983). Moreover, recent findings suggest that relatively intense exercise decreases the activity of phosphofructokinase, which could diminish glucose catabolism temporarily and may be caused in part by elevations in competing substrates (Gibb et al. 2017a; Brookes and Taegtmeyer 2017). A transient decrease in glycolysis during intense exercise appears important for exercise-induced cardiac growth because low phosphofructokinase activity caused by expression of a cardiac-specific, kinase-deficient 6-phosphofructokinase/fructose-2,6-bisphosphatase transgene (Glyco^{Lo} mice) partially phenocopies the exercise-adapted heart and regulates genes (e.g., *Cebpb*, *Cited4* (Bezzerides et al. 2016; Bostrom et al. 2010)) required for myocardial growth (Gibb et al. 2017a). Moreover, the increase in cardiac glycogen levels after intense exercise (Gibb et al. 2017a; Oliveira et al. 2018; Riehle et al. 2014; Segel et al. 1975) seems to be caused by higher levels of circulating FFA, lower cardiac phosphofructokinase activity, and higher glycogen synthase activity (Gibb et al. 2017a; Conlee et al. 1981; Conlee and Tipton 1977; Garcia et al. 2009; Segel and Mason 1978).

In mice, exercise promotes acute fission of cardiac mitochondria, the process by which one mitochondrion separates into two, in a manner dependent on adrenergic signaling, which appears to enhance mitochondrial function (Coronado et al. 2017). Periodic mitochondrial fission may be important for regulating glucose and lipid metabolism (Salabei and Hill 2013; Buck et al. 2016) as well as mitochondrial quality control. For the latter, mitophagy appears to be increased in the heart during and shortly after a bout of exercise (Ogura et al. 2011; Lee et al. 2016) and could cull defective mitochondria via autophagy (Shirihai et al. 2015; Vasquez-Trincado et al. 2016; Hill et al. 2012), a process important for the enhanced oxidative metabolism

Fig. 12.2 (continued) physiological conditions (**b**), the concentrations of which are directly influenced by exercise intensity (**c**). BCAAs, branched chain amino acids; CA, catecholamines; Ca^{2+}, calcium; FFA, free fatty acids; GCG, glucagon

observed in the trained heart. Collectively, these findings suggest that exercise-induced changes in glycolytic rate and mitochondrial metabolism may regulate the cardiac growth program, changes that are likely dependent on the type, intensity, and duration of exercise.

Circulating lactate levels can increase robustly with intense exercise, which is significant because lactate extraction by the heart is dependent on arterial lactate concentration (Gertz et al. 1980). Specifically, intense exercise (e.g., 60–80% of VO_2max) promotes large increases in arterial lactate (Stanley 1991), which can increase five- to tenfold (to >10 mM) during the exercise bout. This increase in lactate is due primarily to lactate extrusion by working skeletal muscle (Fig. 12.2), although several other tissues can influence lactate levels (Jang et al. 2019; Hui et al. 2020). During exercise, lactate oxidation could account for up to 60–90% of substrate utilization (Kaijser and Berglund 1992; Bertrand et al. 1977; Drake et al. 1980; Keul 1971). Although exercise at lower intensities, e.g., 40% of VO_2 max, does not increase circulating lactate remarkably (Gertz et al. 1988), the contribution of lactate oxidation to myocardial oxidative metabolism appears higher than that occurring in the sedentary state (Gertz et al. 1988). Interestingly, why the myocardium has been shown itself to release lactate, net extraction still occurs (Gertz et al. 1981). Of note, lactate may enhance fatty acid oxidation in the heart (Goodwin and Taegtmeyer 2000), which further increases ATP generation under high workloads.

Plasma metabolomic profiling studies show that other critical substrates such as ketone bodies as well as many TCA cycle metabolites are influenced by exercise (Lewis et al. 2010; Morville et al. 2020). Ketone bodies such as β-hydroxybutyrate and acetoacetate are important sources of ATP production for the heart; in the presence of high ketone body concentrations, ketone body oxidation can become the major fuel for the heart (Ho et al. 2021). A critical determinant of cardiac ketone body oxidation rates is their circulating levels (Ho et al. 2019, 2021). Blood ketone body levels can increase with exercise, but less so in trained athletes or in individuals who eat a high-carbohydrate meal before exercise (Koeslag et al. 1980). As ketogenesis occurs in conditions of low glucose, it is possible that ketone body production and utilization may occur following exercise exhaustion or in athletes experiencing reactive hypoglycemia. The most significant exercise-induced rises in circulating ketone body levels appear to occur shortly after a bout of exercise, with 400% higher levels of β-hydroxybutyrate reported to occur 1 h after exercise in humans (Walsh et al. 1998). While some studies suggest that exercise training prevents post-exercise ketosis, other studies indicate that exercise training augments ketone body levels more in trained subjects than in untrained subjects (Koeslag 1982). Recent studies also indicate that β-hydroxybutyrate levels increase more after endurance exercise than after resistance exercise (Morville et al. 2020). Therefore, whether the contribution of cardiac ketone oxidation to ATP production increases during or early after exercise may depend on nutritional state, the level of physical fitness, and the type of exercise, with endurance exercise influencing ketone body utilization more than resistance exercise.

Hormonal Regulation of Cardiac Metabolism with Exercise In addition to circulating substrates, cardiac metabolism is regulated by hormonal changes (Fig. 12.2). For example, moderate acute exercise increases plasma catecholamine concentrations two- to fourfold (Kreisman et al. 2000, 2001), whereas intense exercise can increase plasma epinephrine and norepinephrine levels by 14- to 18-fold (Kjaer et al. 1986; Manzon et al. 1998; Marliss et al. 1991). A primary function of catecholamines in the heart is to increase contractility: catecholamines enhance cytosolic calcium (Ca^{2+}), which is required for excitation-contraction coupling. This increase in cytosolic Ca^{2+} also increases mitochondrial Ca^{2+} uptake through the mitochondrial Ca^{2+} uniporter. Heightened levels of intramitochondrial Ca^{2+} enhance oxidative phosphorylation via the activation of Ca^{2+}-dependent dehydrogenases, which helps match higher energetic demands with increased contractility (Glancy and Balaban 2012; Kwong et al. 2015; Luongo et al. 2015). Exercise-induced increases in catecholamines also stimulate adipose tissue lipolysis (Arner et al. 1990a; Hellstrom et al. 1996), a response mitigated by local adipose tissue β-adrenergic blockade (Arner et al. 1990b; Bulow 1981), and may also augment skeletal muscle lactate production (Febbraio et al. 1998; Qvisth et al. 2008), thereby increasing circulating fatty acid and lactate levels. In the isolated heart, catecholamines increase glucose uptake, glycogenolysis, glycolysis, and glucose oxidation, increasing their contributions to ATP production (Collins-Nakai et al. 1994; Goodwin et al. 1998b); however, the in vivo response may differ because of the elevated levels of competing substrates available to the heart during intense exercise (e.g. FFA, lactate).

Plasma insulin is perhaps the most potent regulator of systemic metabolism, as its primary action is to increase glucose uptake through the activation of Akt2, which promotes Glut4 translocation to the membrane (DeBosch and Muslin 2008). During exercise, insulin levels decrease modestly (Kreisman et al. 2001; Kjaer et al. 1986), a response that may prevent hypoglycemia as muscle glucose uptake positively correlates with exercise intensity (Kjaer et al. 1986; Manzon et al. 1998; Marliss et al. 1991) (see Fig. 12.2). The corresponding decrease in insulin receptor activation may diminish cardiac uptake of glucose acutely; however, increased contraction, such as occurs with exercise, has been shown to be sufficient to promote Glut4 translocation to the cardiomyocyte plasma membrane (Kolter et al. 1992; Till et al. 1997), which may involve both the AMPK- and PKC-mediated signaling (Luiken et al. 2008, 2015). It is also possible that lower levels of insulin contribute to the switch to utilization of fatty acids during exercise. Because insulin is a known inhibitor of adipose tissue lipolysis (Degerman et al. 1998; Duncan et al. 2007), exercise-induced decreases in insulin may prevent inhibition of lipolysis, thereby increasing blood fatty acid levels.

Insulin-like growth factor 1 (IGF1) and its receptor (IGF1r) are well known to regulate cardiac adaptation to exercise. During acute exercise, the anterior pituitary gland promotes release of growth hormone (GH), which increases IGF1 production in the liver (Schwarz et al. 1996; Sutton and Lazarus 1976). Additionally, exercise increases local IGF1 production in several target tissues, including the heart, thereby acting in an autocrine or paracrine fashion (Neri Serneri et al. 2001). Although IGF1 may activate AKT1 rather than AKT2 to control cardiac growth signaling (DeBosch et al. 2006a, b), it nevertheless increases the amplitude of cytosolic Ca^{2+} transients and the peak free Ca^{2+} concentration (Freestone et al. 1996; Stromer et al. 1996) leading to an increase in mitochondrial Ca^{2+} uptake and mitochondrial respiration (Troncoso et al. 2012). Cardiac glucose metabolism also seems affected by IGF1 signaling because glucose uptake and PFK2 phosphorylation are enhanced in response to IGF1 (Pentassuglia et al. 2016; Pozuelo Rubio et al. 2003). As IGF1 is a critical growth factor, it is likely that the metabolic effects of IGF1 develop post-exercise once liver IGF1 production has occurred and tissue protein synthesis takes place.

Acute exercise also stimulates the release of a number of other hormones. This includes stimulation of the renin–angiotensin–aldosterone axis, resulting in increased circulating renin, aldosterone, angiotensin II, and ACTH (Luger et al. 1988; Maeda et al. 2005). Increases in the satiety hormones, peptide tyrosine, glucagon-like peptide 1 (GLP-1), and pancreatic polypeptide are also seen with exercise (although no change in leptin levels is observed) (Hulver and Houmard 2003), as well as a decrease in the oxerigenic hormone, acylated ghrelin (Dorling et al. 2018). Moreover, exercise may also affect levels of thyroid hormones (Caralis et al. 1977; Ciloglu et al. 2005; Deligiannis et al. 1993) and glucagon (Galbo et al. 1975; Gyntelberg et al. 1977). Glucagon is particularly interesting because, along with insulin, it helps to maintain plasma glucose homeostasis (Trefts et al. 2015; Wolfe et al. 1986) and can influence the heart via hormonal signaling (Charron and Vuguin 2015; Petersen et al. 2018) and modulate cardiac function and metabolism (Gonzalez-Munoz et al. 2008; Harney and Rodgers 2008; Karwi et al. 2019). Glucagon also stimulates ketogenesis in diabetics (Liljenquist et al. 1974); however, whether this occurs during exercise when insulin levels are modestly decreased is unknown. Combined, these hormonal changes can contribute to an increase in circulating fatty acids and an insulin-resistant effect at the muscle level (Dorling et al. 2018; Ogihara et al. 2002; Taniyama et al. 2005). The many hormonal changes seen during acute exercise appear to contribute to a rise in circulating fatty acids and lactate, resulting in an increase in their use by the heart.

Direct Myocardial Control of Energy Metabolism During Exercise Increases in contractile activity in the heart are accompanied by increases in glycolysis, mitochondrial TCA cycle activity, and oxidative phosphorylation (Oram et al. 1973). Mitochondrial oxidative metabolism of all energy substrates also increases with cardiac work, although the relative proportion of individual energy substrates can change. Acute increases in contractile activity of the heart (such as seen with exercise) result in a preferential increase in ATP production from glycolysis and glucose oxidation (Goodwin et al. 1998a; Collins-Nakai et al. 1994; Williamson 1964). During exercise or atrial pacing, cardiac glucose uptake and glycolysis are increased, accompanied by an increase in glucose oxidation and an increase in lactate release from the heart (Gertz et al. 1988; Bergman et al. 2009b; Stanley et al. 1988). The rise in glucose oxidation is probably the result of an increase in mitochondrial Ca^{2+}, which occurs during exercise (Dawn et al. 2004; Huang et al. 2007, 2009), resulting in an activation of pyruvate dehydrogenase (PDH) (Collins-

Nakai et al. 1994), the rate-limiting enzyme for glucose oxidation. A rise in cytoplasmic Ca^{2+} may also contribute to the increase in glycolytic rates seen during increased cardiac work (Collins-Nakai et al. 1994).

Despite the increased preference of the heart for glycolysis and glucose oxidation during increases in cardiac work, a rise in circulating fatty acids and lactate can switch cardiac energy substrate preference toward fatty acid oxidation and lactate oxidation during acute exercise. In rat hearts exposed to the high fatty acids and lactate levels seen during exercise, the relative contribution of lactate oxidation to O_2 consumption at high workload increases from 4 to 21%, while fatty acid oxidation increases from 35 to 44% of O_2 consumption (Goodwin and Taegtmeyer 2000). An increase in cardiac work by atrial pacing human subjects also results in an increase in cardiac fatty acid uptake (Bergman et al. 2009a). An increase in myocardial triacylglycerol turnover is also observed, further increasing the relative contribution of fatty acid oxidation to ATP production (Goodwin and Taegtmeyer 2000; Saddik and Lopaschuk 1991). In addition to the increase in circulating levels of lactate seen during exercise, the activation of cardiac PDH due to increased mitochondrial Ca^{2+} levels could also contribute to increases in the contribution of lactate-derived pyruvate for mitochondrial oxidative metabolism.

12.2.3 Chronic Effects of Exercise on the Heart

Exercise can promote mild cardiac hypertrophy or chamber enlargement (Weiner and Baggish 2012; DeMaria et al. 1978; Maillet et al. 2013), which is reversible upon prolonged exercise cessation (Maron et al. 1993; Olah et al. 2017; Waring et al. 2015) (Fig. 12.3a). Exercise intensity appears to determine the degree of physiological cardiac remodeling, with exercise-induced hemodynamic changes providing a stimulus for cardiac adaptation. Isometric or static exercises such as weightlifting result in little change in cardiac output and are associated with mild concentric hypertrophy and a normal to mildly enlarged left atrium. This increase in cardiac wall thickness is caused by the parallel addition of sarcomeres within cardiomyocytes. Prolonged isotonic or dynamic aerobic exercise, such as long-distance running or swimming, requires sustained elevations in cardiac output and promotes eccentric left ventricular hypertrophy, right ventricular dilation, and biatrial enlargement (Weiner and Baggish 2012; Fulghum and Hill 2018; Morganroth et al. 1975; Spence et al. 2011; Mihl et al. 2008). This form of cardiac hypertrophy is associated with addition of cardiomyocyte sarcomeres in series. Endurance training-induced cardiac remodeling has been suggested to be phasic in nature, with one study suggesting initial concentric LV hypertrophy giving way to later eccentric LV hypertrophy (Arbab-Zadeh et al. 2014) and another suggesting increases in chamber size followed later by increases in wall thickness (Weiner et al. 2015).

Regular intense endurance exercise typically decreases resting and submaximal heart rates (DeMaria et al. 1978; Utomi et al. 2013); however, the effects of exercise

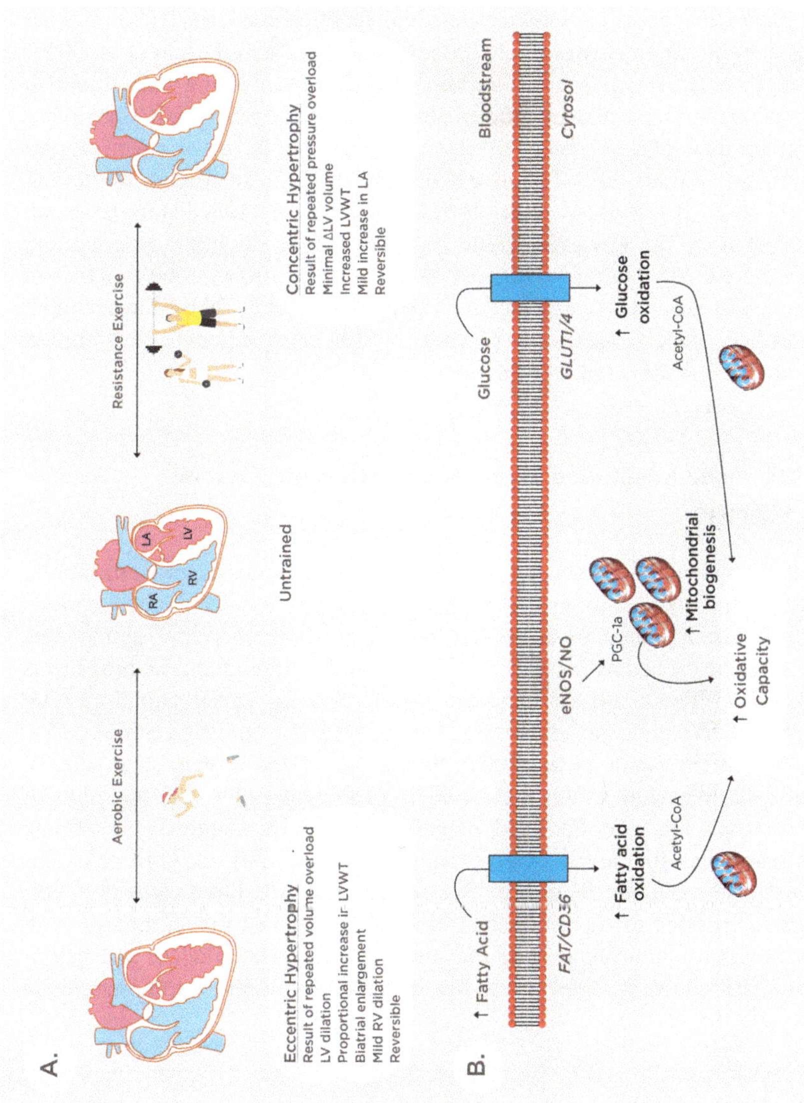

Fig. 12.3 Structural and metabolic cardiac adaptations to exercise training. Exercise-induced remodeling responses in the heart are dependent upon the intensity, duration, and mode of exercise training. **(a)** Prolonged aerobic exercise leads to series addition of sarcomeres that manifests as an eccentric hypertrophic response with proportional increases in LV chamber dimension and LVWT. Resistance training leads to a parallel addition of sarcomeres that manifests as a concentric hypertrophic response with minimal changes in LV chamber dimension, but increased LVWT. **(b)** Exercise-induced metabolic

on other indices of cardiac function such as systolic and diastolic heart function are less noticeable. For example, a meta-analysis of athletes participating in endurance, strength, or combined dynamic and static sports suggests little change in systolic or diastolic function compared with control subjects (Pluim et al. 2000). Nevertheless, several studies identified changes in diastolic function in exercise-adapted subjects. Endurance exercise appears to enhance diastolic function modestly (Utomi et al. 2013; Baggish et al. 2010; Naylor et al. 2005; Caso et al. 2000; Prasad et al. 2007; D'Andrea et al. 2010; Baggish et al. 2008a), and strength training may diminish diastolic function, as suggested by studies showing impairment of left ventricular relaxation in football players (Baggish et al. 2008a). In the rested state, regular exercise does not remarkably change ejection fraction or fractional shortening (Utomi et al. 2013; Bar-Shlomo et al. 1982; Bekaert et al. 1981; Douglas et al. 1986; Gilbert et al. 1977); however, subtle changes captured by tissue Doppler and speckle-tracking echocardiography suggest modestly enhanced systolic function in subjects adapted to exercise (Baggish et al. 2008b; Weiner et al. 2010; Simsek et al. 2013). In addition, exercise appears to promote modest cardiomyocyte proliferation (Waring et al. 2014; Vujic et al. 2018).

12.2.3.1 Metabolic Adaptation of the Heart to Chronic Exercise Regimens

A large body of evidence suggests that chronic exercise promotes adaptive metabolic remodeling in the myocardium (Fig. 12.3b). Chronic exercise-induced cardiac hypertrophy is accompanied by a modest increase in mitochondrial abundance, which appears to be caused by increased eNOS-dependent mitochondrial biogenesis (Vettor et al. 2014). This could result in an increased capacity for mitochondrial ATP synthesis by oxidative phosphorylation (Beer et al. 2008; Bruning and Sturek 2015; Gibb and Hill 2018; Kemi et al. 2007; Seo et al. 2016; Stolen et al. 2003). Nevertheless, the extent to which metabolism changes and even the direction of the changes remain unclear. Perfused mouse heart studies suggest that chronic exercise increases baseline levels of glycolysis (Gibb et al. 2017a), glucose oxidation, and fatty acid oxidation (Riehle et al. 2014). In contrast, basal cardiac glycolysis has been suggested to be diminished in exercise-adapted rats compared with hearts from sedentary controls, despite increases in myocardial glucose and palmitate oxidation (Burelle et al. 2004). As a result, while studies are inconsistent as to

---

Fig. 12.3 (continued) remodeling in the heart includes mitochondrial biogenesis and overall increases in oxidative metabolism (e.g., glucose and FFA oxidation). eNOS, endothelial nitric oxide synthase; FAT/CD36, fatty-acid translocase; GLUT1/4, glucose transporter 1/4; LA, left atrium; LV, left ventricle; LVWT, left ventricular wall thickness; NO, nitric oxide; PGC-1α, peroxisome proliferator-activated receptor gamma coactivator 1-alpha; RA, right atrium; RV, right ventricle

whether glycolysis increases or decreases with chronic exercise, they are consistent in showing an increase in glucose oxidation. This may be explained in part by the increased mitochondrial oxidative capacity following chronic exercise training. In support of the increase in fatty acid oxidation, upon exercise adaptation, genes responsible for fatty acid transport and catabolism are elevated, which integrate to amplify cardiac fatty acid utilization (Strom et al. 2005; Bonen et al. 2000; Jeppesen et al. 2009). These metabolic changes may depend on exercise intensity because a moderate-intensity treadmill regimen showed no effect on myocardial glucose oxidation, palmitate oxidation, or oxygen consumption, whereas a high-intensity, interval style regimen increased glucose oxidation, diminished palmitate oxidation, and decreased resting myocardial oxygen consumption (Hafstad et al. 2011). The reasons for discrepancies are not known but may include the use of different rodent strains or types of exercise, or differences in cardiac perfusion protocols.

Because rate-limiting steps of glycolysis have been suggested to modulate biosynthetic pathways (Gibb and Hill 2018), it is likely that periodic changes in phosphofructokinase activity occurring during exercise regimens may be important for modulating cardiac anabolism and growth. The phosphofructokinase step of glycolysis could be particularly important for regulating the pentose phosphate pathway (PPP), which is important for nucleotide synthesis and redox regulation (Yamamoto et al. 2014; Yi et al. 2012; Boada et al. 2000; Blackmore and Shuman 1982). Indeed, modeling studies in the adult heart demonstrate that phosphofructokinase activity is a key regulator of both the PPP and the polyol pathway (Cortassa et al. 2015). In isolated cardiac myocytes, phosphofructokinase activity also modulates the hexosamine biosynthetic pathway (HBP) and the glycerophospholipid synthesis pathway (GLP), potentially by modulating glycolytic intermediate availability and by indirectly regulating mitochondria-derived molecules important for building block synthesis (e.g., aspartate) (Gibb et al. 2017b). Metabolomic studies indicate that phosphofructokinase activity also influences several amino acid and lipid metabolites in the heart (Gibb et al. 2017a). Although it remains unclear how exercise affects hexokinase and pyruvate kinase in the heart, pyruvate kinase activity has been shown to be elevated in both the hearts of exercise-adapted rats (York et al. 1975) and dogs (Stuewe et al. 2000).

Role for Metabolite Signaling in Cardiac Adaptation Metabolite signaling could also promote cardiac adaptation to exercise. Because glucose-derived metabolites regulate prohypertrophic kinases, such as mammalian target of rapamycin (mTOR) and AMPK (Maillet et al. 2013), it remains possible that exercise-induced changes in glucose 6-phosphate (G6P) and 5-amino-4-imidazolecarboxamide ribonucleotide (AICAR) (Hurlimann et al. 2011) could regulate mTOR and AMPK activity in the heart (Sen et al. 2013; Kundu et al. 2015; Roberts et al. 2014; Sullivan et al. 1994). Predictions from crossover theorem (Chance and Williams 1955; Chance et al. 1955; Chance and Williams 1956; Heinrich and Rapoport 1974) and modeling studies (Cortassa et al. 2015) insinuate that acute decreases in phosphofructokinase activity, such as occurs during exercise (Gibb and Hill 2018), may increase G6P, which may not only activate mTOR but could augment PPP activity and nucleotide biosynthesis, an intermediate of which is AICAR. In addition, recent studies suggest that phosphoglucose isomerase inhibition could also regulate G6P (Karlstaedt et al. 2020).

Circulating metabolites may also play a hormonal role in exercise-induced cardiac growth. Hormone-mediated lipolysis during exercise liberates palmitoleate (C16:1n7), which could stimulate cardiac growth by a receptor-mediated mechanism (Foryst-Ludwig et al. 2015). This palmitoleate-induced hypertrophic effect is equivalent to the fatty acid-induced cardiac hypertrophy observed in pythons (Riquelme et al. 2011). It is interesting that FFAs such as palmitoleate increase not only with acute exercise (Rodahl et al. 1964), but also remain elevated in the exercise-adapted state (Papadopoulos et al. 1969; Monleon et al. 2014), which could indicate that they are lasting metabolic signals that sustain cardiac adaptations. Because numerous metabolites stimulate GPCRs (Husted et al. 2017), it is possible that other metabolites elevated by exercise play important roles in cardiac adaptation. Indeed, recent studies indicate that endurance and resistance exercise differentially influence the levels of numerous circulating metabolites with cognate GPCRs, which could influence tissue adaptation via receptor-mediated signaling (Morville et al. 2020).

12.2.4 Exercise-Induced Cardioprotection: A Clinical Application of Metabolic Adaptations in the Exercised Heart

Among the most dynamic features of metabolic control in the exercised heart is its innate ability to become protected against a host of disease stressors. Clinically relevant physiologic stressors include bioenergetic supply-demand mismatch which occurs during ischemia-reperfusion injury, manifesting as exertional angina and myocardial infarction. The metabolic and anatomical adaptations highlighted earlier in this chapter underpin the metabolic resilience afforded to the exercised heart during ischemic challenges. The term *cardioprotection* is born from both scientific and clinical paradigms and includes attributes which cannot be separated from the bioenergetic processes described within this chapter. Accordingly, cardioprotection more broadly encompasses four key facets: (1) beneficial modification of cardiovascular disease (CVD) risk factors, (2) anatomical modification of ventricular architecture, (3) revascularization of the myocardium, and (4) biochemical upregulation (or allosteric activation) of endogenous mediators of biochemical protection (Quindry 2017; Thijssen et al. 2018). The first three cardioprotective facets represent chronic adaptive responses to exercise training and will be discussed later in this chapter. In contrast, the latter factor, biochemical cardioprotection, reflects an acute benefit of short-term (days to weeks) exposure to exercise, termed "*exercise preconditioning*," and directly relates to acute metabolic alterations central to this chapter.

To understand the role of metabolic control and exercise preconditioning against CVD, one must first consider the experimental approaches which underpin this area of research. Specifically, exercise preconditioning research studies are commonly performed in rat and mouse exercise models that are designed to parallel human exercise prescriptions in terms of the frequency, intensity, time, and type of exercise. Summative descriptions of the experimental exercise training and ischemia reperfusion surgical protocols are provided in Fig. 12.4. To this end, exercise prescriptions are scaled to the animal model in order to parallel human exercise as defined in the American College of Sports Medicine 10th edition of Guidelines for Exercise Testing and Prescription (Riebe et al. 2018). Similarly, animal exercise is then followed by experimental approaches to produce CVD, typically surgically induced ischemia reperfusion in vivo, or global ischemia applied ex vivo, to mimic myocardial infarction in clinical settings. Scientific validity of preclinical exercise preconditioning studies depends upon the clinical relevance of both the infarct challenge and the outcome measures of cardiac pathology which include electrocardiographic (ECG), measures of ventricular pump function (e.g., transthoracic echocardiography), and postmortem markers of tissue death. In aggregate, these types of studies continue to be recognized for the translative potential from animals to humans (Quindry 2017; Thijssen et al. 2018).

Descriptive evidence from exercise preconditioning from the last 30 years indicates that exercise equally protects both males and females against myocardial infarction (Brown et al. 2005a; Chicco et al. 2007). Perhaps of equal importance, given that CVD manifests as a function of advancing age, exercise preconditions the hearts of both young and old animals (Quindry et al. 2005; Starnes et al. 2003). The dose of exercise needed to precondition the heart in all of the aforementioned studies is as little as 1–3 days of moderate-intensity exercise, of any modality (e.g., treadmill running and swimming) performed for at least 30 min (Brown et al. 2005a; Demirel et al. 2001). It is important to recognize that the exercise-preconditioned phenotype is highly reproducible, and represents a threshold-dependent phenomenon, in that both moderate- and high-intensity exercise are equally protective against experimental infarction (Starnes et al. 2003; Lennon et al. 2004a; Miller et al. 2015; Quindry et al. 2012). Moreover, the infarct-sparing effects of a 3-day regimen of moderate-intensity exercise last for at least 9 days following the last bout of structured physical activity (Lennon et al. 2004b).

Translating from animals to humans, these findings are important because they suggest that the "dose" of exercise needed to evoke cardioprotection is relatively low compared to maximal aerobic capacity. Stated differently, it appears that one does not need to be highly exercise trained to have a cardioprotected heart. Indeed, the short time course (days) needed to obtain clinically relevant protection underscores the notion that primary mechanisms of exercise cardiac preconditioning are endogenous biochemical mediators, upregulated independent of cardiac remodeling. Relative to the topic of exercise and myocardial metabolism, this broad conclusion about exercise dose indicates that the acute metabolic demands placed on the heart during bouts of physical activity elicit metabolic alterations (described previously),

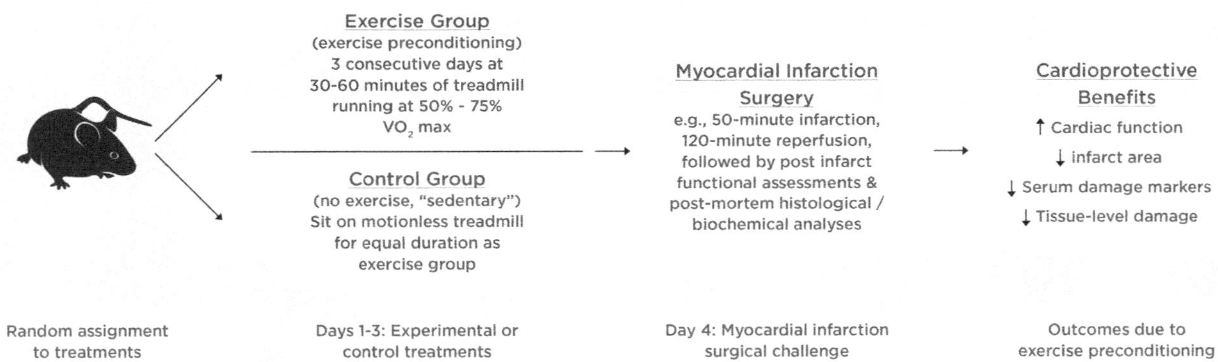

Fig. 12.4 Study design for exercise and cardioprotection research. Anesthetized animals (e.g., rodents) are assigned to either the exercise cardioprotection or sedentary groups. Exercise duration, intensity, and modality parallel human exercise prescriptions. Following exercise or sedentary treatments, typically on study day 4, animals receive a clinically relevant myocardial infarction surgery. Outcome measures include assessment of cardiac function and postmortem assessment of cardioprotective mediators. Exercise-induced cardioprotection confers a protected phenotype against a sustained infarct challenge

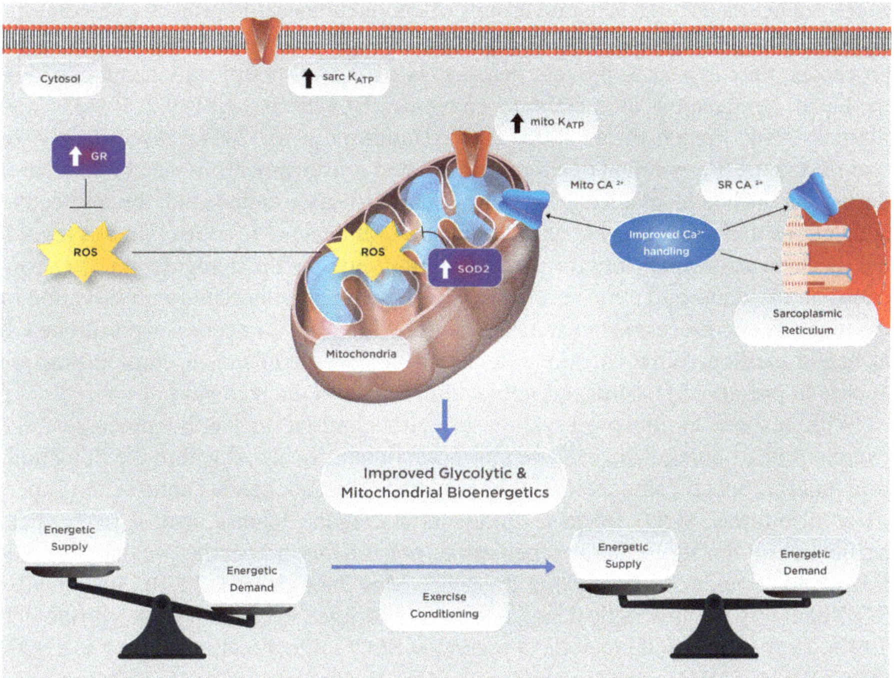

Fig. 12.5 Upregulated mechanisms of exercise-induced cardioprotection against ischemia-reperfusion injury. Exercise upregulates numerous endogenous mechanisms of protection against ischemic injury. Protective factors directly and indirectly mitigate the stress imposed upon cardiac metabolism during ischemia and reperfusion. Protective factors include the quenching of reactive oxygen species (ROS) through the fortification of endogenous antioxidants (GR, glutathione reductase; SOD2, manganese isoform of superoxide dismutase), mitochondrial and sarcolemmal ATP-sensitive K+ channels (K_{ATP}), and preservation of Ca^{2+} control through the protection of Ca^{2+} handling proteins of the mitochondria and sarcoplasmic reticulum. Exercise-induced cardioprotection therefore appears to be due to blunted ROS production and improved Ca^{2+} handling, exerting synergistic protection to improve the energetic supply-demand mismatch occurring during ischemia-reperfusion injury

and a parallel set of biochemical alterations that directly and indirectly protect the heart's ability to sustain bioenergetic production during a supply-demand mismatch.

The mechanisms responsible for exercise preconditioning are pleotropic (Fig. 12.5), a fact that is necessary to counter the multifaceted pathology of ischemia-reperfusion injury. Most prominently, exercise results in notable preservation of bioenergetic control during ischemia reperfusion (Bowles et al. 1992; Bowles and Starnes 1994; Starnes et al. 2007). More specifically, the cellular pathology of myocardial infarction includes cytosolic and mitochondrial Ca^{2+} dyshomeostasis, in addition to free radical overload which results in oxidative stress. Ca^{2+} dysregulation and oxidative stress occur in concert, both secondary to the bioenergetic supply-demand mismatch of ischemia reperfusion (Bolli 1988; Bolli and Marban 1999). Moreover, the aforementioned metabolic adaptations are thereby preserved in exercised hearts through separate groups of endogenous cardioprotective mediators. While these protective factors are not necessarily related directly to metabolic processes, they appear to prevent deleterious oxidative modifications and calcium-mediated degradation of metabolic enzymes (Bowles et al. 1992; Bowles and Starnes 1994; French et al. 2006, 2008; Hamilton et al. 2004). Accordingly, to understand the preservation of exercise-mediated improvements in metabolic control during an ischemic challenge, it is essential to also understand the protective adaptations that either prevent or mitigate loss of bioenergetic control. In this regard, the mechanisms responsible for preconditioning involve the upregulation of endogenous antioxidants and ionic regulation, the latter being important to preservation of the bioenergetic processes in exercised hearts. Moreover, protection is partitioned in ischemic cardiomyocytes to the extent that unique mediators of cardioprotection appear to preserve cytosolic and mitochondrial metabolic pathways.

With respect to the preservation of mitochondrial metabolic processes, the exercised heart upregulates endogenous antioxidants localized within the mitochondrial matrix. Specifically, as shown in Fig. 12.5, the endogenous antioxidant superoxide dismutase (SOD) serves as an essential cellular defense against free radical accumulation during ischemia reperfusion, presumably preventing oxidative modification to components of electron transport (McCullough et al. 2020). Specifically, SOD converts the free radical superoxide to hydrogen peroxide (H_2O_2) (Fridovich 1997). Isoforms of SOD include copper/zinc SOD, mitochondrial manganese SOD (MnSOD or SOD2), and extracellular SOD (Fridovich 1997). SOD2 has been confirmed as being essential to the exercise preconditioning phenotype (French et al. 2008). Using a knockdown approach, via intraperitoneal injections of antisense oligonucleotides directed against SOD2, it has been demonstrated that a modest (20–50%) increase in SOD2 expression is protective against ventricular dysrhythmias and tissue infarct. Interestingly, however SOD2 upregulation is not sufficient for preventing deficits in ventricular pump function (Lennon et al. 2004a; Yamashita et al. 1999; Hamilton et al. 2003). Because endogenous antioxidants typically function as a network defense, SOD2 production of the radical species H_2O_2 must subsequently be converted to H_2O and O_2 by downstream activity of the endogenous enzyme catalase and enzymes of the glutathione system. Preliminary data indicate that catalase, while specific to both mitochondrial and cytosolic spaces, is inconsistently upregulated in exercised hearts (Lennon et al. 2004a, b; French et al. 2008; Dao et al. 2011). These inconsistencies likely reflect the fact that organelle-affiliating catalase has not been examined within the context of a divergent role for the enzyme within. In addition, upregulation of the glutathione system has been found to be essential to complement the effects of SOD2 within the mitochondria. As shown in Fig. 12.5, myocardial glutathione reductase activity appears to be enhanced following exercise and thereby serves as the essential second step for exercise-induced cardioprotection against free radical generation (Frasier et al. 2011).

Preservation of metabolic function during ischemia reperfusion is fundamentally dependent upon mitigation of cytosolic Ca^{2+} overload, which has been shown to directly and indirectly limit glycolytic and mitochondrial bioenergetic energy production (Omar et al. 2010; Tu et al. 2020). Accordingly, improved Ca^{2+} handling during ischemia reperfusion is also central to the exercise preconditioned phenotype (Bowles et al. 1992). While some conflicts exist within published studies on the topic, most evidence indicates that exercise training improves mitochondrial tolerance to Ca^{2+} overload, blunting mitochondrial permeability transition pore (MPTP) opening (Starnes et al. 2007; Marcil et al. 2006; Magalhaes et al. 2014). Moreover, increased mitochondrial antioxidant fortifications limit Ca^{2+} overload (Starnes et al. 2007; Buja 2005), protecting against oxidative modification of Ca^{2+} handling proteins such as L-type Ca^{2+} channels, phospholamban, and sarcoplasmic/endoplasmic reticulum Ca^{2+} ATPase (SERCA), and attenuation of calpain-mediated protein degradation (French et al. 2006, 2008). In regard to the interrelated effects of antioxidant quenching of free radicals and improved Ca^{2+} handling, as shown in Fig. 12.5, it is currently unknown whether these effects are the result of direct antioxidant events in the cytosol or due to upstream quenching of free radical chain reactions (French et al. 2006, 2008). Moreover, to date, the specific role of mitigating both mitochondrial and Ca^{2+} overload has not been directly linked to preservation of cardiac metabolism and requires scientific confirmation.

As discussed previously, exercise preconditioning improves bioenergetic control during ischemia reperfusion through the preservation of mitochondrial function in addition to improved antioxidant fortification and Ca^{2+} regulation (Bowles et al. 1992; Starnes et al. 2007; French et al. 2006, 2008; Marcil et al. 2006; Magalhaes et al. 2014). Adjuvant to these observations is the bolstering of mitochondrial subpopulations within exercised hearts. Specifically, mitochondria subpopulations located at the subsarcolemmal space and intermyofibrillar mitochondria provide distinct, although incompletely understood, facets of protection to the exercised heart (Kavazis et al. 2009). Presumably, subsarcolemmal mitochondria are essential to preserving bioenergetic demand related to ionic regulation, while intermyofibrillar mitochondria generally sustain the metabolic demands related to myocardial contraction. Nonetheless, it does appear that within exercised hearts exposed to ischemia reperfusion, both mitochondrial subpopulations exhibit decreased MPTP opening and preserved bioenergetic function (Kavazis et al. 2008; Lee et al. 2012). In addition to elevated mitochondrial antioxidants such as SOD2, exercised hearts also appear to be protected by downregulation of the enzyme monoamine oxidase (MAO), which catalyzes amine oxidation reactions and the production of reactive oxygen species (Kavazis et al. 2009; Deshwal et al. 2017).

Related to improvements in mitochondrial function, alterations in the expression or activity of the ATP-sensitive potassium channels (K_{ATP}), located on the sarcolemma (sarc K_{ATP}) and inner mitochondrial membranes (mito K_{ATP}), are essential to exercise preconditioning; however, the precise role of these putative mediators of exercise preconditioning remains a topic of debate (Light et al. 2001; Gross and Peart 2003) . Findings from exercised hearts indicated that both sarc K_{ATP} and mito K_{ATP} are essential to the exercise preconditioned phenotype, as presented in Fig. 12.5 (Chicco et al. 2007; Brown et al. 2005b; Johnson et al. 2006). Notably, this mechanism is the only currently known example of sex-specific differences of exercise preconditioning. Specifically, it was observed that as compared with male counterparts, hearts from exercised females overexpress Kir6.2 (a subunit of K_{ATP}

channels) (Chicco et al. 2007; Johnson et al. 2006). Importantly, however, hearts from exercised males are nonetheless protected via K_{ATP} channel activation (Chicco et al. 2007; Johnson et al. 2006). To this end, the mechanisms of exercise-induced cardioprotection do not apply to all aspects of ischemia reperfusion injury. In this instance, mito K_{ATP} channels are essential for preventing ventricular dysrhythmias, while the sarc K_{ATP} channels are essential for preventing post-infarct tissue necrosis and apoptotic processes (Quindry et al. 2010, 2012).

12.3 Summary

Exercise promotes acute and chronic changes in cardiac metabolism by altering the levels of circulating hormones, substrates, and metabolic enzymes. These changes are important not only for structural cardiac adaptation but also for protecting the heart from injury (Fig. 12.6). Interestingly, different types of exercise have disparate effects on cardiac metabolism, remodeling, and function. The goal of future studies is to develop a more thorough understanding of how exercise-induced metabolic changes integrate with the development of cardiac resilience and how metabolism cues transcriptional changes that elicit structural changes and protect the heart from injury.

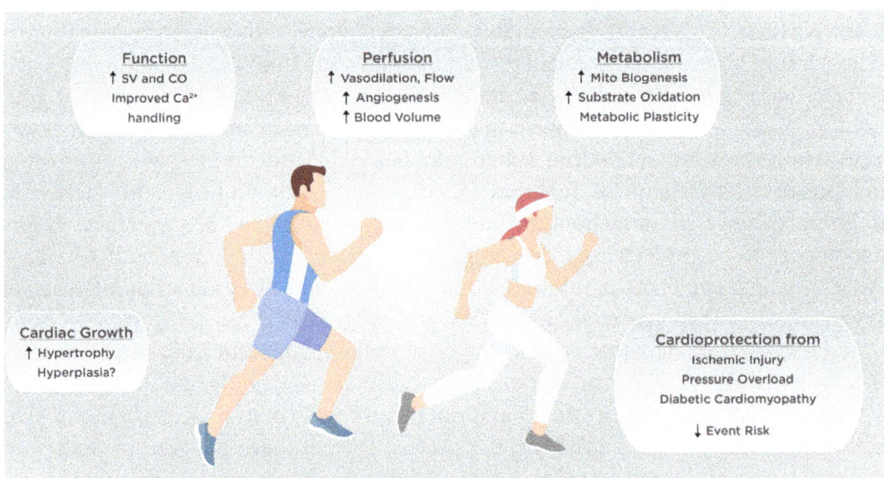

Fig. 12.6 Summary of the cardiovascular benefits of exercise. Exercise elicits a myriad of structural, functional, and metabolic adaptations which result in a cardioprotective state. SV, stroke volume; CO, cardiac output

References

Andrews Portes L et al (2009) Swimming training attenuates remodeling, contractile dysfunction and congestive heart failure in rats with moderate and large myocardial infarctions. Clin Exp Pharmacol Physiol 36(4):394–399

Arbab-Zadeh A et al (2014) Cardiac remodeling in response to 1 year of intensive endurance training. Circulation 130(24):2152–2161

Arner P, Kriegholm E, Engfeldt P (1990a) In situ studies of catecholamine-induced lipolysis in human adipose tissue using microdialysis. J Pharmacol Exp Ther 254(1):284–288

Arner P et al (1990b) Adrenergic regulation of lipolysis in situ at rest and during exercise. J Clin Invest 85(3):893–898

Baggish AL et al (2008a) Training-specific changes in cardiac structure and function: a prospective and longitudinal assessment of competitive athletes. J Appl Physiol (1985) 104(4):1121–1128

Baggish AL et al (2008b) The impact of endurance exercise training on left ventricular systolic mechanics. Am J Physiol Heart Circ Physiol 295(3):H1109–H1116

Baggish AL et al (2010) Differences in cardiac parameters among elite rowers and subelite rowers. Med Sci Sports Exerc 42(6):1215–1220

Barboza CA et al (2013) Ventricular and autonomic benefits of exercise training persist after detraining in infarcted rats. Eur J Appl Physiol 113(5):1137–1146

Bar-Shlomo BZ et al (1982) Left ventricular function in trained and untrained healthy subjects. Circulation 65(3):484–488

Beer M et al (2008) Effects of exercise training on myocardial energy metabolism and ventricular function assessed by quantitative phosphorus-31 magnetic resonance spectroscopy and magnetic resonance imaging in dilated cardiomyopathy. J Am Coll Cardiol 51(19):1883–1891

Bekaert I et al (1981) Non-invasive evaluation of cardiac function in professional cyclists. Br Heart J 45(2):213–218

Bergman BC et al (2009a) Myocardial FFA metabolism during rest and atrial pacing in humans. Am J Physiol Endocrinol Metab 296(2):E358–E366

Bergman BC et al (2009b) Myocardial glucose and lactate metabolism during rest and atrial pacing in humans. J Physiol 587(Pt 9):2087–2099

Bertrand ME et al (1977) Maximal exercise in normal subjects: changes in coronary sinus blood flow, contractility and myocardial extraction of FFA and lactate. Eur J Cardiol 5(6):481–491

Bezucha GR et al (1982) Comparison of hemodynamic responses to static and dynamic exercise. J Appl Physiol Respir Environ Exerc Physiol 53(6):1589–1593

Bezzerides VJ et al (2016) CITED4 induces physiologic hypertrophy and promotes functional recovery after ischemic injury. JCI Insight 1(9)

Blackmore PF, Shuman EA (1982) Regulation of hepatic altro heptulose 1,7-bisphosphate levels and control of flux through the pentose pathway by fructose 2,6-bisphosphate. FEBS Lett 142(2):255–259

Blair SN et al (1989) Physical fitness and all-cause mortality. A prospective study of healthy men and women. JAMA 262(17):2395–2401

Blair SN et al (1996) Influences of cardiorespiratory fitness and other precursors on cardiovascular disease and all-cause mortality in men and women. JAMA 276(3):205–210

Boada J et al (2000) Cells overexpressing fructose-2,6-bisphosphatase showed enhanced pentose phosphate pathway flux and resistance to oxidative stress. FEBS Lett 480(2-3):261–264

Bolli R (1988) Oxygen-derived free radicals and postischemic myocardial dysfunction ("stunned myocardium"). J Am Coll Cardiol 12(1):239–249

Bolli R, Marban E (1999) Molecular and cellular mechanisms of myocardial stunning. Physiol Rev 79(2):609–634

Bonen A et al (2000) Acute regulation of fatty acid uptake involves the cellular redistribution of fatty acid translocase. J Biol Chem 275(19):14501–14508

Bostrom P et al (2010) C/EBPbeta controls exercise-induced cardiac growth and protects against pathological cardiac remodeling. Cell 143(7):1072–1083

Bowles DK, Starnes JW (1994) Exercise training improves metabolic response after ischemia in isolated working rat heart. J Appl Physiol (1985) 76(4):1608–1614
Bowles DK, Farrar RP, Starnes JW (1992) Exercise training improves cardiac function after ischemia in the isolated, working rat heart. Am J Physiol 263(3 Pt 2):H804–H809
Brookes PS, Taegtmeyer H (2017) Metabolism: a direct link between cardiac structure and function. Circulation 136(22):2158–2161
Brown DA et al (2005a) Susceptibility of the heart to ischaemia-reperfusion injury and exercise-induced cardioprotection are sex-dependent in the rat. J Physiol 564(Pt 2):619–630
Brown DA et al (2005b) Cardioprotection afforded by chronic exercise is mediated by the sarcolemmal, and not the mitochondrial, isoform of the KATP channel in the rat. J Physiol 569(Pt 3):913–924
Bruning RS, Sturek M (2015) Benefits of exercise training on coronary blood flow in coronary artery disease patients. Prog Cardiovasc Dis 57(5):443–453
Buck MD et al (2016) Mitochondrial dynamics controls T cell fate through metabolic programming. Cell 166(1):63–76
Buja LM (2005) Myocardial ischemia and reperfusion injury. Cardiovasc Pathol 14(4):170–175
Bulow J (1981) Human adipose tissue blood flow during prolonged exercise, III. Effect of beta-adrenergic blockade, nicotinic acid and glucose infusion. Scand J Clin Lab Invest 41(4): 415–424
Burelle Y et al (2004) Regular exercise is associated with a protective metabolic phenotype in the rat heart. Am J Physiol Heart Circ Physiol 287(3):H1055–H1063
Caralis DG, Edwards L, Davis PJ (1977) Serum total and free thyroxine and triiodothyronine during dynamic muscular exercise in man. Am J Physiol 233(2):E115–E118
Caso P et al (2000) Pulsed Doppler tissue imaging in endurance athletes: relation between left ventricular preload and myocardial regional diastolic function. Am J Cardiol 85(9):1131–1136
Chance B, Williams GR (1955) Respiratory enzymes in oxidative phosphorylation. III. The steady state. J Biol Chem 217(1):409–427
Chance B, Williams GR (1956) Respiratory enzymes in oxidative phosphorylation. VI. The effects of adenosine diphosphate on azide-treated mitochondria. J Biol Chem 221(1):477–489
Chance B et al (1955) Respiratory enzymes in oxidative phosphorylation. V. A mechanism for oxidative phosphorylation. J Biol Chem 217(1):439–451
Charron MJ, Vuguin PM (2015) Lack of glucagon receptor signaling and its implications beyond glucose homeostasis. J Endocrinol 224(3):R123–R130
Chicco AJ et al (2007) Sex-specific and exercise-acquired cardioprotection is abolished by sarcolemmal KATP channel blockade in the rat heart. Am J Physiol Heart Circ Physiol 292(5): H2432–H2437
Ciloglu F et al (2005) Exercise intensity and its effects on thyroid hormones. Neuro Endocrinol Lett 26(6):830–834
Clausen JP (1976) Circulatory adjustments to dynamic exercise and effect of physical training in normal subjects and in patients with coronary artery disease. Prog Cardiovasc Dis 18(6): 459–495
Collins-Nakai RL, Noseworthy D, Lopaschuk GD (1994) Epinephrine increases ATP production in hearts by preferentially increasing glucose metabolism. Am J Physiol 267(5 Pt 2):H1862–H1871
Conlee RK, Tipton CM (1977) Cardiac glycogen repletion after exercise: influence of synthase and glucose 6-phosphate. J Appl Physiol Respir Environ Exerc Physiol 42(2):240–244
Conlee RK, Dalsky GP, Robinson KC (1981) The influence of free fatty acids on glycogen recovery in rat heart after exercise. Eur J Appl Physiol Occup Physiol 47(4):377–383
Coronado M et al (2017) Physiologic mitochondrial fragmentation is a normal cardiac adaptation to increased energy demand. Circ Res 122(2):282–295
Cortassa S et al (2015) From metabolomics to fluxomics: a computational procedure to translate metabolite profiles into metabolic fluxes. Biophys J 108(1):163–172
Coyle EF (2000) Physical activity as a metabolic stressor. Am J Clin Nutr 72(2 Suppl):512S–520S

Coyle EF, Gonzalez-Alonso J (2001) Cardiovascular drift during prolonged exercise: new perspectives. Exerc Sport Sci Rev 29(2):88–92

D'Andrea A et al (2010) Left ventricular myocardial velocities and deformation indexes in top-level athletes. J Am Soc Echocardiogr 23(12):1281–1288

Dao VT et al (2011) Catalase activity prevents exercise-induced up-regulation of vasoprotective proteins in venous tissue. J Cell Mol Med 15(11):2326–2334

Dawn B et al (2004) IL-6 plays an obligatory role in late preconditioning via JAK-STAT signaling and upregulation of iNOS and COX-2. Cardiovasc Res 64(1):61–71

DeBosch BJ, Muslin AJ (2008) Insulin signaling pathways and cardiac growth. J Mol Cell Cardiol 44(5):855–864

DeBosch B et al (2006a) Akt2 regulates cardiac metabolism and cardiomyocyte survival. J Biol Chem 281(43):32841–32851

DeBosch B et al (2006b) Akt1 is required for physiological cardiac growth. Circulation 113(17): 2097–2104

de Cabo R et al (2014) The search for antiaging interventions: from elixirs to fasting regimens. Cell 157(7):1515–1526

Degerman E et al (1998) Phosphorylation and activation of hormone-sensitive adipocyte phosphodiesterase type 3B. Methods 14(1):43–53

de Groot MJ et al (1993) Substrate-induced changes in the lipid content of ischemic and reperfused myocardium. Its relation to hemodynamic recovery. Circ Res 72(1):176–186

Deligiannis A et al (1993) Plasma TSH, T3, T4 and cortisol responses to swimming at varying water temperatures. Br J Sports Med 27(4):247–250

DeMaria AN et al (1978) Alterations in ventricular mass and performance induced by exercise training in man evaluated by echocardiography. Circulation 57(2):237–244

Demirel HA et al (2001) Short-term exercise improves myocardial tolerance to in vivo ischemia-reperfusion in the rat. J Appl Physiol (1985) 91(5):2205–2212

Deshwal S et al (2017) Emerging role of monoamine oxidase as a therapeutic target for cardiovascular disease. Curr Opin Pharmacol 33:64–69

Dorling J et al (2018) Acute and chronic effects of exercise on appetite, energy intake, and appetite-related hormones: the modulating effect of adiposity, sex, and habitual physical activity. Nutrients 10(9)

Douglas PS et al (1986) Left ventricular structure and function by echocardiography in ultraendurance athletes. Am J Cardiol 58(9):805–809

Drake AJ, Haines JR, Noble MI (1980) Preferential uptake of lactate by the normal myocardium in dogs. Cardiovasc Res 14(2):65–72

Duncan RE et al (2007) Regulation of lipolysis in adipocytes. Annu Rev Nutr 27:79–101

Egan B, Zierath JR (2013) Exercise metabolism and the molecular regulation of skeletal muscle adaptation. Cell Metab 17(2):162–184

Febbraio MA et al (1998) Effect of epinephrine on muscle glycogenolysis during exercise in trained men. J Appl Physiol (1985) 84(2):465–470

Foryst-Ludwig A et al (2015) Adipose tissue lipolysis promotes exercise-induced cardiac hypertrophy involving the lipokine C16:1n7-palmitoleate. J Biol Chem 290(39):23603–23615

Frasier CR et al (2011) Short-term exercise preserves myocardial glutathione and decreases arrhythmias after thiol oxidation and ischemia in isolated rat hearts. J Appl Physiol (1985) 111(6):1751–1759

Freestone NS, Ribaric S, Mason WT (1996) The effect of insulin-like growth factor-1 on adult rat cardiac contractility. Mol Cell Biochem 163–164:223–229

French JP et al (2006) Ischemia-reperfusion-induced calpain activation and SERCA2a degradation are attenuated by exercise training and calpain inhibition. Am J Physiol Heart Circ Physiol 290(1):H128–H136

French JP et al (2008) Exercise-induced protection against myocardial apoptosis and necrosis: MnSOD, calcium-handling proteins, and calpain. FASEB J 22(8):2862–2871

Fridovich I (1997) Superoxide anion radical (O2-.), superoxide dismutases, and related matters. J Biol Chem 272(30):18515–18517
Fulghum K, Hill BG (2018) Metabolic mechanisms of exercise-induced cardiac remodeling. Front Cardiovasc Med 5:127
Galbo H, Holst JJ, Christensen NJ (1975) Glucagon and plasma catecholamine responses to graded and prolonged exercise in man. J Appl Physiol 38(1):70–76
Garcia M et al (2009) Phosphofructo-1-kinase deficiency leads to a severe cardiac and hematological disorder in addition to skeletal muscle glycogenosis. PLoS Genet 5(8):e1000615
Gertz EW et al (1980) Myocardial lactate extraction: multi-determined metabolic function. Circulation 61(2):256–261
Gertz EW et al (1981) Myocardial lactate metabolism: evidence of lactate release during net chemical extraction in man. Circulation 63(6):1273–1279
Gertz EW et al (1988) Myocardial substrate utilization during exercise in humans. Dual carbon-labeled carbohydrate isotope experiments. J Clin Invest 82(6):2017–2025
Gibb AA, Hill BG (2018) Metabolic coordination of physiological and pathological cardiac remodeling. Circ Res 123:107–128
Gibb AA et al (2017a) Exercise-induced changes in glucose metabolism promote physiological cardiac growth. Circulation 136(22):2144–2157
Gibb AA et al (2017b) Integration of flux measurements to resolve changes in anabolic and catabolic metabolism in cardiac myocytes. Biochem J 474(16):2785–2801
Gilbert CA et al (1977) Echocardiographic study of cardiac dimensions and function in the endurance-trained athlete. Am J Cardiol 40(4):528–533
Glancy B, Balaban RS (2012) Role of mitochondrial Ca2+ in the regulation of cellular energetics. Biochemistry 51(14):2959–2973
Gonzalez-Munoz C et al (2008) Glucagon increases contractility in ventricle but not in atrium of the rat heart. Eur J Pharmacol 587(1–3):243–247
Goodwin GW, Taegtmeyer H (2000) Improved energy homeostasis of the heart in the metabolic state of exercise. Am J Physiol Heart Circ Physiol 279(4):H1490–H1501
Goodwin GW, Taylor CS, Taegtmeyer H (1998a) Regulation of energy metabolism of the heart during acute increase in heart work. J Biol Chem 273(45):29530–29539
Goodwin GW et al (1998b) Energy provision from glycogen, glucose, and fatty acids on adrenergic stimulation of isolated working rat hearts. Am J Physiol 274(4):H1239–H1247
Gross GJ, Peart JN (2003) KATP channels and myocardial preconditioning: an update. Am J Physiol Heart Circ Physiol 285(3):H921–H930
Gyntelberg F et al (1977) Effect of training on the response of plasma glucagon to exercise. J Appl Physiol Respir Environ Exerc Physiol 43(2):302–305
Hafstad AD et al (2011) High intensity interval training alters substrate utilization and reduces oxygen consumption in the heart. J Appl Physiol (1985) 111(5):1235–1241
Hamilton KL et al (2003) Exercise, antioxidants, and HSP72: protection against myocardial ischemia/reperfusion. Free Radic Biol Med 34(7):800–809
Hamilton KL et al (2004) MnSOD antisense treatment and exercise-induced protection against arrhythmias. Free Radic Biol Med 37(9):1360–1368
Harney JA, Rodgers RL (2008) Insulin-like stimulation of cardiac fuel metabolism by physiological levels of glucagon: involvement of PI3K but not cAMP. Am J Physiol Endocrinol Metab 295(1):E155–E161
Heinonen I et al (2014) Organ-specific physiological responses to acute physical exercise and long-term training in humans. Physiology (Bethesda) 29(6):421–436
Heinrich R, Rapoport TA (1974) A linear steady-state treatment of enzymatic chains. Critique of the crossover theorem and a general procedure to identify interaction sites with an effector. Eur J Biochem 42(1):97–105
Hellstrom L, Blaak E, Hagstrom-Toft E (1996) Gender differences in adrenergic regulation of lipid mobilization during exercise. Int J Sports Med 17(6):439–447

Hill DW, Butler SD (1991) Haemodynamic responses to weightlifting exercise. Sports Med 12(1): 1–7
Hill BG et al (2012) Integration of cellular bioenergetics with mitochondrial quality control and autophagy. Biol Chem 393(12):1485–1512
Ho KL et al (2019) Increased ketone body oxidation provides additional energy for the failing heart without improving cardiac efficiency. Cardiovasc Res 115(11):1606–1616
Ho KL et al (2021) Ketones can become the major fuel source for the heart but do not increase cardiac efficiency. Cardiovasc Res 117(4):1178–1187
Huang MH et al (2007) Mediating delta-opioid-initiated heart protection via the beta2-adrenergic receptor: role of the intrinsic cardiac adrenergic cell. Am J Physiol Heart Circ Physiol 293(1): H376–H384
Huang MH et al (2009) Reducing ischaemia/reperfusion injury through delta-opioid-regulated intrinsic cardiac adrenergic cells: adrenopeptidergic co-signalling. Cardiovasc Res 84(3): 452–460
Hui S et al (2020) Quantitative fluxomics of circulating metabolites. Cell Metab 32(4):676–688 e4
Hulver MW, Houmard JA (2003) Plasma leptin and exercise: recent findings. Sports Med 33(7): 473–482
Hurlimann HC et al (2011) Physiological and toxic effects of purine intermediate 5-amino-4-imidazolecarboxamide ribonucleotide (AICAR) in yeast. J Biol Chem 286(35):30994–31002
Husted AS et al (2017) GPCR-mediated signaling of metabolites. Cell Metab 25(4):777–796
Jang C et al (2019) Metabolite exchange between mammalian organs quantified in pigs. Cell Metab 30(3):594–606 e3
Jeppesen J et al (2009) Contractions but not AICAR increase FABPpm content in rat muscle sarcolemma. Mol Cell Biochem 326(1-2):45–53
Johnson MS, Moore RL, Brown DA (2006) Sex differences in myocardial infarct size are abolished by sarcolemmal KATP channel blockade in rat. Am J Physiol Heart Circ Physiol 290(6): H2644–H2647
Joyner MJ, Green DJ (2009) Exercise protects the cardiovascular system: effects beyond traditional risk factors. J Physiol 587(Pt 23):5551–5558
Kaijser L, Berglund B (1992) Myocardial lactate extraction and release at rest and during heavy exercise in healthy men. Acta Physiol Scand 144(1):39–45
Karlstaedt A et al (2020) Glucose 6-phosphate accumulates via phosphoglucose isomerase inhibition in heart muscle. Circ Res 126(1):60–74
Karwi QG et al (2019) Targeting the glucagon receptor improves cardiac function and enhances insulin sensitivity following a myocardial infarction. Cardiovasc Diabetol 18(1):1
Kavazis AN et al (2008) Exercise induces a cardiac mitochondrial phenotype that resists apoptotic stimuli. Am J Physiol Heart Circ Physiol 294(2):H928–H935
Kavazis AN et al (2009) Exercise training induces a cardioprotective phenotype and alterations in cardiac subsarcolemmal and intermyofibrillar mitochondrial proteins. Am J Physiol Heart Circ Physiol 297(1):H144–H152
Kemi OJ et al (2007) Exercise training restores aerobic capacity and energy transfer systems in heart failure treated with losartan. Cardiovasc Res 76(1):91–99
Kemppainen J et al (2002) Myocardial and skeletal muscle glucose uptake during exercise in humans. J Physiol 542(Pt 2):403–412
Keul J (1971) Myocardial metabolism in athletes. Adv Exp Med Biol 11:447–467
Kjaer M et al (1986) Increased epinephrine response and inaccurate glucoregulation in exercising athletes. J Appl Physiol (1985) 61(5):1693–1700
Koeslag JH (1982) Post-exercise ketosis and the hormone response to exercise: a review. Med Sci Sports Exerc 14(5):327–334
Koeslag JH, Noakes TD, Sloan AW (1980) Post-exercise ketosis. J Physiol 301:79–90
Kolter T et al (1992) Contraction-induced translocation of the glucose transporter Glut4 in isolated ventricular cardiomyocytes. Biochem Biophys Res Commun 189(2):1207–1214

Kreisman SH et al (2000) Epinephrine infusion during moderate intensity exercise increases glucose production and uptake. Am J Physiol Endocrinol Metab 278(5):E949–E957

Kreisman SH et al (2001) Norepinephrine infusion during moderate-intensity exercise increases glucose production and uptake. J Clin Endocrinol Metab 86(5):2118–2124

Kuhl JE et al (2006) Exercise training decreases the concentration of malonyl-CoA and increases the expression and activity of malonyl-CoA decarboxylase in human muscle. Am J Physiol Endocrinol Metab 290(6):E1296–E1303

Kundu BK et al (2015) Remodeling of glucose metabolism precedes pressure overload-induced left ventricular hypertrophy: review of a hypothesis. Cardiology 130(4):211–220

Kwong JQ et al (2015) The mitochondrial calcium uniporter selectively matches metabolic output to acute contractile stress in the heart. Cell Rep 12(1):15–22

Lassers BW et al (1971) Effect of prolonged exercise on myocardial metabolism in man. Br Heart J 33(4):609

Lassers BW et al (1972a) Effect of nicotinic acid on myocardial metabolism in man at rest and during exercise. J Appl Physiol 33(1):72–80

Lassers BW, Kaijser L, Carlson LA (1972b) Myocardial lipid and carbohydrate metabolism in healthy, fasting men at rest: studies during continuous infusion of 3 H-palmitate. Eur J Clin Invest 2(5):348–358

Lee Y et al (2012) Exercise protects cardiac mitochondria against ischemia-reperfusion injury. Med Sci Sports Exerc 44(3):397–405

Lee Y et al (2016) Cardiac kinetophagy coincides with activation of anabolic signaling. Med Sci Sports Exerc 48(2):219–226

Lennon SL et al (2004a) Exercise and myocardial tolerance to ischaemia-reperfusion. Acta Physiol Scand 182(2):161–169

Lennon SL et al (2004b) Loss of exercise-induced cardioprotection after cessation of exercise. J Appl Physiol (1985) 96(4):1299–1305

Lewis GD et al (2010) Metabolic signatures of exercise in human plasma. Sci Transl Med 2(33): 33ra37

Light PE et al (2001) Distinct myoprotective roles of cardiac sarcolemmal and mitochondrial KATP channels during metabolic inhibition and recovery. FASEB J 15(14):2586–2594

Liljenquist JE et al (1974) Effects of glucagon on lipolysis and ketogenesis in normal and diabetic men. J Clin Invest 53(1):190–197

Lombardo TA et al (1953) The effect of exercise on coronary blood flow, myocardial oxygen consumption and cardiac efficiency in man. Circulation 7(1):71–78

Lopaschuk GD et al (2010) Myocardial fatty acid metabolism in health and disease. Physiol Rev 90(1):207–258

Luger A et al (1988) Acute exercise stimulates the renin-angiotensin-aldosterone axis: adaptive changes in runners. Horm Res 30(1):5–9

Luiken JJ et al (2008) Identification of protein kinase D as a novel contraction-activated kinase linked to GLUT4-mediated glucose uptake, independent of AMPK. Cell Signal 20(3):543–556

Luiken JJ, Glatz JF, Neumann D (2015) Cardiac contraction-induced GLUT4 translocation requires dual signaling input. Trends Endocrinol Metab 26(8):404–410

Luongo TS et al (2015) The mitochondrial calcium uniporter matches energetic supply with cardiac workload during stress and modulates permeability transition. Cell Rep 12(1):23–34

MacDougall JD et al (1985) Arterial blood pressure response to heavy resistance exercise. J Appl Physiol (1985) 58(3):785–790

MacDougall JD et al (1992) Factors affecting blood pressure during heavy weight lifting and static contractions. J Appl Physiol (1985) 73(4):1590–1597

Maeda S et al (2005) Acute exercise causes an enhancement of tissue renin-angiotensin system in the kidney in rats. Acta Physiol Scand 185(1):79–86

Magalhaes J et al (2014) Modulation of cardiac mitochondrial permeability transition and apoptotic signaling by endurance training and intermittent hypobaric hypoxia. Int J Cardiol 173(1):40–45

Maillet M, van Berlo JH, Molkentin JD (2013) Molecular basis of physiological heart growth: fundamental concepts and new players. Nat Rev Mol Cell Biol 14(1):38–48
Manzon A et al (1998) Glucose infusion partially attenuates glucose production and increases uptake during intense exercise. J Appl Physiol (1985) 85(2):511–524
Marcil M et al (2006) Exercise training induces respiratory substrate-specific decrease in Ca2+-induced permeability transition pore opening in heart mitochondria. Am J Physiol Heart Circ Physiol 290(4):H1549–H1557
Marliss EB et al (1991) Glucoregulatory and hormonal responses to repeated bouts of intense exercise in normal male subjects. J Appl Physiol (1985) 71(3):924–933
Maron BJ et al (1993) Reduction in left ventricular wall thickness after deconditioning in highly trained Olympic athletes. Br Heart J 69(2):125–128
McCullough DJ et al (2020) Endurance exercise training in pulmonary hypertension increases skeletal muscle electron transport chain supercomplex assembly. Pulm Circ 10(2): 2045894020925762
Mercken EM et al (2012) Of mice and men: the benefits of caloric restriction, exercise, and mimetics. Ageing Res Rev 11(3):390–398
Mihl C, Dassen WR, Kuipers H (2008) Cardiac remodelling: concentric versus eccentric hypertrophy in strength and endurance athletes. Neth Heart J 16(4):129–133
Miller LE et al (2015) Involvement of the delta-opioid receptor in exercise-induced cardioprotection. Exp Physiol 100(4):410–421
Monleon D et al (2014) Metabolomic analysis of long-term spontaneous exercise in mice suggests increased lipolysis and altered glucose metabolism when animals are at rest. J Appl Physiol (1985) 117(10):1110–1119
Mora S et al (2007) Physical activity and reduced risk of cardiovascular events: potential mediating mechanisms. Circulation 116(19):2110–2118
Morganroth J et al (1975) Comparative left ventricular dimensions in trained athletes. Ann Intern Med 82(4):521–524
Morris JN et al (1953) Coronary heart-disease and physical activity of work. Lancet 265(6796): 1111–1120. Concl
Morville T et al (2020) Plasma metabolome profiling of resistance exercise and endurance exercise in humans. Cell Rep 33(13):108554
Musi N et al (2005) Functional role of AMP-activated protein kinase in the heart during exercise. FEBS Lett 579(10):2045–2050
Myers J et al (2002) Exercise capacity and mortality among men referred for exercise testing. N Engl J Med 346(11):793–801
Naylor LH et al (2005) Reduced ventricular flow propagation velocity in elite athletes is augmented with the resumption of exercise training. J Physiol 563(Pt 3):957–963
Neri Serneri GG et al (2001) Increased cardiac sympathetic activity and insulin-like growth factor-I formation are associated with physiological hypertrophy in athletes. Circ Res 89(11):977–982
No Authors (1967) Adolph Fick (1829-1901), mathematician, physicist, physiologist. JAMA 202(12):1100–1101
Nystoriak MA, Bhatnagar A (2018) Cardiovascular effects and benefits of exercise. Front Cardiovasc Med 5:135
Ogihara T et al (2002) Angiotensin II-induced insulin resistance is associated with enhanced insulin signaling. Hypertension 40(6):872–879
Ogura Y et al (2011) Single bout of running exercise changes LC3-II expression in rat cardiac muscle. Biochem Biophys Res Commun 414(4):756–760
Olah A et al (2017) Complete reversion of cardiac functional adaptation induced by exercise training. Med Sci Sports Exerc 49(3):420–429
Oliveira LDC et al (2018) Excessive treadmill training enhances the insulin signaling pathway and glycogen deposition in mice hearts. J Cell Biochem 120(2):1304–1317

Olver TD, Ferguson BS, Laughlin MH (2015) Molecular mechanisms for exercise training-induced changes in vascular structure and function: skeletal muscle, cardiac muscle, and the brain. Prog Mol Biol Transl Sci 135:227–257
Omar MA, Wang L, Clanachan AS (2010) Cardioprotection by GSK-3 inhibition: role of enhanced glycogen synthesis and attenuation of calcium overload. Cardiovasc Res 86(3):478–486
Oram JF, Bennetch SL, Neely JR (1973) Regulation of fatty acid utilization in isolated perfused rat hearts. J Biol Chem 248(15):5299–5309
Paffenbarger RS Jr et al (1986) Physical activity, all-cause mortality, and longevity of college alumni. N Engl J Med 314(10):605–613
Papadopoulos NM, Bloor CM, Standefer JC (1969) Effects of exercise and training on plasma lipids and lipoproteins in the rat. J Appl Physiol 26(6):760–763
Pentassuglia L et al (2016) Neuregulin-1beta promotes glucose uptake via PI3K/Akt in neonatal rat cardiomyocytes. Am J Physiol Endocrinol Metab 310(9):E782–E794
Petersen KM et al (2018) Hemodynamic effects of glucagon: a literature review. J Clin Endocrinol Metab 103(5):1804–1812
Plowman SA, Smith DL (2017) Exercise physiology for health, fitness, and performance, 5th edn. Wolters Kluwer, Philadelphia. xix, 771 pages
Pluim BM et al (2000) The athlete's heart. A meta-analysis of cardiac structure and function. Circulation 101(3):336–344
Podbielski B, Palmer WK (1989) Cardiac triacylglycerol content and lipase activity during recovery from exercise. J Appl Physiol (1985) 66(3):1099–1103
Pozuelo Rubio M et al (2003) 14-3-3s regulate fructose-2,6-bisphosphate levels by binding to PKB-phosphorylated cardiac fructose-2,6-bisphosphate kinase/phosphatase. EMBO J 22(14): 3514–3523
Prasad A et al (2007) The effects of aging and physical activity on Doppler measures of diastolic function. Am J Cardiol 99(12):1629–1636
Quindry JC (2017) Exercise-induced cardioprotection and the therapeutic potential of RIPC. J Cardiovasc Pharmacol Ther 22(5):397–403
Quindry J et al (2005) Exercise training provides cardioprotection against ischemia-reperfusion induced apoptosis in young and old animals. Exp Gerontol 40(5):416–425
Quindry JC et al (2010) Mitochondrial KATP channel inhibition blunts arrhythmia protection in ischemic exercised hearts. Am J Physiol Heart Circ Physiol 299(1):H175–H183
Quindry JC et al (2012) Ischemia reperfusion injury, KATP channels, and exercise-induced cardioprotection against apoptosis. J Appl Physiol (1985) 113(3):498–506
Qvisth V et al (2008) Catecholamine regulation of local lactate production in vivo in skeletal muscle and adipose tissue: role of -adrenoreceptor subtypes. J Clin Endocrinol Metab 93(1):240–246
Riebe D et al (2018) ACSM's guidelines for exercise testing and prescription, 10th edn. Wolters Kluwer, Philadelphia. xxx, 472 pages
Riehle C et al (2014) Insulin receptor substrates are essential for the bioenergetic and hypertrophic response of the heart to exercise training. Mol Cell Biol 34(18):3450–3460
Riquelme CA et al (2011) Fatty acids identified in the Burmese python promote beneficial cardiac growth. Science 334(6055):528–531
Roberts DJ et al (2014) Hexokinase-II positively regulates glucose starvation-induced autophagy through TORC1 inhibition. Mol Cell 53(4):521–533
Rodahl K, Miller HI, Issekutz B Jr (1964) Plasma free fatty acids in exercise. J Appl Physiol 19: 489–492
Rowell LB (1986) Human circulation : regulation during physical stress. Oxford University Press, New York. x, 416 p
Rowell LB et al (1969) Human cardiovascular adjustments to rapid changes in skin temperature during exercise. Circ Res 24(5):711–724
Rowland T (2008) Echocardiography and circulatory response to progressive endurance exercise. Sports Med 38(7):541–551

Saddik M, Lopaschuk GD (1991) Myocardial triglyceride turnover and contribution to energy substrate utilization in isolated working rat hearts. J Biol Chem 266(13):8162–8170

Saha AK, Kurowski TG, Ruderman NB (1995) A malonyl-CoA fuel-sensing mechanism in muscle: effects of insulin, glucose, and denervation. Am J Physiol 269(2 Pt 1):E283–E289

Salabei JK, Hill BG (2013) Mitochondrial fission induced by platelet-derived growth factor regulates vascular smooth muscle cell bioenergetics and cell proliferation. Redox Biol 1(1): 542–551

Schwarz AJ et al (1996) Acute effect of brief low- and high-intensity exercise on circulating insulin-like growth factor (IGF) I, II, and IGF-binding protein-3 and its proteolysis in young healthy men. J Clin Endocrinol Metab 81(10):3492–3497

Segel LD, Mason DT (1978) Effects of exercise and conditioning on rat heart glycogen and glycogen synthase. J Appl Physiol Respir Environ Exerc Physiol 44(2):183–189

Segel LD et al (1975) Cardiac glycogen in long-evans rats: diurnal pattern and response to exercise. Am J Physiol 229(2):398–401

Sen S et al (2013) Glucose regulation of load-induced mTOR signaling and ER stress in mammalian heart. J Am Heart Assoc 2(3):e004796

Seo DY et al (2016) Age-related changes in skeletal muscle mitochondria: the role of exercise. Integr Med Res 5(3):182–186

Shirihai OS, Song M, Dorn GW 2nd (2015) How mitochondrial dynamism orchestrates mitophagy. Circ Res 116(11):1835–1849

Simsek Z et al (2013) Speckle tracking echocardiographic analysis of left ventricular systolic and diastolic functions of young elite athletes with eccentric and concentric type of cardiac remodeling. Echocardiography 30(10):1202–1208

Spence AL et al (2011) A prospective randomised longitudinal MRI study of left ventricular adaptation to endurance and resistance exercise training in humans. J Physiol 589(Pt 22): 5443–5452

Stanley WC (1991) Myocardial lactate metabolism during exercise. Med Sci Sports Exerc 23(8): 920–924

Stanley WC et al (1988) Glucose and lactate interrelations during moderate-intensity exercise in humans. Metabolism 37(9):850–858

Starnes JW, Taylor RP, Park Y (2003) Exercise improves postischemic function in aging hearts. Am J Physiol Heart Circ Physiol 285(1):H347–H351

Starnes JW, Barnes BD, Olsen ME (2007) Exercise training decreases rat heart mitochondria free radical generation but does not prevent Ca2+-induced dysfunction. J Appl Physiol (1985) 102(5):1793–1798

Stolen KQ et al (2003) Exercise training improves biventricular oxidative metabolism and left ventricular efficiency in patients with dilated cardiomyopathy. J Am Coll Cardiol 41(3): 460–467

Stolen TO et al (2009) Interval training normalizes cardiomyocyte function, diastolic Ca2+ control, and SR Ca2+ release synchronicity in a mouse model of diabetic cardiomyopathy. Circ Res 105(6):527–536

Strom CC et al (2005) Expression profiling reveals differences in metabolic gene expression between exercise-induced cardiac effects and maladaptive cardiac hypertrophy. FEBS J 272(11):2684–2695

Stromer H et al (1996) Exogenously administered growth hormone and insulin-like growth factor-I alter intracellular Ca2+ handling and enhance cardiac performance. In vitro evaluation in the isolated isovolumic buffer-perfused rat heart. Circ Res 79(2):227–236

Stuewe SR et al (2000) Exercise training enhances glycolytic and oxidative enzymes in canine ventricular myocardium. J Mol Cell Cardiol 32(6):903–913

Sullivan JE et al (1994) Inhibition of lipolysis and lipogenesis in isolated rat adipocytes with AICAR, a cell-permeable activator of AMP-activated protein kinase. FEBS Lett 353(1):33–36

Sutton J, Lazarus L (1976) Growth hormone in exercise: comparison of physiological and pharmacological stimuli. J Appl Physiol 41(4):523–527

Takala TE, Ruskoaho HJ, Hassinen IE (1983) Transmural distribution of cardiac glucose uptake in rat during physical exercise. Am J Physiol 244(1):H131–H137
Taniyama Y et al (2005) Mechanisms of reactive oxygen species-dependent downregulation of insulin receptor substrate-1 by angiotensin II. Arterioscler Thromb Vasc Biol 25(6):1142–1147
Thijssen DHJ et al (2018) Association of exercise preconditioning with immediate cardioprotection: a review. JAMA Cardiol 3(2):169–176
Till M, Kolter T, Eckel J (1997) Molecular mechanisms of contraction-induced translocation of GLUT4 in isolated cardiomyocytes. Am J Cardiol 80(3A):85A–89A
Trefts E, Williams AS, Wasserman DH (2015) Exercise and the regulation of hepatic metabolism. Prog Mol Biol Transl Sci 135:203–225
Troncoso R et al (2012) Energy-preserving effects of IGF-1 antagonize starvation-induced cardiac autophagy. Cardiovasc Res 93(2):320–329
Tu CC, Wan BY, Zeng Y (2020) STIM2 knockdown protects against ischemia/reperfusion injury through reducing mitochondrial calcium overload and preserving mitochondrial function. Life Sci 247:116560
Utomi V et al (2013) Systematic review and meta-analysis of training mode, imaging modality and body size influences on the morphology and function of the male athlete's heart. Heart 99(23): 1727–1733
Vasquez-Trincado C et al (2016) Mitochondrial dynamics, mitophagy and cardiovascular disease. J Physiol 594(3):509–525
Vettor R et al (2014) Exercise training boosts eNOS-dependent mitochondrial biogenesis in mouse heart: role in adaptation of glucose metabolism. Am J Physiol Endocrinol Metab 306(5):E519–E528
Vujic A et al (2018) Exercise induces new cardiomyocyte generation in the adult mammalian heart. Nat Commun 9(1):1659
Walsh NP et al (1998) The effects of high-intensity intermittent exercise on the plasma concentrations of glutamine and organic acids. Eur J Appl Physiol Occup Physiol 77(5):434–438
Waring CD et al (2014) The adult heart responds to increased workload with physiologic hypertrophy, cardiac stem cell activation, and new myocyte formation. Eur Heart J 35(39):2722–2731
Waring CD et al (2015) Cardiac adaptations from 4 weeks of intensity-controlled vigorous exercise are lost after a similar period of detraining. Physiol Rep 3(2)
Weiner RB, Baggish AL (2012) Exercise-induced cardiac remodeling. Prog Cardiovasc Dis 54(5): 380–386
Weiner RB et al (2010) The impact of endurance exercise training on left ventricular torsion. JACC Cardiovasc Imaging 3(10):1001–1009
Weiner RB et al (2015) Exercise-induced left ventricular remodeling among competitive athletes: a phasic phenomenon. Circ Cardiovasc Imaging 8(12)
Williamson JR (1964) Metabolic effects of epinephrine in the isolated, perfused rat heart. I. Dissociation of the glycogenolytic from the metabolic stimulatory effect. J Biol Chem 239:2721–2729
Wisneski JA et al (1987) Myocardial metabolism of free fatty acids. Studies with 14C-labeled substrates in humans. J Clin Invest 79(2):359–366
Wolfe RR et al (1986) Role of changes in insulin and glucagon in glucose homeostasis in exercise. J Clin Invest 77(3):900–907
Xu T et al (2015) Exercise preconditioning attenuates pressure overload-induced pathological cardiac hypertrophy. Int J Clin Exp Pathol 8(1):530–540

Yamamoto T et al (2014) Reduced methylation of PFKFB3 in cancer cells shunts glucose towards the pentose phosphate pathway. Nat Commun 5:3480
Yamashita N et al (1999) Exercise provides direct biphasic cardioprotection via manganese superoxide dismutase activation. J Exp Med 189(11):1699–1706
Yi W et al (2012) Phosphofructokinase 1 glycosylation regulates cell growth and metabolism. Science 337(6097):975–980
York JW, Penney DG, Oscai LB (1975) Effects of physical training on several glycolytic enzymes in rat heart. Biochim Biophys Acta 381(1):22–27

Chapter 13
Metabolism in the Brain During Exercise in Humans

Hannah G. Caldwell, Lasse Gliemann, and Philip N. Ainslie

Abstract This chapter discusses the regulation of cerebral metabolism and fuel utilization at rest and during dynamic whole-body exercise in humans. The relative contributions of cerebral metabolic rates of key substrates (oxygen, glucose, lactate, ketone bodies) are outlined with respect to rest and exercise. A brief overview of the current gold-standard techniques to assess cerebral metabolism during dynamic exercise in humans is also provided, and future research areas are highlighted throughout.

Keywords Cerebral metabolism · Cerebral blood flow · Exercise · Glucose · Lactate · Glycogen · Ketone bodies

13.1 Background

Regulation of the human circulation during exercise is astonishingly complex and is dictated by a balance between systemic sympathetic vasoconstriction and localized vasodilation. The tight regulation of blood flow and pressure prioritizes both (1) systemic arterial blood pressure to provide adequate organ-specific perfusion pressure and (2) sufficient delivery of oxygen and metabolic substrates to satisfy the increased tissue metabolism and the subsequent removal of metabolic end-products. Although

H. G. Caldwell (✉)
Centre for Heart, Lung and Vascular Health, School of Health and Exercise Sciences, University of British Columbia, Kelowna, BC, Canada

Department of Nutrition, Exercise and Sports, The August Krogh Section for Human Physiology, University of Copenhagen, Copenhagen, Denmark
e-mail: hannah.caldwell@ubc.ca

L. Gliemann
Department of Nutrition, Exercise and Sports, The August Krogh Section for Human Physiology, University of Copenhagen, Copenhagen, Denmark

P. N. Ainslie
Centre for Heart, Lung and Vascular Health, School of Health and Exercise Sciences, University of British Columbia, Kelowna, BC, Canada

© The Author(s), under exclusive license to Springer Nature Switzerland AG 2022

G. McConell (ed.), *Exercise Metabolism*, Physiology in Health and Disease,
https://doi.org/10.1007/978-3-030-94305-9_13

Table 13.1 Cerebral blood flow and global metabolic rate at rest and during exercise to exhaustion

	Per 100 g brain tissue	Per whole brain (1400 g)	Relative to whole body	Max increase from rest with exercise
Cerebral blood flow (ml/min)	57	798	15% cardiac output	+15–25%
Cerebral O_2 consumption (ml/min)	3.5	49	20% total VO_2	+30–35%
Cerebral glucose utilization (mg/min)	5.5	77	15% total glucose	+35–40%

[a]Reviewed in Clarke and Sokoloff (1999) with data derived from the literature published by Sokoloff (1960). Relative increases during exercise reviewed by Smith and Ainslie (2017)

the human brain is reflective of <1/50th of the body mass, it accounts for approximately 15–20% of the available cardiac output, body glucose, and oxygen consumption in the resting condition (Table 13.1). Cerebral oxygen consumption is approximately 3.5 ml/100 g brain/min; therefore, an average-sized brain, weighing approximately 1400 g, requires a rate of oxygen uptake ($\dot{V}O_2$) of 49 ml O_2/min (Sokoloff 1960). As the majority of oxygen in the brain is utilized for carbohydrate oxidation, the energy equivalent of the total cerebral metabolic rate is approximately 20 watts or 0.29 kcal/min. This high cerebral metabolic rate, paired with a remarkably restricted capacity for substrate storage (e.g., glycogen <10 μmol/g) and inability to utilize fat as a fuel substrate, necessitates tight regulation of cerebral blood flow. Loss of consciousness occurs within 10–20 seconds of cerebral ischemia, and brain death can ensue if adequate restoration of blood supply is not made. As an example, if the supply of glucose to the brain were theoretically abolished, the brain's total substrate stores—from free glucose, lactate, and astrocytic glycogen—would only sustain cerebral metabolism for approximately 12 min (Barros and Deitmer 2010). As such, the interplay between cerebral substrate delivery (via cerebral blood flow and arterial substrate content) and trans-cerebral exchange is essential to support increases in fuel utilization in the brain during exercise.

13.2 Cerebral Metabolism and Fuel Utilization at Rest

Cerebral metabolism is regulated by blood flow and the arterial to venous transcerebral exchange of primary substrates (i.e., oxygen, glucose, lactate, and ketones) across the cerebral circulation. The cerebral storage capacity of oxygen is exceptionally small relative to its rate of utilization; therefore, continuous cerebral circulation is essential to support oxidative metabolism. The regulation of cerebral substrate delivery is achieved via the neurovascular unit, involving neurons, glial cells, and cortical penetrating arterioles, which relays vascular signaling to tightly regulate cerebral blood flow in response to alterations in local cerebral metabolism (Fig. 13.1). Effective regional cerebral blood flow control is requisite for coordinated substrate delivery and local paracrine regulation in response to alterations in cerebral

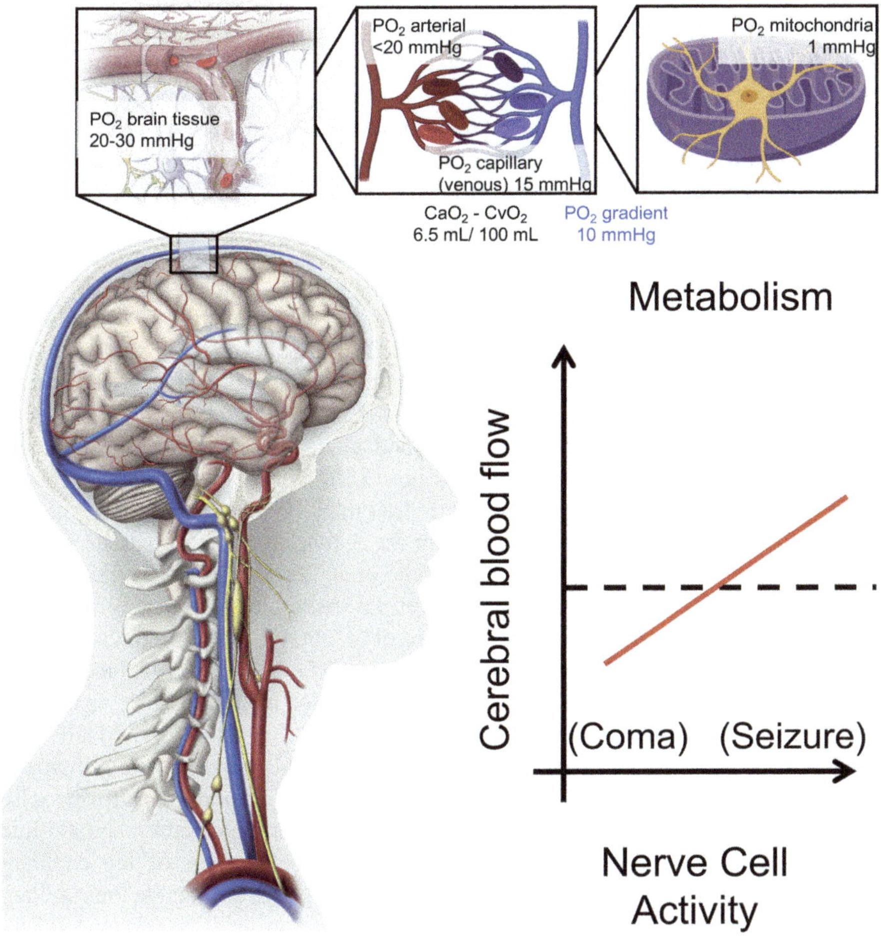

Fig. 13.1 Regulation of cerebral blood flow by metabolism. The cerebral oxygen cascade highlights the necessity for tightly controlled cerebral blood flow regulation. Regulation of cerebral blood flow is affected by changes in localized neuronal activity and metabolism. As an example, the linear relationship between cerebral blood flow and CMRO2 is evidenced by very low to high levels of nerve cell activity, e.g., coma/deep sleep/anesthesia (Madsen and Vorstrup 1991) and exercise/seizure (Siesjö 1978). Figure adapted from (Bailey 2019). Created in BioRender.com

metabolism. At least in pre-clinical models, novel in vivo evidence utilizing the cranial window technique suggests that contractile pericytes may facilitate the regulation of capillary blood flow (Gonzales et al. 2020); however, little is known about whether this occurs during exercise.

At rest, the brain relies almost exclusively on oxygen and glucose. Although lactate in the arterial blood is relatively low at rest (e.g., <1 mmol/l), lactate utilization reportedly contributes to approximately 8% of cerebral metabolism (van Hall et al. 2009). It is noteworthy that the brain does not rely on fatty acid oxidation (Mitchell et al. 2011; Schönfeld and Reiser 2013). The reasons for this are likely fourfold: (1) the rate-limiting passage of fatty acids across the blood-brain barrier; (2) the rate of adenosine triphosphate (ATP) production derived from fatty acids that is slower and less efficient than glucose utilization with respect to oxygen requirements; (3) relatively low enzymatic capacity of necessary fatty acid β-oxidation enzymes (e.g., carnitine palmitolytransferase 1, 3-ketoacyl-CoA thiolase, acyl-CoA dehydrogenase, enoyl-CoA hydratase); and (4) the β-oxidation of fatty acids that generates superoxide which can contribute to severe oxidative stress. However, while fatty acid oxidation does not contribute substantially to ATP production, recent in vitro cultured cell work and in vivo PET imaging evidence support a role for localized insulin-sensitive astrocytes in cerebral fatty acid uptake and metabolism in humans (Rebelos et al. 2020; Heni et al. 2020).

At rest, trans-cerebral uptake of oxygen and glucose each contributes approximately 50% to cerebral metabolic rate (Fig. 13.2). The contribution of cerebral lactate utilization to cerebral metabolism is less than 10% at rest. At rest, oxygen is utilized in the brain almost exclusively for the oxidation of carbohydrates consistent with the respiratory quotient of 0.97 (Nybo et al. 2003; Dalsgaard et al. 2004a). The stable resting cerebral metabolic ratio for oxidative glucose metabolism (often termed OGI) is approximately 5.7 (Dalsgaard et al. 2004b; Smith et al. 2014). This value indicates near complete oxidation of glucose: $6\ O_2 + C_6\ H_{12}O_6$ (glucose) $\rightarrow 6\ H_2O + 6\ CO_2$; approximately 5–10% of the glucose taken up undergoes glycolysis as evidenced by a slight lactate efflux at rest (Dalsgaard et al. 2004b; van Hall et al. 2009). Energy-yielding substrates within the arterial blood enter the brain principally via the blood-brain barrier (Fig. 13.3), albeit other less important pathways also include extracellular fluid, choroid plexuses, and interstitial/cerebrospinal fluid exchange. The blood-brain barrier consists of endothelial cells of the capillary wall, astrocyte end-feet surrounding the capillary, and pericytes outside the capillary basement membrane (Edvinsson and Krause 2002; Simard et al. 2003) (Fig. 13.4). Gases like oxygen and carbon dioxide readily cross the blood-brain barrier via passive diffusion. Larger molecules like glucose, however, cross the blood-brain barrier via specialized transporters within the endothelial cells (e.g., glucose transporter; GLUT- 1, 3, and 5). The glucose transporter GLUT1 is also found on astrocytes and the choroid plexus (Simpson et al. 2007), GLUT3 is found on neurons and to some extent the endothelium (Simpson et al. 2007), and GLUT5 is localized to the microglia. The majority of glucose entering the brain is distributed via the interstitial fluid where it is taken up via facilitated diffusion by neurons (GLUT3) and astrocytes (GLUT1). The rate of glucose transport is higher in the neuron versus astrocytes due to larger catalytic capacity of GLUT3 compared to GLUT1. The regulation of glucose transport at the blood-brain barrier has been recently reviewed in detail elsewhere (Patching 2017; Koepsell 2020).

At rest, circulating ketone bodies, principally acetoacetate and its reduced form β-hydroxybutyrate (βHB), are <0.1 mmol/l, whereas, following prolonged aerobic exercise in the glycogen depleted state, post-exercise ketosis can arise depending on

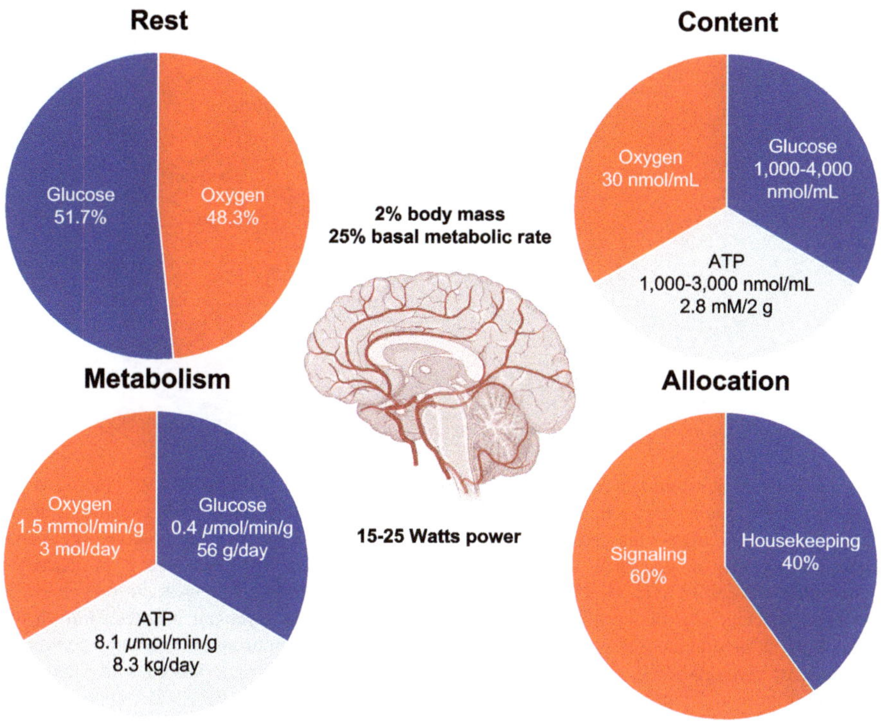

Fig. 13.2 Summary of the bioenergetic requirements of the human brain, emphasizing its severely restricted energy stores of oxygen, glucose, and ATP, paired with high rates of metabolism. Figure adapted from (Bailey 2019). The majority of signaling-related ATP costs involve the Na^+/K^+ ATP pump to regulate synaptic ion fluxes and maintain the resting membrane potential as well as conversion of glutamate into glutamine via astrocytes. Non-signaling "housekeeping" tasks include ATP-dependent actin treadmilling, microtubule dynamic instability, protein synthesis, lipid turnover, microglial motility, mitochondrial proton transfer, and cytoskeletal rearrangements (Attwell and Laughlin 2001; Engl and Attwell 2015). Created in BioRender.com

the intensity and duration of exercise as well as aerobic fitness and nutritional status in the range of 0.3–2.0 mmol/l (Evans et al. 2016). During conditions of hypoglycemia, e.g., fasting/starvation (Owen et al. 1967), post-exhaustive exercise (Askew et al. 1975), and ketogenic diet (Yancy et al. 2004), increases in systemic ketone bodies act as an alternative fuel source for cerebral metabolism. The human brain can readily uptake and oxidize βHB linearly with increases in the systemic circulation (Fig. 13.5) (Owen et al. 1967; Mikkelsen et al. 2015). Acute ketone infusion increases cerebral blood flow by approximately 30–40% and reduces cerebral metabolic rate of glucose by 30% with no change in oxygen utilization in humans (Hasselbalch et al. 1996; Svart et al. 2018). These ketone-induced increases in cerebral blood flow are perhaps in part due to changes in the brain cytosolic NAD^+/NADH ratio (Vlassenko et al. 2006; Xin et al. 2018), reductions in oxidative stress (Shimazu et al. 2013), and/or direct effects of ketones on the vasculature

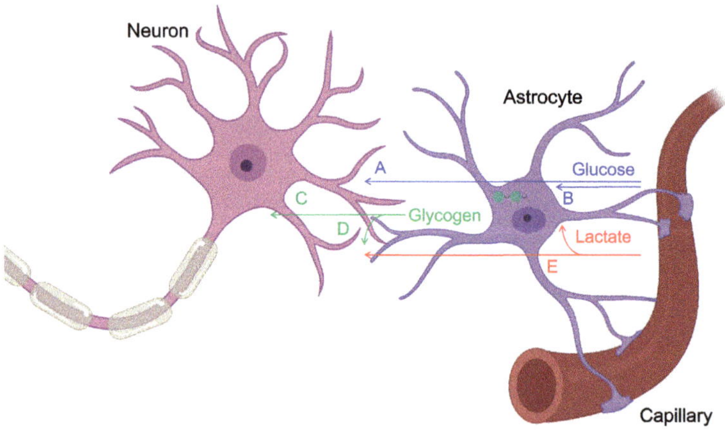

Fig. 13.3 Brain substrate metabolism of glucose, glycogen, and lactate. (**a**) At rest, glucose is the primary cerebral substrate as the uptake ratio O_2/glucose is close to 6. (**b**) Additionally, glucose fuels glycolysis in astrocytes producing lactate that is apparently oxidized by neurons. (**c**) Astrocytic breakdown of glycogen supports astrocyte metabolism, (**d**) particularly in astrocyte enveloping end-feet of synaptic clefts that are devoid of mitochondria. The spatial arrangement of the neurovascular unit may explain why lactate production from glycogen is separate from the breakdown of blood glucose. (**e**) Increases in cerebral lactate uptake during exercise may contribute to astrocytic glycogen-derived lactate oxidation during hypoglycemia experienced with exhaustive exercise. Figure and caption adapted from (Dalsgaard 2006). Created in BioRender.com

(Kimura et al. 2011; Han et al. 2018; Wu et al. 2020). The majority of βHB oxidized by the human brain is supplied by the liver via the arterial blood (Robinson and Williamson 1980; Laffel 1999); however, βHB can be synthesized to a lesser extent endogenously in astrocytes via fatty acid oxidation (Edmond et al. 1987). With reductions in carbohydrate availability—and lower insulin levels—elevations in free fatty acid release and utilization result in excess acetyl-CoA in hepatocytes, where it will be converted to ketone bodies. Although, as discussed above, the brain does not typically rely on fatty acid oxidation at rest, the dependency and utilization of ketone bodies as a substrate during/following prolonged exercise is related to elevations in acetyl-CoA and upregulated free fatty acid availability.

13.2.1 Summary

At rest, the brain relies almost exclusively on aerobic oxidative metabolism as reflected by the large 50% contributions of oxygen and glucose trans-cerebral uptake to the total cerebral metabolic rate. The effective localized control of cerebral blood flow, in part achieved by the highly sensitive pial arteriolar capillary network and tightly regulated blood-brain barrier, allows for coordinated cerebral substrate

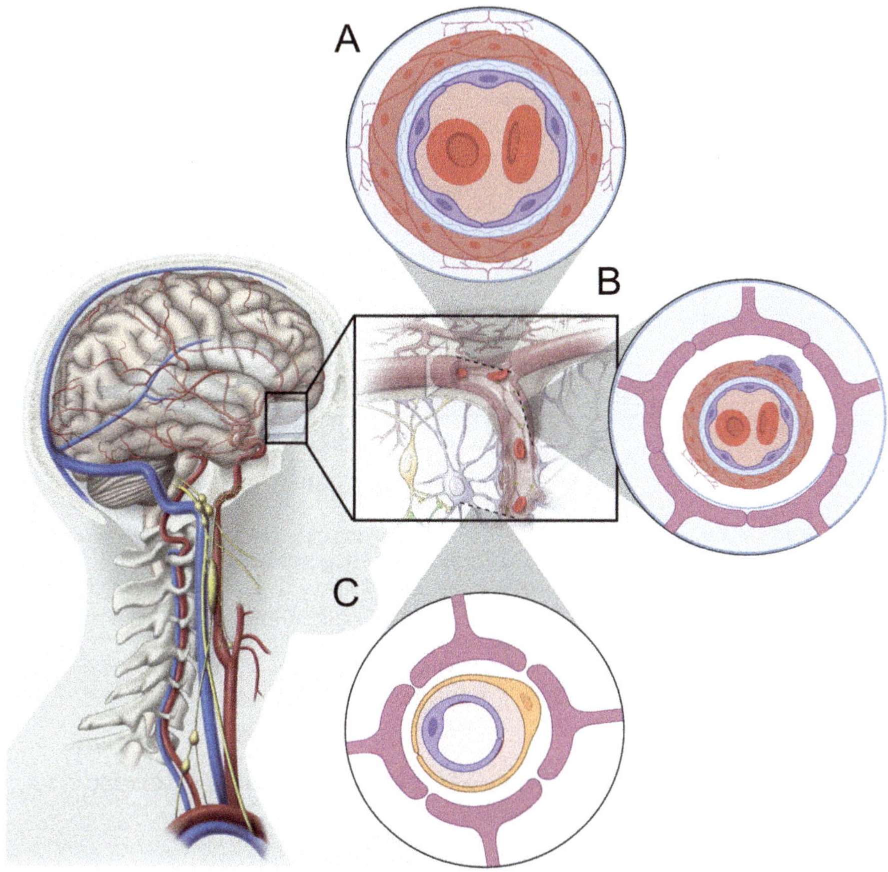

Fig. 13.4 Anatomy of the blood-brain barrier and cerebrovascular tree. At the brain surface, there is a highly vascularized capillary network of pial arterioles that are responsive to vasoactive stimuli (e.g., cerebral perfusion pressure, PO_2/CaO_2, $PCO_2/[H^+]$). (**a**) Pial arterioles are surrounded by the subarachnoid space and are innervated by perivascular nerves acting on vascular smooth muscle cells (VSMC). (**b**) Pial arterioles penetrate the pia mater through the Virchow-Robin space where they are encapsulated by astrocytic glial-like processes termed end-feet as well as perivascular macrophages. (**c**) Intraparenchymal arterioles and branching capillaries are covered by neural processes and pericytes within the basement membrane of the endothelium. Figure and caption adapted from (Caldwell et al. 2021). Created in BioRender.com

delivery. As discussed next, the regulation of cerebral blood flow and substrate delivery is critically important to support increases in total cerebral metabolism during exercise.

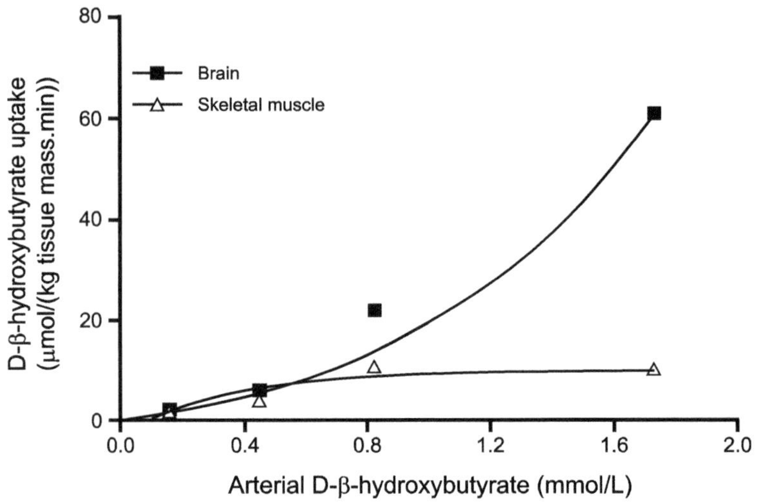

Fig. 13.5 Relationship between the arterial D-β-hydroxybutyrate concentration and cerebral and leg skeletal muscle uptake rate per kg of tissue in healthy males. Brain ketone uptake and subsequent oxidation was linearly related to the arterial ketone concentration. Resting skeletal muscle ketone uptake was markedly less and showed saturation kinetics. These findings substantiate the selective utilization of ketone bodies as a fuel for cerebral metabolism with increased systemic availability. It is noteworthy that the brain readily uptakes D-β-hydroxybutyrate approximately tenfold higher than skeletal muscle at rest. Figure redrawn from (Mikkelsen et al. 2015)

13.3 Regulation of Cerebral Metabolism During Exercise

At rest and during exercise, adequate cerebral oxygen delivery (CDO_2) is necessary to support an appropriate cerebral oxygen reserve (i.e., capillary oxygen volume sufficient to match cerebral metabolic rate of oxygen; Fig. 13.6).

$$CDO_2\ (\text{ml/min}) = CBF \times CaO_2 \tag{13.1}$$

The cerebral metabolic rate of oxygen ($CMRO_2$) reflects the oxidative metabolic rate of the brain and is derived from the product of oxygen delivery (i.e., cerebral blood flow) and the cerebral oxygen extraction (i.e., arterial-venous oxygen content difference; CaO_2–CvO_2).

$$CMRO_2\ (\text{ml/min}) = CBF/100 \times CaO_2 - CvO_2 \tag{13.2}$$

Typically, the resting oxygen extraction fraction (OEF) of the brain is within 20–30% (Gjedde 2005; Ainslie et al. 2014) and is dictated by the fixed oxygen conductivity of cerebral tissue (e.g., capillary surface area and distance for diffusion) and tissue metabolism for a given PO_2 gradient between the tissues and capillaries (Gjedde 2005).

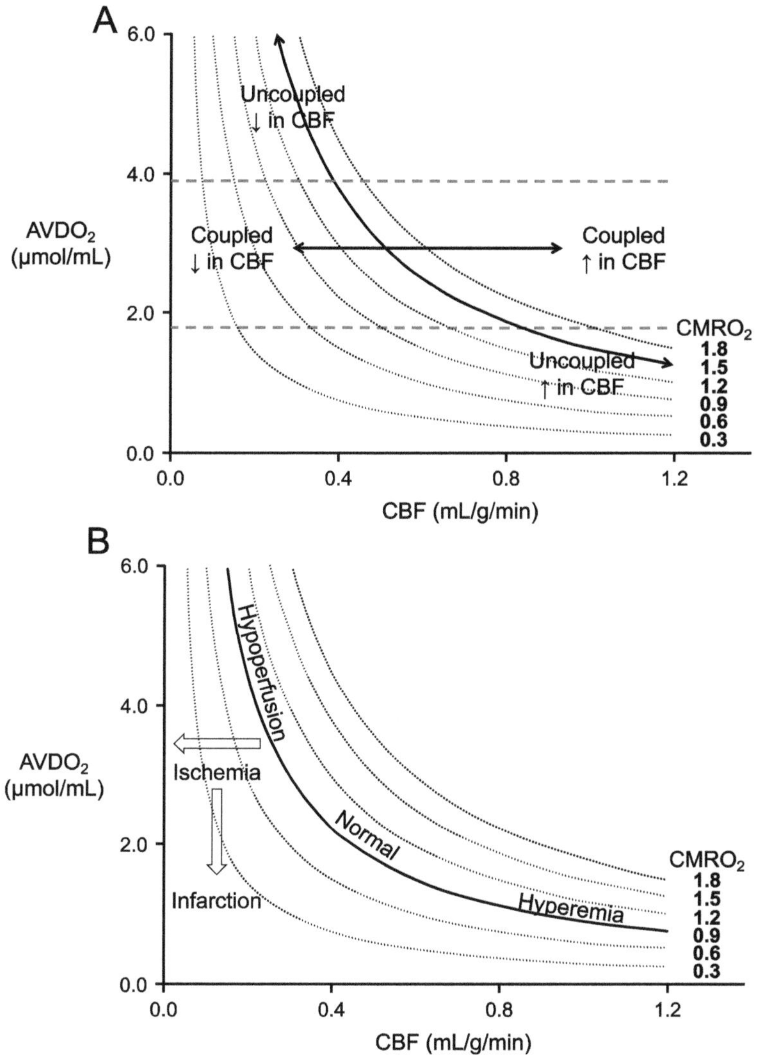

Fig. 13.6 Relationship between arterial-venous oxygen difference ($AVDO_2$) and cerebral blood flow (CBF) at rest. (**a**) If a coupled change in cerebral metabolic rate of oxygen ($CMRO_2$) and CBF occurs, then $AVDO_2$ remains unchanged, and the relationship between CBF and $AVDO_2$ shifts to a new $CMRO_2$ curve *(horizontal arrows)*. If $CMRO_2$ remains constant, then changes in $AVDO_2$ reflect uncoupled changes in CBF *(curved arrows)*. (**b**) Model diagramming the relationship between CBF and cerebral metabolism in comatose patients. In the absence of cerebral ischemia, the $AVDO_2$ and CBF have the relationship illustrated by the solid curve, with $CMRO_2$ averaging 0.9 μmol/g/min. In the presence of cerebral ischemia/infarction *(open arrows)*, $AVDO_2$ and CBF have an unpredictable relationship. Figure and caption adapted from (Robertson et al. 1989)

$$\mathrm{OEF}(\%) = \mathrm{CaO_2} - \mathrm{CvO_2}/\mathrm{CaO_2} \times 100\% \tag{13.3}$$

It is noteworthy that the brain (e.g., unlike the muscle) does not increase capillary recruitment and, therefore, the diffusive surface area for oxygen does not change (Kuschinsky and Paulson 1992; Williams et al. 1993), necessitating large increases in cerebral blood flow and/or oxygen extraction to maintain or elevate $CMRO_2$ (Buxton and Frank 1997; Mintun et al. 2001). As such, increases in cerebral oxygen delivery and metabolism may be described by an exponential relationship with increases in cerebral blood flow in excess of relative increases in $CMRO_2$ (Buxton and Frank 1997). In support of this, the cerebral oxygen extraction is inversely proportional to cerebral blood flow when metabolism is unaltered and directly proportional to metabolism when cerebral blood flow is unaltered. However, although the brain has the capacity to somewhat alter cerebral extraction throughout submaximal cycling exercise to exhaustion, e.g., <5% change in fractional O_2 extraction (Fig. 13.7), it is noteworthy that this response is markedly less than other organs such as the heart or skeletal muscle where oxygen extractions up to 80–90% have been reported (Proctor et al. 1998; Calbet et al. 2004; Lundby et al. 2006). As such, the brain requires sufficient increases in cerebral oxygen delivery (cerebral blood flow × CaO_2) to support elevated $CMRO_2$ at maximal exercise.

With progressive increases in exercise intensity up to approximately 60–70%, maximal oxygen uptake ($\dot{V}O_{2max}$), cerebral blood flow increases steadily (+10–20%) (Sato and Sadamoto 2010) to regulate cerebral substrate delivery (Ide and Secher 2000; Fisher et al. 2013) (Fig. 13.8). The increases in cerebral blood flow during submaximal exercise are mediated via relative alveolar hypoventilation (i.e., small elevations in arterial PCO_2) and increases in cerebral oxidative metabolism ($CMRO_2$) (Nybo et al. 2002; Smith et al. 2014). With heavy exercise intensity (e.g., >70% $\dot{V}O_{2max}$), cerebral blood flow is attenuated via hyperventilatory-induced reductions in arterial PCO_2 and resultant cerebral vasoconstriction (Larsen et al. 2008; Smith et al. 2014). Relative increases in arterial oxygen content (CaO_2) during progressive exercise (Ekblom et al. 1975; Smith et al. 2014) likely offset countervailing reductions in cerebral blood flow to stabilize cerebral oxygen delivery at maximal exercise intensities. Throughout various exercise intensities, the $CMRO_2$ response is coupled to cerebral oxygen delivery (cerebral blood flow × CaO_2) and oxygen extraction (CaO_2–CvO_2) rather than cerebral blood flow per se. This is further supported by the nonlinear relationship between cerebral blood flow/oxygen delivery and $CMRO_2$ (Fox and Raichle 1986; Fox et al. 1988).

13.3.1 Summary

The brain's inability to increase capillary recruitment and relatively small capacity to increase oxygen extraction necessitates large increases in blood flow to support elevations in $CMRO_2$. The increases in cerebral blood flow during submaximal

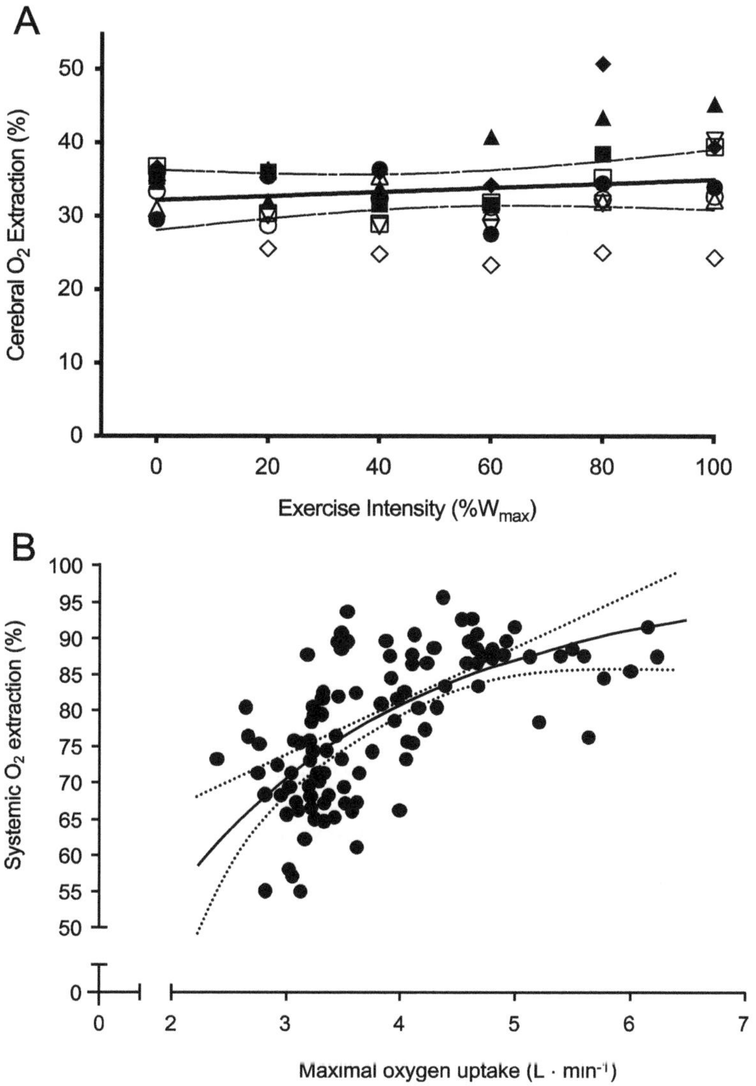

Fig. 13.7 Alterations in cerebral and systemic oxygen extraction during aerobic exercise to exhaustion. (**a**) The relationship between cerebral oxygen extraction—via arterial-jugular venous O_2 difference—throughout incremental cycling exercise to exhaustion (expressed as % maximal workload; $\%W_{max}$). Each symbol represents individual data from $n = 9$ with data from (Smith et al. 2014) with permission. Note the relatively small change in O_2 extraction across a large range of exercise intensities, e.g., $<5\%$ change in fractional O_2 extraction (%). (**b**) The relationship between systemic oxygen extraction (calculated via the Fick equation) and the respective pulmonary maximal oxygen uptake; note the very high attainable systemic O_2 extraction values at maximal exercise. Data are individual values from $n = 115$ healthy young men at whole-body maximal exercise across 17 studies summarized by (Skattebo et al. 2020). Figure redrawn from (Skattebo et al. 2020)

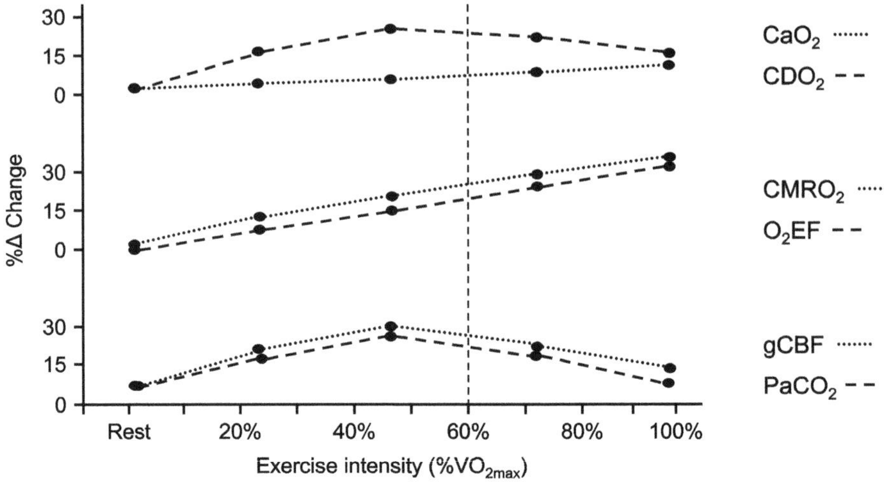

Fig. 13.8 Systemic and cerebrovascular variables affecting cerebral blood flow regulation during aerobic exercise to exhaustion. Generalized relative changes in arterial oxygen content (CaO_2) and cerebral oxygen delivery (CDO_2) (top), cerebral metabolic rate of oxygen ($CMRO_2$) and oxygen extraction fraction (O_2EF) (middle), and global cerebral blood flow (gCBF) and partial pressure of arterial carbon dioxide ($PaCO_2$) (bottom) responses to incremental exercise. The vertical dotted line indicates the point at which ventilatory threshold is typically achieved during exercise. Note the related temporal responses of CBF and $PaCO_2$ during exercise. Figure adapted from (Smith and Ainslie 2017)

exercise are principally attributed to temporal changes in arterial PCO_2 and $CMRO_2$. With maximal exercise, the $CMRO_2$ response is closely related to cerebral oxygen delivery (cerebral blood flow × CaO_2) and oxygen extraction (CaO_2–CvO_2) rather than cerebral blood flow per se.

13.3.2 Techniques to Assess Cerebral Metabolism During Dynamic Exercise in Humans

The first experiments to accurately measure cerebral blood flow and metabolism during exercise in humans utilized the nitrous oxide dilution technique developed by Seymour Kety and Carl Schmidt in 1945 (Kety and Schmidt 1945; Kety and Schmidt 1948). With this approach, inhalation/intravenous infusion of a freely diffusible tracer (e.g., nitrous oxide, xenon, hydrogen) and serial arterial and internal jugular venous blood sampling (i.e., Fick Principle) allows for calculation of absolute cerebral blood flow. The cerebral arterial-venous difference of inert tracer indicates the rate of appearance and clearance of the tracer from the cerebral circulation, respectively, and is used to calculate cerebral blood flow. Cerebral substrate metabolism can thus be calculated as cerebral metabolic rate (CMR) = cerebral blood flow × arterial-venous content difference of oxygen/glucose/lactate/

ketones (Eq. 13.2). Throughout the 1950s, this reliable and robust technique was recognized as the gold-standard method of quantifying cerebral blood flow; however, notably, there are three key considerations: (1) the Kety-Schmidt technique has poor temporal resolution as each steady-state measurement requires serial blood sampling across 10 minutes; (2) there is no indication of regional contributions to total cerebral blood flow; and (3) this technique assumes that venous outflow is symmetrical between both jugular veins, although this assumption is likely inappropriate (Lichtenstein et al. 2001). Within the last decade, Duplex ultrasound has been utilized to quantify regional volumetric blood flow in the extracranial circulation (e.g., carotid and vertebral arteries). Duplex ultrasound has excellent temporal resolution and facilitates synchronized beat-by-beat arterial diameter and pulse-wave blood velocity measurements to calculate volumetric blood flow in the anterior and posterior cerebral circulations, responsible for approximately 70% and 30% of total cerebral blood flow, respectively (Thomas et al. 2015). To address the previously described drawbacks of the Kety-Schmidt technique, cerebral metabolic rate of substrates can be calculated with this noninvasive ultrasound-derived cerebral blood flow—now inclusive of regional differences in flow—paired with direct arterial-venous blood sampling (via the Fick Principle). Table 13.2 outlines the strengths, limitations, and utility of various techniques for the measurement of cerebral metabolism in humans. It is important to note that the Kety-Schmidt technique and the Fick Principle paired with Duplex ultrasound-derived cerebral blood flow are the only techniques available to quantify cerebral substrate metabolism during whole-body exercise involving large muscle mass.

13.4 Cerebral Substrate Oxidation During Exercise

The absolute change in cerebral metabolic rate and relative changes in oxidative cerebral fuel utilization during incremental exercise have been elegantly reviewed by Smith and Ainslie (2017) and are summarized in Fig. 13.9. With incremental exercise to exhaustion, the cerebral metabolic ratios for oxidative glucose and oxidative carbohydrate metabolism (i.e., OGI and OCI, respectively) are progressively attenuated from their stable resting values of approximately 5.7 by up to 40–50% (Dalsgaard et al. 2004b; Smith et al. 2014).

$$\mathrm{OGI} = (\mathrm{CaO_2} - \mathrm{CvO_2})/(\mathrm{Glu_a} - \mathrm{Glu_v}) \tag{13.4}$$

$$\mathrm{OCI} = (\mathrm{CaO_2} - \mathrm{CvO_2})/(\mathrm{Glu_a} - \mathrm{Glu_v}) + \frac{1}{2}(\mathrm{Lac_a} - \mathrm{Lac_v}) \tag{13.5}$$

The reduction in OGI and OCI indicates part of the glucose/lactate was not fully oxidized suggesting an increase in anaerobic non-oxidative metabolism, i.e., larger increases in the respective glucose and carbohydrate (glucose + ½ lactate) uptakes relative to oxygen (Dalsgaard et al. 2004b). This reduction in OGI and OCI during

Table 13.2 Strengths, limitations, and utility of various techniques for the measurement of cerebral metabolism in humans

Method	Strengths	Limitations	Utility	Exercise?
Kety-Schmidt	1. Good reliability 2. Robust technique 3. Volumetric flow measurement	1. Poor temporal resolution (requires steady-state) 2. Invasive 3. Requires highly trained personnel	Measure arterial-venous cerebral metabolism and global CBF at rest and during steady-state whole-body exercise	Yes
Fick principle with Duplex ultrasound	1. Excellent temporal resolution 2. High spatial resolution 3. Volumetric flow measurement	1. Reliability largely based on sonography experience 2. Difficulty getting ultrasound measures at higher exercise intensities 3. Extracranial assessment 4. Invasive catheterizations	Measure arterial-venous cerebral metabolism and regional dynamic CBF responses at rest and during whole-body exercise	Yes
MRI	1. Excellent reliability 2. High spatial resolution 2. Noninvasive	1. Fair temporal resolution 2. Influenced by changes in $CMRO_2$ 3. No volumetric flow measurement	High-quality dynamic (BOLD) and steady-state (ASL) volumetric, regional, and global CBF at rest	No
PET	1. Excellent reliability 2. High-quality images	1. Poor temporal resolution 2. Invasive 3. Unable to measure during whole-body dynamic exercise	Optimum technique for regional changes in CBF and cerebral metabolism at rest	No
NMR spectroscopy	1. Very high reproducibility 2. Wide application for clinical diagnosis 3. Noninvasive	1. Low sensitivity	Measure rates of substrate transport and metabolism in the brain	No
Microdialysis	1. Direct cerebral metabolism within the interstitial space	1. Highly invasive	Only applicable in critical care patients (e.g., stroke, TBI)	No

exercise appears to be unaffected by changes in arterial lactate availability (Volianitis et al. 2011); however, both OGI and OCI are reportedly altered with severe high-altitude hypoxia (e.g., 5050 m) (Smith et al. 2014). Collectively, these

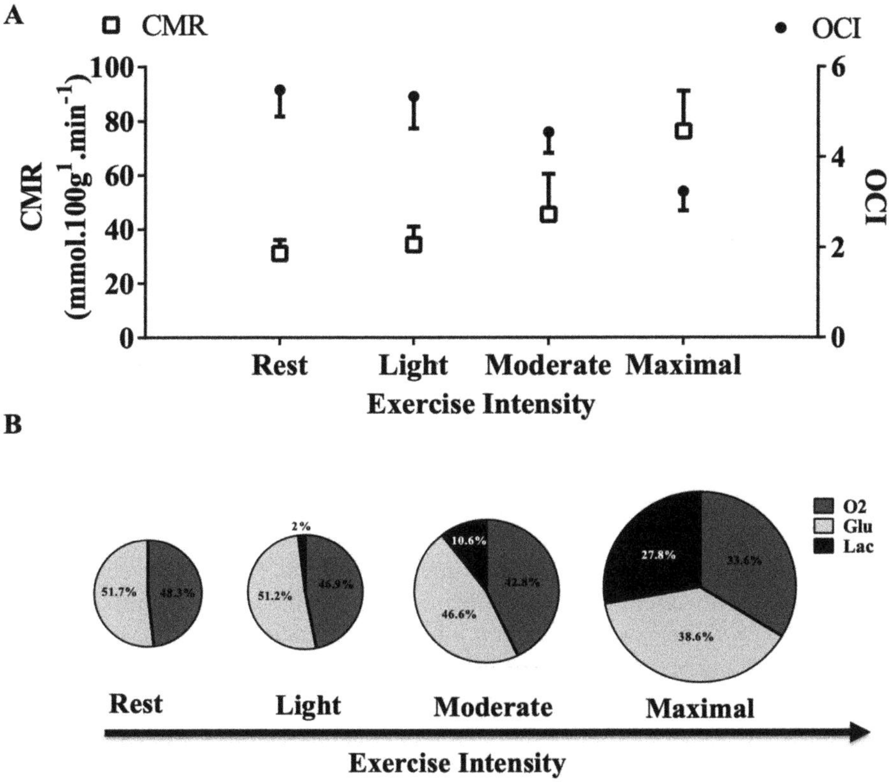

Fig. 13.9 Average cerebral metabolic rate and oxidative cerebral fuel utilization during incremental exercise. (**a**) The cumulative oxidative (CMR_{glu} + ½ CMR_{lac}) and non-oxidative (1/6 $CMRO_2$–(CMR_{glu} + ½ CMR_{lac}) cerebral metabolic ratio (CMR; oxidative plus non-oxidative in glucose equivalent units) and cerebral oxidative carbohydrate ratio (OCI) during rest, light (20–40% W_{max}), moderate (40–60% W_{max}), and maximal (100% W_{max}) cycling exercise. An increase in CMR indicates an increase in global cerebral metabolism, whereas a reduction in OCI indicates a reduction in oxidative metabolism. (**b**) The percentage contribution of oxygen, glucose, and lactate trans-cerebral uptake to the CMR. Sizes of the circles are proportional to the percentage difference in CMR from rest. All values are means and SDs calculated from the eight studies quantifying CMR during incremental exercise (Ide et al. 2000; Nybo et al. 2003; Larsen et al. 2008; Rasmussen et al. 2010b; Brassard et al. 2010; Rasmussen et al. 2010a; Fisher et al. 2013; Trangmar et al. 2014). Figure and caption from (Smith and Ainslie 2017)

data indicate that the contribution of oxidative carbohydrate metabolism to cerebral substrate utilization is influenced by the balance between exercise intensity and arterial lactate availability (Smith and Ainslie 2017). With very intensive rowing exercise, elevated arterial levels of lactate (e.g., >15 mmol/l) contribute upward of 30–40% of total cerebral metabolism (Ide and Secher 2000; Volianitis et al. 2008). Pre-clinical studies indicate astrocytic glycogen-derived lactate may contribute to cerebral metabolism via lactate oxidation during hypoglycemia experienced with exhaustive exercise (Brown and Ransom 2007; Matsui et al. 2017). To date, unlike at rest (Fig. 13.5 above), no in vivo study in humans has investigated the role of ketone bodies as a substrate for cerebral metabolism during exercise.

13.4.1 Incremental Versus Steady-State Exercise and Cerebral Metabolism

The cerebral metabolic ratio (O_2 uptake/glucose + ½ lactate) decreases during prolonged exhaustive exercise where arterial lactate remains low; however, progressive vigorous exercise above lactate threshold facilitates higher cerebral lactate uptake and oxidation due to increases in systemic lactate availability. Glucose is the preferential substrate at rest; yet, oxidation of monocarboxylic acids such as lactate, pyruvate, and ketone bodies can contribute to cerebral metabolism when glucose utilization is restricted and/or when these substrates become available during exhaustive exercise. For example, as shown by Nybo et al. (2003), exercise-induced hypoglycemia during prolonged submaximal exercise (via 3 hours of cycling at 60% VO_{2max} with and without glucose supplementation) reduces cerebral glucose uptake and $CMRO_2$ with increases in cerebral β-hydroxybutyrate uptake (Fig. 13.10). It is noteworthy that systemic free fatty acids increased by up to threefold after 3 hours of cycling exercise without glucose supplementation; however, this did not result in an increase in cerebral arterial-venous free fatty acid exchange (Nybo et al. 2003). These results substantiate that the brain does not rely on fatty acid oxidation to contribute substantially to ATP production (Mitchell et al. 2011; Schönfeld and

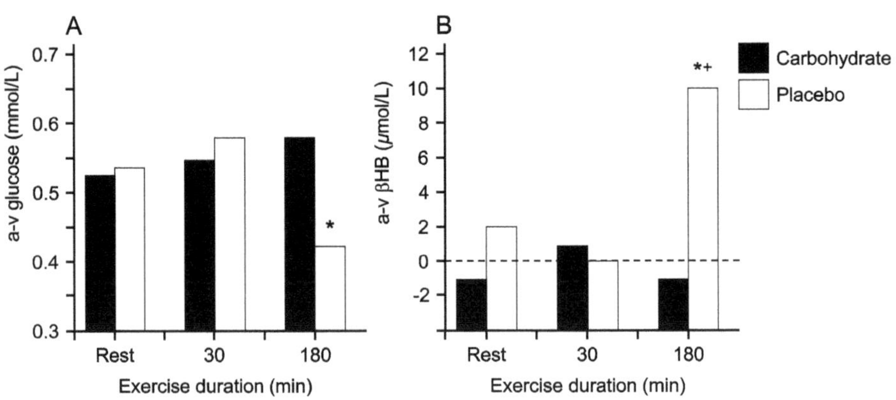

Fig. 13.10 The influence of carbohydrate availability on cerebral substrate utilization during prolonged submaximal cycling exercise. (**a**) Cerebral glucose uptake is reduced with exercise-induced hypoglycemia during prolonged submaximal cycling exercise at 60% VO_{2max}. (**b**) At the same time, cerebral ketone uptake is markedly increased during exercise with restricted carbohydrate availability. βHB, β-hydroxybutyrate. $^{*}P < 0.05$ versus resting value; $^{+}P < 0.05$ versus carbohydrate trial. Figure redrawn with data from (Nybo et al. 2003)

Reiser 2013). Overall, the relative contribution of cerebral substrates to fuel utilization during progressive versus steady-state exercise is markedly affected by systemic carbohydrate availability, exercise intensity, and nutritional status. The majority of studies to date have investigated trans-cerebral metabolism during acute (15-- 30 minutes) incremental/steady-state exercise; therefore, it is difficult to extrapolate these findings to prolonged continuous exercise where marked changes in systemic fuel utilization occur.

13.4.2 Cerebral Metabolic Rate of Oxygen During Exercise

Collectively, progressive relative increases in $CMRO_2$ throughout various exercise intensities occur in parallel with elevated cerebral oxygen delivery (i.e., blood flow $\times$ CaO_2) and cerebral O_2 extraction (Smith and Ainslie 2017). At maximal exercise intensities, $CMRO_2$ is linearly increased by approximately 30%—albeit with considerable reported variability—to prioritize substrate utilization in the face of reductions in cerebral blood flow. This relates to an increase in $CMRO_2$ in excess of cerebral blood flow and emphasizes the importance of adequate cerebral oxygen delivery and extraction at maximal exercise intensities. It is noteworthy that the relative increases in $CMRO_2$ are reflective of the total increase in cerebral metabolism during exercise. The contribution of trans-cerebral oxygen uptake to total cerebral metabolism is reduced by 30% at maximal exercise versus rest, thus emphasizing the increased contribution of anaerobic non-oxidative substrate utilization and a promising role for partial compensation via ketone body oxidation when carbohydrates are restricted.

13.4.3 Cerebral Metabolic Rate of Glucose During Exercise

Taken together, in concert with the elevated oxidative metabolism ($CMRO_2$) described above, the cerebral metabolic rate of glucose (CMR_{glu}) is markedly increased from rest to maximal exercise (Fig. 13.9). Importantly, the relative increase in glucose utilization is a result of increases in total cerebral metabolism, whereas the relative contribution of glucose to trans-cerebral uptake and total fuel oxidation is reduced with progressive exercise to exhaustion. The data presented in Fig. 13.9 summarize eight studies utilizing trans-cerebral arterial venous glucose values (Ide et al. 2000; Nybo et al. 2003; Larsen et al. 2008; Rasmussen et al. 2010b; Brassard et al. 2010; Rasmussen et al. 2010a; Fisher et al. 2013; Trangmar et al. 2014) as well as calculated cerebral blood flow (via Duplex vascular ultrasound) and/or the relative change in cerebral blood velocity-derived estimates of blood flow during incremental exercise. It is noteworthy that the elevated CMR_{glu} from rest to maximal exercise was principally attributable to an increased trans-cerebral exchange of glucose (i.e., higher glucose extraction) as these changes were unrelated to the cerebral glucose delivery (Dalsgaard et al. 2004a; Rasmussen et al. 2011). Indeed, previous studies consistently report glucose delivery at rest and during various exercise intensities (submaximal to exhaustion) to be in excess of cerebral glucose demand (Rasmussen et al. 2011; Fisher et al. 2013; Trangmar et al. 2014; Smith et al. 2014); therefore, quite unlike lactate availability, these data indicate that glucose uptake does not depend on arterial glucose levels.

13.4.4 Cerebral Metabolic Rate of Lactate During Exercise

At rest, the $<10\%$ utilization of lactate does not appreciably contribute to cerebral metabolism; however, cerebral lactate uptake can increase almost tenfold during progressive exercise to exhaustion (Dalsgaard et al. 2004b; Smith et al. 2014). This exponential rise in cerebral lactate uptake during incremental exercise is closely related to the elevated systemic arterial concentration (Larsen et al. 2008; Fisher et al. 2013). In support of this, van Hall et al. (2009) reported a highly significant relationship between arterial lactate availability and lactate extraction at rest and during 30 minutes of cycling at 75% maximal exercise intensity when systemic lactate was previously elevated via intravenous infusion (<1 mmol/l vs. 7 mmol/l). The arterial-venous lactate gradient changes from a small net "release" at rest to a 10 and 17% "uptake" during exercise and recovery, respectively. These increases in cerebral lactate uptake/oxidation occur with corresponding reductions in cerebral glucose uptake, thereby acting to maintain total carbohydrate oxidation (van Hall et al. 2009).

Although lactate uptake is related to systemic lactate availability, previous studies show that total cerebral lactate uptake is not affected by the maximal arterial lactate value achieved during exhaustive exercise. Smith et al. (2014) showed that arterial lactate availability was approximately 50% higher at high altitude versus sea level for a given absolute exercise intensity; however, cerebral lactate metabolism was not different at matched absolute workloads throughout progressive submaximal cycling exercise to exhaustion. These findings are further substantiated by Volianitis et al. (2011) who reported that total lactate uptake is not altered when both arterial pH and lactate values (approx. +10 mmol/l) were elevated (via intravenous sodium bicarbonate infusion) during a 2000-meter maximal rowing time trial. Lastly, pre-clinical evidence shows that exercise induces cerebral-specific vascular endothelial growth factor (VEGF) via lactate signaling with direct relevance to exercise-mediated neovascularization in the brain (Morland et al. 2017).

13.4.5 Ketone Utilization During Exercise?

Ketone body utilization as an alternative substrate for cerebral metabolism during exercise has received little attention with no in vivo human study to date investigating the relative contribution of cerebral metabolic rate of ketones during exercise. Influential work by Cahill and colleagues have established that at rest over 60% of the metabolic energy needs of the brain can be supplied by ketone bodies rather than glucose with situations of carbohydrate restriction (Owen et al. 1967; Cahill 1976); these results are further substantiated by the observation that ketone infusion during acute controlled experimental hypoglycemia (<6 hours) lowers counter-regulatory hormone responses and neuroglycopenic symptoms (Amiel et al. 1991; Veneman et al. 1994). As such, ketones may also contribute as a key alternative fuel substrate to glucose during prolonged exhaustive exercise alongside progressive reductions in cerebral glucose oxidation. As previously described, circulating ketones can reach 1–3 mmol/l during 3–6 days fasting (Owen et al. 1969; Garber et al. 1974; Haymond et al. 1982), prolonged exercise (Johnson et al. 1969; Volek et al. 2016), and very high fat ketogenic diet (Langfort et al. 1996; Johnston et al. 2006). Within these conditions of ketosis, ketones can supply between 50 and 80% of brain energy requirements (Hasselbalch et al. 1994; Hasselbalch et al. 1995; Blomqvist et al. 1995; Hasselbalch et al. 1996; Pan et al. 2000). Nevertheless, the existing literature on fuel utilization during exercise has focused almost exclusively on skeletal muscle, which can resynthesize ATP from ketone bodies, and shows that ketone body disposal in skeletal muscle is elevated up to fivefold during exercise (Hagenfeldt and Wahren 1968; Hagenfeldt and Wahren 1971; Balasse et al. 1978; Féry and Balasse 1983; Wahren et al. 1984; Féry and Balasse 1986; Féry and Balasse 1988). The ketogenic response to fasting prioritizes skeletal muscle protein, which would otherwise be catabolized via gluconeogenesis to produce glucose for cerebral substrate utilization (Veech 2004); this glycogen-sparing effect of ketone metabolism is also advantageous following exhaustive exercise. Lastly, exercise-induced βHB oxidation reportedly improves brain-derived neurotropic factor expression (Sleiman et al. 2016; Marosi et al. 2016); therefore, ketone body utilization may contribute to the therapeutic/neuroprotective effects of exercise (Nay et al. 2021).

13.4.6 Summary

The cerebral metabolic ratios for oxidative glucose and oxidative carbohydrate metabolism (i.e., OGI and OCI, respectively) are progressively attenuated from their stable resting values of approximately 5.7 by up to 40–50% with incremental exercise to exhaustion; such changes indicate an increase in anaerobic non-oxidative metabolism. As a compensatory action, increases in cerebral lactate uptake/oxidation occur with corresponding reductions in cerebral glucose uptake, thereby acting to maintain total carbohydrate oxidation. At rest, cerebral ketone body utilization increases with situations of carbohydrate restriction and/or increases in systemic ketone levels; therefore, it is likely that ketones may also contribute as a key alternative fuel substrate to glucose during prolonged exhaustive exercise alongside progressive reductions in cerebral glucose oxidation; however, this has yet to be experimentally tested in vivo in humans.

13.5 Summary

The interaction between cerebral substrate delivery (via cerebral blood flow and arterial substrate content) and trans-cerebral exchange is essential to support fuel utilization in the brain during exercise. The high cerebral metabolic rate, paired with a remarkably restricted capacity for substrate storage (e.g., glycogen <10 μmol/g), insufficient ability to increase oxygen extraction, and inability to utilize fat as substrate, necessitates tight regulation of cerebral blood flow to support oxidative metabolism. Future work is needed to ascertain the relative contribution of ketone utilization and/or astrocytic glycogen-derived lactate oxidation during hypoglycemia experienced with exhaustive exercise in vivo in humans.

References

Ainslie PN, Shaw AD, Smith KJ et al (2014) Stability of cerebral metabolism and substrate availability in humans during hypoxia and hyperoxia. Clin Sci 126:661–670

Amiel SA, Archibald HR, Chusney G et al (1991) Ketone infusion lowers hormonal responses to hypoglycaemia: evidence for acute cerebral utilization of a non-glucose fuel. Clin Sci 81:189–194

Askew EW, Dohm GL, Huston RL (1975) Fatty acid and ketone body metabolism in the rat: response to diet and exercise. J Nutr 105:1422–1432

Attwell D, Laughlin SB (2001) An energy budget for signaling in the grey matter of the brain. J Cereb Blood Flow Metab 21:1133–1145

Bailey DM (2019) Oxygen and brain death; back from the brink. Exp Physiol 104:1769–1779

Balasse EO, Féry F, Neef MA (1978) Changes induced by exercise in rates of turnover and oxidation of ketone bodies in fasting man. J Appl Physiol Respir Environ Exerc Physiol 44: 5–11

Barros LF, Deitmer JW (2010) Glucose and lactate supply to the synapse. Brain Res Rev 63:149–159

Blomqvist G, Thorell JO, Ingvar M et al (1995) Use of R-beta-[1-11C]hydroxybutyrate in PET studies of regional cerebral uptake of ketone bodies in humans. Am J Physiology-Legacy Content 269:E948–E959

Brassard P, Seifert T, Wissenberg M et al (2010) Phenylephrine decreases frontal lobe oxygenation at rest but not during moderately intense exercise. J Appl Physiol 108:1472–1478

Brown AM, Ransom BR (2007) Astrocyte glycogen and brain energy metabolism. Glia 55:1263–1271

Buxton RB, Frank LR (1997) A model for the coupling between cerebral blood flow and oxygen metabolism during neural stimulation. J Cereb Blood Flow Metab 17:64–72

Cahill GF (1976) Starvation in man. Clin Endocrinol Metab 5:397–415

Calbet JAL, Jensen-Urstad M, van Hall G et al (2004) Maximal muscular vascular conductances during whole body upright exercise in humans. J Physiol 558:319–331

Caldwell HG, Carr JMJR, Minhas JS et al (2021) Acid-base balance and cerebrovascular regulation. J Physiol 599:5337–5359
Clarke DD, Sokoloff L (1999) Substrates of cerebral metabolism. In: Siegel GJ, Agranoff BW, Albers RW et al (eds) Basic neurochemistry: molecular, cellular and medical aspects, 6th edn. Lippincott-Raven, Philadelphia. Available from https://www.ncbi.nlm.nih.gov/books/NBK28048/
Dalsgaard MK (2006) Fuelling cerebral activity in exercising man. J Cereb Blood Flow Metab 26: 731–750
Dalsgaard MK, Ogoh S, Dawson EA et al (2004a) Cerebral carbohydrate cost of physical exertion in humans. Am J Physiol Regul Integr Comp Physiol 287:R534–R540
Dalsgaard MK, Quistorff B, Danielsen ER et al (2004b) A reduced cerebral metabolic ratio in exercise reflects metabolism and not accumulation of lactate within the human brain. J Physiol 554:571–578
Edmond J, Robbins RA, Bergstrom JD et al (1987) Capacity for substrate utilization in oxidative metabolism by neurons, astrocytes, and oligodendrocytes from developing brain in primary culture. J Neurosci Res 18:551–561
Edvinsson LM, Krause DN (2002) Cerebral blood flow and metabolism. Lippincott, Williams & Wilkins, Philadelphia
Ekblom B, Huot R, Stein EM, Thorstensson AT (1975) Effect of changes in arterial oxygen content on circulation and physical performance. J Appl Physiol 39:71–75
Engl E, Attwell D (2015) Non-signalling energy use in the brain. J Physiol 593:3417–3429
Evans M, Cogan KE, Egan B (2016) Metabolism of ketone bodies during exercise and training: physiological basis for exogenous supplementation. J Physiol 595:2857–2871
Féry F, Balasse EO (1983) Ketone body turnover during and after exercise in overnight-fasted and starved humans. Am J Physiol 245:E318–E325
Féry F, Balasse EO (1986) Response of ketone body metabolism to exercise during transition from postabsorptive to fasted state. Am J Physiol 250:E495–E501
Féry F, Balasse EO (1988) Effect of exercise on the disposal of infused ketone bodies in humans. J Clin Endocrinol Metab 67:245–250
Fisher JP, Hartwich D, Seifert T et al (2013) Cerebral perfusion, oxygenation and metabolism during exercise in young and elderly individuals. J Physiol 591:1859–1870
Fox PT, Raichle ME (1986) Focal physiological uncoupling of cerebral blood flow and oxidative metabolism during somatosensory stimulation in human subjects. Proc Natl Acad Sci 83:1140–1144
Fox PT, Raichle ME, Mintun MA, Dence C (1988) Nonoxidative glucose consumption during focal physiologic neural activity. Science 241:462–464
Garber AJ, Menzel PH, Boden G, Owen OE (1974) Hepatic ketogenesis and gluconeogenesis in humans. J Clin Investig 54:981–989
Gjedde A (2005) The pathways of oxygen in brain. I. Delivery and metabolism of oxygen. Adv Exp Med Biol 566:269–275
Gonzales AL, Klug NR, Moshkforoush A et al (2020) Contractile pericytes determine the direction of blood flow at capillary junctions. Proc Natl Acad Sci U S A 117:27022–27033
Hagenfeldt L, Wahren J (1968) Human forearm muscle metabolism during exercise III uptake, release and oxidation of β-hydroxybutyrate and observations on the β-hydroxybutyrate/acetoacetate ratio. Scand J Clin Lab Invest 21:314–320
Hagenfeldt L, Wahren J (1971) Human forearm muscle metabolism during exercise. VI. Substrate utilization in prolonged fasting. Scand J Clin Lab Invest 27:299–306
Han Y-M, Bedarida T, Ding Y et al (2018) β-Hydroxybutyrate prevents vascular senescence through hnRNP A1-mediated upregulation of Oct4. Mol Cell 71:1064–1078.e5
Hasselbalch SG, Knudsen GM, Jakobsen J et al (1994) Brain metabolism during short-term starvation in humans. J Cereb Blood Flow Metab 14:125–131

Hasselbalch SG, Knudsen GM, Jakobsen J et al (1995) Blood-brain barrier permeability of glucose and ketone bodies during short-term starvation in humans. Am J Physiology-Legacy Content 268:E1161–E1166

Hasselbalch SG, Madsen PL, Hageman LP et al (1996) Changes in cerebral blood flow and carbohydrate metabolism during acute hyperketonemia. Am J Physiology-Endocrinology Metabolism 270:E746–E751

Haymond MW, Karl IE, Clarke WL et al (1982) Differences in circulating gluconeogenic substrates during short-term fasting in men, women, and children. Metabolism 31:33–42

Heni M, Eckstein SS, Schittenhelm J et al (2020) Ectopic fat accumulation in human astrocytes impairs insulin action. R Soc Open Sci 7:200701

Ide K, Secher NH (2000) Cerebral blood flow and metabolism during exercise. Prog Neurobiol 61: 397–414

Ide K, Schmalbruch IK, Quistorff B et al (2000) Lactate, glucose and O2 uptake in human brain during recovery from maximal exercise. J Physiol (Lond) 522(Pt 1):159–164

Johnson RH, Walton JL, Krebs HA, Williamson DH (1969) Metabolic fuels during and after severe exercise in athletes and non-athletes. Lancet 2:452–455

Johnston CS, Tjonn SL, Swan PD et al (2006) Ketogenic low-carbohydrate diets have no metabolic advantage over nonketogenic low-carbohydrate diets. Am J Clin Nutr 83:1055–1061

Kety SS, Schmidt CF (1945) The determination of cerebral blood flow in man by the use of nitrous oxide in low concentrations. Am J Physiol 143:53–66

Kety SS, Schmidt CF (1948) The nitrous oxide method for the quantitative determination of cerebral blood flow in man: theory, procedure and normal values. J Clin Investig 27:476–483

Kimura I, Inoue D, Maeda T et al (2011) Short-chain fatty acids and ketones directly regulate sympathetic nervous system via G protein-coupled receptor 41 (GPR41). Proc Natl Acad Sci U S A 108:8030–8035

Koepsell H (2020) Glucose transporters in brain in health and disease. Pflugers Arch 472:1299–1343

Kuschinsky W, Paulson OB (1992) Capillary circulation in the brain. Cerebrovasc Brain Metab Rev 4:261–286

Laffel L (1999) Ketone bodies: a review of physiology, pathophysiology and application of monitoring to diabetes. Diabetes Metab Res Rev 15:412–426

Langfort J, Pilis W, Zarzeczny R et al (1996) Effect of low-carbohydrate-ketogenic diet on metabolic and hormonal responses to graded exercise in men. J Physiol Pharmacol 47:361–371

Larsen TS, Rasmussen P, Overgaard M et al (2008) Non-selective β-adrenergic blockade prevents reduction of the cerebral metabolic ratio during exhaustive exercise in humans. J Physiol 586: 2807–2815

Lichtenstein D, Saïfi R, Augarde R et al (2001) The internal jugular veins are asymmetric. Usefulness of ultrasound before catheterization. Intensive Care Med 27:301–305

Lundby C, Sander M, van Hall G et al (2006) Maximal exercise and muscle oxygen extraction in acclimatizing lowlanders and high altitude natives. J Physiol 573:535–547

Madsen PL, Vorstrup S (1991) Cerebral blood flow and metabolism during sleep. Cerebrovasc Brain Metab Rev 3:281–296

Marosi K, Kim SW, Moehl K et al (2016) 3-Hydroxybutyrate regulates energy metabolism and induces BDNF expression in cerebral cortical neurons. J Neurochem 139:769–781

Matsui T, Omuro H, Liu Y-F et al (2017) Astrocytic glycogen-derived lactate fuels the brain during exhaustive exercise to maintain endurance capacity. Proc Natl Acad Sci U S A 114:6358–6363

Mikkelsen KH, Seifert T, Secher NH et al (2015) Systemic, cerebral and skeletal muscle ketone body and energy metabolism during acute hyper-D-β-hydroxybutyratemia in post-absorptive healthy males. J Clin Endocrinol Metab 100:636–643

Mintun MA, Lundstrom BN, Snyder AZ et al (2001) Blood flow and oxygen delivery to human brain during functional activity: theoretical modeling and experimental data. Proc Natl Acad Sci 98:6859–6864

Mitchell RW, On NH, Del Bigio MR et al (2011) Fatty acid transport protein expression in human brain and potential role in fatty acid transport across human brain microvessel endothelial cells. J Neurochem 117:735–746
Morland C, Andersson KA, Haugen ØP et al (2017) Exercise induces cerebral VEGF and angiogenesis via the lactate receptor HCAR1. Nat Commun 8:15557–15559
Nay K, Smiles WJ, Kaiser J et al (2021) Molecular mechanisms underlying the beneficial effects of exercise on brain function and neurological disorders. Int J Mol Sci 22:4052
Nybo L, Møller K, Volianitis S et al (2002) Effects of hyperthermia on cerebral blood flow and metabolism during prolonged exercise in humans. J Appl Physiol 93:58–64
Nybo L, Møller K, Pedersen BK et al (2003) Association between fatigue and failure to preserve cerebral energy turnover during prolonged exercise. Acta Physiol Scand 179:67–74
Owen OE, Morgan AP, Kemp HG et al (1967) Brain metabolism during fasting. J Clin Investig 46: 1589–1595
Owen OE, Felig P, Morgan AP et al (1969) Liver and kidney metabolism during prolonged starvation. J Clin Investig 48:574–583
Pan JW, Rothman TL, Behar KL et al (2000) Human brain beta-hydroxybutyrate and lactate increase in fasting-induced ketosis. J Cereb Blood Flow Metab 20:1502–1507
Patching SG (2017) Glucose transporters at the blood-brain barrier: function, regulation and gateways for drug delivery. Mol Neurobiol 54:1046–1077
Proctor DN, Shen PH, Dietz NM et al (1998) Reduced leg blood flow during dynamic exercise in older endurance-trained men. J Appl Physiol 85:68–75
Rasmussen P, Nielsen J, Overgaard M et al (2010a) Reduced muscle activation during exercise related to brain oxygenation and metabolism in humans. J Physiol 588:1985–1995
Rasmussen P, Overgaard A, Bjerre AF et al (2010b) The effects of normoxia, hypoxia, and hyperoxia on cerebral haemoglobin saturation using near infrared spectroscopy during maximal exercise. Int J Ind Ergon 40:190–196
Rasmussen P, Wyss MT, Lundby C (2011) Cerebral glucose and lactate consumption during cerebral activation by physical activity in humans. FASEB J 25:2865–2873
Rebelos E, Mari A, Bucci M et al (2020) Brain substrate metabolism and ß-cell function in humans: a positron emission tomography study. Endocrinol Diabetes Metab 3:e00136
Robertson CS, Narayan RK, Gokaslan ZL et al (1989) Cerebral arteriovenous oxygen difference as an estimate of cerebral blood flow in comatose patients. J Neurosurg 70:222–230
Robinson AM, Williamson DH (1980) Physiological roles of ketone bodies as substrates and signals in mammalian tissues. Physiol Rev 60:143–187
Sato K, Sadamoto T (2010) Different blood flow responses to dynamic exercise between internal carotid and vertebral arteries in women. J Appl Physiol 109:864–869
Schönfeld P, Reiser G (2013) Why does brain metabolism not favor burning of fatty acids to provide energy? Reflections on disadvantages of the use of free fatty acids as fuel for brain. J Cereb Blood Flow Metab 33:1493–1499
Shimazu T, Hirschey MD, Newman J et al (2013) Suppression of oxidative stress by β-hydroxybutyrate, an endogenous histone deacetylase inhibitor. Science 339:211–214
Siesjö BK (1978) Brain metabolism and anaesthesia. Acta Anaesthesiol Scand Suppl 70:56–59
Simard M, Arcuino G, Takano T et al (2003) Signaling at the gliovascular interface. J Neurosci 23: 9254–9262
Simpson IA, Carruthers A, Vannucci SJ (2007) Supply and demand in cerebral energy metabolism: the role of nutrient transporters. J Cereb Blood Flow Metab 27:1766–1791
Skattebo Ø, Calbet JAL, Rud B et al (2020) Contribution of oxygen extraction fraction to maximal oxygen uptake in healthy young men. Acta Physiol (Oxf) 230:e13486
Sleiman SF, Henry J, Al-Haddad R et al (2016) Exercise promotes the expression of brain derived neurotrophic factor (BDNF) through the action of the ketone body β-hydroxybutyrate. elife
Smith KJ, Ainslie PN (2017) Regulation of cerebral blood flow and metabolism during exercise. Exp Physiol 102(11):1356–1371

Smith KJ, MacLeod D, Willie CK et al (2014) Influence of high altitude on cerebral blood flow and fuel utilization during exercise and recovery. J Physiol 592:5507–5527
Sokoloff L (1960) The metabolism of the central nervous system *in vivo*. In: Handbook of physiology–neurophysiology, vol 3. American Physiological Society, Washington, DC, pp 1843–1864
Svart M, Gormsen LC, Hansen J et al (2018) Regional cerebral effects of ketone body infusion with 3-hydroxybutyrate in humans: reduced glucose uptake, unchanged oxygen consumption and increased blood flow by positron emission tomography. A randomized, controlled trial. PLoS One 13:e0190556
Thomas KN, Lewis NCS, Hill BG, Ainslie PN (2015) Technical recommendations for the use of carotid duplex ultrasound for the assessment of extracranial blood flow. Am J Physiol Regul Integr Comp Physiol 309:1–7
Trangmar SJ, Chiesa ST, Stock CG et al (2014) Dehydration affects cerebral blood flow but not its metabolic rate for oxygen during maximal exercise in trained humans. J Physiol 592:3143–3160
van Hall G, Strømstad M, Rasmussen P et al (2009) Blood lactate is an important energy source for the human brain. J Cereb Blood Flow Metab 29:1121–1129
Veech RL (2004) The therapeutic implications of ketone bodies: the effects of ketone bodies in pathological conditions: ketosis, ketogenic diet, redox states, insulin resistance, and mitochondrial metabolism. Prostaglandins Leukot Essent Fatty Acids 70:309–319
Veneman T, Mitrakou A, Mokan M et al (1994) Effect of hyperketonemia and hyperlacticacidemia on symptoms, cognitive dysfunction, and counterregulatory hormone responses during hypoglycemia in normal humans. Diabetes Care 43:1311–1317
Vlassenko AG, Rundle MM, Raichle ME, Mintun MA (2006) Regulation of blood flow in activated human brain by cytosolic NADH/NAD+ ratio. Proc Natl Acad Sci 103:1964–1969
Volek JS, Freidenreich DJ, Saenz C et al (2016) Metabolic characteristics of keto-adapted ultra-endurance runners. Metabolism 65:100–110
Volianitis S, Fabricius-Bjerre A, Overgaard A et al (2008) The cerebral metabolic ratio is not affected by oxygen availability during maximal exercise in humans. J Physiol 586:107–112
Volianitis S, Rasmussen P, Seifert T et al (2011) Plasma pH does not influence the cerebral metabolic ratio during maximal whole body exercise. J Physiol 589:423–429
Wahren J, Sato Y, Ostman J et al (1984) Turnover and splanchnic metabolism of free fatty acids and ketones in insulin-dependent diabetics at rest and in response to exercise. J Clin Investig 73: 1367–1376
Williams JL, Shea M, Jones SC (1993) Evidence that heterogeneity of cerebral blood flow does not involve vascular recruitment. Am J Physiology-Legacy Content 264:H1740–H1743
Wu X, Miao D, Liu Z et al (2020) β-Hydroxybutyrate antagonizes aortic endothelial injury by promoting generation of VEGF in diabetic rats. Tissue Cell 64:101345
Xin L, Ipek Ö, Beaumont M et al (2018) Nutritional ketosis increases NAD+/NADH ratio in healthy human brain: an in vivo study by 31P-MRS. Front Nutr 5:62
Yancy WS Jr, Olsen MK, Guyton JR et al (2004) A low-carbohydrate, ketogenic diet versus a low-fat diet to treat obesity and hyperlipidemia. Ann Intern Med 140:769–777

Chapter 14
Effects of Age on Exercise Metabolism

J. Matthew Hinkley, Paul M. Coen, and Bret H. Goodpaster

Abstract Exercise training is an important intervention to improve metabolic health in older adults. The phenotypic adaptations provoked by chronic training are initiated by the metabolic response to an acute bout of exercise; however, how aging affects this initial response is unclear. In this chapter, we discuss the effects of aging on the metabolic response to acute bouts of resistance and endurance exercise and how age-related diseased conditions may further impair this process. Additionally, we describe various lifestyle interventions that improve the metabolic response to exercise.

Keywords Aging · Resistance exercise · Endurance exercise · Diabetes · Cardiovascular disease · Heart failure

14.1 Introduction

It is projected that by 2030, more than 20% of the population in the United States will be over 65 years of age (Ortman et al. 2014). This represents a major healthcare challenge as aging is the strongest risk factor for many common diseases, including cardiovascular disease and type 2 diabetes. Maladaptations in metabolism greatly contribute to the increased risk of cardiometabolic disease in older adults and are caused to a large degree by reduced exercise and physical activity behaviors. For example, cardiorespiratory fitness (VO_2peak) decreases with age (Fleg et al. 2005) and is associated with an increased risk of cardiometabolic diseases and all-cause mortality (Kodama et al. 2009). Older adults are also susceptible to developing impaired glucose tolerance and insulin resistance (Kalyani and Egan 2013) which underly the development of type 2 diabetes. There is a higher prevalence of cardiovascular mortality in older adults with impaired glucose tolerance compared to those with normal glucose tolerance (Kokkinos et al. 2009). Furthermore, the

J. M. Hinkley · P. M. Coen · B. H. Goodpaster (✉)
AdventHealth Translational Research Institute, Orlando, FL, USA
e-mail: Bret.Goodpaster@AdventHealth.com

© The Author(s), under exclusive license to Springer Nature Switzerland AG 2022

G. McConell (ed.), *Exercise Metabolism*, Physiology in Health and Disease,
https://doi.org/10.1007/978-3-030-94305-9_14

age-related loss of muscle mass and function (i.e., sarcopenia) leads to an increased risk of falls, fractures, mobility disability, and loss of functional independence. The healthcare burden that the aging population presents elevates the significance of developing effective therapeutic strategies for common diseases and disorders associated with aging.

Exercise is a powerful way to improve metabolic health, muscle mass, and strength. However, evidence suggests that aging can impact the metabolic response to acute and chronic exercise, an important consideration if exercise is to be utilized as a frontline strategy to promote healthy aging. In this chapter, we will discuss how sedentary aging affects the metabolic response to an acute bout of exercise and how this may relate to the impaired adaptive response to chronic exercise training. Specifically, we will consider the metabolic response to resistance and endurance exercise and how it differs between young and older sedentary adults and discuss potential strategies that may improve the metabolic effect of exercise in older adults. Additionally, we will consider how cardiometabolic diseases (diabetes, cardiovascular disease, and heart failure) and commonly prescribed drugs also impact exercise metabolism in older adults. Finally, we will make the argument that lifelong exercise may represent the most potent strategy for extending health span and negating the impact of aging on the metabolic response to exercise.

14.2 Metabolic Response to Resistance Exercise

Skeletal muscle plays an important role in physical and metabolic function, and the loss of muscle mass and strength contributes to various chronic conditions associated with aging, including frailty, sarcopenia, insulin resistance, and osteoporosis (Wolfe 2006). Progressive resistance exercise training is a valuable intervention to maintain or increase muscle mass and function. Older adults well into their 90s appear to greatly benefit from resistance exercise (Fiatarone et al. 1994); short-term training interventions increase muscle size and strength in elderly men and women (Stec et al. 2017; Taaffe et al. 1996; Trappe et al. 2001; Trappe et al. 2000). However, the benefits of resistance exercise training appear to be attenuated with advanced age in comparison to younger counterparts (Greig et al. 2011; Rivas et al. 2014). Specifically, improvements in whole-muscle mass and function were blunted in both men and women >80 years of age following 12 weeks of progressive resistance exercise training (Raue et al. 2009; Slivka et al. 2008). A recent meta-analysis concluded that older age was associated with a smaller increase in type 1 and 2 myofiber size following resistance exercise training in both men and women (Straight et al. 2020). Those in the ninth decade of life (80–89 years of age) had the lowest effect size for myofiber hypertrophy. In addition to size, single muscle fiber contractile function is not improved in both type 1 and 2 myofibers in men and women >80 years of age following resistance exercise training (Raue et al. 2009; Slivka et al. 2008).

Resistance exercise induced improvements in muscle mass and strength are associated with an accrual of muscle protein, most notably an increase in myofibrillar proteins. With each acute bout of resistance exercise, there is an increase in protein turnover (i.e., protein synthesis and degradation) in muscle. In young adults, muscle protein synthesis rates increase within the first hour following exercise and remains elevated up to 48 hours after the acute bout of exercise (Dreyer et al. 2006; Phillips et al. 1997; Walker et al. 2011). The increase in protein synthesis appears to stem from enhanced mRNA translation through activation of the mammalian/mechanistic target of rapamycin complex (mTORC1) (Walker et al. 2011). While the importance of mTORC1 on muscle hypertrophy was first described in rodents (Bodine et al. 2001), human clinical trials have shown that pharmacological inhibition of mTORC1 by rapamycin can reduce protein synthetic rates in young adults following a bout of resistance exercise (Drummond et al. 2009). Interestingly, protein degradation rates also increase following acute resistance exercise in young adults (Phillips et al. 1997). However, net protein balance becomes less negative following resistance exercise, suggesting a greater protein synthetic response following exercise in young adults (Phillips et al. 1997). Over time, the greater balance toward protein synthesis leads to an accumulation of new myofibrillar proteins, eventually resulting in a larger muscle mass (Terzis et al. 2008).

The attenuated hypertrophic response to resistance exercise training may stem from an inability to adapt to an acute stimulus. Following an acute bout of resistance exercise, older adults appear to have a blunted anabolic response in comparison to younger counterparts (Kumar et al. 2009; Walker et al. 2011; Welle et al. 1995). The differences in anabolic sensitivity are evident across a spectrum of exercise intensities in older adults (Kumar et al. 2009). Further, the protein synthetic response appears to be delayed in older adults, peaking at 3–6 hours post-exercise in comparison to young adults, which occurs 1–3 hours following exercise (Drummond et al. 2008). In line with the impaired protein synthetic response, older adults present a blunted anabolic signaling response, as evident by lower mTORC1 activation (Fry et al. 2011; Kumar et al. 2009). Interestingly, Kumar et al. have shown that doubling of exercise volume (3 sets vs. 6 sets) increased protein synthesis rates and downstream signaling targets of mTORC1 (p70S6K phosphorylation) in older adults (Kumar et al. 2012). When comparing young and older adults that completed an acute bout of resistance exercise at 75% of 1-RM, doubling of exercise volume led to a similar anabolic response between groups (Kumar et al. 2012), suggesting that an increase in exercise volume may negate the age-related anabolic resistance.

Along with exercise volume, other strategies have been implemented to improve anabolic sensitivity to resistance exercise in older adults. Ingestion of protein supplementation stimulates skeletal muscle protein synthesis and can exacerbate that anabolic effect of resistance exercise in young adults (Walker et al. 2011). Similar to resistance exercise, supplemental protein, particularly essential amino acids (EAA) like leucine, appear to stimulate protein synthesis via activation of mTORC1 (Dreyer et al. 2008). With feeding alone, protein supplemental can increase skeletal muscle protein synthetic rates in older adults to a similar level as young counterparts when levels of EAA are sufficient (leucine content ~3 g) (Katsanos et al. 2006). When combined with resistance exercise, protein supplementation can also act synergistically to increase protein synthesis rates in aged-skeletal muscle when total protein content exceeds 20 grams, with a plateau at ~40 grams of protein (Drummond et al. 2008; Symons et al. 2011; Yang et al. 2012). This is significantly more than young adults, who only need ~10 grams of protein to stimulate protein synthesis rates following resistance exercise, with a plateau response occurring at ~20 grams of protein (Moore et al. 2009; Tang et al. 2007). Delivery of amino acids is reduced in older adults, providing a possible explanation for larger bolus of protein that is needed to stimulate skeletal muscle protein synthesis (Dillon et al. 2011; Moro et al. 2016; Timmerman et al. 2010). However, despite this impairment, these findings suggest the combination of resistance exercise and protein supplementation may be sufficient to improve anabolic sensitivity in skeletal muscle of older adults (Breen and Phillips 2011).

Low-load blood flow restriction (BFR) has been suggested to be an important tool to induce muscle hypertrophy (Loenneke et al. 2012). The advantages of this type of training include reduced mechanical stress on joints and bones, which is particularly beneficial for an older adult population that suffers from osteoarthritis and osteoporosis (Centner et al. 2019). Previous work in young adults revealed that low-load BFR can increase the protein synthetic response in skeletal muscle (Fujita et al. 2007). Interestingly, while previous studies have shown a reduced anabolic response to low-load resistance exercise (Kumar et al. 2009), protein synthesis rates and mTORC1 signaling are enhanced in skeletal muscle of older adults following resistance exercise combined with BFR (Fry et al. 2010). The acute effects on protein metabolism appear to lead to increased muscle quality, as older adults engaged in chronic low-load BFR training present increase in muscle mass and strength (Centner et al. 2019; Karabulut et al. 2010). These data suggest low-load BFR training may enhance both the acute and chronic effects of resistance exercise on skeletal muscle protein metabolism.

14.3 Metabolic Response to Endurance Exercise

Endurance-type exercise interventions have been shown to improve multiple endurance-related endpoints in older adults, including faster 400 m walk time, reduced fatigue, and elevated aerobic capacity (American College of Sports Medicine C-Z et al. 2009; Santanasto et al. 2017). In general, endurance exercise training results in 1–2 metabolic equivalent (MET) increase in VO_2peak. This improvement is significant, as a 1 metabolic equivalent (MET) increase in VO_2peak reduces the risk of mortality by ~25% (Kodama et al. 2009). Elevated cardiovascular fitness results in a greater physiological reserve above the frailty threshold (~5 METs) in older active adults, which may be beneficial when acute and chronic illnesses later in life further reduce fitness; this reduced fitness and physiological reserve puts these adults on the precipice of frailty (Myers et al. 2002). Additionally, and contrary to common dogma, structured endurance exercise interventions may improve myofiber size and strength in older adults, which translates to elevated muscle mass and physical function (Harber et al. 2012).

The high-frequency, low-load muscle contractions provoked by endurance exercise can increase the rate of both fat and glucose oxidation by nearly tenfold to meet the increased energetic demands of prolonged exercise (Goodpaster and Sparks 2017; Klein et al. 1994; Mittendorfer and Klein 2001). Substrate availability is dependent on delivery both from the plasma and from intracellular stores. Once inside the cell, glucose can be converted to ATP through either anaerobic or aerobic metabolism, while fat is primarily converted to ATP by aerobic metabolism. Oxidative metabolism within the mitochondria provides the majority of energy during prolonged moderate-intensity exercise (25–60% VO_2peak) (Klein et al. 1994; Mittendorfer and Klein 2001). Chronic endurance exercise improves the ability to convert carbohydrate and lipid fuels to energy, which appears to be mediated by an increase in mitochondrial content and function in skeletal muscle (Oliveira et al. 2021).

Measuring gas exchange during exercise (i.e., volume of oxygen consumed (VO_2) and volume of carbon dioxide expelled (VCO_2) via indirect calorimetry provides rates of fat and glucose oxidation. With an increase in exercise intensity, the respiratory exchange ratio (RER; VCO_2/VO_2) is gradually increased, suggesting a greater reliance of carbohydrate oxidation. Interestingly, for a given absolute exercise intensity, RER is higher in older adults, suggesting a greater reliance on carbohydrate oxidation (and less reliance on fat) during endurance exercise (Julius et al. 1967; Montoye 1982; Sial et al. 1998). In addition to gas exchange measures, the use of isotopically labeled substrates can aid to identify and quantify the contribution of exogenous and endogenous sources of substrate oxidation. At a similar absolute exercise intensity, the contribution of fat and carbohydrates on total energy expenditure in young adults was ~50% each, while the majority (two-thirds) of total energy came from carbohydrate sources in older adults, with a greater reliance of muscle glycogen stores (Sial et al. 1998). As muscle glycogen levels are reduced with age (Meredith et al. 1989), a greater reliance of muscle glycogen levels can lead to an impaired ability to continue exercise in older adults, i.e., can lead to fatigue. The lower fat oxidation does not appear to be related to reduced substrate availability, as the rate of fatty acid appearance during exercise is greater in older adults than younger counterparts (Sial et al. 1998). Additionally, it has been shown that the increased appearance of plasma free fatty acids observed in older adults reduces the reliance on intracellular fat stores during exercise (Boon et al. 2007; Chee et al. 2016), leading to increased accumulation of lipids within the subsarcolemmal regions of muscle (Chee et al. 2016). This can have severe implications beyond exercise capacity, as elevated subsarcolemmal lipids are associated with insulin resistance (Chee et al. 2016).

Cardiorespiratory fitness (VO_2peak) starts to decline 3–6% each decade starting at the fourth decade of life, with greater loss (>20% per decade) after the age of 70 years (Fleg et al. 2005). Thus, the greater reliance of carbohydrate oxidation when exercising at a similar absolute exercise intensity may simply be due to older adults exercising at a greater percentage of their VO_2peak. When utilizing a similar relative exercise intensity (i.e., percent of VO_2peak), findings are conflicting, with results showing older adults had higher (Sial et al. 1998; Silverman and Mazzeo 1996), lower (Hagberg et al. 1988), or no difference (Tankersley et al. 1991) in RER in comparison to younger counterparts. The discrepancy between studies may be due to phenotypic characteristics of the participants, including physical activity levels (discussed later in this chapter). When phenotypically similar young and older adults exercised at a similar relative intensity (i.e., 50% of VO_2peak), the age-related suppression of fat oxidation remains (Sial et al. 1998). In contrast to exercise at a similar absolute exercise intensity, reduction in fat oxidation in older adults when exercising at a similar relative exercise intensity may potentially be due to an inability to stimulate lipolysis, as indicated by a suppression in the rate of free fatty acid appearance during exercise (Sial et al. 1998).

Along with a reduction in fat oxidation, carbohydrate oxidation was also suppressed in older adults when exercising at a similar relative exercise intensity as younger adults (Sial et al. 1998). This would suggest the oxidative capacity of skeletal muscle is lower in older adults. Indeed, mitochondrial enzyme activity is lower in skeletal muscle of older adults (Coggan et al. 1992). The reduction in mitochondrial enzyme activity is a product of a lower mitochondrial number and impairments in mitochondrial function in older adults (Mittendorfer and Klein 2001). Mitochondrial function can be assessed in vivo using magnetic resonance spectroscopy (31P-MRS) which measures changes in phosphocreatine levels following an acute bout of muscle contractions. In response to contractile activity, the ability to resynthesize phosphocreatine is lower in sedentary older adults (Braganza et al. 2019; Conley et al. 2000). Additionally, the balance between ATP hydrolysis and ATP generation with acute exercise is greater in older adults, suggesting an impaired ability of the mitochondria to resynthesis ATP in response to contractile activity (Coggan et al. 1993). Along with in vivo measures, mitochondrial function can be assessed ex vivo using muscle biopsy and isolated blood cells obtained from participants using high-resolution respirometry. In line with in vivo assessments, skeletal muscle and blood-based mitochondrial energetics is impaired in older adults (Braganza et al. 2019; Short et al. 2005). The changes in mitochondrial function can impact whole-body oxidative capacity, with lower mitochondrial energetics associated with reduced VO_2peak and higher levels of fatigue.

As mitochondria play an important role in facilitating substrate utilization during exercise, various therapies have been suggested to improve mitochondrial function. Most notably, carnitine and nitrates have been purported to improve fat metabolism and mitochondrial energetics, respectively, and may improve substrate metabolism during exercise (Chee et al. 2021; Larsen et al. 2011). Fat oxidation is facilitated by the ability of fatty acids to enter the mitochondria via the rate limiting enzyme carnitine palmitoyltransferase 1 (CPT1) (Fritz and McEwen 1959). Carnitine is critical to facilitate fatty acid flux into the mitochondria, and previous work in young adults has shown that carnitine supplementation increases fat oxidation during exercise at a relative intensity (50% VO_2peak), as well as a reduction in glycogen utilization during exercise (Stephens et al. 2013; Wall et al. 2011). Similar findings have been shown in older adults, as 25 weeks of carnitine supplementation, which increased muscle carnitine levels by 20%, led to an increase in fat oxidation during exercise at 50% VO_2peak (Chee et al. 2021). Along with carnitine, dietary inorganic nitrate supplementation has recently been identified to improve endurance performance (Jones 2014). Specifically, as little as 3 days of nitrate supplementation has been shown to reduce oxygen consumption and increase exercise efficiency (wattage/oxygen consumed) during an acute bout of endurance exercise (50% VO_2peak) (Larsen et al. 2011). Mechanistically, dietary nitrates improve energy transfer and efficiency in skeletal muscle mitochondria (Larsen et al. 2011). The improvements in mitochondrial efficiency have implications for exercise, as the improvement in mitochondrial efficiency is associated with lower oxygen consumption during an acute bout of endurance exercise, as well as related to a higher exercise efficiency during the bout (Larsen et al. 2011). While the role of dietary nitrates has not been explored in older adults, these data suggest dietary nitrates may be able to improve the capacity of skeletal muscle mitochondria to increase ATP production during energetic stress (i.e., exercise).

14.4 Considerations for Understanding the Effect of Age on Exercise Metabolism

14.4.1 Impact of Cardiometabolic Disease and Age on Exercise Metabolism

Along with the negative impact of aging on the body, older adults also need to contend with an increased risk of chronic diseases. In the United States, nearly 85% of older adults have at least one chronic disease condition and 60% having at least two conditions (Fong 2019). Most notably, non-communicable cardiometabolic disorders (i.e., cardiovascular disease, heart failure, and type 2 diabetes) are the leading cause of disability and death in the older adult population in the United States (Lopez et al. 2014). The presence of chronic disease conditions can severely impact quality of life in older adults, causing an earlier and steeper decline in functional capacity (Fong 2019). While higher levels of physical activity levels appear to protect against functional or mobility limitations seen in sedentary older adults (Brach et al. 2004), exercise training interventions reveal patients with chronic disease conditions may not respond as well or as completely. Specifically, there may be less improvement in mitochondrial energetics and insulin sensitivity following an endurance exercise training intervention in individuals with impaired glucose tolerance or type 2 diabetes (Kacerovsky-Bielesz et al. 2009; Malin and Kirwan 2012; Solomon et al. 2013). Further, older adults with heart failure are exercise intolerant, as indicated by a decrease in VO_2peak, and do not appear to respond favorably to exercise training interventions (Coats et al. 2017; Kitzman et al. 2002; Tucker et al. 2018).

While the acute effects of exercise on metabolism have not been examined fully in older adults with chronic disorders, possible physiological adaptations provoked by cardiometabolic diseases may lead to a reduced capacity to perform exercise in older adults. Of note, a reduction in cardiorespiratory fitness (VO_2peak) is associated with the development of many cardiometabolic disorders and remains lower when diagnosed with the condition (Kitzman et al. 2002; Kokkinos et al. 2009). Previous work has shown that a reduction in VO_2peak may explain the reduction in fat oxidation in older sedentary adults (Sial et al. 1998). In line with this, it has been shown that older obese adults, who are at a higher risk of cardiometabolic disorders, have a reduction in fat oxidation in response to acute endurance exercise in comparison to lean counterparts (Chee et al. 2016). The impairments in cardiorespiratory fitness and fat oxidation may stem from disease-associated reductions in mitochondrial density and function in comparison to sedentary yet healthy counterparts (Chomentowski et al. 2011; Drexler et al. 1992; Ritov et al. 2010; Ritov et al. 2005; Tucker et al. 2018). In addition to a reduction in cardiorespiratory fitness, older adults with cardiometabolic disorders present a reduction in skeletal muscle mass and strength in comparison to sedentary but otherwise healthy counterparts (Cicoira et al. 2001; Mancini et al. 1992; Minotti et al. 1991; Park et al. 2009; Park et al. 2006). Collectively, reduced cardiorespiratory fitness and muscle quality provoked by cardiometabolic diseases may impair the ability to perform exercise in these individuals, leading to an aberrant metabolic response to exercise and reduced adaptive ability to respond to chronic exercise training. More research is needed to examine the acute response to resistance and endurance exercise to determine how exercise metabolism is altered in a growing demographic of older adults with cardiometabolic diseases.

14.4.2 Exercise-Drug Interaction

Many older adults take at least one prescribed drug to treat various disease conditions, including diabetes and cardiovascular disease. In fact, more individuals take a prescribed drug (44% of population) than those who exercise (16.4% of adults >65 years of age) (Miller and Thyfault 2020). With many physicians recommending exercise in addition to drug treatment, it is important to understand exercise-pharmacological interactions to determine how commonly prescribed drugs may benefit or deter the positive adaptations to exercise. Two drugs that have recently been scrutinized for their role on exercise adaptations in older adults are metformin and statins (Miller and Thyfault 2020). Metformin is the fourth most commonly prescribed drug in the United States and is a first-line defense against type 2 diabetes (American Diabetes Association 2018). While metformin has been proposed to treat other diseased states associated with aging (i.e., cancer, cognitive decline, and cardiovascular), recent clinical trials have been implemented to test whether the drug can prevent the onset of many age-related diseases (Barzilai et al. 2016). In the case of type 2 diabetes, metformin and exercise are commonly prescribed together; yet recent evidence has shown that drug treatment may attenuate the positive benefits to exercise. In insulin-resistant individuals, metformin has been shown to abolish the acute effects of endurance exercise on muscle insulin sensitivity (Sharoff et al.

2010), as well blunting the chronic effects of endurance exercise training on insulin sensitivity, cardiovascular risk, and VO_2max (Braun et al. 2008; Malin and Braun 2013; Malin et al. 2013). The negative effects of metformin on endurance exercise training have been extended to healthy older adults, as it has recently been shown that a combination of 12 weeks of endurance exercise with metformin attenuated exercise-induced increases in whole-body and skeletal muscle oxidative capacity, as well as insulin sensitivity (Konopka et al. 2019). In addition to endurance exercise, metformin has been shown to attenuate exercise-induced increases in skeletal muscle mass following 14 weeks of progressive exercise training in older healthy adults (Walton et al. 2019), which may stem from blunting of transcript pathway-associated muscle hypertrophy (Kulkarni et al. 2020). Together, these findings suggest metformin may mitigate the positive benefits of both endurance and resistance exercise training in older adults.

Statins are a potent cholesterol-lowering drug, with nearly 40 million adults in the United States prescribed the drug to reduce the risk of cardiovascular disease (Salami et al. 2017). While beneficial toward cholesterol lowering, statins appear to induce several negative side effects, including impaired insulin sensitivity and muscle pain (myalgia) (Larsen et al. 2018; Morville et al. 2019). Similar to metformin, many physicians recommend exercise along with statins to treat hypercholesterolemia. Interestingly, middle-aged individuals with high fitness taking statins lowers mortality risk by 34%, which is greater than each treatment alone (Kokkinos et al. 2014). Further, when comparing statin users only, those with a higher exercise capacity (>8.5 METs) was associated with 52% lower mortality (Kokkinos et al. 2014). This would suggest that improvements in fitness through exercise training in conjunction with statin treatment may reduce mortality risk in older adults. However, recent work has suggested that, similar to metformin, statin treatment may blunt the positive effects of exercise. In a recent review (Miller and Thyfault 2020), the authors described a preliminary study that examined the effects of 12 weeks of statin treatment with moderate-intensity endurance exercise (60–75% heart rate reserve, 45 minutes/day, 5 days/week) on aerobic capacity and muscle mitochondrial content in overweight/obese adults. The study revealed that statin treatment abolished exercise-induced increases in VO_2max and muscle citrate synthase activity. A possible explanation for impaired response to exercise may be due to an increased prevalence of muscle pain (i.e., myalgia) in individuals treated with statins which may augment muscle strength, endurance, and overall performance (Noyes and Thompson 2017). Unlike metformin, how the combined effects of aging and statin treatment affect exercise-induced adaptations have not been clearly defined. However, based on the negative adaptations of statins alone, as well as emerging data combining drug and exercise treatment in obese adults, it appears that statins may lead to negative adaptations at the whole-body and individual tissue level.

14.4.3 Healthy Aging

While aging is typically associated with a sedentary lifestyle, cardiometabolic disease, and frailty, recent evidence has suggested that lifelong exercise training may negate many detrimental effects of aging. The late 1960s triggered the beginning of the exercise boom in the United States, in which men and women began engaging in lifelong structured physical activity, including recreational activities as well as training for competition, to maintain physical health. Many of these individuals have maintained their active lifestyle through the eighth decade of life, providing a unique group to examine whether exercise can negate the detrimental effects of age on whole-body health. Previous work has shown that cardiorespiratory fitness in male and female lifelong exercisers is elevated when compared to older healthy but sedentary counterparts (Gries et al. 2018; Heath et al. 1981; Pollock et al. 1997; Trappe et al. 2013). Using data required from the Fitness Registry and the Importance of Exercise Database (FRIEND) (Kaminsky et al. 2017), recent work has shown that VO_2peak values for lifelong exercisers were the equivalent to individuals 15–35 years younger (Gries et al. 2018). Vigorous training leads to further improvements in cardiorespiratory fitness, with older men who exercise for performance (i.e., master athletes) having a nearly 40% greater relative VO_2peak in comparison to those who exercise for fitness and health (Gries et al. 2018). Lifelong exercisers have increased left ventricular mass and diastolic filling, which equates to increased oxygen pulse, a surrogate for stroke volume, as well as increased muscle capillarization (Gries et al. 2018). Additionally, older adults engaged in >5 years of endurance exercise (swimming, running, and cycling) have elevated in vivo and ex vivo mitochondrial energetics (Distefano et al. 2018; Lanza et al. 2008). Along with improved endurance parameters, lifelong exercise training can improve skeletal muscle mass and function (Chambers et al. 2020; Distefano et al. 2018; Gries et al. 2019). While the effects of lifelong exercise on the acute metabolic response to exercise have not been fully explored, these data suggest that chronic physical activity attenuates the age-related declines in cardiorespiratory fitness and muscle quality. As both these parameters aid in exercise responsiveness, it is possible that the aging athlete may respond positively to acute exercise in comparison to sedentary counterparts.

14.5 Conclusion

While the effects of exercise training on improving many age-related complications have been well described, the impact of aging on the acute metabolic response to a single bout of exercise has not been thoroughly discussed. In this chapter, we provide evidence that aging appears to negatively impact the acute metabolic responses to both resistance and endurance exercise (Fig. 14.1), potentially leading to an impaired ability to positively respond to exercise training interventions. The

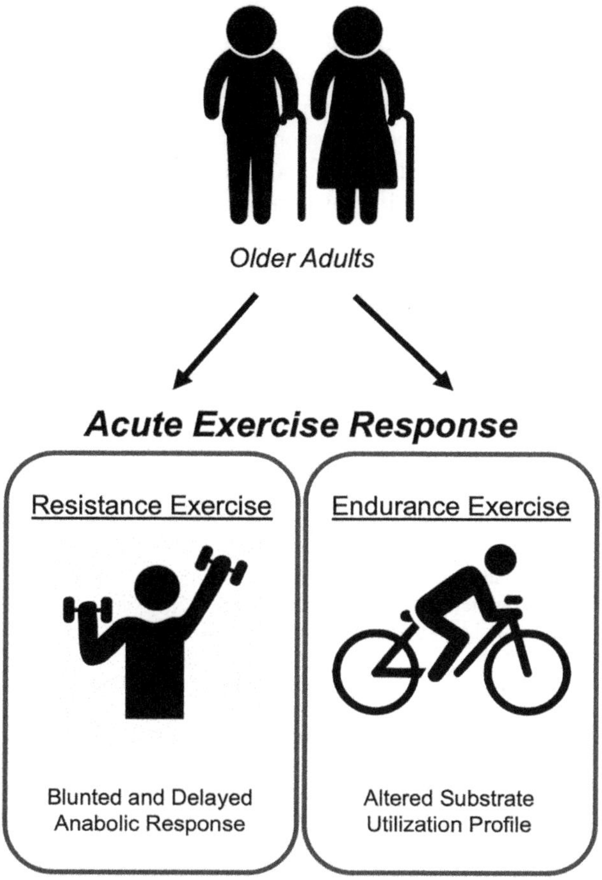

Fig. 14.1 Overview of the metabolic response to an acute bout of exercise in older adults. Sedentary older adults (>65 years old) participating in a single bout of either resistance or endurance exercise have an impaired metabolic response in comparison to younger counterparts. Specifically, following a bout of resistance exercise, the anabolic response (i.e., protein synthetic rate) is lower in aged skeletal muscle. Further, the anabolic response is delayed when compared to younger adults completing a similar relative intensity bout. Following endurance exercise, older adults have an altered substrate utilization profile than young adults. This includes (1) higher respiratory quotient ratio (RER) when performing exercise at similar absolute and relative intensities, (2) lower fatty acid oxidation in skeletal muscle when performing exercise at similar absolute and relative intensities, and (3) lower lipolytic rate in adipose tissue when performing exercise at a similar relative intensity

majority of studies that have studied the impact of aging on exercise metabolism have compared older sedentary adults to younger individuals. However, aging should be considered along a spectrum rather than a singular event, spanning older sedentary adults suffering from cardiometabolic diseases (unhealthy aging) to master athletes competing at a high level (healthy aging). It is important for future work to understand how these different aging phenotypes can affect exercise metabolism and, importantly, the phenotypic characteristics that either impair or improve the ability to respond to an acute bout of exercise. Further, the use of multi-omics platforms (i.e., transcriptomics, proteomics, phospho-proteomics, metabolomics, lipidomics, etc.) could yield important information about the molecular mediators that define the impaired metabolic response to acute exercise in older adults. Regardless, despite the impaired response to acute bouts of exercise, chronic exercise training is still sufficient to improve the negative aspects of aging (i.e., lower cardiorespiratory fitness, reduced muscle mass/function) and is an important intervention to improve the quality of life in the growing older adult population.

References

American College of Sports Medicine C-Z, Chodzko-Zajko WJ, Proctor DN, Fiatarone Singh MA, Minson CT, Nigg CR, Salem GJ, Skinner JS (2009) American College of Sports Medicine position stand. Exercise and physical activity for older adults. Med Sci Sports Exerc 41:1510–1530

American Diabetes Association (2018) 8. Pharmacologic approaches to glycemic treatment: standards of medical care in diabetes-2018. Diabetes Care 41:S73–S85

Barzilai N, Crandall JP, Kritchevsky SB, Espeland MA (2016) Metformin as a tool to target aging. Cell Metab 23:1060–1065

Bodine SC, Stitt TN, Gonzalez M, Kline WO, Stover GL, Bauerlein R, Zlotchenko E, Scrimgeour A, Lawrence JC, Glass DJ, Yancopoulos GD (2001) Akt/mTOR pathway is a crucial regulator of skeletal muscle hypertrophy and can prevent muscle atrophy in vivo. Nat Cell Biol 3:1014–1019

Boon H, Jonkers RAM, Koopman R, Blaak EE, Saris WHM, Wagenmakers AJM, van Loon LJC (2007) Substrate source use in older, trained males after decades of endurance training. Med Sci Sports Exerc 39:2160–2170

Brach JS, Simonsick EM, Kritchevsky S, Yaffe K, Newman AB, Health, Aging and Body Composition Study Research Group (2004) The association between physical function and lifestyle activity and exercise in the health, aging and body composition study. J Am Geriatr Soc 52:502–509

Braganza A, Corey CG, Santanasto AJ, Distefano G, Coen PM, Glynn NW, Nouraie S-M, Goodpaster BH, Newman AB, Shiva S (2019) Platelet bioenergetics correlate with muscle energetics and are altered in older adults. JCI insight 5:e128248

Braun B, Eze P, Stephens BR, Hagobian TA, Sharoff CG, Chipkin SR, Goldstein B (2008) Impact of metformin on peak aerobic capacity. Appl Physiol Nutr Metab 33:61–67

Breen L, Phillips SM (2011) Skeletal muscle protein metabolism in the elderly: interventions to counteract the "anabolic resistance" of ageing. Nutr Metab (Lond) 8:68

Centner C, Wiegel P, Gollhofer A, König D (2019) Effects of blood flow restriction training on muscular strength and hypertrophy in older individuals: a systematic review and meta-analysis. Sports Med 49:95–108

Chambers TL, Burnett TR, Raue U, Lee GA, Finch WH, Graham BM, Trappe TA, Trappe S (2020) Skeletal muscle size, function, and adiposity with lifelong aerobic exercise. J Appl Physiol 128:368–378

Chee C, Shannon CE, Burns A, Selby AL, Wilkinson D, Smith K, Greenhaff PL, Stephens FB (2016) Relative contribution of Intramyocellular lipid to whole-body fat oxidation is reduced with age but Subsarcolemmal lipid accumulation and insulin resistance are only associated with overweight individuals. Diabetes 65:840–850

Chee C, Shannon CE, Burns A, Selby AL, Wilkinson D, Smith K, Greenhaff PL, Stephens FB (2021) Increasing skeletal muscle carnitine content in older individuals increases whole-body fat oxidation during moderate-intensity exercise. Aging Cell 20:e13303

Chomentowski P, Coen PM, Radiková Z, Goodpaster BH, Toledo FGS (2011) Skeletal muscle mitochondria in insulin resistance: differences in intermyofibrillar versus subsarcolemmal subpopulations and relationship to metabolic flexibility. J Clin Endocrinol Metab 96:494–503

Cicoira M, Zanolla L, Franceschini L, Rossi A, Golia G, Zamboni M, Tosoni P, Zardini P (2001) Skeletal muscle mass independently predicts peak oxygen consumption and ventilatory response during exercise in noncachectic patients with chronic heart failure. J Am Coll Cardiol 37:2080–2085

Coats AJS, Forman DE, Haykowsky M, Kitzman DW, McNeil A, Campbell TS, Arena R (2017) Physical function and exercise training in older patients with heart failure. Nat Rev Cardiol 14: 550–559

Coggan AR, Spina RJ, King DS, Rogers MA, Brown M, Nemeth PM, Holloszy JO (1992) Histochemical and enzymatic comparison of the gastrocnemius muscle of young and elderly men and women. J Gerontol 47:B71–B76

Coggan AR, Abduljalil AM, Swanson SC, Earle MS, Farris JW, Mendenhall LA, Robitaille PM (1993) Muscle metabolism during exercise in young and older untrained and endurance-trained men. J Appl Physiol 75:2125–2133

Conley KE, Jubrias SA, Esselman PC (2000) Oxidative capacity and ageing in human muscle. J Physiol 526:203–210

Dillon EL, Casperson SL, Durham WJ, Randolph KM, Urban RJ, Volpi E, Ahmad M, Kinsky MP, Sheffield-Moore M (2011) Muscle protein metabolism responds similarly to exogenous amino acids in healthy younger and older adults during NO-induced hyperemia. Am J Physiol Regul Integr Comp Physiol 301:R1408–R1417

Distefano G, Standley RA, Zhang X, Carnero EA, Yi F, Cornnell HH, Coen PM (2018) Physical activity unveils the relationship between mitochondrial energetics, muscle quality, and physical function in older adults. J Cachexia Sarcopenia Muscle 9:279–294

Drexler H, Riede U, Münzel T, König H, Funke E, Just H (1992) Alterations of skeletal muscle in chronic heart failure. Circulation 85:1751–1759

Dreyer HC, Fujita S, Cadenas JG, Chinkes DL, Volpi E, Rasmussen BB (2006) Resistance exercise increases AMPK activity and reduces 4E-BP1 phosphorylation and protein synthesis in human skeletal muscle. J Physiol 576:613–624

Dreyer HC, Drummond MJ, Pennings B, Fujita S, Glynn EL, Chinkes DL, Dhanani S, Volpi E, Rasmussen BB (2008) Leucine-enriched essential amino acid and carbohydrate ingestion following resistance exercise enhances mTOR signaling and protein synthesis in human muscle. Am J Physiol Endocrinol Metab 294:E392–E400

Drummond MJ, Dreyer HC, Pennings B, Fry CS, Dhanani S, Dillon EL, Sheffield-Moore M, Volpi E, Rasmussen BB (2008) Skeletal muscle protein anabolic response to resistance exercise and essential amino acids is delayed with aging. J Appl Physiol 104:1452–1461

Drummond MJ, Fry CS, Glynn EL, Dreyer HC, Dhanani S, Timmerman KL, Volpi E, Rasmussen BB (2009) Rapamycin administration in humans blocks the contraction-induced increase in skeletal muscle protein synthesis. J Physiol 587:1535–1546

Fiatarone MA, O'Neill EF, Ryan ND, Clements KM, Solares GR, Nelson ME, Roberts SB, Kehayias JJ, Lipsitz LA, Evans WJ (1994) Exercise training and nutritional supplementation for physical frailty in very elderly people. N Engl J Med 330:1769–1775

Fleg JL, Morrell CH, Bos AG, Brant LJ, Talbot LA, Wright JG, Lakatta EG (2005) Accelerated longitudinal decline of aerobic capacity in healthy older adults. Circulation 112:674–682

Fong JH (2019) Disability incidence and functional decline among older adults with major chronic diseases. BMC Geriatr 19:323

Fritz IB, McEwen B (1959) Effects of carnitine on fatty-acid oxidation by muscle. Science 129: 334–335

Fry CS, Glynn EL, Drummond MJ, Timmerman KL, Fujita S, Abe T, Dhanani S, Volpi E, Rasmussen BB (2010) Blood flow restriction exercise stimulates mTORC1 signaling and muscle protein synthesis in older men. J Appl Physiol 108:1199–1209

Fry CS, Drummond MJ, Glynn EL, Dickinson JM, Gundermann DM, Timmerman KL, Walker DK, Dhanani S, Volpi E, Rasmussen BB (2011) Aging impairs contraction-induced human skeletal muscle mTORC1 signaling and protein synthesis. Skelet Muscle 1:11

Fujita S, Abe T, Drummond MJ, Cadenas JG, Dreyer HC, Sato Y, Volpi E, Rasmussen BB (2007) Blood flow restriction during low-intensity resistance exercise increases S6K1 phosphorylation and muscle protein synthesis. J Appl Physiol 103:903–910

Goodpaster BH, Sparks LM (2017) Metabolic flexibility in health and disease. Cell Metab 25: 1027–1036

Greig CA, Gray C, Rankin D, Young A, Mann V, Noble B, Atherton PJ (2011) Blunting of adaptive responses to resistance exercise training in women over 75y. Exp Gerontol 46:884–890

Gries KJ, Raue U, Perkins RK, Lavin KM, Overstreet BS, D'Acquisto LJ, Graham B, Finch WH, Kaminsky LA, Trappe TA, Trappe S (2018) Cardiovascular and skeletal muscle health with lifelong exercise. J Appl Physiol 125:1636–1645

Gries KJ, Minchev K, Raue U, Grosicki GJ, Begue G, Finch WH, Graham B, Trappe TA, Trappe S (2019) Single-muscle fiber contractile properties in lifelong aerobic exercising women. J Appl Physiol 127:1710–1719

Hagberg JM, Seals DR, Yerg JE, Gavin J, Gingerich R, Premachandra B, Holloszy JO (1988) Metabolic responses to exercise in young and older athletes and sedentary men. J Appl Physiol 65:900–908

Harber MP, Konopka AR, Undem MK, Hinkley JM, Minchev K, Kaminsky LA, Trappe TA, Trappe S (2012) Aerobic exercise training induces skeletal muscle hypertrophy and age-dependent adaptations in myofiber function in young and older men. J Appl Physiol 113: 1495–1504

Heath GW, Hagberg JM, Ehsani AA, Holloszy JO (1981) A physiological comparison of young and older endurance athletes. J Appl Physiol 51:634–640

Jones AM (2014) Dietary nitrate supplementation and exercise performance. Sports Med 44(Suppl 1):S35–S45

Julius S, Amery A, Whitlock LS, Conway J (1967) Influence of age on the hemodynamic response to exercise. Circulation 36:222–230

Kacerovsky-Bielesz G, Chmelik M, Ling C, Pokan R, Szendroedi J, Farukuoye M, Kacerovsky M, Schmid AI, Gruber S, Wolzt M, Moser E, Pacini G, Smekal G, Groop L, Roden M (2009) Short-term exercise training does not stimulate skeletal muscle ATP synthesis in relatives of humans with type 2 diabetes. Diabetes 58:1333–1341

Kalyani RR, Egan JM (2013) Diabetes and altered glucose metabolism with aging. Endocrinol Metab Clin N Am 42:333–347

Kaminsky L, Imboden M, Arena R, Myers J (2017) Reference standards for cardiorespiratory fitness measured with cardiopulmonary exercise testing using cycle Ergometry: data from the fitness registry and the importance of exercise National Database (FRIEND) registry. Mayo Clin Proc 92:228–233

Karabulut M, Abe T, Sato Y, Bemben MG (2010) The effects of low-intensity resistance training with vascular restriction on leg muscle strength in older men. Eur J Appl Physiol 108:147–155

Katsanos CS, Kobayashi H, Sheffield-Moore M, Aarsland A, Wolfe RR (2006) A high proportion of leucine is required for optimal stimulation of the rate of muscle protein synthesis by essential amino acids in the elderly. Am J Physiol Endocrinol Metab 291:E381–E387

Kitzman DW, Little WC, Brubaker PH, Anderson RT, Hundley WG, Marburger CT, Brosnihan B, Morgan TM, Stewart KP (2002) Pathophysiological characterization of isolated diastolic heart failure in comparison to systolic heart failure. JAMA 288:2144–2150

Klein S, Coyle EF, Wolfe RR (1994) Fat metabolism during low-intensity exercise in endurance-trained and untrained men. Am J Phys 267:E934–E940

Kodama S, Saito K, Tanaka S, Maki M, Yachi Y, Asumi M, Sugawara A, Totsuka K, Shimano H, Ohashi Y, Yamada N, Sone H (2009) Cardiorespiratory fitness as a quantitative predictor of all-cause mortality and cardiovascular events in healthy men and women: a meta-analysis. JAMA 301:2024–2035

Kokkinos P, Myers J, Nylen E, Panagiotakos DB, Manolis A, Pittaras A, Blackman MR, Jacob-Issac R, Faselis C, Abella J, Singh S (2009) Exercise capacity and all-cause mortality in African American and Caucasian men with type 2 diabetes. Diabetes Care 32:623–628

Kokkinos P, Faselis C, Myers J, Kokkinos JP, Doumas M, Pittaras A, Kheirbek R, Manolis A, Panagiotakos D, Papademetriou V, Fletcher R (2014) Statin therapy, fitness, and mortality risk in middle-aged hypertensive male veterans. Am J Hypertens 27:422–430

Konopka AR, Laurin JL, Schoenberg HM, Reid JJ, Castor WM, Wolff CA, Musci RV, Safairad OD, Linden MA, Biela LM, Bailey SM, Hamilton KL, Miller BF (2019) Metformin inhibits mitochondrial adaptations to aerobic exercise training in older adults. Aging Cell 18:e12880

Kulkarni AS, Peck BD, Walton RG, Kern PA, Mar JC, Windham ST, Bamman MM, Barzilai N, Peterson CA (2020) Metformin alters skeletal muscle transcriptome adaptations to resistance training in older adults. Aging (Albany NY) 12:19852–19866

Kumar V, Selby A, Rankin D, Patel R, Atherton P, Hildebrandt W, Williams J, Smith K, Seynnes O, Hiscock N, Rennie MJ (2009) Age-related differences in the dose-response relationship of muscle protein synthesis to resistance exercise in young and old men. J Physiol 587: 211–217

Kumar V, Atherton PJ, Selby A, Rankin D, Williams J, Smith K, Hiscock N, Rennie MJ (2012) Muscle protein synthetic responses to exercise: effects of age, volume, and intensity. J Gerontol A Biol Sci Med Sci 67:1170–1177

Lanza IR, Short DK, Short KR, Raghavakaimal S, Basu R, Joyner MJ, McConnell JP, Nair KS (2008) Endurance exercise as a countermeasure for aging. Diabetes 57:2933–2942

Larsen FJ, Schiffer TA, Borniquel S, Sahlin K, Ekblom B, Lundberg JO, Weitzberg E (2011) Dietary inorganic nitrate improves mitochondrial efficiency in humans. Cell Metab 13:149–159

Larsen S, Vigelsø A, Dandanell S, Prats C, Dela F, Helge JW (2018) Simvastatin-induced insulin resistance may be linked to decreased lipid uptake and lipid synthesis in human skeletal muscle: the LIFESTAT study. J Diabetes Res 2018:9257874

Loenneke JP, Wilson JM, Marín PJ, Zourdos MC, Bemben MG (2012) Low intensity blood flow restriction training: a meta-analysis. Eur J Appl Physiol 112:1849–1859

Lopez AD, Williams TN, Levin A, Tonelli M, Singh JA, Burney PGJ, Rehm J, Volkow ND, Koob G, Ferri CP (2014) Remembering the forgotten non-communicable diseases. BMC Med 12:200

Malin SK, Braun B (2013) Effect of metformin on substrate utilization after exercise training in adults with impaired glucose tolerance. Appl Physiol Nutr Metab 38:427–430

Malin SK, Kirwan JP (2012) Fasting hyperglycaemia blunts the reversal of impaired glucose tolerance after exercise training in obese older adults. Diabetes Obes Metab 14:835–841

Malin SK, Nightingale J, Choi S-E, Chipkin SR, Braun B (2013) Metformin modifies the exercise training effects on risk factors for cardiovascular disease in impaired glucose tolerant adults. Obesity (Silver Spring) 21:93–100

Mancini DM, Walter G, Reichek N, Lenkinski R, McCully KK, Mullen JL, Wilson JR (1992) Contribution of skeletal muscle atrophy to exercise intolerance and altered muscle metabolism in heart failure. Circulation 85:1364–1373

Meredith CN, Frontera WR, Fisher EC, Hughes VA, Herland JC, Edwards J, Evans WJ (1989) Peripheral effects of endurance training in young and old subjects. J Appl Physiol 66:2844–2849

Miller BF, Thyfault JP (2020) Exercise-pharmacology interactions: metformin, statins, and Healthspan. Physiology (Bethesda) 35:338–347

Minotti JR, Christoph I, Oka R, Weiner MW, Wells L, Massie BM (1991) Impaired skeletal muscle function in patients with congestive heart failure. Relationship to systemic exercise performance. J Clin Invest 88:2077–2082

Mittendorfer B, Klein S (2001) Effect of aging on glucose and lipid metabolism during endurance exercise. Int J Sport Nutr Exerc Metab 11(Suppl):S86–S91
Montoye HJ (1982) Age and oxygen utilization during submaximal treadmill exercise in males. J Gerontol 37:396–402
Moore DR, Robinson MJ, Fry JL, Tang JE, Glover EI, Wilkinson SB, Prior T, Tarnopolsky MA, Phillips SM (2009) Ingested protein dose response of muscle and albumin protein synthesis after resistance exercise in young men. Am J Clin Nutr 89:161–168
Moro T, Ebert SM, Adams CM, Rasmussen BB (2016) Amino acid sensing in skeletal muscle. Trends Endocrinol Metab 27:796–806
Morville T, Dohlmann T, Kuhlman AB, Monberg T, Torp M, Hartmann B, Holst JJ, Larsen S, Helge JW, Dela F (2019) Glucose homeostasis in statin users-the LIFESTAT study. Diabetes Metab Res Rev 35:e3110
Myers J, Prakash M, Froelicher V, Do D, Partington S, Atwood JE (2002) Exercise capacity and mortality among men referred for exercise testing. N Engl J Med 346:793–801
Noyes AM, Thompson PD (2017) The effects of statins on exercise and physical activity. J Clin Lipidol 11:1134–1144
Oliveira AN, Richards BJ, Slavin M, Hood DA (2021) Exercise is muscle mitochondrial Medicine. Exerc Sport Sci Rev 49:67–76
Ortman JM, Velkoff VA, Hogan H (2014) An aging nation: the older population in the United States, current population reports, P25-1140. US Census Bureau, Washington, DC
Park SW, Goodpaster BH, Strotmeyer ES, de Rekeneire N, Harris TB, Schwartz AV, Tylavsky FA, Newman AB (2006) Decreased muscle strength and quality in older adults with type 2 diabetes: the health, aging, and body composition study. Diabetes 55:1813–1818
Park SW, Goodpaster BH, Lee JS, Kuller LH, Boudreau R, de Rekeneire N, Harris TB, Kritchevsky S, Tylavsky FA, Nevitt M, Cho Y, Newman AB, Health, Aging, and Body Composition Study (2009) Excessive loss of skeletal muscle mass in older adults with type 2 diabetes. Diabetes Care 32:1993–1997
Phillips SM, Tipton KD, Aarsland A, Wolf SE, Wolfe RR (1997) Mixed muscle protein synthesis and breakdown after resistance exercise in humans. Am J Phys 273:E99–E107
Pollock ML, Mengelkoch LJ, Graves JE, Lowenthal DT, Limacher MC, Foster C, Wilmore JH (1997) Twenty-year follow-up of aerobic power and body composition of older track athletes. J Appl Physiol 82:1508–1516
Raue U, Slivka D, Minchev K, Trappe S (2009) Improvements in whole muscle and myocellular function are limited with high-intensity resistance training in octogenarian women. J Appl Physiol 106:1611–1617
Ritov VB, Menshikova EV, He J, Ferrell RE, Goodpaster BH, Kelley DE (2005) Deficiency of subsarcolemmal mitochondria in obesity and type 2 diabetes. Diabetes 54:8–14
Ritov VB, Menshikova EV, Azuma K, Wood R, Toledo FGS, Goodpaster BH, Ruderman NB, Kelley DE (2010) Deficiency of electron transport chain in human skeletal muscle mitochondria in type 2 diabetes mellitus and obesity. Am J Physiol Endocrinol Metab 298:E49–E58
Rivas DA, Lessard SJ, Rice NP, Lustgarten MS, So K, Goodyear LJ, Parnell LD, Fielding RA (2014) Diminished skeletal muscle microRNA expression with aging is associated with attenuated muscle plasticity and inhibition of IGF-1 signaling. FASEB J 28:4133–4147
Salami JA, Warraich H, Valero-Elizondo J, Spatz ES, Desai NR, Rana JS, Virani SS, Blankstein R, Khera A, Blaha MJ, Blumenthal RS, Lloyd-Jones D, Nasir K (2017) National Trends in statin use and expenditures in the US adult population from 2002 to 2013: insights from the medical expenditure panel survey. JAMA Cardiol 2:56–65
Santanasto AJ, Glynn NW, Lovato LC, Blair SN, Fielding RA, Gill TM, Guralnik JM, Hsu F-C, King AC, Strotmeyer ES, Manini TM, Marsh AP, McDermott MM, Goodpaster BH, Pahor M, Newman AB (2017) LIFE study group. Effect of physical activity versus health education on physical function, grip strength and mobility. J Am Geriatr Soc 65:1427–1433

Sharoff CG, Hagobian TA, Malin SK, Chipkin SR, Yu H, Hirshman MF, Goodyear LJ, Braun B (2010) Combining short-term metformin treatment and one bout of exercise does not increase insulin action in insulin-resistant individuals. Am J Physiol Endocrinol Metab 298:E815–E823

Short KR, Bigelow ML, Kahl J, Singh R, Coenen-Schimke J, Raghavakaimal S, Nair KS (2005) Decline in skeletal muscle mitochondrial function with aging in humans. Proc Natl Acad Sci U S A 102:5618–5623

Sial S, Coggan AR, Hickner RC, Klein S (1998) Training-induced alterations in fat and carbohydrate metabolism during exercise in elderly subjects. Am J Phys 274:E785–E790

Silverman HG, Mazzeo RS (1996) Hormonal responses to maximal and submaximal exercise in trained and untrained men of various ages. J Gerontol A Biol Sci Med Sci 51:B30–B37

Slivka D, Raue U, Hollon C, Minchev K, Trappe S (2008) Single muscle fiber adaptations to resistance training in old (>80 yr) men: evidence for limited skeletal muscle plasticity. Am J Physiol Regul Integr Comp Physiol 295:R273–R280

Solomon TPJ, Malin SK, Karstoft K, Haus JM, Kirwan JP (2013) The influence of hyperglycemia on the therapeutic effect of exercise on glycemic control in patients with type 2 diabetes mellitus. JAMA Intern Med 173:1834–1836

Stec MJ, Thalacker-Mercer A, Mayhew DL, Kelly NA, Tuggle SC, Merritt EK, Brown CJ, Windham ST, Dell'Italia LJ, Bickel CS, Roberts BM, Vaughn KM, Isakova-Donahue I, Many GM, Bamman MM (2017) Randomized, four-arm, dose-response clinical trial to optimize resistance exercise training for older adults with age-related muscle atrophy. Exp Gerontol 99: 98–109

Stephens FB, Wall BT, Marimuthu K, Shannon CE, Constantin-Teodosiu D, Macdonald IA, Greenhaff PL (2013) Skeletal muscle carnitine loading increases energy expenditure, modulates fuel metabolism gene networks and prevents body fat accumulation in humans. J Physiol 591: 4655–4666

Straight CR, Fedewa MV, Toth MJ, Miller MS (2020) Improvements in skeletal muscle fiber size with resistance training are age-dependent in older adults: a systematic review and meta-analysis. J Appl Physiol 129:392–403

Symons TB, Sheffield-Moore M, Mamerow MM, Wolfe RR, Paddon-Jones D (2011) The anabolic response to resistance exercise and a protein-rich meal is not diminished by age. J Nutr Health Aging 15:376–381

Taaffe DR, Pruitt L, Pyka G, Guido D, Marcus R (1996) Comparative effects of high- and low-intensity resistance training on thigh muscle strength, fiber area, and tissue composition in elderly women. Clin Physiol 16:381–392

Tang JE, Manolakos JJ, Kujbida GW, Lysecki PJ, Moore DR, Phillips SM (2007) Minimal whey protein with carbohydrate stimulates muscle protein synthesis following resistance exercise in trained young men. Appl Physiol Nutr Metab 32:1132–1138

Tankersley CG, Smolander J, Kenney WL, Fortney SM (1991) Sweating and skin blood flow during exercise: effects of age and maximal oxygen uptake. J Appl Physiol 71:236–242

Terzis G, Georgiadis G, Stratakos G, Vogiatzis I, Kavouras S, Manta P, Mascher H, Blomstrand E (2008) Resistance exercise-induced increase in muscle mass correlates with p70S6 kinase phosphorylation in human subjects. Eur J Appl Physiol 102:145–152

Timmerman KL, Lee JL, Fujita S, Dhanani S, Dreyer HC, Fry CS, Drummond MJ, Sheffield-Moore M, Rasmussen BB, Volpi E (2010) Pharmacological vasodilation improves insulin-stimulated muscle protein anabolism but not glucose utilization in older adults. Diabetes 59: 2764–2771

Trappe S, Williamson D, Godard M, Porter D, Rowden G, Costill D (2000) Effect of resistance training on single muscle fiber contractile function in older men. J Appl Physiol 89:143–152

Trappe S, Godard M, Gallagher P, Carroll C, Rowden G, Porter D (2001) Resistance training improves single muscle fiber contractile function in older women. Am J Physiol Cell Physiol 281:C398–C406

Trappe S, Hayes E, Galpin A, Kaminsky L, Jemiolo B, Fink W, Trappe T, Jansson A, Gustafsson T, Tesch P (2013) New records in aerobic power among octogenarian lifelong endurance athletes. J Appl Physiol 114:3–10

Tucker WJ, Haykowsky MJ, Seo Y, Stehling E, Forman DE (2018) Impaired exercise tolerance in heart failure: role of skeletal muscle morphology and function. Curr Heart Fail Rep 15:323–331

Walker DK, Dickinson JM, Timmerman KL, Drummond MJ, Reidy PT, Fry CS, Gundermann DM, Rasmussen BB (2011) Exercise, amino acids, and aging in the control of human muscle protein synthesis. Med Sci Sports Exerc 43:2249–2258

Wall BT, Stephens FB, Constantin-Teodosiu D, Marimuthu K, Macdonald IA, Greenhaff PL (2011) Chronic oral ingestion of L-carnitine and carbohydrate increases muscle carnitine content and alters muscle fuel metabolism during exercise in humans. J Physiol 589:963–973

Walton RG, Dungan CM, Long DE, Tuggle SC, Kosmac K, Peck BD, Bush HM, Villasante Tezanos AG, McGwin G, Windham ST, Ovalle F, Bamman MM, Kern PA, Peterson CA (2019) Metformin blunts muscle hypertrophy in response to progressive resistance exercise training in older adults: a randomized, double-blind, placebo-controlled, multicenter trial: the MASTERS trial. Aging Cell 18:e13039

Welle S, Thornton C, Statt M (1995) Myofibrillar protein synthesis in young and old human subjects after three months of resistance training. Am J Phys 268:E422–E427

Wolfe RR (2006) The underappreciated role of muscle in health and disease. Am J Clin Nutr 84: 475–482

Yang Y, Breen L, Burd NA, Hector AJ, Churchward-Venne TA, Josse AR, Tarnopolsky MA, Phillips SM (2012) Resistance exercise enhances myofibrillar protein synthesis with graded intakes of whey protein in older men. Br J Nutr 108:1780–1788

Chapter 15
Sex-Specific Effects on Exercise Metabolism

Anne-Marie Lundsgaard, Andreas M. Fritzen, and Bente Kiens

Abstract Women and men exhibit different anthropometric and physiologic characteristics, along with a sex-specific morphologic and metabolic imprint of skeletal muscle. These sex differences integrate to impact on metabolism during exercise. Men have a greater maximal exercise capacity than equally trained women. Besides this, a remarkable sex difference is a greater fatty acid oxidation in women than men at the same relative exercise intensity. The greater fatty acid oxidation may lead to less amino acid oxidation during exercise and potentially muscle glycogen sparing in women compared with men. Several sex-specific morphologic and molecular features of skeletal muscle appear to explain the differences in substrate utilization during exercise in women and men. Here, factors such as muscle fiber type composition, capillarization, and substrate availability within skeletal muscle will be discussed in a sex-comparative manner. The influence of sex on mitochondria—specifically, the energy generating pathways as beta-oxidation and glycolysis, the *tricarboxylic acid cycle* (TCA) cycle, and electron transport chain capacities—will also be reviewed.

Keywords Fat oxidation · Muscle-triacylglycerol · Muscle lipolysis · Muscle fiber types

15.1 Introduction

Women and men exhibit sex-specific anthropometric, physiologic, morphologic, and metabolic skeletal muscle characteristics, which together integrate to impact on substrate metabolism during physical activity. The marked differences in skeletal muscle metabolism between sexes are emphasized by metabolomics analysis showing that in serum, obtained in the fasting state from ~1800 men and women,

A.-M. Lundsgaard · A. M. Fritzen · B. Kiens (✉)
The Lipid Research Group, August Krogh Section for Molecular Physiology, Department of Nutrition, Exercise and Sports, Faculty of Science, University of Copenhagen, Copenhagen, Denmark
e-mail: bkiens@nexs.ku.dk

© The Author(s), under exclusive license to Springer Nature Switzerland AG 2022
G. McConell (ed.), *Exercise Metabolism*, Physiology in Health and Disease,
https://doi.org/10.1007/978-3-030-94305-9_15

one-third of the metabolites showed sex-specific expression, with pathway analysis revealing strong differences within glucose, fatty acid, as well as amino acid metabolism (Krumsiek et al. 2015). Evidence from gene microarray analysis suggests a distinct sexual dimorphism in the metabolic machinery in human skeletal muscle. In biopsies from the vastus lateralis muscle obtained in the resting, fasting state from 45 men and women, sex differences were observed for one-fourth of the identified genes (Lindholm et al. 2014), and sex was reported to have a stronger influence on gene expression in skeletal muscle than training status (Roth et al. 2002). More specifically, gene microarray analyses have revealed enrichment of genes in fatty acid metabolism in female muscle and in genes associated with pathways related to protein catabolism in male muscle (Chapman et al. 2020; Lindholm et al. 2014).

Sex differences in exercise metabolism are ascribed to both sex chromosomes and sex hormones. Sex-dependent differences in chromosomes and hormones result in differences in physiological variables, as body fat content and distribution, as well as skeletal muscle characteristics, in turn exerting an impact on sex-specific exercise metabolism. To accurately investigate the effect of sex on metabolism during exercise, proper matching of men and women at the level of exercise capacity, habitual physical activity, and training history is essential. The sexes must also be compared at exercise eliciting a similar intensity in workload. In many sex-comparative studies, it has become the norm that women are subjected to experiments in the follicular phase, a time point where differences between sexes in circulating levels of sex hormones as estrogen and progesterone are minimized. Menstrual cycle phase and the use of oral contraceptives have been taken into account in some—but not all studies comparing exercise metabolism between sexes. These two factors can have a role in substrate metabolism during exercise in women and need to be taken into account when designing and interpreting scientific studies within this area (Sims and Heather 2018).

In this chapter, sex differences in body composition and the capacity for oxygen uptake will be introduced first, followed by the role of sex on substrate metabolism during acute aerobic exercise. The most remarkable sex-dependent exercise phenotype appears to be the greater oxidation of fatty acids during aerobic exercise in women compared with men. The plethora of potential mechanisms involved in the greater fat oxidative capacity in women will be discussed, with specific consideration given to the role of skeletal muscle composition and the mitochondrial machinery. At the end of the chapter, an overview of anaerobic capacity and sex-specific regulation of glycolytic capacity will be included. The evidence presented herein is derived from studies in premenopausal women and men, unless otherwise stated.

15.2 Body Composition and Maximal Oxygen Uptake

After adjustment for body weight, women have approximately two-thirds the skeletal muscle mass of their male counterparts, as measured by magnetic resonance imaging (MRI) scanning of ~470 men and women (Janssen et al. 2000). This sex difference in skeletal muscle mass is attributed to the male-specific Y chromosome (i.e., genetics), hormonal differences (e.g., testosterone levels), and differences in physical activity patterns in men and women (i.e., environment), which over time jointly contribute to sex differences in muscle mass (Fig. 15.1).

Women generally have a greater body fat mass than men, independently of training status, and the mass of adipose tissue in women is typically equal to or even greater than their skeletal muscle mass. Varying within age (but not taking physical activity into consideration), 6–12% higher body fat mass was observed in women, when a large cohort of 16,000 12 to 80-year-old men and women were analyzed by bioelectrical impedance (Chumlea et al. 2002). One of the most important sex-specific features is the different body fat distribution, known as the android and gynoid distribution pattern. Men have a higher amount of visceral adipose tissue, whereas women have more subcutaneous fat, in particular in the gluteo-femoral region, as measured by computed tomography (CT) scanning and MRI (Kvist et al. 1988; Lemieux et al. 1993). Independent of sex, the subcutaneous fat depot comprises the majority (~80%) of total body fat.

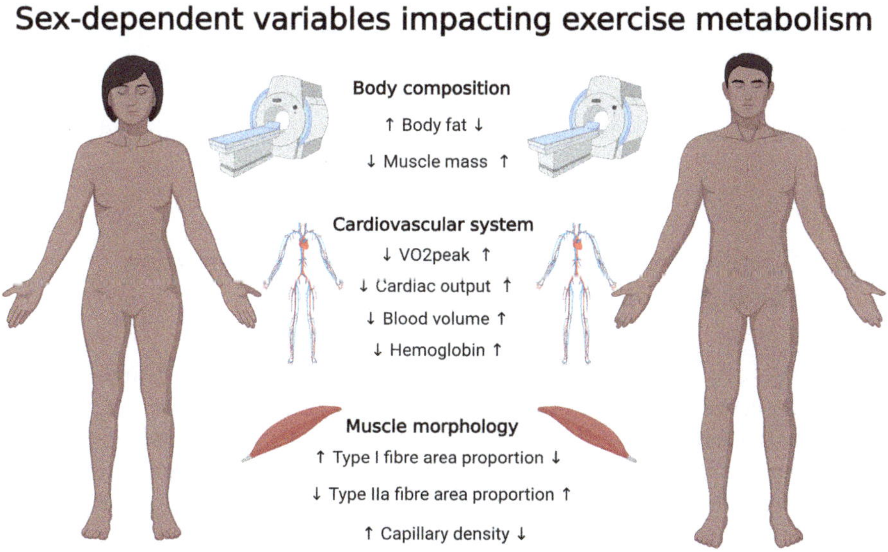

Fig. 15.1 Sex-dependent variables impacting exercise metabolism. Summary of the sex-dependent variables at whole-body and skeletal muscle level, which in the main text are discussed to have impact on exercise substrate metabolism

The maximal oxygen uptake (VO_2peak) is determined by the maximal cardiac output and the maximal peripheral (mainly skeletal muscle) oxygen extraction. When VO_2peak is expressed as liters per min, a ~ 30–40% higher VO_2peak is typically observed in trained men compared with equally trained women (based on training history), but when expressed relative to absolute body mass, VO_2peak is still ~ 10–20% higher in men than women matched for training status and history (Charkoudian and Joyner 2004). This sex difference is ascribed mainly to different dimensions of the cardiovascular oxygen transport systems and inherent differences muscle mass between women and men. Regarding the cardiovascular contribution to sex differences in VO_2peak, a lower maximal cardiac output is present in women compared with men. The maximal heart rate is either similar (Joyner and Casey 2015) or slightly higher in men than in women (Loe et al. 2013; Sydó et al. 2014), while the maximal stroke volume is approximately 30% lower in women (Joyner and Casey 2015), primarily due to a lower left ventricular mass of the female heart (even after correction for body size). Together with a 20–30% lower total blood volume and the ~10–15% lower blood hemoglobin concentration in women than in men (Murphy 2014), these physiological factors contribute to a skeletal muscle lower oxygen delivery in women than in men. It appears, however, that the cardiovascular sex differences may be compensated to some extent by increased peripheral oxygen extraction in female skeletal muscle (Beltrame et al. 2017), as discussed in a subsequent paragraphs.

In addition to lower oxygen delivery in women than men, the lower total muscle mass in women than men also plays an important role in the sex difference in maximal oxygen uptake. This is illustrated by the findings that when women and men are carefully matched at training status, with similar maximal oxygen uptake expressed relative to lean body mass, the sex difference in VO_2peak became non-significant (Steffensen et al. 2002). Matching women and men at VO_2peak per lean body mass is thus required to investigate the impact of sex per se and in particular when sex-specific metabolism in skeletal muscle is investigated.

15.3 Greater Contribution of Fatty Acids to Oxidation During Aerobic Exercise in Women

It is well documented that women have a greater resting, fasting plasma fatty acid concentration than men (Høeg et al. 2009; Karpe et al. 2011; Koutsari et al. 2011). Tracer studies have shown that this results from a greater fasting plasma fatty acid rate of appearance in women compared with men, which seems to be due to a greater fat mass rather than a greater lipolytic activity per unit of fat mass (Mittendorfer et al. 2009). Despite the greater circulating fatty acid availability in women than men, the whole-body fatty acid oxidation rate has been shown to be similar during resting, fasting conditions between sexes in the majority of studies (Carter et al. 2001a; Høeg et al. 2011; Horton et al. 1998; McKenzie et al. 2000; Roepstorff et al. 2006a).

However, a greater resting fatty oxidation in women than in men was found in one study (Høeg et al. 2009). The greater circulating fatty acid concentrations in women may lead to the greater intramyocellular triacylglycerol (IMTG) content that has consistently been found in female compared with male skeletal muscle, as a positive association has been found between circulating fatty acid concentrations and IMTG content across sexes (Haugaard et al. 2009). In contrast to resting, fasting conditions, a greater relative contribution of fatty acids to energy metabolism in women is well-documented during submaximal aerobic exercise. Accordingly, during prolonged submaximal aerobic exercise at the same relative workload, a greater fatty acid oxidation was obtained in women compared with matched men (Roepstorff et al. 2006a). These findings are supported by indirect calorimetry data from 25 studies, where the respiratory exchange ratio (RER) consistently indicated a greater fatty acid oxidation in women than in men during submaximal aerobic exercise performed for 60–120 min at an intensity in the range of 35–75% of VO_2peak (Tarnopolsky 2008). The lower RER values during exercise in women than in men are observed for both untrained and trained individuals and are maintained when untrained women and men complete a similar training intervention (Carter et al. 2001a; Friedlander et al. 1998).

The absolute fatty acid oxidation rate can be calculated from indirect calorimetry measures. During exercise, the fatty acid oxidation rate increases from low to moderate intensities, from where the rate decreases slightly with increasing exercise intensities. When the maximal rate of fatty acid oxidation was calculated from indirect calorimetry during submaximal, moderate intensity incremental exercise in 300 women and men, it was demonstrated to be higher in women than in men (Venables et al. 2005). Interestingly, maximal rate of fatty acid oxidation is also elicited at a greater relative intensity of VO_2peak in women than in men (52% of VO_2peak for women compared with 45% of VO_2peak in men) (Venables et al. 2005). This indicates that women are able to maintain the favoring of fatty acid oxidation at a higher exercise intensity than men, whereas men shift to predominantly glucose oxidation at lower levels of submaximal exercise intensity than women.

Despite the greater use of fatty acids as substrate during aerobic exercise in women, glucose disposal during exercise appears to be similar in women and men. Hence, removal of glucose from the circulation, representing uptake mainly in skeletal muscle, was found to be similar in untrained men and women during moderate-intensity exercise, both before and after a training period (Carter et al. 2001a). Likewise, using a unilateral leg extension exercise model, no sex difference in glucose uptake across the exercising leg was observed during moderate-intensity exercise in untrained men and women when expressed per unit of lean leg mass (Mittendorfer et al. 2002). Even during more intense whole-body exercise (14–15 min at 88% of VO_2peak), glucose removal from the blood expressed relative to lean body mass was found to be similar between trained women and men (Marliss et al. 2000). Friedlander et al. (1998) studied glucose kinetics in matched, trained men and women and showed similar whole-body glucose rate of removal relative to body mass during moderate-intensity exercise, despite lower RER values in women, which gave rise to the hypothesis that there is greater muscle glycogen-sparing during exercise in women than men (Friedlander et al. 1998). Future studies needs to further clarify whether glycogen breakdown is actually lower in women than men during a matched exercise bout.

15.4 A Role of Muscle Fiber Type Composition in the Greater Exercise Fatty Acid Oxidation in Women

When biopsies from the vastus lateralis muscle were obtained before and after moderate intensity aerobic exercise, it was shown that, concomitantly with the higher fatty acid oxidation in women than men, the content of free AMP and the activation of the energy sensor, AMP- activated protein kinase (AMPK), were less in female than male skeletal muscle (Roepstorff et al. 2006a). This could indicate a better maintenance of intramyocellular energy balance during submaximal exercise in skeletal muscle of women compared with men, suggesting an increased potential for oxidative (fatty acid) metabolism in women.

The morphologic and enzymatic properties of the individual muscle fibers depend on the myosin heavy chain (MHC) expression, which in humans comprises primarily type I and IIA, and some IIX. Sex differences in muscle fiber type composition have mainly been investigated in the vastus lateralis muscle. When investigated by gene microarray in recreationally active women and men, female muscle (vastus lateralis) expressed 35% more MHCI mRNA and 30% and 15% less MHCIIA and MHCIIX mRNA, respectively, than male muscle (Welle et al. 2008). There is thus a greater encoding of type I muscle fibers in female muscle. By histochemical techniques like the myosin ATPase-, silver-, or immuno-staining, the fiber composition can be expressed relative to the muscle biopsy area. By histochemical analysis, *the area proportion* of type I fibers has been described to be 22–35% greater in women, while a corresponding greater area proportion of type IIA fibers has been observed in men (Carter et al. 2001a; Høeg et al. 2009; Roepstorff et al. 2006a; Staron et al. 2000; Steffensen et al. 2002; Esbjörnsson-Liljedahl et al. 1999; Yasuda et al. 2005). A smaller *individual cross-sectional area* of type II fibers (Carter et al. 2001a; Esbjörnsson-Liljedahl et al. 1999; Høeg et al. 2009; Miller et al. 1993; Roepstorff et al. 2006a; Simoneau and Bouchard 1989; Staron et al. 2000; Yasuda et al. 2005), and in some cases type I fibers (Simoneau and Bouchard 1989; Staron et al. 2000), has been observed in women compared with men. Also, a smaller cross-sectional area of both type I and IIA have been found when investigated in isolated single fibers (Jeon et al. 2019; Trappe et al. 2003). Thus, there is a greater total muscle area covered by type I fibers in women, and this is independent of training status (Carter et al. 2001a; Steffensen et al. 2002). Interestingly, the content of satellite cells, the precursor stem cells of skeletal muscle, is greater in type I than type II fibers in female muscle while being greater in type II fibers of men (Horwath et al. 2020). This could also contribute to the sex-specific muscle fiber type composition.

The number of capillaries surrounding each type I or II muscle fiber is reported to be similar in women and men, but due to a smaller individual fiber area (in particular, in type II fibers), a greater capillary density per given muscle area is observed in women (Høeg et al. 2009; Roepstorff et al. 2006a). This influences the perfusion of the muscle while reducing oxygen and substrate diffusion distances to individual muscle fibers. This may contribute to explain observations of more efficient skeletal muscle oxygen extraction during submaximal exercise in women compared with men (Beltrame et al. 2017). Notably, a positive correlation has been found between fatty acid oxidation during exercise and both the proportion of type I fibers and capillary density (Roepstorff et al. 2006a). This indicates that sex-specific differences in muscle morphology in women may indeed contribute to the observation of higher fatty acid oxidation during submaximal exercise.

15.5 Sex Differences in Mitochondrial Capacity and Function

To understand the greater capacity for fatty acid oxidation in female skeletal muscles, it is worth considering the potential contribution of sex-specific regulation of mitochondrial capacity and function, in addition to the effects of sex on muscle fiber type and capillarization. When examining sex differences in exercise metabolism, the aerobic fitness of women and men must be carefully matched, as mitochondrial quantity and the mitochondrial capacities for beta-oxidation, TCA cycle, and oxidative phosphorylation in the electron transport chain all correlate highly with VO_2peak (Jacobs and Lundby 2013). Therefore, VO_2peak/kg lean body mass should be similar between sexes for proper muscle mitochondrial comparisons.

When assessed by electron microscopy, muscle mitochondrial density was similar in recreationally active women compared with men matched at VO_2peak/kg lean body mass (Tarnopolsky et al. 2007) or matched at weekly physical activity level (Crane et al. 2010; Hoppeler et al. 1985). In addition, a similar increase in muscle mitochondrial density was found in women and men following 6 and 7 weeks of aerobic exercise training (Hoppeler et al. 1985; Tarnopolsky et al. 2007). One study showed greater mitochondrial density in moderately trained women than men (Montero et al. 2018). However, in this study the sexes were matched at VO_2peak/kg body mass, and hence the women might have been more highly trained at the level of skeletal muscle than the men. It seems thus likely that the volume of mitochondria in skeletal muscle is not influenced by sex, despite well-established differences of muscle fiber type composition.

Along with this, measurement of mitochondrial respiration capacities in muscle samples from women and men have also shown no sex differences (Karakelides et al. 2010; Miotto et al. 2018; Thompson et al. 2013). More specifically, maximal

ATP production rates measured in isolated mitochondria from recreationally active women and men were similar (Karakelides et al. 2010). Similar maximal mitochondrial respiration rates were also measured in muscle bundles or permeabilized muscle fibers from sedentary overweight women and men (Thompson et al. 2013) and from moderately trained women and men (Miotto et al. 2018), respectively. In these studies of mitochondrial respiration, different substrates as pyruvate, palmitoyl-carnitine, or different TCA metabolites were applied to provide fuel for the mitochondrial respiration, and respiration was related to measures of mitochondrial protein content. Thus, no sex differences seem to be apparent for the intrinsic mitochondrial oxidative capacity.

The overall capacity to generate ATP from glucose- or fatty acid-derived substrates in mitochondria is likely similar in female and male skeletal muscle. When the beta-oxidation, TCA cycle, and electron transport chain are investigated at the enzyme or protein level, sex differences seem limited to the proximal part of the beta-oxidation (Fig. 15.2).

Carnitine palmitoyl-transferase 1 (CPT1) is a key regulator of mitochondrial fatty acid import and hence fatty acid oxidation in skeletal muscle. Fatty acyl-CoAs must be converted to acyl-carnitine to cross the outer mitochondrial membrane, and this reaction is catalyzed by CPT1. CPT1 protein content and maximal activity, measured in isolated mitochondria from muscle biopsies of both untrained and trained women and men, were found to be similar (Berthon et al. 1998; Costill et al. 1979; Miotto et al. 2018). This suggests that it is not a greater capacity for CTP1 flux in female muscle that contributes to the sex differences in fatty acid oxidation during exercise.

Within the mitochondria, fatty acyl-CoA enters beta-oxidation. In this pathway, very long- and medium-chain acyl-CoA dehydrogenase (VLCAD and MCAD) protein contents are higher in muscle of moderately trained women than in men (Fu et al. 2009; Maher et al. 2010). This is also the case for the content of trifunctional protein α (TFPα) (Fu et al. 2009; Maher et al. 2009, 2010), which catalyzes the second and third beta-oxidation reaction. The hydroxy acyl-CoA dehydrogenase (HAD) enzyme forms part of TFPα and catalyzes the third reaction, which leads to NADH. Contrary to the sex difference in TFPα, the maximal activity of HAD has been shown to be similar in untrained and trained women and men (Carter et al. 2001a; Roepstorff et al. 2005). Women and men thus have a similar capacity for mitochondrial fatty acid import, and while women may express some of the proximal beta-oxidation enzymes to a greater extent, the studies which have assessed respiration with palmitoyl-carnitine as substrate suggest a similar overall beta-oxidation capacity in women and men.

Acetyl-CoAs from beta-oxidation as well as from glycolysis enter the TCA cycle, which produces NADH and $FADH_2$ coenzymes for the electron transport chain. The maximal activity of citrate synthase (CS) is similar in skeletal muscle from both untrained and trained women and men (Chapman et al. 2020; Høeg et al. 2009; Roepstorff et al. 2005; Thompson et al. 2013; McKenzie et al. 2000). Moreover, maximal CS activity increased to a similar extent in women and men after 7 weeks of aerobic training (Carter et al. 2001a), further supporting a similar metabolic capacity

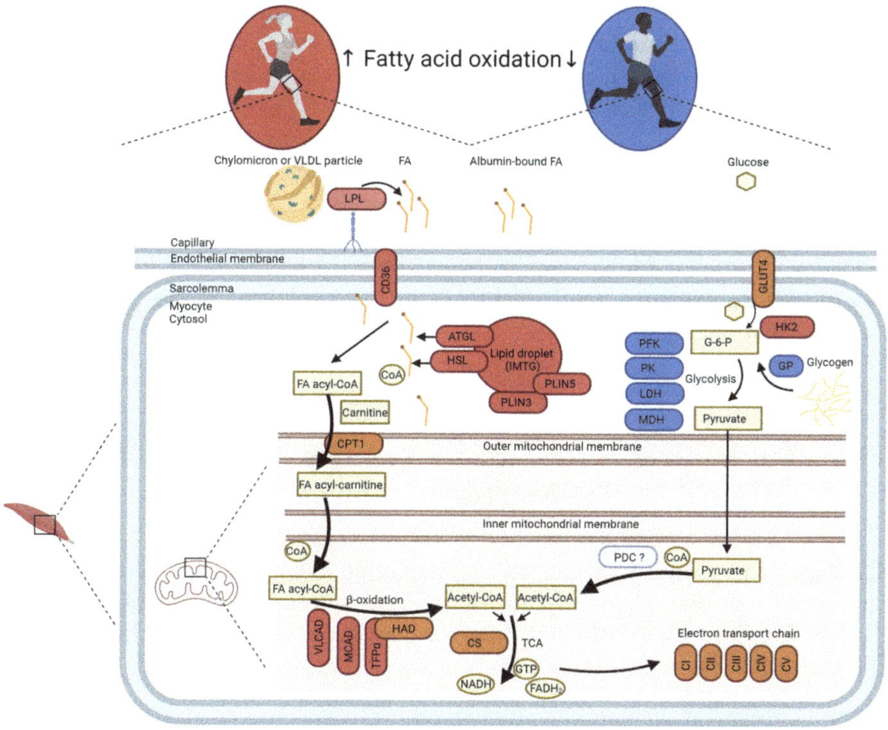

Fig. 15.2 Sex-dependent molecular imprint involved in substrate metabolism in skeletal muscle during exercise. Illustration of proteins and enzymes involved in substrate metabolism in skeletal muscle, of which the protein expression or maximal activity are investigated in gender-comparative muscle biopsy studies. Red color illustrates a higher expression in female compared with male muscle; blue color illustrates a higher expression in male than female muscle. Proteins or enzymes shown to be similar in expression are depicted in orange. Lipoprotein lipase is shown as pink due to findings of higher LPL mRNA in female compared with male muscle, with no available data on protein expression. In women, lipid metabolic proteins related to uptake of fatty acids from the circulation, intramyocellular release of fatty acids from triacylglycerol, and the proximal β-oxidation are more abundant in women. Resting skeletal muscle intramyocellular triacylglycerol (IMTG) content are shown to be higher in women, while resting glycogen content are shown to be similar in female and male muscle. Mitochondrial proteins in the TCA cycle and electron transport chain are equally expressed in skeletal muscle of women and men and hence not likely to explain differences in substrate metabolism. In male skeletal muscle, a greater molecular capacity for glycolysis and glycogen breakdown is obtained. Abbreviations: *ATGL* adipose triglyceride lipase, *CI-V* electron transport chain complex I–V, *CD36* cluster of differentiation 36 or SR/B3, *CS* citrate synthase, *FA* fatty acid, *GP* glycogen phosphorylase, *GLUT4* glucose transporter 4, *HAD* beta-hydroxyacyl-CoA dehydrogenase, *HK2* hexokinase 2, *HSL* hormone sensitive lipase, *LPL* lipoprotein lipase, *MCAD* medium-chain acyl-CoA dehydrogenase, *PLIN3 and –5* perilipin 3 and 5, *PDC* pyruvate dehydrogenase complex, *PFK* phosphofructokinase, *PK* pyruvate kinase, *TCA* tricarboxylic acid cycle, *TFPα* trifunctional protein α, *VLCAD* very long chain acyl-CoA dehydrogenase

at the level of CS between sexes. At the electron transport chain complexes, the maximal activity of enzymes at the level of complex I to V is reported to be similar between in muscle from moderately trained women and men (Carter et al. 2001a; Crane et al. 2010; McKenzie et al. 2000). In support, protein contents of electron transport chain complex I to V were similar in moderately trained women and men (Miotto et al. 2018).

The summarized findings of similar capacities for acetyl-CoA flux through TCA and ATP production in the electron transport chain suggest that the higher fatty acid oxidation in women during exercise potentially may be regulated at the proximal part of the beta-oxidation, but it is more likely to be regulated by the availability of intramuscular fatty acid substrates.

15.6 The Sources of the Greater Fraction of Fatty Acids Oxidized During Exercise in Women

The fatty acids oxidized in skeletal muscle during exercise can originate from the circulation, as either adipose tissue-derived fatty acids or fatty acids liberated from hydrolysis of circulating triacylglycerol, or they can be derived from intramuscular sources as IMTGs.

15.6.1 Plasma Free Fatty Acids and Fatty Acids Derived from VLDL-TG

During exercise, there is an increase in circulating catecholamines, which is one of the major mechanisms stimulating lipolysis in adipose tissue via adrenoreceptors. The whole-body lipolytic rate can be determined following labeled glycerol infusion, and from such studies, it has been shown that plasma glycerol rate of appearance can be increased by up to four fold compared with resting values during submaximal moderate-intensity exercise (Romijn et al. 2000; Wolfe et al. 1990). When compared during exercise at the same relative moderate exercise intensity, a higher plasma glycerol and fatty acid concentration have been observed in untrained and moderately trained women compared with matched men (Carter et al. 2001a; Davis et al. 2000; Roepstorff et al. 2006b). This could indicate that in individuals that are not endurance trained, a greater plasma fatty acid availability in women than men could contribute to the higher fatty acid oxidation. On the other hand, in endurance trained women and men, a number of studies have shown similar plasma glycerol and fatty acid concentration between sexes, as well as a similar entry rate of fatty acids in the bloodstream during exercise (Burguera et al. 2000; Roepstorff et al. 2002). Thus, in the highly trained state, the liberation of fatty acids from adipose tissue lipolysis appears to be the same. Accordingly, when uptake and oxidation of plasma fatty acids were measured across the exercising leg in endurance trained individuals, no sex differences in the uptake and oxidation of plasma fatty acids were obtained irrespective of whether the exercise was performed at low, moderate, or high intensities (Burguera et al. 2000; Roepstorff et al. 2002; Romijn et al. 2000; Steffensen et al. 2002). While untrained to moderately trained women may achieve greater free fatty acid availability in the circulation during submaximal exercise, the availability and utilization of free fatty acids from the circulation are not likely to be a main factor in the sex difference in exercise fatty acid oxidation observed in more highly trained men and women.

Plasma VLDL-TG-derived fatty acids may also contribute to fatty acid utilization during exercise, though the relative contribution of this substrate has been considered minor. Muscle lipoprotein lipase (LPL) is important for hydrolysis of circulating (VLDL)-TG. There is no sex difference in LPL activity in skeletal muscle (mLPL activity) in the resting, fasted state (Kiens et al. 2004). It has been shown that following moderate-intensity exercise, mLPL activity was similar to pre-exercise values in women, while it increased 56% compared with resting levels in men (Perreault et al. 2004). Some studies thus suggest that VLDL-TG might comprise a more significant energy substrate in men rather than women during exercise (Roepstorff et al. 2002); however, tracer studies have revealed similar VLDL-TG clearance and oxidation rates in recreationally active women and men during 90 min exercise at 50% of VO_2peak (Sondergaard et al. 2011).

Altogether, these data indicate that VLDL-TG derived fatty acids are not oxidized to a greater extent in women than men during exercise. Together with the data showing similar oxidation of fatty acids from adipose tissue lipolysis, at least in the more trained state, these findings point to a sex-specific utilization of intramyocellular fatty acids.

15.6.2 Intramyocellular Triacylglycerol Utilization

It has consistently been demonstrated that IMTG content is higher in women compared with men, by either biochemical analysis, histochemical staining, or noninvasively by ^{1}H-MRS (Devries et al. 2007; Høeg et al. 2009; Miller et al. 1993; Moro et al. 2009; Roepstorff et al. 2002; Roepstorff et al. 2006a; Steffensen et al. 2002; Tarnopolsky et al. 2007). As type I muscle fibers contain more IMTG than type II fibers (Essén et al. 1975), the greater proportion of type I fibers in female muscle could contribute to explaining these sex difference in IMTG content.

A higher skeletal muscle protein content of the fatty acid translocase cluster of differentiation 36/SR-B2 (CD36) has been documented in both untrained and trained women compared with men (Kiens et al. 2004; Miotto et al. 2018). The greater IMTG content in female muscle may be associated with their greater capacity for CD36-mediated FA uptake, in combination with the higher basal fasting plasma fatty acid concentration in women than in men. The amount of body fat (which seems to contribute to sex differences in resting fasting plasma fatty acid concentrations) has been shown to be an important predictor of IMTG content (Moro et al. 2009).

During moderate-intensity exercise, women utilized IMTG to a greater extent than men, irrespective of training status, as IMTG were reduced by ~25–35% in women following exercise, with a non-significant reduction of IMTG in men (Roepstorff et al. 2002; Roepstorff et al. 2006b; Steffensen et al. 2002). The decline in IMTG during exercise also correlates positively with the basal IMTG content, as shown across sexes (Roepstorff et al. 2006b). Electron microscopy has revealed that IMTGs in female muscle are comprised of a high number of smaller lipid droplets compared with men (Tarnopolsky et al. 2007), which may increase the potential for IMTG hydrolysis by lipases and other proteins involved in the regulation of lipid droplet lipolysis. To this end, lipid droplets in skeletal muscle of women are found to be located closer to mitochondria after an exercise bout (Devries et al. 2007), a location which may increase the potential for IMTG-derived fatty acid oxidation.

A higher TG hydrolase activity by adipose triglyceride lipase (ATGL), the first enzyme in IMTG lipolysis, has been shown in the vastus lateralis muscle of untrained women compared with men (Moro et al. 2009). On the other hand, DAG hydrolase activity, primarily mediated by hormone sensitive lipase (HSL), is not different between sexes, even though both HSL mRNA expression and protein content are higher in skeletal muscle of women than men (Roepstorff et al. 2006b). The expressions of the lipid-droplet coating proteins perilipin 2, 3, 4, and 5 were all 1.5- to 2-fold higher in untrained women than in men (Peters et al. 2012). This could also contribute to a greater female skeletal muscle capacity for lipid droplet lipolysis and interaction between lipid droplets and mitochondria during exercise (Covington et al. 2014; Granneman et al. 2011; Wang et al. 2011).

The greater contribution of IMTG as energy source during exercise in women may thus be associated with greater stores, smaller and more abundant lipid droplets in proximity to mitochondria, and an increased lipolytic capacity for TG hydrolysis compared with men.

The greater fatty acid oxidation during exercise in women may primarily result from differences in IMTG-derived fatty acid oxidation.

15.7 Protein Catabolism and Anabolism During and Following Exercise

When the rates of muscle protein synthesis and breakdown were investigated by labeled amino acid infusion and subsequent measurement of tracer incorporation into muscle proteins in samples from medial vastus lateralis biopsies, men and women showed similar resting muscle protein fractional synthesis and breakdown rate in the overnight-fasted state (Dreyer et al. 2010; Fujita et al. 2007; Markofski and Volpi 2011), suggesting no apparent sex differences in basal protein turnover in skeletal muscle.

Urea excretion can be used as a marker of whole-body protein breakdown. It has been shown that urinary urea excretion is lower in women than men on a day with 90 min moderate-intensity exercise at 60% of VO_2peak, when a similar diet was ingested (McKenzie et al. 2000). This was found to be concomitant with leucine oxidation being 118% greater in men than in women during submaximal exercise, when measured by infusion of labeled amino acids (McKenzie et al. 2000). Likewise in other studies, leucine oxidation during aerobic exercise at moderate intensity was also ~70% greater in trained men compared with women (Lamont et al. 2001; Phillips et al. 1993). Lower amino acid oxidation in women seems to coincide with higher fatty acid oxidation.

Following acute resistance exercise, the increase in post-exercise muscle protein synthesis was similar between women and men, when investigated in either the fasted or fed state (Areta et al. 2014; Dreyer et al. 2010; West et al. 2012). However, when the response to acute sprint interval exercise was evaluated in recreationally active men and women using deuterium oxide, mixed muscle protein synthesis was shown to be ~35% lower in women than men at 48 h post-exercise (Scalzo et al. 2014). This discrepancy could relate to the longer study period in the latter study. It remains to be established whether a lower total muscle mass and the lower amino acid utilization during exercise contribute to lower protein requirements in female compared with male athletes. However, the sum of existing studies suggests that protein requirements of female athletes (when expressed per kg body mass) are in the similar range as those in males (Mercer et al. 2020).

15.8 Capacity for Glucose Metabolism and Glycolysis

During muscle contractions, intracellular signaling leads to translocation of glucose transporter 4 (GLUT4) to the plasma membrane leading to increased glucose uptake. GLUT4 protein content is similar in skeletal muscle of moderately trained women and men (Høeg et al. 2009). Also important for glucose uptake into skeletal muscle is the hexokinase II (HKII) protein, which facilitates the gradient for glucose transport into muscle by intracellular phosphorylation of glucose to glucose-6-phosphate (G6P). For HKII, a 56% higher protein content has been demonstrated in women compared with men (Høeg et al. 2011). Via this increased capacity for intracellular phosphorylation of glucose, women seem to have a greater potential for glucose uptake, which, however, does not appear to come in play during submaximal aerobic exercise (Carter et al. 2001b; Friedlander et al. 1998; Marliss et al. 2000; Mittendorfer et al. 2002).

In the resting, fasted state, skeletal muscle glycogen content is not different between untrained or trained women and men during conditions of controlled diet and optimal energy intake (Roepstorff et al. 2002; Roepstorff et al. 2006a; Tarnopolsky et al. 1990). During submaximal exercise (treadmill running for 90 min at 63% of VO_2peak), a lower glycogen breakdown was observed in well-trained women compared with men (Tarnopolsky et al. 1990), while, in endurance trained female and male athletes, 90 min of bicycle exercise at 58% of VO_2peak induced a similar glycogen breakdown in men and women (Roepstorff et al. 2002). The discrepancy between findings may be due to the different exercise modalities and intensities. When it comes to more intense exercise, like 30 s bicycle sprinting, glycogen depletion was described to be 42% less in type I muscle fibers in women than in men, with similar glycogen depletion in type II fibers (Esbjörnsson-Liljedahl et al. 1999). There may thus be a sex difference in glycogen utilization during more intense exercise.

In support of a greater capacity for glycogenolysis in men, a higher maximal activity of glycogen phosphorylase (GP) has been reported in muscle homogenates from untrained men compared with women (Green et al. 1984). In skeletal muscle, G6P derived from either glycogen breakdown or glucose taken up from plasma is substrate for glycolysis. In the glycolysis, there are several reports of higher maximal activity of glycolytic enzymes in male compared with female muscle. Higher activities of phosphofructokinase (PFK), pyruvate kinase (PK), lactate dehydrogenase (LDH), and malate dehydrogenase (MDH) have been demonstrated in skeletal muscle from recreationally active men compared with women (Green et al. 1984; Jaworowski et al. 2002; Simoneau and Bouchard 1989). These findings clearly indicate that men have a higher capacity for glycogenolysis and glycolytic flux, which might make them superior compared with women, when it comes to supporting the energy demands of intense exercise. Indicative of a higher potential for glycolysis rather than beta-oxidation in muscle from males, a lower ratio between HAD activity and glycolytic enzyme activity in skeletal muscle was observed in men compared with women (Green et al. 1984). Considering that type II fibers have a higher glycolytic potential compared with type I fibers (Borges and Essén-gustavsson 1989), the sex difference in glycolytic capacity could, at least in part, be related to the greater contribution of type II fibers to muscle mass in men.

15.9 Greater Anaerobic Capacity in Men than Women

Along with the higher capacity for glycolysis in skeletal muscle of men, several indices point toward men demonstrating a more pronounced anaerobic component during high-intensity exercise. Such measures must be performed and interpreted carefully with regard to the amount of work performed and the anaerobic work capacity of the two sexes. Besides, the work performed must be related to body weight and/or even better active muscle mass. During a single 30 s sprint with gender-adjusted resistance, leading to 30% lower total work in women than men (0.086 and 0.095 kg/kg body mass), it has been shown that there is a 35% lower anaerobic contribution in women when corrected for body mass (Hill and Smith 1993). The anaerobic performance during high-intensity exercise in men has been associated with the greater glycolytic activity in muscle and is correlated with the higher proportion of type II fibers (Esbjörnsson et al. 1993).

Greater post-exercise disturbances of the blood lactate concentrations have accordingly been observed in men compared with equally trained women after several 30 s sprints on a cycle ergometer performed at 0.075 kp/kg body mass (Esbjörnsson-Liljedahl et al. 2002). It has also been shown that intramuscular lactate accumulation was greater in men than women during both 10 and 30 s sprint exercise performed at 4.9 J/kg body mass (Jacobs et al. 1983). However, this study did find that muscle lactate accumulation was similar, when adjusted for the greater work in men. Finally, one study applied ^{31}P-MRS to skeletal muscle of women and men during intense, maximal 60s contractions and demonstrated higher glycolytic flux in male compared with female muscle despite similar oxidative energy contribution during this period (Russ et al. 2005). Again, men also produced more power during these intense contractions making it difficult to interpret whether men had a higher anaerobic component relative to the power produced. Together, this clearly illustrates the potential bias of the greater muscular work in men and emphasizes the difficulties in comparing women and men at high-intensity anaerobic exercise protocols, due to differences in both muscle mass and strength/anaerobic capacity. A recent study indicated that women and men fatigued and recruited anaerobic energy to a similar extent during repeated supramaximal exercise bouts at 120% of VO_2peak until exhaustion interspaced by 20 sec recovery periods, when differences in muscle mass were considered (Martin-Rincon et al. 2021). Men, however, demonstrated higher anaerobic capacity than women, even after normalization to the lower extremities lean mass, but this advantage was only manifested during the first bout of supramaximal exercise (Martin-Rincon et al. 2021).

Future studies carefully matching men and women and also the work performed need to further clarify potential sex differences in anaerobic exercise metabolism during high-intense exercise.

15.10 Concluding Remarks: Summarizing Sex Differences

When women and men are matched carefully on aerobic training status, women have lower absolute maximal oxygen uptake than men, which relates to their smaller muscle mass, lower maximal cardiac output, and lower hemoglobin concentration.

It remains unequivocal that there is a greater fatty acid oxidation in women than men at the same relative exercise intensity and that the maximal fatty acid oxidation rate is higher in women. Several important sex differences at the level of the morphological and molecular imprint of skeletal muscle contribute to explain this. There is a greater proportion of type I fibers per given muscle area in women, and a smaller fiber size is also reported, which leads to a greater capillary density per given muscle area in women. There is generally a greater total IMTG content in skeletal muscle of women, and lipid droplet morphology favors oxidation via smaller lipid droplets which are in closer association with mitochondria, as well as greater content of lipid droplet-associated proteins involved in lipolysis. Altogether, this likely leads to a greater contribution of IMTG-derived fatty acids to fatty acid oxidation during exercise in women compared with men. Concomitant with the greater fatty acid oxidation in women than men, a similar uptake of glucose into muscle is observed between sexes. It might therefore seems intuitive that muscle glycogen use during exercise would be less in female muscle, but that needs to be determined. As compensation for the lower contribution of fat to energy metabolism, men exhibit a greater amino acid oxidation during exercise than women.

At the mitochondrial level, TCA and electron transport chain capacities appear to be similar in women and men. This is why the greater fatty acid oxidation during exercise takes its basis in the substrate (fatty acid) delivery to muscle mitochondria in women and potentially also in a greater capacity of the proximal part of the beta-oxidation pathway in female muscle. There is solid evidence showing a greater molecular capacity for glycolysis in male than female muscle, and hence there are several indications of a greater exercise capacity and anaerobic component at supramaximal exercise intensities in men.

References

Areta JL, Burke LM, Camera DM, West DWD, Crawshay S, Moore DR, Stellingwerff T, Phillips SM, Hawley JA, Coffey VG (2014) Reduced resting skeletal muscle protein synthesis is rescued by resistance exercise and protein ingestion following short-term energy deficit. Am J Physiol-Endocrinol Metabolism 306(8):E989–E997

Beltrame T, Villar R, Hughson RL (2017) Sex differences in the oxygen delivery, extraction, and uptake during moderate-walking exercise transition. Appl Physiol Nutr Metab 42(9):994–1000

Berthon PM, Howlett RA, Heigenhauser GJF, Spriet LL (1998) Human skeletal muscle carnitine palmitoyltransferase I activity determined in isolated intact mitochondria. J Appl Physiol 85(1): 148–153

Borges O, Essén-gustavsson B (1989) Enzyme activities in type I and II muscle fibres of human skeletal muscle in relation to age and torque development. Acta Physiol Scand 136(1):29–36

Burguera B, Proctor D, Dietz N, Guo Z, Joyner M, Jensen MD (2000) Leg free fatty acid kinetics during exercise in men and women. Am J Physiol-Endocrinol Metab 278(1):E113–E117

Carter SL, Rennie CD, Hamilton SJ, Tarnopolsky MA (2001a) Changes in skeletal muscle in males and females following endurance training. Can J Physiol Pharmacol 79(5):386–392

Carter SL, Rennie C, Tarnopolsky MA (2001b) Substrate utilization during endurance exercise in men and women after endurance training. Am J Physiol-Endocrinol Metab 280(6):E898–E907

Chapman MA, Arif M, Emanuelsson EB, Reitzner SM, Lindholm ME, Mardinoglu A, Sundberg CJ (2020) Skeletal muscle transcriptomic comparison between long-term trained and untrained men and women. Cell Rep 31(12):107808

Charkoudian N, Joyner MJ (2004) Physiologic considerations for exercise performance in women. Clin Chest Med 25(2):247–255

Chumlea W, Guo S, Kuczmarski R, Flegal K, Johnson C, Heymsfield S, Lukaski H, Friedl K, Hubbard V (2002) Body composition estimates from NHANES III bioelectrical impedance data. Int J Obes 26(12):1596–1609

Costill DL, Fink WJ, Getchell LH, Ivy JL, Witzmann FA (1979) Lipid metabolism in skeletal muscle of endurance-trained males and females. J Appl Physiol 47(4):787–791

Covington JD, Galgani JE, Moro C, LaGrange JM, Zhang Z, Rustan AC, Ravussin E, Bajpeyi S (2014) Skeletal muscle Perilipin 3 and Coatomer proteins are increased following exercise and are associated with fat oxidation. PLoS One 9(3):e91675

Crane JD, Devries MC, Safdar A, Hamadeh MJ, Tarnopolsky MA (2010) The effect of aging on human skeletal muscle mitochondrial and Intramyocellular lipid ultrastructure. J Gerontol Ser A Biol Med Sci 65A(2):119–128

Davis SN, Galassetti P, Wasserman DH, Tate D (2000) Effects of gender on neuroendocrine and metabolic Counterregulatory responses to exercise in Normal man. J Clin Endocrinol Metabol 85(1):224–230

Devries MC, Lowther SA, Glover AW, Hamadeh MJ, Tarnopolsky MA (2007) IMCL area density, but not IMCL utilization, is higher in women during moderate-intensity endurance exercise, compared with men. Am J Physiol-Regul Integr Com Physiol 293(6):R2336–R2342

Dreyer HC, Fujita S, Glynn EL, Drummond MJ, Volpi E, Rasmussen BB (2010) Resistance exercise increases leg muscle protein synthesis and mTOR signalling independent of sex. Acta Physiol 199(1):71–81

Esbjörnsson M, Sylvén C, Holm I, Jansson E (1993) Fast twitch Fibres may predict anaerobic performance in both females and males. Int J Sports Med 14(05):257–263

Esbjörnsson-Liljedahl M, Sundberg CJ, Norman B, Jansson E (1999) Metabolic response in type I and type II muscle fibers during a 30-s cycle sprint in men and women. J Appl Physiol 87(4): 1326–1332

Esbjörnsson-Liljedahl M, Bodin K, Jansson E (2002) Smaller muscle ATP reduction in women than in men by repeated bouts of sprint exercise. J Appl Physiol 93(3):1075–1083

Essén B, Jansson E, Henriksson J, Taylor AW, Saltin B (1975) Metabolic characteristics of fibre types in human skeletal muscle. Acta Physiol Scand 95(2):153–165

Friedlander AL, Casazza GA, Horning MA, Huie MJ, Piacentini MF, Trimmer JK, Brooks GA (1998) Training-induced alterations of carbohydrate metabolism in women: women respond differently from men. J Appl Physiol 85(3):1175–1186

Fu MH, Maher AC, Hamadeh MJ, Ye C, Tarnopolsky MA (2009) Exercise, sex, menstrual cycle phase, and 17β-estradiol influence metabolism-related genes in human skeletal muscle. Physiol Genomics 40(1):34–47

Fujita S, Rasmussen BB, Bell JA, Cadenas JG, Volpi E (2007) Basal muscle intracellular amino acid kinetics in women and men. Am J Physiol-Endocrinol Metab 292(1):E77–E83

Granneman JG, Moore H-PH, Mottillo EP, Zhu Z, Zhou L (2011) Interactions of Perilipin-5 (Plin5) with adipose triglyceride lipase. J Biol Chem 286(7):5126–5135

Green HJ, Fraser IG, Ranney DA (1984) Male and female differences in enzyme activities of energy metabolism in vastus lateralis muscle. J Neurol Sci 65(3):323–331

Haugaard SB, Mu H, Vaag A, Madsbad S (2009) Intramyocellular triglyceride content in man, influence of sex, obesity and glycaemic control. Eur J Endocrinol 161(1):57–64

Hill DW, Smith JC (1993) Gender difference in anaerobic capacity: role of aerobic contribution. Br J Sports Med 27(1):45–48

Høeg L, Roepstorff C, Thiele M, Richter EA, Wojtaszewski JFP, Kiens B (2009) Higher intramuscular triacylglycerol in women does not impair insulin sensitivity and proximal insulin signaling. J Appl Physiol 107(3):824–831

Høeg LD, Sjøberg KA, Jeppesen J, Jensen TE, Frøsig C, Birk JB, Bisiani B, Hiscock N, Pilegaard H, Wojtaszewski JFP, Richter EA, Kiens B (2011) Lipid-induced insulin resistance affects women less than men and is not accompanied by inflammation or impaired proximal insulin signaling. Diabetes 60(1):64–73

Hoppeler H, Howald H, Conley K, Lindstedt SL, Claassen H, Vock P, Weibel ER (1985) Endurance training in humans: aerobic capacity and structure of skeletal muscle. J Appl Physiol 59(2):320–327

Horton TJ, Pagliassotti MJ, Hobbs K, Hill JO (1998) Fuel metabolism in men and women during and after long-duration exercise. J Appl Physiol 85(5):1823–1832

Horwath O, Moberg M, Larsen FJ, Philp A, Apró W, Ekblom B (2020) Influence of sex and fiber type on the satellite cell pool in human skeletal muscle. Scand J Med Sci Sports 31(2):303–312

Jacobs RA, Lundby C (2013) Mitochondria express enhanced quality as well as quantity in association with aerobic fitness across recreationally active individuals up to elite athletes. J Appl Physiol 114(3):344–350

Jacobs I, Tesch PA, Bar-Or O, Karlsson J, Dotan R (1983) Lactate in human skeletal muscle after 10 and 30 s of supramaximal exercise. J Appl Physiol 55(2):365–367

Janssen I, Heymsfield SB, Wang Z, Ross R (2000) Skeletal muscle mass and distribution in 468 men and women aged 18–88 yr. J Appl Physiol 89(1):81–88

Jaworowski Å, Porter MM, Holmbäck AM, Downham D, Lexell J (2002) Enzyme activities in the tibialis anterior muscle of young moderately active men and women: relationship with body composition, muscle cross-sectional area and fibre type composition. Acta Physiol Scand 176(3):215–225

Jeon Y, Choi J, Kim HJ, Lee H, Lim J-Y, Choi S-J (2019) Sex- and fiber-type-related contractile properties in human single muscle fiber. *Journal of exercise*. Rehabilitation 15(4):537–545

Joyner MJ, Casey DP (2015) Regulation of increased blood flow (hyperemia) to muscles during exercise: A hierarchy of competing physiological needs. Physiol Rev 95(2):549–601

Karakelides H, Irving BA, Short KR, O'Brien P, Nair KS (2010) Age, obesity, and sex effects on insulin sensitivity and skeletal muscle mitochondrial function. Diabetes 59(1):89–97

Karpe F, Dickmann JR, Frayn KN (2011) Fatty acids, obesity, and insulin resistance: time for a reevaluation. Diabetes 60(10):2441–2449

Kiens B, Roepstorff C, Glatz JFC, Bonen A, Schjerling P, Knudsen J, Nielsen JN (2004) Lipid-binding proteins and lipoprotein lipase activity in human skeletal muscle: influence of physical activity and gender. J Appl Physiol 97(4):1209–1218

Koutsari C, Basu R, Rizza RA, Nair KS, Khosla S, Jensen MD (2011) Nonoxidative free fatty acid disposal is greater in young women than men. J Clin Endocrinol Metabol 96(2):541–547

Krumsiek J, Mittelstrass K, Do KT, Stückler F, Ried J, Adamski J, Peters A, Illig T, Kronenberg F, Friedrich N, Nauck M, Pietzner M, Mook-Kanamori DO, Suhre K, Gieger C, Grallert H, Theis FJ, Kastenmüller G (2015) Gender-specific pathway differences in the human serum metabolome. Metabolomics 11(6):1815–1833

Kvist H, Chowdhury B, Grangård U, Tylén U, Sjöström L (1988) Total and visceral adipose-tissue volumes derived from measurements with computed tomography in adult men and women: predictive equations. Am J Clin Nutr 48(6):1351–1361

Lamont LS, McCullough AJ, Kalhan SC (2001) Gender differences in leucine, but not lysine, kinetics. J Appl Physiol 91(1):357–362

Lemieux S, Prud'homme D, Bouchard C, Tremblay A, Després JP (1993) Sex differences in the relation of visceral adipose tissue accumulation to total body fatness. Am J Clin Nutr 58(4): 463–467

Lindholm ME, Huss M, Solnestam BW, Kjellqvist S, Lundeberg J, Sundberg CJ (2014) The human skeletal muscle transcriptome: sex differences, alternative splicing, and tissue homogeneity assessed with RNA sequencing. FASEB J 28(10):4571–4581

Loe H, Rognmo Ø, Saltin B, Wisløff U (2013) Aerobic capacity reference data in 3816 healthy men and women 20–90 years. PLoS One 8(5):e64319

Maher AC, Fu MH, Isfort RJ, Varbanov AR, Qu XA, Tarnopolsky MA (2009) Sex differences in global mRNA content of human skeletal muscle. PLoS One 4(7):e6335

Maher AC, Akhtar M, Vockley J, Tarnopolsky MA (2010) Women have higher protein content of β-oxidation enzymes in skeletal muscle than men. PLoS One 5(8):e12025

Markofski MM, Volpi E (2011) Protein metabolism in women and men: similarities and disparities. Curr Opin Clin Nutr Metab Care 14(1):93–97

Marliss EB, Kreisman SH, Manzon A, Halter JB, Vranic M, Nessim SJ (2000) Gender differences in glucoregulatory responses to intense exercise. J Appl Physiol 88(2):457–466

Martin-Rincon M, Gelabert-Rebato M, Perez-Valera M, Galvan-Alvarez V, Morales-Alamo D, Dorado C, Boushel R, Hallen J, Calbet JAL (2021) Functional reserve and sex differences

during exercise to exhaustion revealed by post-exercise ischaemia and repeated supramaximal exercise. J Physiol 599(16):3853–3878
McKenzie S, Phillips SM, Carter SL, Lowther S, Gibala MJ, Tarnopolsky MA (2000) Endurance exercise training attenuates leucine oxidation and BCOAD activation during exercise in humans. Am J Physiol-Endocrinol Metabolism 278(4):E580–E587
Mercer D, Convit L, Condo D, Carr AJ, Hamilton DL, Slater G, Snipe RMJ (2020) Protein requirements of pre-menopausal female athletes: systematic literature review. Nutrients 12(11):3527
Miller AEJ, MacDougall JD, Tarnopolsky MA, Sale DG (1993) Gender differences in strength and muscle fiber characteristics. Eur J Appl Physiol Occup Physiol 66(3):254–262
Miotto PM, McGlory C, Holloway TM, Phillips SM, Holloway GP (2018) Sex differences in mitochondrial respiratory function in human skeletal muscle. Am J Physiol-Regul Integr Comp Physiol 314(6):R909–R915
Mittendorfer B, Horowitz JF, Klein S (2002) Effect of gender on lipid kinetics during endurance exercise of moderate intensity in untrained subjects. Am J Physiol-Endocrinol Metabolism 283(1):E58–E65
Mittendorfer B, Magkos F, Fabbrini E, Mohammed BS, Klein S (2009) Relationship between body fat mass and free fatty acid kinetics in men and women. Obesity 17(10):1872–1877
Montero D, Madsen K, Meinild-Lundby A-K, Edin F, Lundby C (2018) Sexual dimorphism of substrate utilization: differences in skeletal muscle mitochondrial volume density and function. Exp Physiol 103(6):851–859
Moro C, Galgani JE, Luu L, Pasarica M, Mairal A, Bajpeyi S, Schmitz G, Langin D, Liebisch G, Smith SR (2009) Influence of gender, obesity, and muscle lipase activity on Intramyocellular lipids in sedentary individuals. J Clin Endocrinol Metabol 94(9):3440–3447
Murphy WG (2014) The sex difference in haemoglobin levels in adults — mechanisms, causes, and consequences. Blood Rev 28(2):41–47
Perreault L, Lavely JM, Kittelson JM, Horton TJ (2004) Gender differences in lipoprotein lipase activity after acute exercise. Obes Res 12(2):241–249
Peters SJ, Samjoo IA, Devries MC, Stevic I, Robertshaw HA, Tarnopolsky MA (2012) Perilipin family (PLIN) proteins in human skeletal muscle: the effect of sex, obesity, and endurance training. Appl Physiol Nutr Metab 37(4):724–735
Phillips SM, Atkinson SA, Tarnopolsky MA, MacDougall JD (1993) Gender differences in leucine kinetics and nitrogen balance in endurance athletes. J Appl Physiol 75(5):2134–2141
Roepstorff C, Steffensen CH, Madsen M, Stallknecht B, Kanstrup I-L, Richter EA, Kiens B (2002) Gender differences in substrate utilization during submaximal exercise in endurance-trained subjects. Am J Physiol-Endocrinol Metabolism 282(2):E435–E447
Roepstorff C, Schjerling P, Vistisen B, Madsen M, Steffensen CH, Rider MH, Kiens B (2005) Regulation of oxidative enzyme activity and eukaryotic elongation factor 2 in human skeletal muscle: influence of gender and exercise. Acta Physiol Scand 184(3):215–224
Roepstorff C, Donsmark M, Thiele M, Vistisen B, Stewart G, Vissing K, Schjerling P, Hardie DG, Galbo H, Kiens B (2006a) Sex differences in hormone-sensitive lipase expression, activity, and phosphorylation in skeletal muscle at rest and during exercise. Am J Physiol-Endocrinol Metabolism 291(5):E1106–E1114
Roepstorff C, Thiele M, Hillig T, Pilegaard H, Richter EA, Wojtaszewski JFP, Kiens B (2006b) Higher skeletal muscle AMPKalpha2 activation and lower energy charge and fat oxidation in men than in women during submaximal exercise. J Physiol 574(1):125–138
Romijn JA, Coyle EF, Sidossis LS, Rosenblatt J, Wolfe RR (2000) Substrate metabolism during different exercise intensities in endurance-trained women. J Appl Physiol 88(5):1707–1714
Roth SM, Ferrell RE, Peters DG, Metter EJ, Hurley BF, Rogers MA (2002) Influence of age, sex, and strength training on human muscle gene expression determined by microarray. Physiol Genomics 10(3):181–190
Russ DW, Lanza IR, Rothman D, Kent-Braun JA (2005) Sex differences in glycolysis during brief, intense isometric contractions. Muscle Nerve 32(5):647–655

Scalzo RL, Peltonen GL, Binns SE, Shankaran M, Giordano GR, Hartley DA, Klochak AL, Lonac MC, Paris HLR, Szallar SE, Wood LM, Peelor FF, Holmes WE, Hellerstein MK, Bell C, Hamilton KL, Miller BF (2014) Greater muscle protein synthesis and mitochondrial biogenesis in males compared with females during sprint interval training. FASEB J 28(6):2705–2714
Simoneau JA, Bouchard C (1989) Human variation in skeletal muscle fiber-type proportion and enzyme activities. Am J Physiol-Endocrinol Metabolism 257(4):E567–E572
Sims ST, Heather AK (2018) Myths and methodologies: reducing scientific design ambiguity in studies comparing sexes and/or menstrual cycle phases. Exp Physiol 103(10):1309–1317
Sondergaard E, Rahbek I, Sørensen LP, Christiansen JS, Gormsen LC, Jensen MD, Nielsen S (2011) Effects of exercise on VLDL-triglyceride oxidation and turnover. Am J Physiol-Endocrinol Metabolism 300(5):E939–E944
Staron RS, Hagerman FC, Hikida RS, Murray TF, Hostler DP, Crill MT, Ragg KE, Toma K (2000) Fiber type composition of the vastus Lateralis muscle of young men and women. J Histochem Cytochem 48(5):623–629
Steffensen CH, Roepstorff C, Madsen M, Kiens B (2002) Myocellular triacylglycerol breakdown in females but not in males during exercise. Am J Physiol-Endocrinol Metab 282(3):E634–E642
Sydó N, Abdelmoneim SS, Mulvagh SL, Merkely B, Gulati M, Allison TG (2014) Relationship between exercise heart rate and age in men vs women. Mayo Clin Proc 89(12):1664–1672
Tarnopolsky MA (2008) Sex differences in exercise metabolism and the role of 17-Beta estradiol. Med Sci Sports Exerc 40(4):648–654
Tarnopolsky LJ, MacDougall JD, Atkinson SA, Tarnopolsky MA, Sutton JR (1990) Gender differences in substrate for endurance exercise. J Appl Physiol 68(1):302–308
Tarnopolsky MA, Rennie CD, Robertshaw HA, Fedak-Tarnopolsky SN, Devries MC, Hamadeh MJ (2007) Influence of endurance exercise training and sex on intramyocellular lipid and mitochondrial ultrastructure, substrate use, and mitochondrial enzyme activity. Am J Physiol-Regul Integr Comp Physiol 292(3):R1271–R1278
Thompson JR, Swanson SA, Casale GP, Johanning JM, Papoutsi E, Koutakis P, Miserlis D, Zhu Z, Pipinos II (2013) Gastrocnemius mitochondrial respiration: are there any differences between men and women? J Surg Res 185(1):206–211
Trappe S, Gallagher P, Harber M, Carrithers J, Fluckey J, Trappe T (2003) Single muscle fibre contractile properties in young and old men and women. J Physiol 552(1):47–58
Venables MC, Achten J, Jeukendrup AE (2005) Determinants of fat oxidation during exercise in healthy men and women: a cross-sectional study. J Appl Physiol 98(1):160–167
Wang H, Sreenivasan U, Hu H, Saladino A, Polster BM, Lund LM, Gong D, Stanley WC, Sztalryd C (2011) Perilipin 5, a lipid droplet-associated protein, provides physical and metabolic linkage to mitochondria. J Lipid Res 52(12):2159–2168
Welle S, Tawil R, Thornton CA (2008) Sex-related differences in gene expression in human skeletal muscle. PLoS One 3(1):e1385
West DWD, Burd NA, Churchward-Venne TA, Camera DM, Mitchell CJ, Baker SK, Hawley JA, Coffey VG, Phillips SM (2012) Sex-based comparisons of myofibrillar protein synthesis after resistance exercise in the fed state. J Appl Physiol 112(11):1805–1813
Wolfe RR, Klein S, Carraro F, Weber JM (1990) Role of triglyceride-fatty acid cycle in controlling fat metabolism in humans during and after exercise. Am J Physiol Endocrinol Metabolism 258(2):E382–E389
Yasuda N, Glover EI, Phillips SM, Isfort RJ, Tarnopolsky MA (2005) Sex-based differences in skeletal muscle function and morphology with short-term limb immobilization. J Appl Physiol 99(3):1085–1092

Chapter 16
Circadian Rhythms and Exercise Metabolism

Christopher A. Wolff, Stuart J. Hesketh, and Karyn A. Esser

Abstract The goal of this chapter is to review the current state of the emerging field of circadian rhythms and exercise metabolism. Since this area of research is fairly new, we have broken up this chapter into the following sections. (1) The role of circadian rhythms and in maintaining homeostasis; (2) introduction to the circadian clock mechanism and clock output; (3) the role of the circadian clock in regulation of resting fat and carbohydrate metabolism in the skeletal muscle; and (4) interactions between exercise and circadian rhythms. We hope that this chapter can serve as a reference and/or entry point for scientists wanting to integrate circadian concepts in their understanding or research design of exercise metabolism.

Keywords Circadian biology · Exercise physiology · Skeletal muscle · Chronotherapy · Homeostasis · Predictive homeostasis

16.1 Circadian Rhythms Underlie Predictive Homeostasis in Physiology

One of the most common questions we are asked about circadian rhythms and physiology is about how or whether this concept of daily variation has implications for our understanding of exercise and metabolism. Our simple answer is that yes, because circadian rhythms are a fundamental and essential mechanism of cell homeostasis. To help frame the role of circadian clocks in maintaining homeostasis, we provide a historical perspective of the interaction between circadian biology, physiology, and homeostasis. Walter Cannon defined homeostasis as the property of a system in which a variable, such as blood glucose, is actively regulated to remain very nearly constant (Moore-Ede 1986). For many years, the majority of mechanisms in place to support homeostasis were reactive processes. For example, after a

C. A. Wolff · S. J. Hesketh · K. A. Esser (✉)
Department of Physiology and Functional Genomics, Myology Institute, University of Florida, Florida, USA
e-mail: kaesser@ufl.edu

© The Author(s), under exclusive license to Springer Nature Switzerland AG 2022

G. McConell (ed.), *Exercise Metabolism*, Physiology in Health and Disease,
https://doi.org/10.1007/978-3-030-94305-9_16

meal increased blood glucose is sensed, the pancreas releases insulin which functions to facilitate glucose uptake into tissues and bringing blood glucose levels back down. However, in 1986, Dr. Martin Moore-Ede presented the Bowditch Lecture at the American Physiological Society meeting and discussed the concept of "predictive homeostasis" based on several physiological parameters, such as heart rate and cortisol, that were known to change prior to a predictable event such as awakening (Moore-Ede 1986). Dr. Moore-Ede proposed that our systems evolved in an environment with very predictable environmental changes, such as light-dark, and as such there were intrinsic mechanisms in place that anticipate these changes. These predictive homeostatic mechanisms allowed for more rapid adjustments with benefits for survival. It is now clear that the circadian timing mechanism is found in virtually all cells within the body which are critical for predictive homeostasis. Consistent with the physiological benefits of this predictive system, it is now clear that disruption to the circadian timing mechanism, or predictive homeostasis, leads to diminished ability of the system to maintain homeostasis with negative health outcomes. Thus, it is critical to understand that circadian rhythms and the clocks responsible are a fundamental part of daily cell and systems physiology. Additionally, the fact that circadian rhythms have been demonstrated at the subcellular level (e.g., chromatin availability) at the cell and tissue levels (e.g., insulin sensitivity) and systems level (e.g., heart rate and behavior) serves as a reminder that our resting cell, tissue, and system physiology is constantly oscillating. Thus, when we introduce exercise or exercise training, we do so on a moving physiology and metabolic baseline.

16.1.1 A Brief History of Circadian Biology and Exercise Physiology

Exercise physiology and circadian biology are well-established fields with rich histories. The history of exercise physiology is often traced to the musings of ancient Greek physicians (Tipton 2003), while circadian biology can trace roots back around 400 B.C. when Androsthenes, a scribe for Alexander the Great, noted that the leaves of certain plants displayed opening and closing patterns coincident with sunrise and sunset (Persson and Bondke Persson 2019). Despite these histories, integration of these fundamental concepts has been relatively recent, seemingly starting in the late 1940s, but expanding rapidly in the last 10 years.

One of the first circadian and exercise studies looked at time of day as a variable for human physical work capacity (reviewed in: Kleitman 1949). Subsequent studies examined athletic performance at a few times of day (Wahlberg and Astrand 1973; Conroy and O'Brien 1974), while the first experiments to explore circadian variation in exercise tolerance were completed in the late 1960s (Voigt et al. 1968; Crockford and Davies 1969). Contrary to our current understanding, the initial work did not identify circadian variation in exercise performance, though the authors point out this could have been due to the repeated bouts of exercise across the day or large interindividual variability. However, similar to the initial work in chronobiology, the initial integration of circadian concepts into exercise physiology were more systemic (e.g., exercise capacity or heart rate) and not on the molecular scale. Only with the discovery of the molecular components of the circadian timing mechanism has the interest in how exercise and circadian biology work together been identified and explored. For those interested in additional and/or more detailed training in circadian biology, there are numerous online resources, including a digital course directed by Drs. Martha Merrow and Till Roenneberg (https://www.coursera.org/learn/circadian-clocks).

16.2 The Core Molecular Clock Mechanism

Daily rhythms in physiology and metabolism can be considered in a framework of predictive homeostasis, and it is now clear that these daily rhythms are driven by the circadian clock mechanism. In this section we provide an overview of the clock to provide critical framework for integration into exercise metabolism studies. The core molecular clock is a timing mechanism that exists in virtually all cells in the body and is highly conserved across mammals. The basis of the clock is a transcription/translation feedback loop that maintains an approximately 24 h period independent of external inputs (time cues: zeitgebers). The clock is comprised of a set of molecules that form the positive limb (considered transcriptional activators) and a different set of molecules that form the negative limb (considered transcriptional repressors). The components and function of the molecular clock have been presented in significant detail, and reviews of the state of knowledge of the clock mechanism can be found here (Mohawk et al. 2012; Takahashi 2017).

16.2.1 The Components of the Molecular Clock

The Positive Arm of the Molecular Clock In mammalian tissues, the positive limb of the molecular clock is driven by two transcription factors. Brain and muscle ARNT-like 1 (BMAL1, also known as ARNTL) and CLOCK, these are basic helix-loop-helix Per-Arnt-Sim (bHLH-PAS) transcription factors. These transcription factors form a heterodimer and bind to E-box sequences (5'-CANNTG-3′, and 5'-CACGTG-3′ most often) across the genome to activate transcription. Importantly, BMAL1/CLOCK bind the promoters of the core circadian genes Period (Pers) and cryptochrome (Crys), transcriptionally activating expression of these negative arm genes (Fig. 16.1). Seminal work from the Takahashi lab revealed more than 5000 binding sites for both CLOCK and BMAL1 across the liver genome over a 24 h period and over 75% of those targeted genes were transcriptionally active (Koike et al. 2012). These findings highlight two key points in molecular clock function.

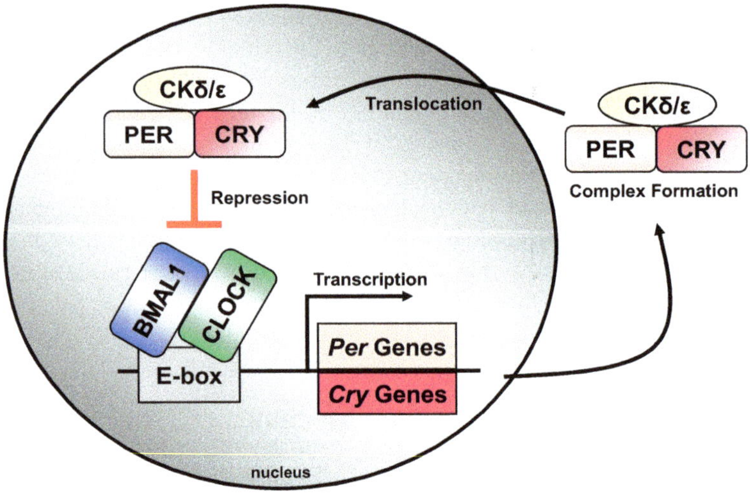

Fig. 16.1 A simplified core circadian clock. A simplified schematic that depicts the core circadian transcription factors BMAL1 and CLOCK which heterodimerize and transcribe the circadian repressors Period (*Pers*) and Cryptochrome (*Crys)* families of genes. The *Pers* and *Crys* mRNAs are translated in the cytoplasm where they form a multimers with casein kinase (CK delta and epsilon). This repressor complex translocates into the nucleus to repress the transcriptional activity of the BMAL1:CLOCK heterodimer. This process takes roughly 24 hours to complete a cycle

First, that BMAL1 and CLOCK directly regulate the Per/Cry genes as a key part of the molecular clock. Second, BMAL1 and CLOCK bind in a time of day-dependent manner to 1000s of sites across the genome, demonstrating that in addition to keeping a 24 h cycle of the Per/Cry families, the core molecular clock is contributing, in a time of day manner, to a diverse transcriptional network within the cell (Takahashi 2017).

The Negative Arm of the Molecular Clock The negative arm of the core clock is comprised of the period (Per1, Per2, and Per3) and cryptochrome (Cry1 and Cry2) gene families. Studies have shown that, the *Per* and *Cry* families of genes are transcribed by BMAL1 and CLOCK in the middle of the rest phase (e.g., afternoon for nocturnal rodents), leading to their protein expression levels peaking in the late rest phase. Upon translation, the PER and CRY proteins interact with each other, as well as with kinases, such as casein kinase 1δ and 1ε (CK1 δ/ε). These protein complexes translocate to the nucleus where they inhibit BMAL1/CLOCK transcriptional activity and thereby decrease their own transcription. As PER/CRY/CK1 represses BMAL1/CLOCK transcription, the abundance of repressive transcripts declines, as does the relative content of the repressor complex proteins, as they have short half-lives due to proteasomal degradation. Once the transcriptional repression subsides, BMAL1/CLOCK transcription increases, beginning the next circadian transcriptional cycle.

Similar to the widespread genomic localization observed for BMAL/CLOCK, the PER/CRY proteins also bind at 1000s of sites across the genome (Koike et al. 2012).

While some of the genomic sites are overlapping between the positive and negative arm, indicative of the expected transcription repression by the negative arm on the positive arm, several hundred sites are not shared between positive and negative arm proteins. The specific role of the PER/CRY families binding to other sites of the genome is currently underexplored, though some data suggest the negative arm proteins serve as mediators of metabolic input into circadian timing (Lamia et al. 2009; Schmutz et al. 2010). Additional work exploring the potential functions of the negative arm proteins independent of the circadian time keeping mechanism is warranted. Together, however, the core molecular clock mechanism is responsible for the daily transcriptional program contributing to the cellular capacity to maintain homeostasis.

16.2.2 Clock Output: The Daily Program of Transcription Underlying the "Moving Baseline" in Cell Physiology

The daily pattern of gene expression outside of the core clock is referred to as the *circadian clock output* (Fig. 16.2). The most common approach to define the circadian clock output for a tissue is to assess the total number of oscillating mRNAs over a 24 h–48 h time course. To date, numerous studies have explored the circadian clock output in muscle as well as other tissues and found the number of

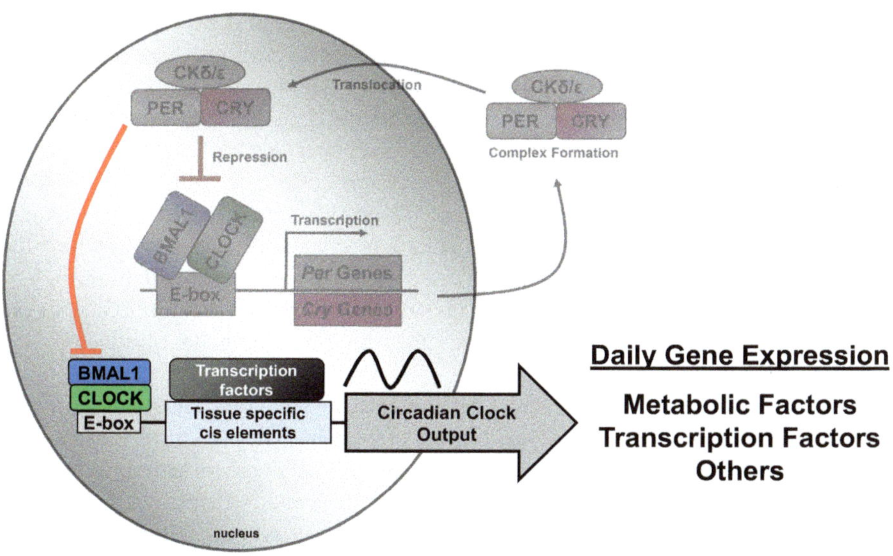

Fig. 16.2 Output from the core circadian clock. The BMAL1:CLOCK heterodimer also contributes to a daily program of gene expression in all cells, and this is called the circadian clock output. The circadian clock output is specific to each tissue and is important for a number of different cellular functions, including transcriptional regulation, metabolism, and homeostasis

oscillating genes represent up to 50% of known protein coding genes (Miller et al. 2007; McCarthy et al. 2007; Pizarro et al. 2012; Zhang et al. 2014; Mure et al. 2018). Restated, the circadian clock output comprises a significant portion of the daily transcriptional program within a tissue, including contributing to daily oscillations in mRNAs associated with metabolism, and general cellular functions. Importantly, the daily variation of metabolic gene expression is not random, but serves to temporally align cellular metabolic gene expression with both rest and active behavior as well as patterns of food availability and intake. In a simple sense, the clock mechanism in muscle functions to temporally segregate periods of fuel oxidation vs. storage. For the purpose of this chapter, the metabolic clusters of genes regulated by the circadian clock in muscle provide the framework for our understanding of intersection of circadian biology with exercise metabolism.

16.2.3 Environmental Inputs Adjust the Timing of the Clock and Clock Output

A fundamental principle of the circadian clock is that the period length is set at ~24 h, but cues from the environment can modify or alter the timing or phase of the clock. The ability to move the phase of the clock allows the timing system and its output to adapt to changes in light/dark cycles. The ability to modulate phase means that the clock and clock output can adjust to the environmental changes and supports the function of predictive homeostasis. To date, a majority of the work exploring the mechanisms through which the molecular clock senses and responds to environmental inputs have focused on how the circadian system responds to the timing of light cues. Research in this area has demonstrated that the clock mechanism is differentially sensitive to the environmental time cue (Aschoff 1965; Pittendrigh and Daan 1976). This means that the exact same cue can induce one of three predictable outcomes, (1) a phase advance, (2) a phase delay, and (3) no change; the outcomes depend on the status of the clock mechanism at the time at which the zeitgeber is received/sensed. The importance of this aspect of circadian biology in relationship to exercise and metabolism is that time of exercise can result in changes in phase of peripheral tissue clocks which then modifies the metabolic factors that are part of clock output.

We now know that both feeding and exercise provide environmental cues that impact the phase settings of clocks in peripheral tissues. In particular, current data have demonstrated that one acute exercise bout is sufficient to alter the phase of peripheral circadian clocks in human (Youngstedt et al. 2019) as well as rodent (Gannon and Rea 1995; Schroeder et al. 2012; Kemler et al. 2020) models. Most recently, studies have shown that exercise, like light, can induce a predictable phase response. This means that if one exercises earlier than when they are normally active the phase of the muscle clock will advance and in contrast, if one exercises later than when they are normally active, the phase of the muscle clock will delay. If one exercises around their normal active period we predict that there would be no shift in the phase of the muscle clock. These predictions are primarily based on the effects of an acute bout of exercise. And as a reminder, if the clock shifts in phase, then this will also shift the clock output (Youngstedt et al. 2019; Kemler et al. 2020). Therefore, it is not surprising that there are time of day specific effects of exercise on transcriptomic and metabolomic responses to the exercise bout (Ezagouri et al. 2019; Sato et al. 2019). The known mechanisms through which exercise influences the timing of the molecular clock output are discussed in Sect. 16.4.1. Since this field has only matured in the last 2 years, the profound diversity of exercise interventions, such as resistance vs. endurance, as well as exercise bout duration and intensity represent a wide range of important new areas of research.

16.3 The Role of the Molecular Clock in Daily Patterns of Muscle Metabolism

The ultimate focus of this chapter is on the intersection of circadian rhythms and exercise metabolism. However to extract exercise-specific effects, one first has to have an understanding of the daily changes in metabolism that occur downstream of the circadian clock at rest. Thus, this section reviews the current understanding of substrate metabolism changes downstream of the circadian clock in the skeletal muscle. In particular, we highlight the interplay between circadian control, glucose, and lipid metabolism during steady-state conditions.

16.3.1 Circadian Control of Glucose Metabolism

Skeletal muscle is the largest insulin sensitive organ in the body, and it plays an essential role in whole-body glucose homeostasis being responsible for ~85% of insulin-stimulated glucose uptake (DeFronzo and Tripathy 2009). Insulin sensitivity in the skeletal muscle also displays robust circadian variations in mice (Dyar et al. 2014; Aras et al. 2019) and humans (de Goede et al. 2018). The circadian variation in skeletal muscle insulin sensitivity, peaking during the beginning of the active/feeding phase, stimulates a rise in tissue glucose uptake and oxidation in anticipation of increased locomotor activity. In contrast, during the resting/fasting phase, the muscle displays reduced insulin sensitivity leading to a decrease in glucose uptake and glycolytic flux in muscle cells. Disruption of the intrinsic muscle clock causes muscle insulin resistance and altered muscle glucose metabolism, showing insulin-dependent glucose uptake in skeletal muscle reduced by muscle-specific inactivation of *Bmal1* (Dyar et al. 2014). One potential explanation for the daily change in muscle insulin sensitivity is the daily oscillation of the muscle glucose transporter GLUT4 displaying highest expression during active/feeding phase (Harfmann et al. 2016).

The circadian control of GLUT4 has been clearly demonstrated as loss of *Bmal1* in skeletal muscle leads to a significant reduction (~45%) in GLUT4 protein levels across the diurnal cycle compared to control (Dyar et al. 2014; Harfmann et al. 2016).

In addition to the circadian variation in insulin sensitivity and glucose uptake, studies have shown that the muscle clock contributes to variation in pyruvate dehydrogenase, PDH activity, as well as the two rate limiting enzymes in glucose metabolism: hexokinase and phosphofructokinase (Dyar et al. 2014). Pizarro and colleagues (2013) report hexokinase-2 expression displays a circadian rhythm that peaks in at the beginning of the active/feeding phase, as well as the genes responsible for phosphofructokinase expression (Pfkfb 1/3/4) which also display circadian rhythmicity and peak in expression during the mid-late resting/fasting phase in humans (Pizarro et al. 2012). However, in muscle in which the circadian clock is stopped (i.e., muscle-specific *Bmal1* knockout mice), the expression of these enzymes (i.e., PDH, Hk2, Pfkfb) is dampened resulting in a reduced protein expression and diminished enzyme activity, indicating impaired glycolytic flux (Dyar et al. 2014; Hodge et al. 2015; Harfmann et al. 2016). Together these data illustrate that the muscle clock modulates both glucose uptake and utilization over time of day in skeletal muscle.

Excess carbohydrate is stored as glycogen in skeletal muscle which accounts for approximately 70–80% of whole-body stores (Ivy et al. 1988; Jensen et al. 2011). Unlike the liver, skeletal muscle glycogen content is not responsible for maintaining blood glucose concentrations, rather it is a rapidly accessible energy store utilized during active contractions (Jensen et al. 2011), with glycogenesis regulated by the enzymatic activity of both glucose-6-phosphate and glycogen synthase (Viijlar-Palasí and Guinovart 1997) and glycogen breakdown regulated by phosphorylase (Howlett et al. 1998; Jensen and Richter 2012). Previous studies have reported that there is a diurnal rhythm of glycogen content in skeletal muscle, displaying highest values during the mid-active/feeding phase (Leighton et al. 1988). Interestingly, muscle-specific knockout of the clock gene, *Bmal1*, is associated with significant increases in muscle glycogen content. Whether this clock-dependent change in muscle glycogen is due to diminished glycolytic flux or alteration in glycogenolysis is not yet know (Harfmann et al. 2016).

In summary, the circadian clock in muscle is important for regulating pathways involved in glucose uptake/insulin sensitivity, glucose utilization, and glucose storage. Thus, independent of exercise there are time of day differences in these important muscle metabolic parameters. A key job of the muscle clock is to prepare the tissue for the activity onset following a fast upon awakening and for changes in substrate availability associated with feeding. These temporal patterns likely have consequences for exercise outcome measures. For example, if the phase of the muscle clock is disrupted, such as during shift work or jet lag, it would be predicted that this would result in impaired exercise performance, in part, through a reduction in glucose uptake and expression of key glycolytic enzymes.

16.3.2 Circadian Control of Lipid Metabolism

The muscle clock also has been shown to contribute to the regulation of skeletal muscle lipid oxidation over time of day. Genes that regulate each step of lipid metabolism, from fatty acid transport to oxidation and back to storage, are known to exhibit circadian rhythms (Zhang et al. 2014; Hodge et al. 2015; Dyar et al. 2018a). Fatty acid transfer into the inner mitochondrial matrix occurs in a time of day manner, peaking in the early to mid-active phase, to support oxidative metabolism with energy availability. Analysis of microarray datasets from circadian time course studies reveals that the gene encoding for acyl-carnitine translocase, *Slc25a20*, reaches peak expression in the middle of the resting/fasting period (Indiveri et al. 2011; Zhang et al. 2014). Further, two genes that encode for lipid transport are fatty acid binding proteins Fabp3 and 4, which are also expressed in a circadian manner, showing highest mRNA expression levels in the early and mid-resting/fasting periods (Syamsunarno et al. 2013; Hodge et al. 2015; Dyar et al. 2018a). Consistent with these data, multiple genes that encode for β-oxidation, such as enoyl CoA hydratase, tri-functional enzyme subunits Hadh α/β, and the acetyl-CoA acyltransferase, have also been identified to be circadian, reaching peak expression around the mid-resting/fasting phase (Zhang et al. 2014). These observations are consistent with predictive homeostasis with important enzymes for oxidative metabolism increasing prior to waking and activity onset. In addition, these observations illustrate that the muscle clock can regulate with malonyl-CoA, known to promote β-oxidation reaching peak expression during similar rest period as that of the circadian β-oxidation genes, aligning lipid metabolism to fit a circadian profile in skeletal muscle (Schmidt and Herpin 1998; Saggerson 2008). In contrast to increases in fat oxidation patterns during rest, the lipogenic genes *Acly*, *Acaca*, and *Fasn* all peak expression at the end of the active/feeding phase (Funai and Semenkovich 2011; Ameer et al. 2014) Thus, the muscle clock contributes to a baseline pattern that temporally segregates fat oxidation versus fat storage over time of day.

Analysis of the circadian time course from metabolomic and lipidomic data from mice in which the core clock gene, *Bmal1*, is knocked out was performed in 2018 by Dyar et al., The major findings were that loss of muscle clock function led to significant perturbations in fatty acid, triglyceride, and phospholipid metabolism. For example, Bmal1 KO mice displayed reduced levels of muscle triglycerides, and this was associated with significant downregulation of *Dgat2* gene expression. *Dgat2* is implicated in fat storage and known to encode for the enzyme responsible for the conversion of diacylglycerols to triglycerides. Interestingly, this finding was concomitant with the upregulation of genes known to be involved in fatty acid synthesis (Dyar et al. 2018b), thus serving to effectively flip the temporal periods of lipid storage and utilization in animals without a functional muscle clock. Since these mice maintain a normal central clock and their activity and feeding patterns are normal, this flipping of substrate storage and utilization in Bmal1 KO mice illustrates the significant contribution that the circadian clock makes to substrate metabolism under resting conditions.

16.4 Exercise, Circadian Rhythms, and Metabolism

The temporal gating of substrate metabolism by circadian clocks is an important component of homeostasis, highlighted by studies in which circadian rhythms are disrupted being linked to a variety of metabolic diseases (Masri and Sassone-Corsi 2018). The majority of daily physiological processes, including many aspects of metabolism, undergo changes about the relative time of day. Many of these metabolic variations, which we refer to as the "moving baseline," are downstream of the circadian clock within each tissue, including skeletal muscle. In this section, we discuss two topics; (1) what is understood about the integration of exercise metabolism and circadian rhythms in skeletal muscle and (2) what happens when we exercise at different times of the day.

It has been recognized for over 10 years that exercise at different times of the day leads to distinct performance outcomes. For example, daily variations in resistance and endurance exercise peak performance have been reported to fluctuate when studied during the active/feeding phase in humans (Souissi et al. 2004; Ab Malik et al. 2020; Mirizio et al. 2020) and rodents (Ezagouri et al. 2019). As such, consistently, studies across species have revealed that variables such as skeletal muscle strength and oxidative capacity demonstrate significant differences in performance outcomes over time of day (Atkinson and Reilly 1996; van Moorsel et al. 2016; de Goede et al. 2018). For example, numerous studies have demonstrated increased maximal isometric strength in the later afternoon versus morning (Douglas et al. 2021), while oxidative capacity peaks in the late evening (Reilly and Waterhouse 2009; van Moorsel et al. 2016). Thus, these exercise performance outcomes provide evidence that the circadian timing of our metabolism throughout the day (i.e., morning vs. afternoon) likely impacts exercise response and in turn performance outcomes such as endurance capacity and maximal strength (Ezagouri et al. 2019; Douglas et al. 2021).

16.4.1 The Influence of Circadian Rhythms on Acute Exercise Responses

Daily variations in exercise capacity likely stem from differences in metabolic starting points derived from the daily changes in the metabolic status that differs between early and late exercisers. For example, exercise during the early active/feeding phase, when hepatic glycogen content is reduced, rather than exercise at the late active/feeding phase, when hepatic glycogen content is increased, results in the rapid depletion of carbohydrate energy stores in skeletal muscle and a shift toward utilization of fatty acid metabolism thereby favoring endurance type exercise. Ezagouri et al. (2019) investigated this concept by assigning animals to sedentary and exercise groups during either the early (ZT14) or late (ZT22) active period, subjecting both wild-type and double Per1/2 knockout mice to moderate-intensity exercise (reported ~55% VO2max). Interestingly, wild-type mice ran ~67% longer at ZT22 than ZT13, whereas the time of day effect on endurance capacity was abolished in the Per1/2 knockouts. These results were also corroborated in healthy, young humans who were subjected to a submaximal constant-load exercise protocol, equivalent to the moderate-intensity exercise protocol of the mice. Each participant performed the exercise protocol on two occasions (0800 and 1800), separated by appropriate wash out period. Strikingly, maximal oxygen consumption (VO_2max) was significantly lower at 1800 than 0800, but the respiratory exchange ratio (RER) was significantly higher at 1800 than 0800, indicating greater use of carbohydrate metabolism in the later exercisers. However, blood glucose levels were reported to be higher at exercise performed at 0800 in comparison to 1800. Previous studies have shown a greater aerobic capacity in the evening compared to the morning hours in humans, with improved capacity associated with a greater reliance on carbohydrates (higher RER) (Drust et al. 2005; Reilly and Waterhouse 2009; Küüsmaa et al. 2016; Thosar et al. 2018). These findings are consistent with the rodent data described above (Ezagouri et al. 2019) and thus suggest that the time of day differences in aerobic capacity are likely orchestrated through the circadian clock impact on substrate metabolism in muscle and likely other metabolic tissues.

Another important question that arises from the "moving baseline concept" is whether the time of exercise elicits a differential molecular response. Analysis of clock output data from rodents and humans provides perspective on the underlying variation of gene expression over time of day (McCarthy et al. 2007; Andrews et al. 2010; Zhang et al. 2014; Perrin et al. 2018). Now, when the exact same bout of exercise is performed at two distinctly different times of day, Sato et al. and Ezagouri et al. (2019) found that the exercise-induced gene expression changes were significantly different. While this may be surprising, when you consider that the exercise intervention is occurring on a moving, and not flat, transcriptomic baseline, this outcome is actually expected (Fig. 16.3). Work from two different labs over the last 2+ years highlights the impact of time of day to an acute response to treadmill exercise. In the study by Ezagouri et al., (2019) they analyzed skeletal muscle gene expression and the metabolite profile from two different exercised groups of mice during the early active vs. late active phase and found that a greater number of significantly regulated transcripts were unique to exercise at the early active phase, ZT14 (343 genes), rather than those unique to the late active phase, ZT22 (125 genes), with a there was a modest overlap of 160 genes (Ezagouri et al. 2019). Functional cluster analysis of the differentially regulated RNAseq data found that genes associated with insulin signaling and glucose metabolism were upregulated when exercise was performed at early active phase, ZT14, but not with later exercise at ZT22. In contrast, genes linked to the FoxO signaling pathway were the more enriched category in the late exercise, ZT22 group. Functional analysis of the genes that changed in common between groups was many nuclear encoded mitochondrial genes.

A similar study was reported in Sato et al. (2019), in which they subjected mice to a 1 h acute bout of treadmill exercise at the early resting/fasting phase (ZT3) or during the early active/feeding phase (ZT15) with respective control groups (Sato

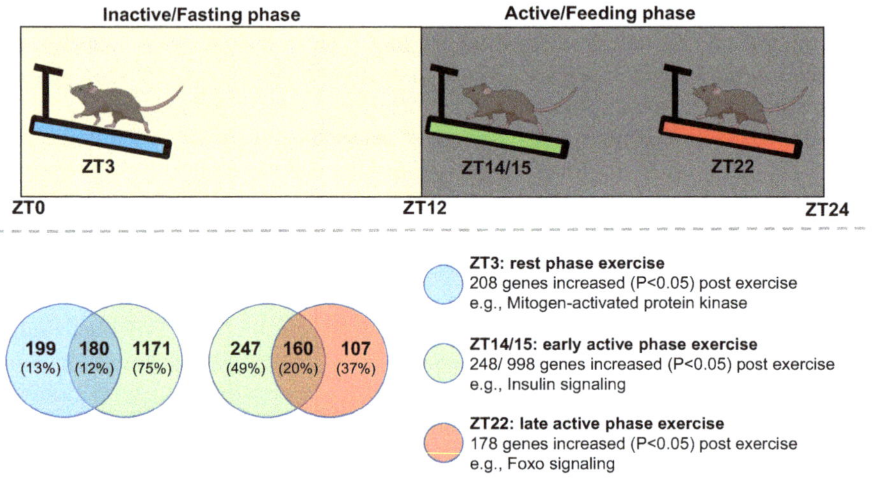

Fig. 16.3 Distinct time of day gene expression responses to treadmill exercise. Treadmill running exercise at the early inactive phase (ZT3) versus the early active phase (ZT15) or at the early active phase (ZT14) compared to the late active phase (ZT22) produces different transcriptional responses. Examples of functional enrichment pathways are provided to demonstrate differential responses based on time of day. The venn diagrams display the upregulated genes only as a percentage of significantly regulated genes, unique to each exercise period: ZT3 vs. ZT15 (blue and green) and ZT14 vs. ZT22 (green and red). All data adapted from Sato et al. (2019) and Ezagouri et al. (2019). Running mouse images were downloaded from BioRender.com

et al. 2019). For reference, both the Sato et al. and Ezagouri et al. studies have one group of mice that exercise in the later active phase (ZT14 or ZT15). However, the Sato et al. (2019) paper had their second group of mice run at ZT3 which is running at 3 hours in the light/rest phase. Transcriptomic analysis of gastrocnemius muscle revealed that both times of exercise demonstrated a significant change in gene expression. However, there was only a small overlap between the number of exercise-responsive transcripts that were upregulated (12%) and downregulated (5%) when comparing the exercise response at ZT3 to ZT15, confirming a large time of day specificity. Moreover, gene ontology clustering of exercise-responsive genes at different times of day also highlighted the transcriptional response to exercise is time of day specific. This was consistent with metabolomic analysis, which showed upregulated and downregulated metabolites also have little overlap, 13% and 1%, respectively, between exercise at different time points. While this does not address potential changes with long-term training, it does highlight the interaction of exercise and the circadian clock, providing evidence that clock output is different over time of day even in response to an acute exercise intervention.

Exercise outcomes within important signaling pathways for metabolism have also been reported to be affected by circadian timing in a time of day-specific manner. For example, the mechanistic target of rapamycin complex 1 (mTORC1) and peroxisome proliferator activated receptor gamma coactivator 1 alpha (PGC1α) are two pathways that have been widely reported in the exercise physiology literature.

However, these exercise stimulated signaling pathways have also been identified as being downstream of the molecular clock, thus providing a mechanism by which circadian timing may modulate exercise responses (Um et al. 2011; Ramanathan et al. 2018). Further, there is evidence to suggest that the core components of the molecular clock can interact with some of these pathways. For example, PER2, which peaks at the end of the resting/fasting phase, reduces mTORC1 activity in the liver (Wu et al. 2019). However, it is not yet clear whether the interaction of clock components and such pathways are conserved across tissues, e.g., skeletal muscle. Nevertheless, the molecular responses of exercise in skeletal muscle exhibit a clear time-dependent effect for acute exercise. For example, resistance exercise performed in humans at the early active/feeding phase leads to acute mTORC1 activation in skeletal muscle, whereas the same exercise in the late active/feeding phase has been documented to produce a more blunted signaling response (Sedliak et al. 2009). Despite this influence of acute resistance exercise bout performed at different times on hypertrophic signaling, training studies have found that when the resistance exercise is performed at different times of day, there is no difference in the magnitude of skeletal muscle hypertrophy (Küüsmaa et al. 2016; Sedliak et al. 2018). However, it is important to note that training (regularly repeated exercise bouts) at either morning or afternoon/evening will lead to shifting of the muscle clock and clock output in phase. Thus, these findings to not rule out that time of exercise does not matter for muscle growth, but rather they reinforce that maintaining a regular exercise training schedule is important. This is a very new area of research, and thus, there is still very much to be learned.

16.4.2 Exercise Can Target the Molecular Clock and Modify the Moving Baseline

Exercise and scheduled physical activity have been established to be environmental time cues, or zeitgeber, that can modify the phase of the muscle clock. The evidence in support of this has been accumulating since the late 1980s and early 1990s when novel wheel access at different times of day was found to be sufficient to shift the phase of circadian behavioral rhythms in mice and hamsters (Edgar et al. 1991; Edgar and Dement 1991). In follow-up work to these original studies, forced treadmill exercise training in rodents (Wolff and Esser 2012; Schroeder et al. 2012) and humans (Youngstedt et al. 2019) further confirmed exercise serves as a zeitgeber. Specifically, depending on the time of exercise, the muscle clock, and its output, will shift in phase in a predictable manner.

The mechanisms through which exercise modifies the phase of the molecular clock and subsequently the clock output are incompletely understood, but have been the target of numerous recent investigations. Emerging data demonstrate that muscle contractions, as a component of exercise, can directly modulate the expression of core clock components in a time of day-dependent manner. Kemler et al. (et al. 2020) subjected PER2::LUC circadian reporter mice to an acute bout of 60 min treadmill exercise at three different times of day (Kemler et al. 2020). Exercise at ZT5 induced a phase advance of the clock, whereas exercise at ZT11 induced a phase delay of the clock. However, exercise in the middle of the active phase, at ZT17, did not alter the muscle clock phase. These time of day-specific responses mirror the anticipated phase response curves. Ex vivo muscle contraction studies showed that the expression of *Per2* is modified acutely, and this results in an altered phase (Small et al. 2020). Small et al. (2020) provided more mechanistic data finding that calcium influx stimulated by contraction leads to the binding of the phosphorylated form of cAMP response element-binding protein (CREB) to the *Per2* promoter. Together, these experiments point to CREB activation as a key mechanism, whereby exercise can alter the phase of the circadian clock in muscle.

While exercise seems to directly affect the expression levels of the core clock genes, well-known exercise-induced kinases have been also shown to modify the phase of the circadian clock. While still early, these associations suggest the potential for these kinases to also contribute to the muscle clock and clock output. For the purpose of this chapter, we will discuss four well-known exercise-induced factors that have been reported to interact with the clock mechanism (Fig. 16.4). Specifically, exercise induces activation of the cAMP response element binding protein (CREB), 5' AMP-activated protein kinase (AMPK), hypoxia inducible factor 1 alpha (HIF-1α), and sirtuin 1 (SIRT1). We review the current understanding of these four exercise and circadian factors in the following paragraphs.

Exercise increases CREB phosphorylation, and CREB plays a role in changing circadian phase in other tissues. Recent data in skeletal muscle following acute exercise or electrical pulse stimulated contractions indicate that there are also CREB-induced increases in *Per2* expression (Small et al. 2020). CREB is activated by increases in cAMP levels and protein kinase A signaling and calcium-specific signaling, as well as through activation of the MAP kinase pathways. It is currently unclear if exercise activates CREB through one or a combination of all of these pathways. Additionally, the time of day-specific effects of exercise on core clock gene expression (Ezagouri et al. 2019; Sato et al. 2019; Kemler et al. 2020) suggest that either (1) exercise-induced CREB activity is different over time of day or (2) the effect of CREB of core clock gene expression varies over time of day. Further work is needed to explore these potential differences and inform the mechanism of CREB-induced changes in core clock gene expression and the subsequent changes in the timing of the circadian phase and output.

AMPK is a well-known exercise-responsive protein (Richter and Ruderman 2009). AMPK is often studied in its context as an energy sensor and thereby metabolic regulator. Therefore the interaction between AMPK and the molecular clock potentially represents the primary interface linking changes in metabolism to change in circadian timing. Current data indicate that AMPK affects the timing of the circadian clock through altering the stability of CRY1 proteins through phosphorylation and subsequent targeted degradation of CRY1 by the proteasome (Lamia et al. 2009). While the endurance exercise-induced activation of AMPK was not different at different times of day (Ezagouri et al. 2019), the AMPK-induced reduction in

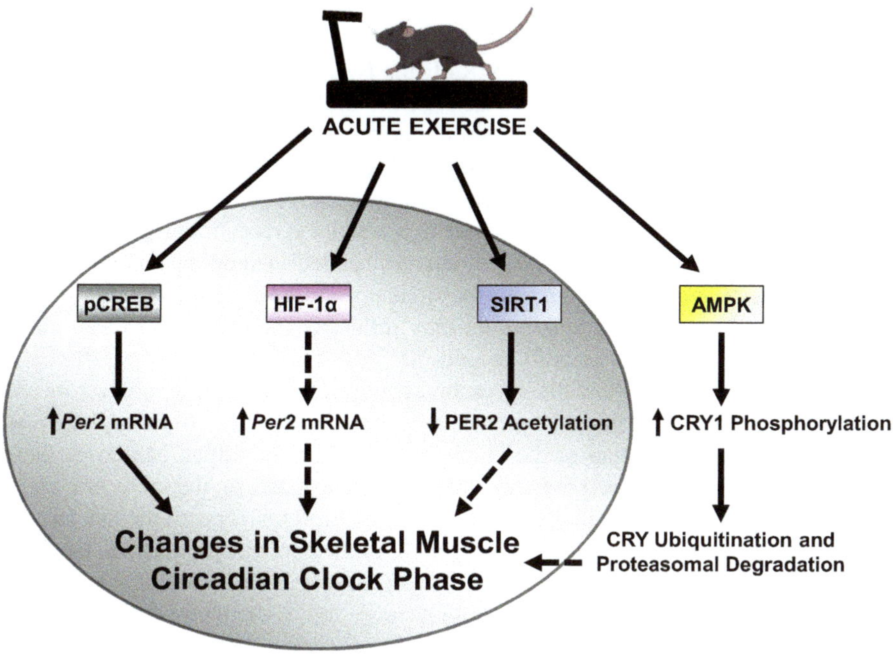

Fig. 16.4 Exercise as an input to the core circadian clock. Exercise is an effector of the core skeletal muscle circadian clock. Bouts of acute exercise increase the phosphorylation of CRE binding protein 1 (CREB) and AMP-activated protein kinase (AMPK) and upregulate hypoxia-inducible factor 1 alpha (HIF-1α) and the NAD-dependent deacetylase sirtuin 1 (SIRT1). Individually, each of these exercise responsive targets has been linked to changes in core clock gene expression or protein levels. The anticipated effect of these changes in core clock gene/protein expression is changes in the timing of the circadian system. Solid lines represent direct experimental evidence, while dotted lines represent anticipated exercise-induced effects. Running mouse images were downloaded from BioRender.com

CRY1 content is beneficial for the upregulation of fatty acid oxidation genes (Jordan et al. 2017). Further, the CRY proteins have a strong repressive effect on these lipid metabolism genes, reiterating the daily variation in metabolic gene expression is influenced directly by the core clock machinery.

Endurance exercise also increases the expression of HIF-1α, a protein with a structure similar to Bmal1 (Wu et al. 2017). And HIF-1α may contribute to the exercise-induced transcription of core clock genes (Adamovich et al. 2014; Peek et al. 2017). Importantly, loss of HIF-1α was required for the molecular clock adaptation to an experimental jet-lag paradigm, suggesting a role of HIF-1α in phase changes, which is specifically remodeling the timing of the molecular clock output. Moreover, there was a time of day-specific effect of an acute exercise bout on both the induction of HIF-1α and the corresponding increase in HIF-1α gene targets (Peek et al. 2017). Recent data also indicate that HIF-1α binds to the Per2 promoter and increases its expression, potentially explaining its role in changing circadian phase (Wu et al. 2017).

Finally, current data suggest that SIRT1, the NAD+ dependent deacetylase, can target both the positive and negative arms of the molecular clock. SIRT1 not only binds to BMAL1:CLOCK to increase amplitude of circadian transcription, but it also deacetylates PER2, increasing its degradation (i.e., repressing the repressor) (Asher et al. 2008; Foteinou et al. 2018). The SIRT1-induced increase in circadian output amplitude may represent a mechanism through which exercise training enhances the function of the circadian clock, though data are needed to support this postulation. Additionally, because SIRT1 contributes to changes in histone acetylation, SIRT1-induced changes in chromatin structure may influence the exercise-induced circadian output, but more data are still needed.

While we have highlighted the interaction between exercise-inducible factors and the molecular clock, additional targets likely exist that will further explain how exercise modifies the timing of the circadian clock and its output. Moreover these data have focused on skeletal muscle, and the exercise effect on the molecular clock in other peripheral tissues remains largely unexplored. Regardless of the mechanism through which exercise is modifying the timing of the core circadian clock, there are clear exercise-induced changes in the timing of the core circadian clock with implications for the timing of its output (e.g., lipid oxidation genes). These exercise-induced changes in circadian clock output timing therefore likely result in differences in daily substrate metabolism, insulin sensitivity, and other processes that contribute to maintaining homeostasis within tissues and across the organism. Finally, most of the data discussed in the context of exercise and the circadian clock are from acute exercise studies. The effect of exercise training on the circadian clock and clock output remains significantly understudied and represents an exciting future direction of the circadian exercise metabolism field.

16.5 Conclusions

In summary, circadian rhythms and the circadian clock are fundamental parts of our cell and systems biology. The circadian clocks function to support homeostatis through their role regulating gene expression in anticipation of predictable changes in the environment and behavior. It is well established that the circadian clock in muscle and other tissues function to regulate gene expression and the tissue metabolome in a time of day manner. A significant component of clock output is genes that are important for substrate metabolism including pathways for fuel oxidation as well as storage. These findings are supported by studies in both humans and rodents and highlight that the metabolic response to an acute bout of exercise will differ based on time of day. While these findings are quite clear, there are still many large gaps in the field. For example, the majority of exercise studies are done with more endurance exercise with very limited data on resistance exercise. We also know very little about how exercise intensity or duration impacts aspects of circadian clock output. The issue of the impact of training, and not just acute exercise, is also not well studied. These large gaps represent exciting new opportunities for both basic exercise science and applied and clinical interventions. In particular there is growing interest in whether time of day exercise strategies could be helpful with preventing or delaying the development of metabolic diseases. Lastly, we want to note that moving forward, future exercise studies in both human and rodent interventions must take care to provide transparent reporting of circadian conditions (e.g., light/dark cycles, feeding status, and habitual activity), as well as robust time of day sampling rates with appropriate controls in order to draw meaningful comparisons and interpretations from the data.

References

Ab Malik Z, Bowden Davies KA, Hall ECR et al (2020) Diurnal differences in human muscle isometric force in vivo are associated with differential phosphorylation of Sarcomeric M-band proteins. Proteomes 8:22

Adamovich Y, Rousso-Noori L, Zwighaft Z et al (2014) Circadian clocks and feeding time regulate the oscillations and levels of hepatic triglycerides. Cell Metab 19:319–330

Ameer F, Scandiuzzi L, Hasnain S et al (2014) De novo lipogenesis in health and disease. Metabolism 63:895–902

Andrews JL, Zhang X, McCarthy JJ et al (2010) CLOCK and BMAL1 regulate MyoD and are necessary for maintenance of skeletal muscle phenotype and function. Proc Natl Acad Sci 107: 19090–19095

Aras E, Ramadori G, Kinouchi K et al (2019) Light entrains diurnal changes in insulin sensitivity of skeletal muscle via ventromedial hypothalamic neurons. Cell Rep 27:2385–2398.e3

Aschoff J (1965) CIRCADIAN RHYTHMS IN MAN. Science 148:1427–1432

Asher G, Gatfield D, Stratmann M et al (2008) SIRT1 regulates circadian clock gene expression through PER2 deacetylation. Cell 134:317–328

Atkinson G, Reilly T (1996) Circadian variation in sports performance. Sport Med 21:292–312

Conroy RT, O'Brien M (1974) Proceedings: diurnal variation in athletic performance. J Physiol 236:51P

Crockford GW, Davies CT (1969) Circadian variations in responses to submaximal exercise on a bicycle ergometer. J Physiol 201:94P–95P

de Goede P, Wefers J, Brombacher EC et al (2018) Circadian rhythms in mitochondrial respiration. J Mol Endocrinol 60:R115–R130

DeFronzo RA, Tripathy D (2009) Skeletal muscle insulin resistance is the primary defect in type 2 diabetes. Diabetes Care 32:S157–S163

Douglas CM, Hesketh SJ, Esser KA (2021) Time of day and muscle strength: a circadian output? Physiology 36:44–51

Drust B, Waterhouse J, Atkinson G et al (2005) Circadian rhythms in sports performance--an update. Chronobiol Int 22:21–44

Dyar KA, Ciciliot S, Wright LE et al (2014) Muscle insulin sensitivity and glucose metabolism are controlled by the intrinsic muscle clock. Mol Metab 3:29–41

Dyar KA, Hubert MJ, Mir AA et al (2018a) Transcriptional programming of lipid and amino acid metabolism by the skeletal muscle circadian clock. PLoS Biol 16:e2005886

Dyar KA, Lutter D, Artati A et al (2018b) Atlas of circadian metabolism reveals system-wide coordination and communication between clocks. Cell 174:1571–1585.e11

Edgar DM, Dement WC (1991) Regularly scheduled voluntary exercise synchronizes the mouse circadian clock. Am J Physiol Integr Comp Physiol 261:R928–R933

Edgar DM, Martin CE, Dement WC (1991) Activity feedback to the mammalian circadian pacemaker: influence on observed measures of rhythm period length. J Biol Rhythm 6:185–199

Ezagouri S, Zwighaft Z, Sobel J et al (2019) Physiological and molecular dissection of daily variance in exercise capacity. Cell Metab 30:78–91.e4

Foteinou PT, Venkataraman A, Francey LJ et al (2018) Computational and experimental insights into the circadian effects of SIRT1. Proc Natl Acad Sci U S A 115:11643–11648

Funai K, Semenkovich CF (2011) Skeletal muscle lipid flux: running water carries no poison. Am J Physiol Metab 301:E245–E251

Gannon RL, Rea MA (1995) Twelve-hour phase shifts of hamster circadian rhythms elicited by voluntary wheel running. J Biol Rhythm 10:196–210

Harfmann BD, Schroder EA, Kachman MT et al (2016) Muscle-specific loss of Bmal1 leads to disrupted tissue glucose metabolism and systemic glucose homeostasis. Skelet Muscle 6:12

Hodge BA, Wen Y, Riley LA et al (2015) The endogenous molecular clock orchestrates the temporal separation of substrate metabolism in skeletal muscle. Skelet Muscle 5:17

Howlett RA, Parolin ML, Dyck DJ et al (1998) Regulation of skeletal muscle glycogen phosphorylase and PDH at varying exercise power outputs. Am J Phys 275:R418–R425

Indiveri C, Iacobazzi V, Tonazzi A et al (2011) The mitochondrial carnitine/acylcarnitine carrier: function, structure and physiopathology. Mol Asp Med 32:223–233

Ivy JL, Katz AL, Cutler CL et al (1988) Muscle glycogen synthesis after exercise: effect of time of carbohydrate ingestion. J Appl Physiol 64:1480–1485

Jensen TE, Richter EA (2012) The journal of physiology regulation of glucose and glycogen metabolism during and after exercise. J Physiol 590:1069–1076

Jensen J, Rustad PI, Kolnes AJ, Lai Y-C (2011) The role of skeletal muscle glycogen breakdown for regulation of insulin sensitivity by exercise. Front Physiol 2:112

Jordan SD, Kriebs A, Vaughan M et al (2017) CRY1/2 selectively repress PPARδ and limit exercise capacity. Cell Metab 26:243–255.e6

Kemler D, Wolff CA, Esser KA (2020) Time-of-day dependent effects of contractile activity on the phase of the skeletal muscle clock. J Physiol 598:3631–3644

Kleitman N (1949) Biological rhythms and cycles. Physiol Rev 29:1–30

Koike N, Yoo S-H, Huang H-C et al (2012) Transcriptional architecture and chromatin landscape of the core circadian clock in mammals. Science 338:349–354

Küüsmaa M, Schumann M, Sedliak M et al (2016) Effects of morning versus evening combined strength and endurance training on physical performance, muscle hypertrophy, and serum hormone concentrations. Appl Physiol Nutr Metab 41:1285–1294

Lamia KA, Sachdeva UM; DiTacchio L et al (2009) AMPK regulates the circadian clock by cryptochrome phosphorylation and degradation. Science 326:437–440

Leighton B, Kowalchuk JM, Challiss RAJ, Newsholme EA (1988) Circadian rhythm in sensitivity of glucose metabolism to insulin in rat soleus muscle. Am J Physiol Metab 255:E41–E45

Masri S, Sassone-Corsi P (2018) The emerging link between cancer, metabolism, and circadian rhythms. Nat Med 24:1795–1803

McCarthy JJ, Andrews JL, McDearmon EL et al (2007) Identification of the circadian transcriptome in adult mouse skeletal muscle. Physiol Genomics 31:86–95

Miller BH, McDearmon EL, Panda S et al (2007) Circadian and CLOCK-controlled regulation of the mouse transcriptome and cell proliferation. Proc Natl Acad Sci U S A 104:3342–3347

Mirizio GG, Nunes RSM, Vargas DA et al (2020) Time-of-day effects on short-duration maximal exercise performance. Sci Rep 10:9485

Mohawk JA, Green CB, Takahashi JS (2012) Central and peripheral circadian clocks in mammals. Annu Rev Neurosci 35:445–462

Moore-Ede MC (1986) Physiology of the circadian timing system: predictive versus reactive homeostasis. Am J Physiol Integr Comp Physiol 250:R737–R752

Mure LS, Le HD, Benegiamo G et al (2018) Diurnal transcriptome atlas of a primate across major neural and peripheral tissues. Science (80-) 359:eaao0318

Peek CB, Levine DC, Cedernaes J et al (2017) Circadian clock interaction with HIF1α mediates oxygenic metabolism and anaerobic glycolysis in skeletal muscle. Cell Metab 25:86–92
Perrin L, Loizides-Mangold U, Chanon S et al (2018) Transcriptomic analyses reveal rhythmic and CLOCK-driven pathways in human skeletal muscle. elife 7:e34114
Persson PB, Bondke Persson A (2019) Circadian rhythms. Acta Physiol 225:e13220
Pittendrigh CS, Daan S (1976) A functional analysis of circadian pacemakers in nocturnal rodents. J comp Physiol ? A 106:333–355
Pizarro A, Hayer K, Lahens NF, Hogenesch JB (2012) CircaDB: a database of mammalian circadian gene expression profiles. Nucleic Acids Res 41:D1009–D1013
Ramanathan C, Kathale ND, Liu D et al (2018) mTOR signaling regulates central and peripheral circadian clock function. PLoS Genet 14:e1007369
Reilly T, Waterhouse J (2009) Sports performance: is there evidence that the body clock plays a role? Eur J Appl Physiol 106:321–332
Richter EA, Ruderman NB (2009) AMPK and the biochemistry of exercise: implications for human health and disease. Biochem J 418:261–275
Saggerson D (2008) Malonyl-CoA, a key signaling molecule in mammalian cells. Annu Rev Nutr 28:253–272
Sato S, Basse AL, Schönke M et al (2019) Time of exercise specifies the impact on muscle metabolic pathways and systemic energy homeostasis. Cell Metab 30:92–110.e4
Schmidt I, Herpin P (1998) Carnitine Palmitoyltransferase I (CPT I) activity and its regulation by Malonyl-CoA are modulated by age and cold exposure in skeletal muscle mitochondria from newborn pigs. J Nutr 128:886–893
Schmutz I, Ripperger JA, Baeriswyl-Aebischer S, Albrecht U (2010) The mammalian clock component PERIOD2 coordinates circadian output by interaction with nuclear receptors. Genes Dev 24:345–357
Schroeder AM, Truong D, Loh DH et al (2012) Voluntary scheduled exercise alters diurnal rhythms of behaviour, physiology and gene expression in wild-type and vasoactive intestinal peptide-deficient mice. J Physiol 590:6213–6226
Sedliak M, Finni T, Cheng S et al (2009) Effect of time-of-day-specific strength training on muscular hypertrophy in men. J Strength Cond Res 23:2451–2457
Sedliak M, Zeman M, Buzgó G et al (2018) Morphological, molecular and hormonal adaptations to early morning versus afternoon resistance training. Chronobiol Int 35:450–464
Small L, Altıntaş A, Laker RC et al (2020) Contraction influences Per2 gene expression in skeletal muscle through a calcium-dependent pathway. J Physiol 598:5739–5752
Souissi N, Gauthier A, Sesboüé B et al (2004) Circadian rhythms in two types of anaerobic cycle leg exercise: force-velocity and 30-s Wingate tests. Int J Sports Med 25:14–19
Syamsunarno MRAA, Iso T, Hanaoka H et al (2013) A critical role of fatty acid binding protein 4 and 5 (FABP4/5) in the systemic response to fasting. PLoS One 8:e79386
Takahashi JS (2017) Transcriptional architecture of the mammalian circadian clock. Nat Rev Genet 18:164–179
Thosar SS, Herzig MX, Roberts SA et al (2018) Lowest perceived exertion in the late morning due to effects of the endogenous circadian system. Br J Sports Med 52:1011–1012
Tipton CM (2003) Exercise physiology. Springer, New York, New York, NY
Um J-H, Pendergast JS, Springer DA et al (2011) AMPK regulates circadian rhythms in a tissue- and isoform-specific manner. PLoS One 6:e18450
van Moorsel D, Hansen J, Havekes B et al (2016) Demonstration of a day-night rhythm in human skeletal muscle oxidative capacity. Mol Metab 5:635–645
Viijlar-Palasí C, Guinovart JJ (1997) The role of glucose 6-phosphate in the control of glycogen synthase. FASEB J 11:544–558
Voigt ED, Engel P, Klein H (1968) On the daily operation of physical work capacity. Int Z Angew Physiol 25:1–12
Wahlberg I, Astrand I (1973) Physical work capacity during the day and at night. Work Environ Health 10:65–68

Wolff G, Esser KA (2012) Scheduled exercise phase shifts the circadian clock in skeletal muscle. Med Sci Sports Exerc 44:1663–1670
Wu Y, Tang D, Liu N et al (2017) Reciprocal regulation between the circadian clock and hypoxia signaling at the genome level in mammals. Cell Metab 25:73–85
Wu R, Dang F, Li P et al (2019) The circadian protein Period2 suppresses mTORC1 activity via recruiting Tsc1 to mTORC1 complex. Cell Metab 29:653–667.e6
Youngstedt SD, Elliott JA, Kripke DF (2019) Human circadian phase–response curves for exercise. J Physiol 597:2253–2268
Zhang R, Lahens NF, Ballance HI et al (2014) A circadian gene expression atlas in mammals: implications for biology and medicine. Proc Natl Acad Sci 111:16219–16224

Chapter 17
Metabolic Factors in Skeletal Muscle Fatigue

Nicolas Place and Håkan Westerblad

Abstract We here discuss metabolic causes of skeletal muscle fatigue with focus on peripheral fatigue, that is, negative transient effects on muscle contractility manifested as decreased force production, reduced shortening speed and/or slowed relaxation. We specifically address the following fatigue-related metabolic changes: (1) [ATP] normally remains almost constant but might decrease to critically low levels in severe fatigue where the accompanying transient increase in [ADP] would reduce shortening speed and hence power output. (2) The increase in inorganic phosphate ions (P_i) during intense exercise has a central role in fatigue by reducing the myofibrillar force generating capacity and Ca^{2+} sensitivity and by attenuating sarcoplasmic reticulum (SR) Ca^{2+} release. (3) Acidosis occurs during intense exercise and may depress myofibrillar contractile function; its importance is currently debated. (4) Increases in reactive oxygen/nitrogen species during intense exercise can induce long-lasting protein modifications that delay the recovery after exercise. (5) The depletion of intramyofibrillar glycogen during prolonged exercise is well correlated with decreased force due to impaired SR Ca^{2+} release. Thus, several metabolic alterations contribute to skeletal muscle fatigue, and the relative importance of these depends on factors such as the type of exercise, muscle fibre composition and the training status of the exercising individual.

Keywords Skeletal muscle fatigue · Inorganic phosphate ions · Lactic acidosis · Reactive oxygen/nitrogen species · Glycogen

Skeletal muscle fatigue develops rather slowly when submaximal muscle contractions are repeated over a long period (e.g. marathon running) or more rapidly during high-intensity efforts (e.g. sprint running). Due to its multifaceted nature, muscle fatigue is difficult to define. The topic has been discussed over decades and the

N. Place
Institute of Sport Sciences, University of Lausanne, Lausanne, Switzerland

H. Westerblad (✉)
Department of Physiology and Pharmacology, Karolinska Institutet, Stockholm, Sweden
e-mail: hakan.westerblad@ki.se

© The Author(s), under exclusive license to Springer Nature Switzerland AG 2022

G. McConell (ed.), *Exercise Metabolism*, Physiology in Health and Disease,
https://doi.org/10.1007/978-3-030-94305-9_17

debate continues as evidenced, for instance, by the rather recent proposal to distinguish between fatigue and fatigability (Kluger et al. 2013). From a general perspective, muscle fatigue can be divided into central fatigue defined as a 'progressive reduction in voluntary activation of muscle during exercise' (Gandevia 2001) and peripheral fatigue caused by functional impairments within the muscle fibres (Allen et al. 2008). Importantly, central and peripheral fatigue should not be considered as independent entities because there are intricate, mutual interactions between the central nervous system and muscles during exercise (for recent reviews, see Taylor et al. 2016; Amann et al. 2020; Brownstein et al. 2021). These interactions can be illustrated by a study where individuals performed 5 km cycling time trials with and without pharmacological inhibition of the neural afferent feedback from exercising muscles (Blain et al. 2016). Without afferent feedback, the power output during the first half of the cycling trial was higher, indicative of non-optimal voluntary activation of the exercising muscles. However, the initial higher power output resulted in exaggerated fatigue, inducing metabolic changes in exercising muscles and declining power output during the second half of the exercise resulting in similar finishing time with and without afferent feedback. In the present review, we will focus on metabolic causes of peripheral fatigue; that is, changes in metabolites, which are caused by the high energy demand during intense or prolonged exercise and which may have negative effects on muscle contractility manifested as decreased force production, reduced shortening speed and/or slowed relaxation. In particular, we will address the effects of increases in the metabolic by-products such as inorganic phosphate ions (P_i), H^+, and reactive oxygen/nitrogen species (ROS/RNS) as well as depletion of intramuscular glycogen stores.

17.1 Skeletal Muscle Activation

Well-controlled skeletal muscle contractions form the basis for all types of physical exercise. In the central nervous system, action potentials are generated at the cortical level and travel to the spinal cord where α-motoneurons connect to their defined set of muscle fibres, that is, the motor units, which constitute the basic entities activated by the central nervous system. Note that all muscle fibres within a specific motor unit are of the same type, i.e. express the same myosin heavy chain (MyHC; see below). Precise activation of motor units is important during most types of physical exercise. Highly accurate muscle activation is, for instance, an absolute requirement from the start and throughout exercise with progressively developing fatigue in a great variety of sports such as soccer, figure skating, and downhill skiing, to give a few examples.

At the muscle fibre level, activation starts with the generation of action potentials at the neuromuscular junction (Dulhunty 2006; Stephenson et al. 1998). Action potentials propagate along the muscle fibre surface membrane and into the t-tubular system, where they activate the t-tubular voltage sensor, the dihydropyridine receptor (DHPR). The DHPR is located close to the sarcoplasmic reticulum (SR) Ca^{2+} channel, the ryanodine receptor 1 (RyR1), and the action potential-mediated DHPR conformation change mechanically triggers the opening of RyR1, and Ca^{2+} is released to the cytosol. Ca^{2+} then binds to and changes the configuration of myofibrillar regulatory proteins, the troponin-tropomyosin protein complex, which allows the myosin heads, the adenosine triphosphate (ATP)-driven cross-bridges, to attach to the actin filament and contraction starts. Ca^{2+} is constantly pumped back into the SR by the SR Ca^{2+}-ATPase (SERCA), and when action potentials cease, cytosolic free Ca^{2+} concentration ($[Ca^{2+}]_i$) rapidly declines, and the muscle fibre relaxes. All these steps of the so-called excitation-contraction coupling might be affected during fatiguing exercise.

17.2 Energy Turnover During Exercise

17.2.1 Energy Consumption

Energy consumption in skeletal muscle can increase by up to 100-fold when going from resting condition to intense exercise (Sahlin et al. 1998), and the metabolic consequences of this dramatically increased energy demand are intimately linked to the development of peripheral muscle fatigue. The energy is mainly consumed by the two major ATPases in muscle fibres, i.e. the force producing cross-bridge and the Ca^{2+} pumping by SERCA (Homsher 1987). Within muscle fibres, both ATPases are constantly active during prolonged contractions, whereas only SERCA is active in the breaks between the repeated short contractions of normal locomotion. This illustrates that the energy utilized by cross-bridges vs. SERCAs varies with the type of exercise, and their relative importance for the total energy consumption is currently debated (Barclay et al. 2007; Szentesi et al. 2001; Zhang et al. 2006a). A recent study on isolated mouse muscle fibres and rat muscles in situ demonstrates the importance of SERCA energy utilization for the total energy consumption (Cheng et al. 2019). By using a pharmacological troponin activator to increase the myofibrillar Ca^{2+} sensitivity, the authors showed that when the same submaximal force could be produced at a lower $[Ca^{2+}]_i$ and hence with a decreased requirement for SR Ca^{2+} pumping, the energy consumption during repeated contractions was markedly reduced, and fatigue was delayed.

17.2.2 Energy Sources

The immediate energy required to drive cross-bridges and ion pumps comes from:

$$\text{ATP} => \text{ADP} + \text{P}_i + \text{energy}$$
$$(\text{ADP} = \text{adenosine diphosphate}; \text{P}_i = \text{inorganic phosphate}).$$

The intracellular store of ATP is small (5–6 mM) and would be depleted within ~2 s in a fully activated muscle fibre (Sahlin et al. 1998). Thus, ADP must be rapidly re-phosphorylated by other metabolic pathways to avoid ATP depletion. These can be divided into anaerobic pathways, which are fast and dominate during high-intensity physical activity, and the slower aerobic pathways that dominate during prolonged submaximal exercise (Sahlin et al. 1998). Peripheral fatigue develops rapidly during high-intensity exercise when a large proportion of the energy demand must be covered by anaerobic metabolism. On the other hand, peripheral fatigue is largely prevented during low-intensity exercise where aerobic metabolism dominates with glucose/glycogen and fat being fully metabolized to CO_2 and H_2O. In this latter case, severe peripheral fatigue will basically not occur until the glycogen stores in muscle are becoming depleted (Coyle et al. 1986; Hermansen et al. 1967).

The dominating anaerobic pathways to regenerate ATP are degradation of phosphocreatine (PCr) and breakdown of muscle glycogen to lactate and hydrogen ions (H^+), although a minor contribution can also come from myokinase, which is considered to be a near-equilibrium reaction in vivo (2 ADP < => ATP + AMP). Adenosine monophosphate (AMP) will subsequently be deaminated to inositol monophosphate (IMP) and NH_4^+.

The PCr store in muscle fibres is ~30–40 mM. PCr can donate its phosphate group to ADP via a reaction catalysed by creatine kinase (CK):

$$\text{PCr} + \text{ADP} <=> \text{Cr} + \text{ATP} \qquad (\text{Cr} = \text{creatine})$$

During intense exercise with high ATP consumption, the reaction is driven to the right with the net effects being a reduction in [PCr] and increases in [Cr] and [P_i], whereas [ATP] remains almost constant (Fig. 17.1a):

$$\text{PCr} + \text{ADP} => \text{Cr} + \text{ATP} => \text{Cr} + \text{ADP} + \text{P}_\text{i} + \text{energy}$$

When [PCr] reaches low levels, [ATP] starts to fall, and [ADP] and [AMP] show transient increases, while inositol monophosphate (IMP) accumulates; thus, IMP can be measured to assess whether transient increases in ADP and AMP occurred during the exercise bout (Zhang et al. 2008). The PCr store is regenerated when exercise becomes less intense or during the recovery period after exercise by energy generated through aerobic metabolism, i.e. the reaction is driven to the left.

17.2.3 Muscle Fibre Types

Our muscles are composed of muscle cells with large differences in metabolic profile, contractile speed, and cellular Ca^{2+} handling properties in order to cope with broadly different challenges, e.g. from long-term upright standing to quick high-power movements during explosive sport activities (Bottinelli and Reggiani 2000; Spangenburg and Booth 2003). The presently dominating classification

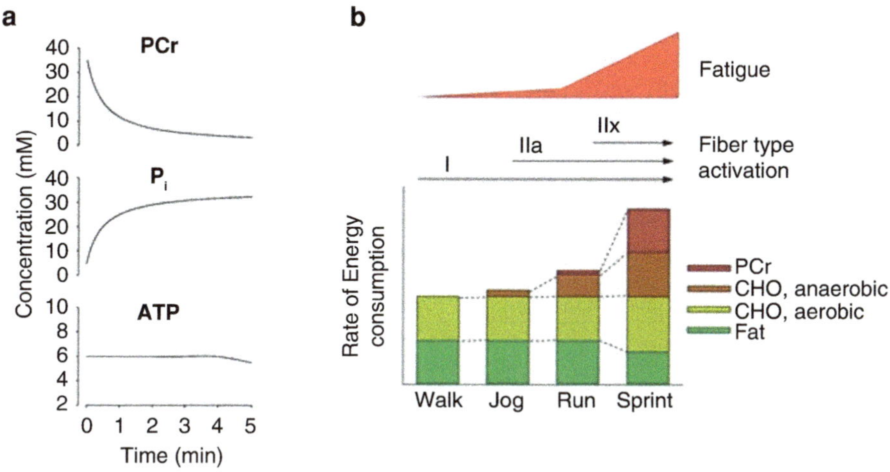

Fig. 17.1 (**a**) Schematic illustration of how the decrease in [PCr] is mirrored by an increase in [P_i], while [ATP] remains virtually constant in skeletal muscle during intense exercise. (**b**) Schematic illustration of the progressing involvement of metabolic pathways to regenerate ATP and activation of the different fibre types during exercise of increasing intensity. Note that the resulting muscle fatigue develops slowly when the energy requirement can be met by aerobic metabolism and much faster when anaerobic metabolism must be added. PCr = phosphocreatine; P_i = inorganic phosphate; CHO = carbohydrate

system for mammalian skeletal muscle is based on the expression of MyHC isoforms and type I, IIa and IIx MyHC are expressed in human muscles. The rate of cross-bridge cycling is determined by the MyHC isoform with type I being the slowest, type IIa intermediate and IIx the fastest. Numerous other protein isoforms also differ between muscle fibres, and their gene expression is controlled by multiple interacting mechanisms (Spangenburg and Booth 2003). Sometimes a pattern of gene co-expression exists in that the slow MyHC type I is co-expressed with 'slow' isoforms of other proteins, but this is not always the case. From a metabolic perspective, the MyHC-based fibre typing is relevant because the cross-bridges of a fast MyHC isoform consume ATP more rapidly than a slow isoform. The other major ATP-consuming protein in skeletal muscles, the SERCAs, also exists in two isoforms, SERCA1 mainly in fast type II fibres and SERCA2 mainly in slow type I fibres, and the density of pumps is much higher in fast than in slow fibres (Bottinelli and Reggiani 2000; Periasamy and Kalyanasundaram 2007). Moreover, fast type II fibres generally have a lower oxidative capacity than slow type I fibres, although this is not always the case (Baldwin et al. 1972). Thus, to sum up:

- Type I fibres have slow MyHC, relatively few SERCAs and a high aerobic capacity and are therefore highly fatigue resistant.
- Type IIa fibres have fast MyHC, many SERCAs and a generally lower aerobic capacity than type I fibres, although this is highly training-dependent, and they fatigue more rapidly than type I fibres.

- Type IIx fibres have the fastest MyHC, many SERCAs, low aerobic capacity and fatigue rapidly.

Motor units are activated according to their size (Henneman and Olson 1965), and this relates to the physiological properties of their respective muscle fibres. Thus, motoneurons activating type I fibres are recruited at low exercise intensity, followed by type IIa fibres and finally type IIx fibres. Figure 17.1b presents a simplified overview of how the major energy metabolic systems are used during locomotion of increasing intensities and how this relates to the activation of motor units with different fibre types and fatigue development.

17.2.4 Metabolic Changes in Various Sport Events

PCr is a primary fuel for the most intense exercise forms lasting a few seconds, such as 60–100 m sprint running and weightlifting. Anaerobic glycogen breakdown contributes substantially to the energy supply for intense efforts lasting ~1–3 min, such as 800 m track and field running and 200 m freestyle swimming (Spriet 1992). On the other end of the spectrum, aerobic degradation of glycogen or triglycerides is the dominating energy source during prolonged endurance exercise, such as marathon running and the increasingly popular ultra-endurance events like mountain running, triathlon and swim-run (Tiller et al. 2021). Furthermore, the fast and slow energy systems alternate over time in many sports, such as during a 90 min soccer match where repeated short sprints requiring anaerobic energy sources are combined with walking/slow running allowing the immediate consequences of anaerobic metabolism to be recovered (i.e. restoration of muscle fibre PCr stores and transport of lactate ions and H^+ out of the cells). Note also that the slower aerobic energy systems are generally activated from the start of any physical activity; for instance, a significant (~40%) contribution of aerobic metabolism was observed during a 30 'all-out' bicycle exercise (Medbø and Tabata 1989). Thus, the metabolic cause (s) of peripheral muscle fatigue varies considerably between different sports, and more than one cause may be of significant importance within a given sport; for instance, a fatigue-induced decrease in muscle force during a soccer match might be related to accumulated effects of the repeated sprints (e.g. transient increases in $[P_i]$ due to PCr usage as an energy source) or glycogen depletion due the long duration (90 min) of the event as such.

17.2.5 Force and $[Ca^{2+}]_i$ During Induction of Fatigue

Cellular causes underlying the impaired contractile function in fatigued muscles have been thoroughly studied in experiments performed on single intact mammalian muscle fibres (Cheng and Westerblad 2017). Such experiments with simultaneous measurements of force and $[Ca^{2+}]_i$ during fatigue induced by repeated tetanic stimulation in easily fatigued mouse, rat and human muscle fibres have revealed three phases of fatigue development: (1) an initial phase where force is reduced while tetanic $[Ca^{2+}]_i$ increases; (2) a second phase with relatively stable tetanic $[Ca^{2+}]_i$ and force; and (3) a final fast decline in both tetanic $[Ca^{2+}]_i$ and force (Westerblad and Allen 1991; Lunde et al. 2001; Olsson et al. 2020). The last phase with rapidly decreasing tetanic $[Ca^{2+}]_i$ can be seen as a safety mechanism to prevent [ATP] to fall to critically low levels where cross-bridges would enter rigour states and SR Ca^{2+} uptake fails (Cheng et al. 2018). Notably, force is much better maintained during repeated tetanic stimulation of isolated fatigue-resistant muscle fibres, and the three phases of fatigue development observed in easily fatigued fibres are not readily distinguishable or absent in highly fatigue resistant fibres (Olsson et al. 2020; Lunde et al. 2006; Zhang et al. 2006b).

17.2.6 Methods to Measure Exercise-Induced Changes in Muscle Metabolites in Humans

The assessment of the metabolic responses to exercise in humans can be performed non-invasively through ^{31}phosphorus magnetic resonance spectroscopy (^{31}P-MRS) or through biochemical measurements on muscle biopsies. A major advantage with ^{31}P-MRS is that metabolites can be quantified with reasonable time resolution throughout the course of an exercise bout and the following recovery period. However, measurements require extensive infrastructure facilities, they are limited to phosphagens (e.g. ATP, PCr, P_i, monophosphoric sugars) and intracellular pH, and global, whole-body exercises cannot be performed since subjects have to lie on a bed in close proximity to the electromagnetic coil (Liu et al. 2017). Due to these practical limitations of the ^{31}P-MRS technique, the vast majority of studies addressing metabolic changes in muscle fatigue use muscle biopsies. There are several drawbacks also with this technique: it is invasive, and individuals may be reluctant to allow several biopsies to be taken; it provides a snapshot of the situation at the time the biopsy was obtained, and therefore the time course of metabolic changes cannot be followed in great detail, and transient metabolic changes can remain undetected; there will be some delay between the end of exercise and the time when the biopsy is taken, and hence rapidly recovering metabolic changes will be missed. Thus, present methods do not allow detailed metabolite measurements at high time resolution throughout an exercise bout and the subsequent recovery period. Nevertheless, available data are of high enough quality to make it possible to correlate metabolic changes to the impaired contractile function. In the following paragraph, we give two examples of sport activities where the energy demand is high and acute peripheral fatigue develops due to a large dependency on anaerobic metabolism.

17.2.7 Exercise with Quasi-isometric Contractions vs. Dynamic Contractions

In alpine skiing, high levels of force are produced, mostly under isometric or eccentric conditions, for 90–120 s (Ferguson 2010). Thus, there is a high proportion of active motor units within the leg muscles, as evidenced by high electromyographic activity levels (Hintermeister et al. 1997), which combined with a limited blood flow due to high intramuscular pressure would lead to a major decrease in [PCr] (Krustrup et al. 2009), and the concomitant increase in [P_i] likely has a central role in the development of peripheral fatigue during the race. Other sports, such as flatwater kayaking, require high-power dynamic contractions mainly in the upper body. Limited blood flow due to prolonged high intramuscular pressure would be less of a problem with dynamic contractions, but acute fatigue-inducing anaerobic metabolism still largely contributes to energy required during 500–1000 m (~2–4 min duration) kayak performance (Bishop 2000; Tesch 1983). Interestingly, muscle biopsies collected in the deltoid muscle before and after simulated 500 m and 10,000 m (~45 min duration) kayak races indicate that glycogen depletion occurred after both types of exercise, while reduced [PCr], and hence a large proportion of anaerobic metabolism, was observed only after the shorter, more intense exercise; notably, [ATP] was unchanged after both exercises (Tesch and Karlsson 1984).

17.3 Effects of Specific Exercise-Induced Changes in Muscle Metabolites on Peripheral Fatigue

17.3.1 Decreased [ATP]/Increased [ADP]

^{31}P-NMR studies generally show well-maintained [ATP] during exercise (Newham and Cady 1990; Miller et al. 1988) unless the exercise is severe and PCr stores become very low (Taylor et al. 1986). In the latter case, biochemical analysis of muscle biopsies shows that large changes can occur in individual muscle fibres. For instance, after maximal cycling exercise, [ATP] was reduced to ~20% of the resting value in fast type IIx fibres, and [IMP] increased from undetectable levels to ~5 mM, which indicates that transient increases in [ADP] and [AMP] had occurred (Karatzaferi et al. 2001).

SR Ca^{2+} release via RyR1 is facilitated by ATP, while ADP and AMP act as weak competitive agonists. In cells, ATP generally exists as MgATP, which means that a decrease in [ATP] is accompanied by an increase in the free cytosolic [Mg^{2+}] ([Mg^{2+}]$_i$), and Mg^{2+} inhibits RyR1 opening (Lamb and Stephenson 1994; Meissner et al. 1986). However, larger changes in [ATP] and [Mg^{2+}]$_i$ than those generally observed during fatiguing contractions are required to induce a substantial inhibition of voltage-activated RyR1 Ca^{2+} release (Westerblad and Allen 1992); hence this mechanism would only affect SR Ca^{2+} release during very intense physical activities (Dahlstedt et al. 2000).

The power output during physical exercise depends both on the muscle capacity to generate force and to shorten (power = force × velocity). Increased [ADP] is known to decrease the shortening velocity of muscle fibres (Cooke and Pate 1985; Metzger 1996). In severe fatigue with depleted PCr stores, [ADP] transients during contractions can reach high enough levels to impose a physiologically significant inhibition of shortening velocity (Westerblad et al. 1998), which adds to the decreased force production to further decrease power output.

17.3.2 Increased [P_i]

In human, it is possible to monitor [P_i] during isometric, single-joint exercise using ^{31}P-MRS, and [P_i] can increase from ~5 mM to ~30 mM during severe exercise (Newham and Cady 1990; Cady et al. 1989; Sundberg et al. 2019). Noteworthy, large interindividual variation in P_i accumulation has been reported; for instance, repeated 30 s maximal voluntary plantar flexions lead to 75% decrease in [PCr] in sprinters, while it was only 40% in endurance-trained runners; thus, the resulting increase in P_i was ~3 times larger in the sprinters (Johansen and Quistorff 2003). It is also interesting to note that the increase in [Pi] is not specific to high intensity exercise as comparable increases were observed at the end of a series of repeated knee extensions performed either as (1) maximum voluntary contractions (MVCs) until reaching a stable low force level or (2) at ~55% MVC torque until task failure (Burnley et al. 2010). However, P_i may not accumulate during repeated contractions performed at low force (e.g. 20% MVC force) unless blood flow is restricted (Sugaya et al. 2011); this enhanced metabolic disturbance is thought to be a trigger for the hypertrophic effects observed with resistance training under blood flow restriction (Pearson and Hussain 2015).

The quantification of phosphagens with ^{31}P-MRS can be used to increase the physiological understanding of issues related to fatigue. For instance, a greater fatigability during dynamic exercise performed by old vs. young adults is a well-described feature, but the underlying mechanisms are not fully understood. A recent ^{31}P-MRS study showed a good correlation between a faster decline in mechanical power and a larger increase in [P_i] (~32 vs. 23 mM) during repeated voluntary knee extensions in old (mean age 76 years) than in young (mean age 23 years) adults (Sundberg et al. 2019), as illustrated in Fig. 17.2a. This was not accompanied by any difference in the [PCr] recovery kinetics between the two groups, which indicates no notable difference in muscle oxidative capacity mitochondrial function (Sundberg et al. 2019). Thus, from an energy metabolism perspective, the superior fatigue resistance in young individuals would be due to energetically more efficient contractions (i.e. less energy consumed) than in old individuals.

Measurements of phosphagens with ^{31}P-MRS have also been used to compare the metabolic response to repeated tetanic contractions induced by neuromuscular

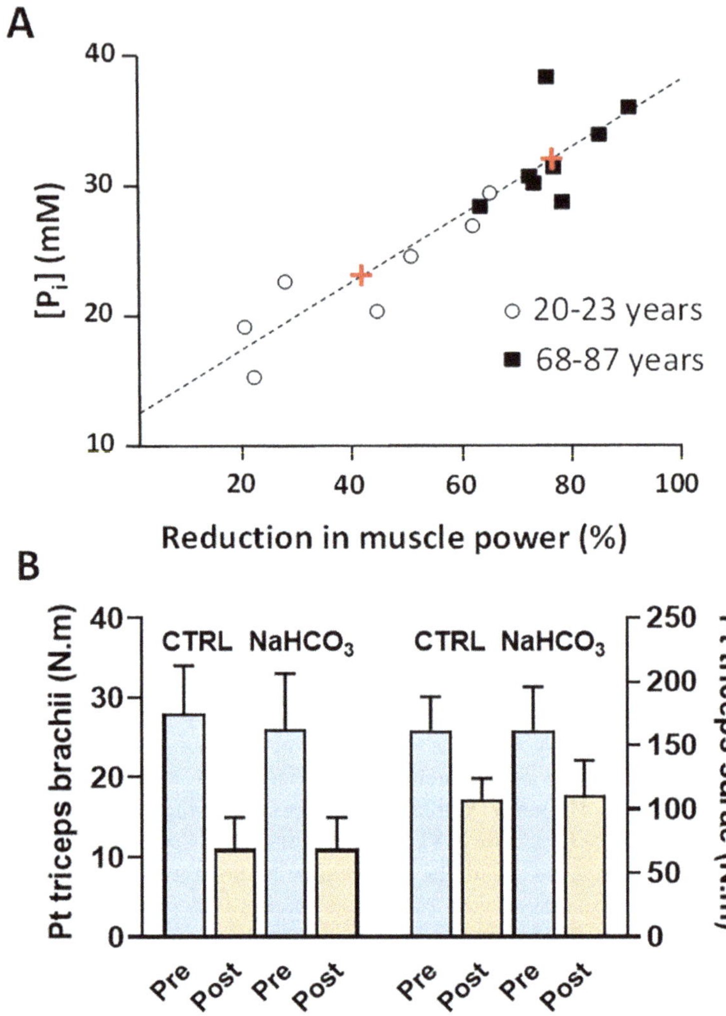

Fig. 17.2 (**a**) Linear relationship ($R^2 = 0.84$) between [P_i] and the reduction in muscle power induced by repeated maximal dynamic knee extensions in young and older individuals (adapted from (Sundberg et al. 2019)). The red crosses correspond to the mean for each group. (**b**) Peak twitch amplitude (mean ± SD) from two human muscles (left, triceps brachii; right, triceps surae) obtained before (pre) and after (post) repeated electrically induced tetanic contractions performed with supplementation of 0.3 g kg^{-1} $NaHCO_3$ or placebo (CTRL) (figure made from data presented in Siegler et al. 2016)

electrical stimulation (NMES) vs. voluntary contractions. An early study showed higher [P_i]/[PCr] ratio and lower pH during NMES than during load-matched voluntary contractions of the knee extensors (Vanderthommen et al. 1999). These data were confirmed in a subsequent study that also showed splitting of the P_i peak in the ^{31}P-MRS spectra with NMES, which was interpreted as activation of two populations of muscle fibres (oxidative vs. glycolytic) during this exercise modality (Jubeau et al. 2015). Thus, ^{31}P-MRS quantification of [Pi] during fatiguing contractions revealed that motor units are recruited in a non-selective manner (i.e. not following the size principle) with NMES (Bickel et al. 2011).

Experiments on skinned muscle fibres show inhibitory effects of increased [P_i] on myofibrillar function, decreasing both the force generating capacity and myofibrillar Ca^{2+} sensitivity (Cooke et al. 1988; Godt and Nosek 1989; Debold et al. 2006). Skinned fibre experiments also show that increased [P_i] might inhibit depolarization-induced SR Ca^{2+} release by acting on the RyR1 (Steele and Duke 2003; Duke and Steele 2001). In addition, P_i might enter the SR during fatigue, which can result in the Ca^{2+}-P_i solubility product being exceeded, precipitation of CaP_i and decreased free Ca^{2+} available for release (Duke and Steele 2001; Fryer et al. 1995; Dutka et al. 2005; Westerblad and Allen 1996; Ferreira et al. 2021). Accordingly, the $[Ca^{2+}]_i$ increase obtained with high doses of caffeine or 4-chloro-m-cresol, which directly stimulate Ca^{2+} release via RyR1, was reduced in fatigued muscle fibres indicating decreased Ca^{2+} available for rapid release (Westerblad and Allen 1991; Kabbara and Allen 1999), and measurements of the SR free $[Ca^{2+}]$ showed a decline during fatiguing stimulation in both isolated toad muscle fibres and mouse muscle studied in situ (Kabbara and Allen 2001; Allen et al. 2011).

Experiments on CK-deficient mouse muscle fibres, which cannot break down PCr, illustrate the important role of PCr energy buffering in high-intensity exercise as well as the inhibitory effect of increased [P_i] during more prolonged fatiguing stimulation (Dahlstedt et al. 2000). These fibres displayed impaired contractile function at the onset of high-intensity stimulation, where PCr breakdown functions as an important energy buffering system, whereas during more prolonged stimulation, they showed neither decreased cross-bridge force production, reduced myofibrillar Ca^{2+} sensitivity, nor decreased SR Ca^{2+} release. Notably all these features partially returned towards the wild-type phenotype after injecting CK into these CK-deficient cells, which allowed PCr breakdown to occur (Dahlstedt et al. 2003).

17.3.3 Acidosis

Physical exercise at an intensity requiring a large proportion of anaerobic metabolism will result in marked accumulation of lactate and hydrogen ions (Ferguson et al. 2018). The resulting acidosis in muscle fibres has classically been viewed as an integral causative factor in peripheral muscle fatigue. In resting human skeletal muscle fibres, intracellular pH (pH_i) at rest is ~7.1. During exhaustive exercise, pH_i has been shown to decrease to as low as ~6.5 (Spriet et al. 1989; Sahlin et al. 1976), whereas in other studies pH_i was only reduced to ~6.8 or 6.9 at exhaustion (Bangsbo et al. 1996; Hogan et al. 1999). This shows that severe fatigue can occur at markedly different degrees of acidosis, and as discussed in this chapter, important roles of several other metabolic alterations in fatigued muscle fibres have been identified. Thus, the importance of acidosis in peripheral muscle fatigue is currently debated (Ferguson et al. 2018). Nevertheless, lactate and hydrogen ions are effectively transported out of muscle fibres (Juel 1997), and the resulting increase in blood lactate is easy to measure and constitutes a good indicator of the usage of anaerobic metabolism by muscles during exercise.

One way to study the importance of acidosis as a cause of fatigue during physical exercise is to apply pH buffering agents, such as sodium bicarbonate ($NaHCO_3$). The results of studies investigating effects of $NaHCO_3$ ingestion on exercise performance generally show minor improvements or no effect, and the effect depends on many factors, such as the type of exercise, dose and duration of supplementation, training status and gender (Saunders et al. 2021). For instance, a meta-analysis showed slightly increased mean power (~2%) during high-intensity races of short duration with $NaHCO_3$ supplementation, but the underlying mechanisms remained unclear (Carr et al. 2011). Conversely, the authors of a recent systematic review concluded that it is unclear whether supplementation with $NaHCO_3$ enhances performance or simply balances deficits in buffering capacity between athletes (Hadzic et al. 2019). The evidence for a role of $NaHCO_3$ supplementation on peripheral fatigue is scarce. In a recent study, repeated electrically induced tetanic contractions of two muscle groups (triceps surae and triceps brachii with predominately slow- and fast-twitch fibres, respectively) did not reveal any difference in central or peripheral fatigue after $NaHCO_3$ intake vs. placebo despite slightly higher blood pH (7.48 vs. 7.42) before exercise in the $NaHCO_3$ group (Siegler et al. 2016), as illustrated in Fig. 17.2b.

Low pH has been shown to inhibit the activation of isolated RyR1 incorporated in planar lipid bilayers (Laver et al. 2000; Ma et al. 1988). However, the physiological action potential-induced SR Ca^{2+} release via RyR1 is little affected by low pH (Lamb and Stephenson 1994; Lamb et al. 1992), and increased tetanic $[Ca^{2+}]_i$ has been observed in acidified mouse and human muscle fibres (Olsson et al. 2020; Westerblad and Allen 1993). Thus, current evidence implies that acidosis does not impair muscle function during fatiguing exercise by inhibiting SR Ca^{2+} release.

Studies on skinned fibre experiments performed at room temperature (~20 °C) showed that a reduction in pH resulted in marked decreases in cross-bridge force generating capacity, myofibrillar force production and shortening velocity (Fabiato and Fabiato 1978; Metzger and Moss 1987). On the other hand, experiments conducted at more physiological temperatures (~30 °C) showed little effect of acidosis on maximum force production and shortening velocity in skinned rabbit muscle fibres (Pate et al. 1995; Chase and Kushmerick 1988), intact mouse muscle fibres (Westerblad et al. 1997), whole mouse muscles (Wiseman et al. 1996) and intact human muscle fibres (Olsson et al. 2020). Furthermore, acidification induced by elevating the bath CO_2 concentration prior to fatigue induced by repeated tetanic contractions had little or no effect on the rate of fatigue development in either isolated intact mouse or human muscle fibres (Olsson et al. 2020; Bruton et al. 1998). Nevertheless, recent experiments on mammalian skinned muscle fibres showed that the combination of acidosis and increased $[P_i]$ can induce large inhibitory effects on myofibrillar force production, Ca^{2+} sensitivity and shortening velocity even at physiological temperatures (Debold et al. 2016). Thus, experiments on skinned and intact muscle fibres provide apparently conflicting results regarding the importance of acidosis on myofibrillar function at physiological temperatures; hence the causative role of acidosis in mammalian muscle fatigue is still debated (Fitts 2016; Westerblad 2016).

17.3.4 Reactive Oxygen/Nitrogen Species

Another consequence of the increased energy consumption during fatiguing stimulation is accelerated ROS/RNS production (Powers and Jackson 2008). The dominant ROS in cells is superoxide ($O_2^{\bullet-}$) and its downstream derivatives, such as hydrogen peroxide (H_2O_2). Classically, mitochondria have been considered as the major site for $O_2^{\bullet-}$ production in muscle fibres during repeated contractions, and there are recent studies supporting this notion (Wei et al. 2011; Cheng et al. 2015). On the other hand, recent studies indicate that the enzyme NADPH oxidase 2 (NOX2) might be the major source for contraction-mediated $O_2^{\bullet-}$ (Michaelson et al. 2010; Pal et al. 2013; Sakellariou et al. 2013). Thus, the importance of different sources of $O_2^{\bullet-}$ in contracting muscle is still uncertain.

As it is relatively difficult to quantify ROS production in response to exercise in humans and most studies only show modest increases in ROS/RNS in muscle during exercise (Cheng et al. 2016), many studies have focused on the potential efficacy of antioxidant treatment on muscle fatigue and recovery. Almost 30 years ago, Reid et al. published a pioneer human study showing that prior infusion *N*-acetylcysteine (NAC, a thiol donor with antioxidant properties) improved force production of the *tibialis anterior* muscle by ~15% during repeated tetanic contractions (Reid et al. 1994). Interestingly, the beneficial effect of NAC was observed when submaximal forces (10 Hz) were evoked, while there was no influence when fatigue was evoked at near maximal (40 Hz) contractions. Later studies suggest that NAC is effective at improving exercise performance at unsaturated $[Ca^{2+}]_i$ levels during single-joint (Matuszczak et al. 2005) and whole-body (Medved et al. 2004; McKenna et al. 2006; Slattery et al. 2014) exercise.

While prolonged increases in ROS have multiple have adverse effects on muscle function and contribute to muscle dysfunction in several pathological conditions, such as rheumatoid arthritis and cancer (Yamada et al. 2017; Abrigo et al. 2018), transient increases in ROS production have important roles in normal cellular signalling, and effective antioxidant treatment can blunt positive training-induced muscle adaptations (Ristow et al. 2009; Paulsen et al. 2014; Gomez-Cabrera et al. 2008; Strobel et al. 2011). Our own studies focusing on fatigue-induced effects of increased ROS in muscle fibres generally show little effect during the actual induction of fatigue, whereas large ROS-mediated effects are seen during the subsequent recovery period (Cheng et al. 2016; Place et al. 2009). Fatigued muscle frequently enters a prolonged state of severely depressed submaximal force, i.e. prolonged low-frequency force depression (PLFFD) (Allen et al. 2008; Skurvydas et al. 2016); note that the effects of antioxidants *during* human exercise are also preferentially seen with submaximal contractions (see above). At the muscle fibre level, depressed submaximal force can be due to decreased SR Ca^{2+} release and/or reduced myofibrillar Ca^{2+} sensitivity. Antioxidants do not prevent PLFFD, but they can change the dominant underlying mechanism from decreased SR Ca^{2+} release to reduced myofibrillar Ca^{2+} sensitivity (Cheng et al. 2015; Bruton et al. 2008), which may have implications on the response to endurance training. Impaired SR Ca^{2+} release caused by redox modifications of RyR1 is associated with increased SR Ca^{2+} leak at rest, and the resulting increase in resting $[Ca^{2+}]_i$ can stimulate mitochondrial biogenesis and thereby improve muscle endurance (Wright et al. 2007; Bruton et al. 2010; Ivarsson et al. 2019; Zanou et al. 2021). Conversely, decreased myofibrillar Ca^{2+} sensitivity is unlikely to induced major adaptations. Thus, ROS-induced changes in RyR1 structure and function likely play a central role in the triggering of beneficial muscular adaptations in response to endurance training, and prevention of these RyR1 modifications with effective antioxidant treatment would then hamper these beneficial effects.

The central RNS in cells is nitric oxide ($NO^•$). $NO^•$ is generated via enzymatic reactions of nitric oxide synthases (NOS), and the rate at which it is produced in muscle fibres increases during repeated contractions (Cheng et al. 2015; Pye et al. 2007). $NO^•$ can also be formed from the inorganic anions nitrate (NO_3^-) and nitrite (NO_2^-), and the $NO^•$ produced in this manner might increase in the hypoxic and acidic conditions of muscles during intense exercise (Larsen et al. 2010). Human and mouse exercise performance can be improved by dietary NO_3^- supplementation (Larsen et al. 2010; Jones et al. 2018; Ivarsson et al. 2017; Larsen et al. 2007). A functionally effective extra intake of NO_3^- can be achieved by increasing the amount of green leafy vegetables or beetroot in the diet (Hord et al. 2009). Briefly, NO_3^- can be reduced to NO_2^- in the mouth, which then enhances $NO^•$ bioavailability. This nitrate-nitrite-$NO^•$ pathway has been shown to affect many physiological functions that can improve exercise performance (Lundberg et al. 2018; Jones et al. 2021), as well as attenuate fatigue development in isolated mouse muscle fibres (Bailey et al. 2019). Interestingly, NO_3^- supplementation for 7 consecutive days in humans resulted in improved muscle function similar to that observed with acute exposure to NAC, i.e. increase in knee extensor electrically evoked force at low ($\leq$ 20 Hz) stimulation frequencies (Whitfield et al. 2017; Haider and Folland 2014). Thus, in simplified terms, increasing the bioactivity of the RNS $NO^•$ induces changes in muscle function similar to those obtained by decreasing ROS with antioxidant treatment, which highlights the multifaceted and incompletely understood effects of ROS/RNS in muscle during physical exercise.

17.3.5 Glycogen

Seminal human studies performed more than 50 years ago reported that muscle glycogen stores are depleted during endurance exercise and a good correlation between muscle glycogen content and endurance performance was observed (Hermansen et al. 1967; Bergström et al. 1967). It is now clearly established that carbohydrate supplementation before/during endurance exercise improves performance (Vandenbogaerde and Hopkins 2011). Recent studies also show that carbohydrate ingestion between consecutive bouts of exercise performed with short recovery periods (2–4 hours) improves performance due to increased muscle glycogen repletion (Alghannam et al. 2016; McCarthy and Spriet 2020; Cheng et al. 2017). Altogether, these results confirm that muscle glycogen stores influence the extent and kinetics of muscle fatigue.

Within muscle fibres, glycogen is preferentially located in three distinct subcellular compartments: subsarcolemmal, intermyofibrillar (i.e. between myofibrils and close to the mitochondria and SR) and intramyofibrillar (i.e. within the contracting myofibrils) (Marchand et al. 2002; Nielsen et al. 2009; Ørtenblad et al. 2013). Studies using electron microscopy show a preferential depletion of intramyofibrillar glycogen in fatigued muscle fibres from humans (Marchand et al. 2007; Ørtenblad et al. 2011; Nielsen et al. 2011; Jensen et al. 2020) and rodents (Nielsen et al. 2009; Nielsen et al. 2014). Moreover, a good correlation between reduced SR Ca^{2+} release and depletion in intramyofibrillar glycogen was observed in mouse muscle fibres fatigued by repeated tetanic contractions (Nielsen et al. 2014). Interestingly, despite being fatigued to 30% of the original force and showing similar decline in intramyofibrillar glycogen content, reduced tetanic $[Ca^{2+}]_i$ could explain all of the force decrease when fibres were submitted to repeated 350 ms tetanic stimulations at long (10 s) intervals, whereas decreased myofibrillar Ca^{2+} sensitivity contributed to the force decrease at short (2 s) intervals (Nielsen et al. 2014). This difference might be taken as evidence for metabolic microenvironments within muscle fibres where the low-intensity stimulation resulted in metabolic disturbance in regions important for SR Ca^{2+} release, while myofibrils were little affected, whereas metabolic disturbances occurred at both sites with the high-intensity stimulation.

17.4 Conclusion

The importance and complexity of energy metabolism in muscle make it highly unlikely that fatigue would depend on a single metabolic mechanism (Booth and Laye 2009). Accordingly, in this chapter we describe mechanisms by which changes in several metabolites can contribute to the decline in contractile function in fatigued muscle as summarized in Fig. 17.3. The relative importance of different metabolites is multifaceted and depends on several factors, such as the type of exercise and the age and training status of the exercising individual. Nevertheless, the balance between the intensity of exercise and the aerobic capacity of the exercising individual has a key role in muscle fatigue, which inevitably occurs at faster rate when a large extent of anaerobic metabolism is required.

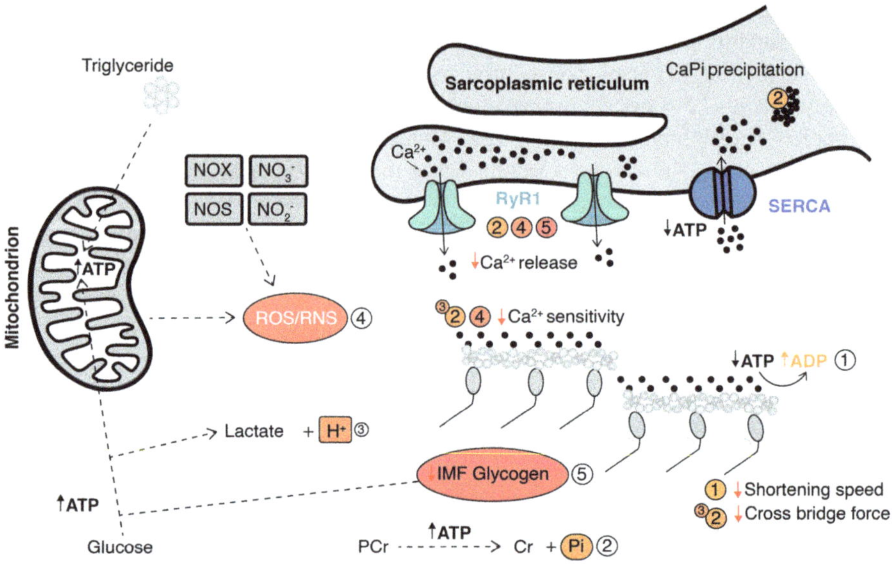

Fig. 17.3 Summary of metabolic factors contributing to muscle fatigue. For clarity purpose each metabolite acting on a given process (reduced SR Ca^{2+} release, decreased myofibrillar Ca^{2+} sensitivity, impaired cross-bridge force and/or decreased shortening speed) is indicated with a number: 1 = increased [ADP]; 2 = increased [P_i]; 3 = acidosis; 4 = increased [ROS/RNS]; 5 = decreased intramyofibrillar (IMF) glycogen content. Abbreviations are the same as those in the text

Acknowledgements We thank Chris Donnelly for assistance in summarizing Fig. 17.3. The work was supported by grants from the Swedish Research Council, the Swedish Research Council for Sport Science and the Swiss Sports Medicine Society.

References

Abrigo J, Elorza AA, Riedel CA, Vilos C, Simon F, Cabrera D et al (2018) Role of oxidative stress as key regulator of muscle wasting during cachexia. Oxidative Med Cell Longev 2018:2063179

Alghannam AF, Jedrzejewski D, Tweddle MG, Gribble H, Bilzon J, Thompson D et al (2016) Impact of muscle glycogen availability on the capacity for repeated exercise in man. Med Sci Sports Exerc 48(1):123–131

Allen DG, Lamb GD, Westerblad H (2008) Skeletal muscle fatigue: cellular mechanisms. Physiol Rev 88(1):287–332

Allen DG, Clugston E, Petersen Y, Röder IV, Chapman B, Rudolf R (2011) Interactions between intracellular calcium and phosphate in intact mouse muscle during fatigue. J Appl Physiol 111(2):358–366

Amann M, Wan HY, Thurston TS, Georgescu VP, Weavil JC (2020) On the influence of group III/IV muscle afferent feedback on endurance exercise performance. Exerc Sport Sci Rev 48(4): 209–216

Bailey SJ, Gandra PG, Jones AM, Hogan MC, Nogueira L (2019) Incubation with sodium nitrite attenuates fatigue development in intact single mouse fibres at physiological P_{O2}. J Physiol 597(22):5429–5443

Baldwin KM, Klinkerfuss GH, Terjung RL, Mole PA, Holloszy JO (1972) Respiratory capacity of white, red, and intermediate muscle: adaptative response to exercise. Am J Phys 222(2): 373–378

Bangsbo J, Madsen K, Kiens B, Richter EA (1996) Effect of muscle acidity on muscle metabolism and fatigue during intense exercise in man. J Physiol 495(2):587–596

Barclay CJ, Woledge RC, Curtin NA (2007) Energy turnover for Ca^{2+} cycling in skeletal muscle. J Muscle Res Cell Motil 28(4–5):259–274

Bergström J, Hermansen L, Hultman E, Saltin B (1967) Diet, muscle glycogen and physical performance. Acta Physiol Scand 71(2):140–150

Bickel CS, Gregory CM, Dean JC (2011) Motor unit recruitment during neuromuscular electrical stimulation: a critical appraisal. Eur J Appl Physiol 111(10):2399–2407

Bishop D (2000) Physiological predictors of flat-water kayak performance in women. Eur J Appl Physiol 82(1–2):91–97

Blain GM, Mangum TS, Sidhu SK, Weavil JC, Hureau TJ, Jessop JE et al (2016) Group III/IV muscle afferents limit the intramuscular metabolic perturbation during whole body exercise in humans. J Physiol 594(18):5303–5315

Booth FW, Laye MJ (2009) Lack of adequate appreciation of physical exercise's complexities can pre-empt appropriate design and interpretation in scientific discovery. J Physiol 587(23): 5527–5539

Bottinelli R, Reggiani C (2000) Human skeletal muscle fibres: molecular and functional diversity. Prog Biophys Mol Biol 73(2–4):195–262

Brownstein CG, Millet GY, Thomas K (2021) Neuromuscular responses to fatiguing locomotor exercise. Acta Physiol (Oxf) 231(2):e13533

Bruton JD, Lännergren J, Westerblad H (1998) Effects of CO_2-induced acidification on the fatigue resistance of single mouse muscle fibers at 28 °C. J Appl Physiol 85(2):478–483

Bruton JD, Place N, Yamada T, Silva JP, Andrade FH, Dahlstedt AJ et al (2008) Reactive oxygen species and fatigue-induced prolonged low-frequency force depression in skeletal muscle fibres of rats, mice and SOD2 overexpressing mice. J Physiol 586(1):175–184

Bruton JD, Aydin J, Yamada T, Shabalina IG, Ivarsson N, Zhang SJ et al (2010) Increased fatigue resistance linked to Ca^{2+}-stimulated mitochondrial biogenesis in muscle fibres of cold-acclimated mice. J Physiol 588(21):4275–4288

Burnley M, Vanhatalo A, Fulford J, Jones AM (2010) Similar metabolic perturbations during all-out and constant force exhaustive exercise in humans: a ^{31}P magnetic resonance spectroscopy study. Exp Physiol 95(7):798–807

Cady EB, Jones DA, Lynn J, Newham DJ (1989) Changes in force and intracellular metabolites during fatigue of human skeletal muscle. J Physiol 418(1):311–325

Carr AJ, Hopkins WG, Gore CJ (2011) Effects of acute alkalosis and acidosis on performance: a meta-analysis. Sports Med 41(10):801–814

Chase PB, Kushmerick MJ (1988) Effects of pH on contraction of rabbit fast and slow skeletal muscle fibers. Biophys J 53:935–946

Cheng AJ, Westerblad H (2017) Mechanical isolation, and measurement of force and myoplasmic free [Ca^{2+}] in fully intact single skeletal muscle fibers. Nat Protoc 12(9):1763–1776

Cheng AJ, Bruton JD, Lanner JT, Westerblad H (2015) Antioxidant treatments do not improve force recovery after fatiguing stimulation of mouse skeletal muscle fibres. J Physiol 593(2): 457–472

Cheng AJ, Yamada T, Rassier D, Andersson DC, Westerblad H, Lanner JT (2016) Reactive oxygen/nitrogen species and contractile function in skeletal muscle during fatigue and recovery. J Physiol 594(18):5149–5160

Cheng AJ, Willis SJ, Zinner C, Chaillou T, Ivarsson N, Ørtenblad N et al (2017) Post-exercise recovery of contractile function and endurance in humans and mice is accelerated by heating and slowed by cooling skeletal muscle. J Physiol 595(24):7413–7426
Cheng AJ, Place N, Westerblad H (2018) Molecular basis for exercise-induced fatigue: the importance of strictly controlled cellular Ca^{2+} handling. Cold Spring Harbor Perspect Med 8(2):a029710
Cheng AJ, Hwee DT, Kim LH, Durham N, Yang HT, Hinken AC et al (2019) Fast skeletal muscle troponin activator CK-2066260 increases fatigue resistance by reducing the energetic cost of muscle contraction. J Physiol 597(17):4615–4625
Cooke R, Pate E (1985) The effects of ADP and phosphate on the contraction of muscle fibers. Biophys J 48:789–798
Cooke R, Franks K, Luciani GB, Pate E (1988) The inhibition of rabbit skeletal muscle contraction by hydrogen ions and phosphate. J Physiol 395:77–97
Coyle EF, Coggan AR, Hemmert MK, Ivy JL (1986) Muscle glycogen utilization during prolonged strenuous exercise when fed carbohydrate. J Appl Physiol 61(1):165–172
Dahlstedt AJ, Katz A, Wieringa B, Westerblad H (2000) Is creatine kinase responsible for fatigue? Studies of skeletal muscle deficient of creatine kinase. FASEB J 14:982–990
Dahlstedt AJ, Katz A, Tavi P, Westerblad H (2003) Creatine kinase injection restores contractile function in creatine-kinase-deficient mouse skeletal muscle fibres. J Physiol 547(2):395–403
Debold EP, Romatowski J, Fitts RH (2006) The depressive effect of P_i on the force-pCa relationship in skinned single muscle fibers is temperature dependent. Am J Physiol Cell Physiol 290(4): C1041–C1C50
Debold EP, Fitts RH, Sundberg CW, Nosek TM (2016) Muscle fatigue from the perspective of a single crossbridge. Med Sci Sports Exerc 48(11):2270–2280
Duke AM, Steele DS (2001) Mechanisms of reduced SR Ca^{2+} release induced by inorganic phosphate in rat skeletal muscle fibers. Am J Physiol Cell Physiol 281(2):C418–CC29
Dulhunty AF (2006) Excitation-contraction coupling from the 1950s into the new millennium. Clin Exp Pharmacol Physiol 33(9):763–772
Dutka TL, Cole L, Lamb GD (2005) Calcium-phosphate precipitation in the sarcoplasmic reticulum reduces action potential-mediated Ca^{2+} release in mammalian skeletal muscle. Am J Physiol Cell Physiol 289:C1502–C1C12
Fabiato A, Fabiato F (1978) Effects of pH on the myofilaments and the sarcoplasmic reticulum of skinned cells from cardiac and skeletal muscles. J Physiol 276(1):233–255
Ferguson RA (2010) Limitations to performance during alpine skiing. Exp Physiol 95(3):404–410
Ferguson BS, Rogatzki MJ, Goodwin ML, Kane DA, Rightmire Z, Gladden LB (2018) Lactate metabolism: historical context, prior misinterpretations, and current understanding. Eur J Appl Physiol 118(4):691–728
Ferreira JJ, Pequera G, Launikonis BS, Rios E, Brum G (2021) A chloride channel blocker prevents the suppression by inorganic phosphate of the cytosolic calcium signals that control muscle contraction. J Physiol 599(1):157–170
Fitts RH (2016) The role of acidosis in fatigue: pro perspective. Med Sci Sports Exerc 48:2335–2338
Fryer MW, Owen VJ, Lamb GD, Stephenson DG (1995) Effects of creatine phosphate and P_i on Ca^{2+} movements and tension development in rat skinned skeletal muscle fibres. J Physiol 482(1):123–140
Gandevia SC (2001) Spinal and supraspinal factors in human muscle fatigue. Physiol Rev 81(4): 1725–1789
Godt RE, Nosek TM (1989) Changes of intracellular milieu with fatigue or hypoxia depress contraction of skinned rabbit skeletal and cardiac muscle. J Physiol 412:155–180
Gomez-Cabrera MC, Domenech E, Romagnoli M, Arduini A, Borras C, Pallardo FV et al (2008) Oral administration of vitamin C decreases muscle mitochondrial biogenesis and hampers training-induced adaptations in endurance performance. Am J Clin Nutr 87(1):142–149

Hadzic M, Eckstein ML, Schugardt M (2019) The impact of sodium bicarbonate on performance in response to exercise duration in athletes: a systematic review. J Sports Sci Med 18(2):271–281
Haider G, Folland JP (2014) Nitrate supplementation enhances the contractile properties of human skeletal muscle. Med Sci Sports Exerc 46(12):2234–2243
Henneman E, Olson CB (1965) Relations between structure and function in the design of skeletal muscles. J Neurophysiol 28:581–598
Hermansen L, Hultman E, Saltin B (1967) Muscle glycogen during prolonged severe exercise. Acta Physiol Scand 71(2):129–139
Hintermeister RA, O'Connor DD, Lange GW, Dillman CJ, Steadman JR (1997) Muscle activity in wedge, parallel, and giant slalom skiing. Med Sci Sports Exerc 29(4):548–553
Hogan MC, Richardson RS, Haseler LJ (1999) Human muscle performance and PCr hydrolysis with varied inspired oxygen fractions: a 31P-MRS study. J Appl Physiol 86(4):1367–1373
Homsher E (1987) Muscle enthalpy production and its relationship to actomyosin ATPase. Annu Rev Physiol 49:673–690
Hord NG, Tang Y, Bryan NS (2009) Food sources of nitrates and nitrites: the physiologic context for potential health benefits. Am J Clin Nutr 90(1):1–10
Ivarsson N, Schiffer TA, Hernandez A, Lanner JT, Weitzberg E, Lundberg JO et al (2017) Dietary nitrate markedly improves voluntary running in mice. Physiol Behav 168:55–61
Ivarsson N, Mattsson CM, Cheng AJ, Bruton JD, Ekblom B, Lanner JT et al (2019) SR Ca^{2+} leak in skeletal muscle fibers acts as an intracellular signal to increase fatigue resistance. J Gen Physiol 151(4):567–577
Jensen R, Ørtenblad N, Stausholm MH, Skjaerbaek MC, Larsen DN, Hansen M et al (2020) Heterogeneity in subcellular muscle glycogen utilisation during exercise impacts endurance capacity in men. J Physiol 598(19):4271–4292
Johansen L, Quistorff B (2003) ^{31}P-MRS characterization of sprint and endurance trained athletes. Int J Sports Med 24(3):183–189
Jones AM, Thompson C, Wylie LJ, Vanhatalo A (2018) Dietary nitrate and physical performance. Annu Rev Nutr 38:303–328
Jones AM, Vanhatalo A, Seals DR, Rossman MJ, Piknova B, Jonvik KL (2021) Dietary nitrate and nitric oxide metabolism: mouth, circulation, skeletal muscle, and exercise performance. Med Sci Sports Exerc 53(2):280–294
Jubeau M, Fur YLE, Duhamel G, Wegrzyk J, Confort-Gouny S, Vilmen C et al (2015) Localized metabolic and t2 changes induced by voluntary and evoked contractions. Med Sci Sports Exerc 47(5):921–930
Juel C (1997) Lactate-proton cotransport in skeletal muscle. Physiol Rev 77(2):321–358
Kabbara AA, Allen DG (1999) The role of calcium stores in fatigue of isolated single muscle fibres from the cane toad. J Physiol 519:169–176
Kabbara AA, Allen DG (2001) The use of fluo-5N to measure sarcoplasmic reticulum calcium in single muscle fibres of the cane toad. J Physiol 534:87–97
Karatzaferi C, de Haan A, Ferguson RA, van Mechelen W, Sargeant AJ (2001) Phosphocreatine and ATP content in human single muscle fibres before and after maximum dynamic exercise. Pflügers Arch. 442(3):467–474
Kluger BM, Krupp LB, Enoka RM (2013) Fatigue and fatigability in neurologic illnesses: proposal for a unified taxonomy. Neurology 80(4):409–416
Krustrup P, Söderlund K, Relu MU, Ferguson RA, Bangsbo J (2009) Heterogeneous recruitment of quadriceps muscle portions and fibre types during moderate intensity knee-extensor exercise: effect of thigh occlusion. Scand J Med Sci Sports 19(4):576–584
Lamb GD, Stephenson DG (1994) Effects of intracellular pH and [Mg^{2+}] on excitation-contraction coupling in skeletal muscle fibres of the rat. J Physiol 478:331–339
Lamb GD, Recupero E, Stephenson DG (1992) Effect of myoplasmic pH on excitation-contraction coupling in skeletal muscle fibres of the toad. J Physiol 448:211–224
Larsen FJ, Weitzberg E, Lundberg JO, Ekblom B (2007) Effects of dietary nitrate on oxygen cost during exercise. Acta Physiol (Oxf) 191(1):59–66

Larsen FJ, Weitzberg E, Lundberg JO, Ekblom B (2010) Dietary nitrate reduces maximal oxygen consumption while maintaining work performance in maximal exercise. Free Radic Biol Med 48(2):342–347

Laver DR, Eager KR, Taoube L, Lamb GD (2000) Effects of cytoplasmic and luminal pH on Ca^{2+} release channels from rabbit skeletal muscle. Biophys J 78(4):1835–1851

Liu Y, Gu Y, Yu X (2017) Assessing tissue metabolism by phosphorous-31 magnetic resonance spectroscopy and imaging: a methodology review. Quant Imaging Med Surg 7(6):707–726

Lundberg JO, Carlström M, Weitzberg E (2018) Metabolic effects of dietary nitrate in health and disease. Cell Metab 28(1):9–22

Lunde PK, Dahlstedt AJ, Bruton JD, Lännergren J, Thorén P, Sejersted OM et al (2001) Contraction and intracellular Ca^{2+} handling in isolated skeletal muscle of rats with congestive heart failure. Circ Res 88:1299–1305

Lunde PK, Sejersted OM, Thorud HM, Tonnessen T, Henriksen UL, Christensen G et al (2006) Effects of congestive heart failure on Ca^{2+} handling in skeletal muscle during fatigue. Circ Res 98(12):1514–1519

Ma J, Fill M, Knudson CM, Campbell KP, Coronado R (1988) Ryanodine receptor of skeletal muscle is a gap junction-type channel. Science 242(4875):99–102

Marchand I, Chorneyko K, Tarnopolsky M, Hamilton S, Shearer J, Potvin J et al (2002) Quantification of subcellular glycogen in resting human muscle: granule size, number, and location. J Appl Physiol 93(5):1598–1607

Marchand I, Tarnopolsky M, Adamo KB, Bourgeois JM, Chorneyko K, Graham TE (2007) Quantitative assessment of human muscle glycogen granules size and number in subcellular locations during recovery from prolonged exercise. J Physiol 580(2):617–628

Matuszczak Y, Farid M, Jones J, Lansdowne S, Smith MA, Taylor AA et al (2005) Effects of *N*-acetylcysteine on glutathione oxidation and fatigue during handgrip exercise. Muscle Nerve 32(5):633–638

McCarthy DG, Spriet LL (2020) Performance effects of carbohydrate ingestion between bouts of intense aerobic interval exercise. Int J Sports Physiol Perform 15(2):262–267

McKenna MJ, Medved I, Goodman CA, Brown MJ, Bjorksten AR, Murphy KT et al (2006) *N*-acetylcysteine attenuates the decline in muscle Na^{+},K^{+}-pump activity and delays fatigue during prolonged exercise in humans. J Physiol 576(Pt 1):279–288

Medbø JI, Tabata I (1989) Relative importance of aerobic and anaerobic energy release during short-lasting exhausting bicycle exercise. J Appl Physiol 67(5):1881–1886

Medved I, Brown MJ, Bjorksten AR, Murphy KT, Petersen AC, Sostaric S et al (2004) *N*-acetylcysteine enhances muscle cysteine and glutathione availability and attenuates fatigue during prolonged exercise in endurance-trained individuals. J Appl Physiol 97(4):1477–1485

Meissner G, Darling E, Eveleth J (1986) Kinetics of rapid Ca^{2+} release by sarcoplasmic reticulum. Effects of Ca^{2+}, Mg^{2+}, and adenine nucleotides. Biochemistry (Mosc) 25(1):236–244

Metzger JM (1996) Effects of phosphate and ADP on shortening velocity during maximal and submaximal calcium activation of the thin filament in skeletal muscle fibers. Biophys J 70(1): 409–417

Metzger JM, Moss RL (1987) Greater hydrogen ion-induced depression of tension and velocity in skinned single fibres of rat fast than slow muscles. J Physiol 393:727–742

Michaelson LP, Shi G, Ward CW, Rodney GG (2010) Mitochondrial redox potential during contraction in single intact muscle fibers. Muscle Nerve 42:522–529

Miller RG, Boska MD, Moussavi RS, Carson PJ, Weiner MW (1988) ^{31}P nuclear magnetic resonance studies of high energy phosphates and pH in human muscle fatigue. Comparison of aerobic and anaerobic exercise. J Clin Invest 81(4):1190–1196

Newham DJ, Cady EB (1990) A ^{31}P study of fatigue and metabolism in human skeletal muscle with voluntary, intermittent contractions at different forces. NMR Biomed 3(5):211–219

Nielsen J, Schrøder HD, Rix CG, Ørtenblad N (2009) Distinct effects of subcellular glycogen localization on tetanic relaxation time and endurance in mechanically skinned rat skeletal muscle fibres. J Physiol 587(Pt 14):3679–3690

Nielsen J, Holmberg HC, Schrøder HD, Saltin B, Ørtenblad N (2011) Human skeletal muscle glycogen utilization in exhaustive exercise: role of subcellular localization and fibre type. J Physiol 589(11):2871–2885

Nielsen J, Cheng AJ, Ørtenblad N, Westerblad H (2014) Subcellular distribution of glycogen and decreased tetanic Ca^{2+} in fatigued single intact mouse muscle fibres. J Physiol 592:2003–2012

Olsson K, Cheng AJ, Al-Ameri M, Wyckelsma VL, Rullman E, Westerblad H et al (2020) Impaired sarcoplasmic reticulum Ca^{2+} release is the major cause of fatigue-induced force loss in intact single fibres from human intercostal muscle. J Physiol 598(4):773–787

Ørtenblad N, Nielsen J, Saltin B, Holmberg HC (2011) Role of glycogen availability in sarcoplasmic reticulum Ca^{2+} kinetics in human skeletal muscle. J Physiol 589(3):711–725

Ørtenblad N, Westerblad H, Nielsen J (2013) Muscle glycogen stores and fatigue. J Physiol 591(18):4405–4413

Pal R, Basu Thakur P, Li S, Minard C, Rodney GG (2013) Real-time imaging of NADPH oxidase activity in living cells using a novel fluorescent protein reporter. PLoS One 8(5):e63989

Pate E, Bhimani M, Franks-Skiba K, Cooke R (1995) Reduced effect of pH on skinned rabbit psoas muscle mechanics at high temperatures: implications for fatigue. J Physiol 486(3):689–694

Paulsen G, Cumming KT, Holden G, Hallen J, Ronnestad BR, Sveen O et al (2014) Vitamin C and E supplementation hampers cellular adaptation to endurance training in humans: a double-blind, randomised, controlled trial. J Physiol 592(8):1887–1901

Pearson SJ, Hussain SR (2015) A review on the mechanisms of blood-flow restriction resistance training-induced muscle hypertrophy. Sports Med 45(2):187–200

Periasamy M, Kalyanasundaram A (2007) SERCA pump isoforms: their role in calcium transport and disease. Muscle Nerve 35(4):430–442

Place N, Yamada T, Zhang SJ, Westerblad H, Bruton JD (2009) High temperature does not alter fatigability in intact mouse skeletal muscle fibres. J Physiol 587(Pt 19):4717–4724

Powers SK, Jackson MJ (2008) Exercise-induced oxidative stress: cellular mechanisms and impact on muscle force production. Physiol Rev 88(4):1243–1276

Pye D, Palomero J, Kabayo T, Jackson MJ (2007) Real-time measurement of nitric oxide in single mature mouse skeletal muscle fibres during contractions. J Physiol 581(1):309–318

Reid MB, Stokic DS, Koch SM, Khawli FA, Leis AA (1994) *N*-acetylcysteine inhibits muscle fatigue in humans. J Clin Invest 94(6):2468–2474

Ristow M, Zarse K, Oberbach A, Klöting N, Birringer M, Kiehntopf M et al (2009) Antioxidants prevent health-promoting effects of physical exercise in humans. Proc Natl Acad Sci U S A 106(21):8665–8670

Sahlin K, Harris RC, Nylind B, Hultman E (1976) Lactate content and pH in muscle obtained after dynamic exercise. Pflügers Arch 367(2):143–149

Sahlin K, Tonkonogi M, Söderlund K (1998) Energy supply and muscle fatigue in humans. Acta Physiol Scand 162:261–266

Sakellariou GK, Vasilaki A, Palomero J, Kayani A, Zibrik L, McArdle A et al (2013) Studies of mitochondrial and nonmitochondrial sources implicate nicotinamide adenine dinucleotide phosphate oxidase(s) in the increased skeletal muscle superoxide generation that occurs during contractile activity. Antioxid Redox Signal 18(6):603–621

Saunders B, de Oliveira LF, Dolan E, Durkalec-Michalski K, McNaughton L, Artioli GG et al (2021) Sodium bicarbonate supplementation and the female athlete: a brief commentary with small scale systematic review and meta-analysis. Eur J Sport Sci 22(5):745–754

Siegler JC, Mudie K, Marshall P (2016) The influence of sodium bicarbonate on maximal force and rates of force development in the triceps surae and brachii during fatiguing exercise. Exp Physiol 101(11):1383–1391

Skurvydas A, Mamkus G, Kamandulis S, Dudoniene V, Valanciene D, Westerblad H (2016) Mechanisms of force depression caused by different types of physical exercise studied by direct electrical stimulation of human quadriceps muscle. Eur J Appl Physiol 116:2251–2224

Slattery KM, Dascombe B, Wallace LK, Bentley DJ, Coutts AJ (2014) Effect of *N*-acetylcysteine on cycling performance after intensified training. Med Sci Sports Exerc 46(6):1114–1123

Spangenburg EE, Booth FW (2003) Molecular regulation of individual skeletal muscle fibre types. Acta Physiol Scand 178(4):413–424

Spriet LL (1992) Anaerobic metabolism in human skeletal muscle during short-term, intense activity. Can J Physiol Pharmacol 70(1):157–165

Spriet LL, Lindinger MI, McKelvie RS, Heigenhauser GJ, Jones NL (1989) Muscle glycogenolysis and H^+ concentration during maximal intermittent cycling. J Appl Physiol 66(1):8–13

Steele DS, Duke AM (2003) Metabolic factors contributing to altered Ca^{2+} regulation in skeletal muscle fatigue. Acta Physiol Scand 179(1):39–48

Stephenson DG, Lamb GD, Stephenson GM (1998) Events of the excitation-contraction-relaxation (E-C-R) cycle in fast- and slow-twitch mammalian muscle fibres relevant to muscle fatigue. Acta Physiol Scand 162(3):229–245

Strobel NA, Peake JM, Matsumoto A, Marsh SA, Coombes JS, Wadley GD (2011) Antioxidant supplementation reduces skeletal muscle mitochondrial biogenesis. Med Sci Sports Exerc 43(6): 1017–1024

Sugaya M, Yasuda T, Suga T, Okita K, Abe T (2011) Change in intramuscular inorganic phosphate during multiple sets of blood flow-restricted low-intensity exercise. Clin Physiol Funct Imaging 31(5):411–413

Sundberg CW, Prost RW, Fitts RH, Hunter SK (2019) Bioenergetic basis for the increased fatigability with ageing. J Physiol 597(19):4943–4957

Szentesi P, Zaremba R, van Mechelen W, Stienen GJ (2001) ATP utilization for calcium uptake and force production in different types of human skeletal muscle fibres. J Physiol 531(2):393–403

Taylor JL, Amann M, Duchateau J, Meeusen R, Rice CL (2016) Neural contributions to muscle fatigue: from the brain to the muscle and back again. Med Sci Sports Exerc 48(11):2294–2306

Taylor DJ, Styles P, Matthews PM, Arnold DA, Gadian DG, Bore P et al (1986) Energetics of human muscle: exercise-induced ATP depletion. Magn Reson Med 3(1):44–54

Tesch PA (1983) Physiological characteristics of elite kayak paddlers. Can J Appl Sport Sci 8(2): 87–91

Tesch PA, Karlsson J (1984) Muscle metabolite accumulation following maximal exercise. A comparison between short-term and prolonged kayak performance. Eur J Appl Physiol Occup Physiol 52(2):243–246

Tiller NB, Elliott-Sale KJ, Knechtle B, Wilson PB, Roberts JD, Millet GY (2021) Do sex differences in physiology confer a female advantage in ultra-endurance sport? Sports Med 51(5):895–915

Vandenbogaerde TJ, Hopkins WG (2011) Effects of acute carbohydrate supplementation on endurance performance: a meta-analysis. Sports Med 41(9):773–792

Vanderthommen M, Gilles R, Carlier P, Ciancabilla F, Zahlan O, Sluse F et al (1999) Human muscle energetics during voluntary and electrically induced isometric contractions as measured by ^{31}P NMR spectroscopy. Int J Sports Med 20(5):279–283

Wei L, Salahura G, Boncompagni S, Kasischke KA, Protasi F, Sheu SS et al (2011) Mitochondrial superoxide flashes: metabolic biomarkers of skeletal muscle activity and disease. FASEB J 25(9):3068–3078

Westerblad H (2016) Acidosis is not a significant cause of skeletal muscle fatigue. Med Sci Sports Exerc 48:2339–2342

Westerblad H, Allen DG (1991) Changes of myoplasmic calcium concentration during fatigue in single mouse muscle fibers. J Gen Physiol 98(3):615–635

Westerblad H, Allen DG (1992) Myoplasmic free Mg^{2+} concentration during repetitive stimulation of single fibres from mouse skeletal muscle. J Physiol 453:413–434

Westerblad H, Allen DG (1993) The influence of intracellular pH on contraction, relaxation and $[Ca^{2+}]_i$ in intact single fibres from mouse muscle. J Physiol 466(1):611–628

Westerblad H, Allen DG (1996) The effects of intracellular injections of phosphate on intracellular calcium and force in single fibres of mouse skeletal muscle. Pflügers Arch. 431(6):964–970

Wiseman RW, Beck TW, Chase PB (1996) Effect of intracellular pH on force development depends on temperature in intact skeletal muscle from mouse. Am J Phys 271(3):C878–CC86

Westerblad H, Bruton JD, Lännergren J (1997) The effect of intracellular pH on contractile function of intact, single fibres of mouse muscle declines with increasing temperature. J Physiol 500(1): 193–204

Westerblad H, Dahlstedt AJ, Lännergren J (1998) Mechanisms underlying reduced maximum shortening velocity during fatigue of intact, single fibres of mouse muscle. J Physiol 510(1): 269–277

Whitfield J, Gamu D, Heigenhauser GJF, VAN Loon LJC, Spriet LL, Tupling AR et al (2017) Beetroot juice increases human muscle force without changing Ca^{2+}-handling proteins. Med Sci Sports Exerc 49(10):2016–2024

Wright DC, Geiger PC, Han DH, Jones TE, Holloszy JO (2007) Calcium induces increases in peroxisome proliferator-activated receptor gamma coactivator-1α and mitochondrial biogenesis by a pathway leading to p38 mitogen-activated protein kinase activation. J Biol Chem 282(26): 18793–18799

Yamada T, Steinz MM, Kenne E, Lanner JT (2017) Muscle weakness in reumatoid arthritis: the role of Ca^{2+} and free radical signaling. EBioMedicine 23:12–19

Zhang SJ, Andersson DC, Sandström ME, Westerblad H, Katz A (2006a) Cross-bridges account for only 20% of total ATP consumption during submaximal isometric contraction in mouse fast-twitch skeletal muscle. Am J Physiol Cell Physiol 291:C147–CC54

Zhang SJ, Bruton JD, Katz A, Westerblad H (2006b) Limited oxygen diffusion accelerates fatigue development in mouse skeletal muscle. J Physiol 572(2):551–559

Zhang SJ, Sandström ME, Aydin J, Westerblad H, Wieringa B, Katz A (2008) Activation of glucose transport and AMP-activated protein kinase during muscle contraction in adenylate kinase-1 knockout mice. Acta Physiol (Oxf) 192(3):413–420

Zanou N, Dridi H, Reiken S, Imamura de Lima T, Donnelly C, De Marchi U, Ferrini M, Vidal J, Sittenfeld L, Feige JN, Garcia-Roves PM, Lopez-Mejia IC, Marks AR, Auwerx J, Kayser B, Place N (2021) Acute RyR1 Ca^{2+} leak enhances NADH-linked mitochondrial respiratory capacity. Nat Commun 12(1):7219

GPSR Compliance
The European Union's (EU) General Product Safety Regulation (GPSR) is a set of rules that requires consumer products to be safe and our obligations to ensure this.

If you have any concerns about our products, you can contact us on

ProductSafety@springernature.com

In case Publisher is established outside the EU, the EU authorized representative is:

Springer Nature Customer Service Center GmbH
Europaplatz 3
69115 Heidelberg, Germany

www.ingramcontent.com/pod-product-compliance
Ingram Content Group UK Ltd.
Pitfield, Milton Keynes, MK11 3LW, UK
UKHW021410091026
15624UKWH00009B/183
* 9 7 8 3 0 3 0 9 4 3 0 4 2 *